BASIC ALGEBRA
A GUIDED APPROACH

Second Edition

BASIC ALGEBRA
A GUIDED APPROACH
Second Edition

Robert A. Carman
Santa Barbara City College

Marilyn J. Carman
Santa Barbara City Schools

1807 1982
175 YEARS OF PUBLISHING

John Wiley & Sons
New York Chichester Brisbane Toronto Singapore

A teacher who can arouse a feeling for one single good action . . . accomplishes more than he who fills our memory with rows on rows of natural objects, classified with name and form. —Goethe

This book is dedicated to our first teachers, our mothers, Nellie and Mary

Library of Congress Cataloging in Publication Data

Carman, Robert A.
   Basic algebra.

  Includes index.
   1. Algebra. I. Carman, Marilyn J. II. Title.
QA152.2.C37 1982   512.9   81-11601
ISBN 0-471-04174-2   AACR2

Printed in the United States of America

10 9 8 7 6

PREFACE

This book is intended for use in a first course in elementary algebra. Specifically, it is designed for the many students with little or no background in algebra and a very limited general ability in mathematics. Most of these students have had little success in mathematics, some openly fear it, and all need a very carefully guided approach with emphasis on understanding and explanation rather than abstraction and formalism. They need instructional materials designed to meet them at their own level of ability and to develop gradually the concepts and skills of elementary algebra. This book is intended for such students.

Those who have difficulty with mathematics will find in this book several special features designed to make it most effective for them. These include the following.

- Careful attention has been given to readability, and reading specialists have helped plan both the written text and the visual organization.
- An optional diagnostic pretest and performance objectives keyed to the text are given at the beginning of each unit. These clearly indicate the content of each unit and provide the student with a sense of direction.
- Each unit ends with a self-test covering the work of the unit.
- The format is clear and easy to follow. It respects the individual needs of each reader, providing immediate feedback at each step to assure understanding and continued attention.
- The *why* as well as the *how* of every concept and operation are carefully explained. Algebra concepts are presented as natural extensions of arithmetic concepts and operations.
- The emphasis is on *explaining* algebra concepts rather than *presenting* them. This book is not a bag of equations or a set of formal proofs.
- Throughout the book special attention has been given to word problems, including specific instruction in an effective strategy for solving such problems. This is the most difficult part of elementary algebra for most students and therefore receives our most careful attention.
- Both routine drill and more imaginative and challenging problems are included. Answers to all of these problems are given in the back of the book.
- Supplementary problem sets are included at the end of each unit, and additional problem sets are available in the accompanying teacher's manual. Answers to all supplementary problem sets are given in the teacher's manual.
- Arithmetic reviews are provided for those students who need to work through the arithmetic concepts and operations generally assumed as prerequisite to the course.
- A light, lively conversational style of writing and a pleasant, easy-to-understand visual approach are used. The use of humor and historical topics are designed to appeal to students who have in the past found mathematics to be dry and uninteresting.

This book has been used in both classroom and individualized instruction settings and carefully field-tested with hundreds of students representing a wide range of interests and ability levels, from junior high school to college graduates. Students who have used the book tell us it is helpful, interesting, even fun to work through. More important, it works—they learn algebra, many of them experiencing success in mathematics for the first time.

Flexibility of use was a major criterion in the design of the book, and field-testing indicates that the book can be used successfully in a variety of course formats. It can be used as a textbook in traditional lecture-oriented courses. It is very effective in situations where an instructor wishes to modify a traditional course by devoting a portion of class time to independent study. The book is especially useful in programs of individualized or self-paced instruction, whether in a learning lab situation, with tutors, with audiotapes, or in totally independent study.

An accompanying teacher's resource book and test manual provides

- information on a variety of self-paced and individualized course formats that may be used;
- multiple forms of all unit tests, brief quizzes, and final examinations; additional problem sets;
- answers to all problems in the unit tests, quizzes, exams, and supplementary problem sets;
- additional references and information on material included in the book.

During the writing of preliminary versions of this book and in its revision we had the good fortune of having it tested under classroom conditions by a very perceptive and capable teacher. To Susanne Culler goes our gratitude for her generous help and encouragement.

This revised second edition includes many improvements suggested by students and teachers who used the first edition. We are especially grateful to the following teachers for their generous advice and valuable assistance: Colin Godfrey, University of Massachusetts; Ronald A. Stoltenberg, Sam Houston State University; Dena Patterson, Santa Monica College; Teri Y. H. Chiang, Mission College; Grace DeVelbiss, Sinclair Community College; Frank Hammons, Sinclair Community College; Sue Myers, Sinclair Community College; John Pfetzing, Sinclair Community College; A. H. Tellez, Glendale Community College; Bryn Gary, Oscar Rose Junior College; Leonard Orman, University of Southern Colorado; and John Loughlin, Lane Community College; and William Whicher, Parks College, St. Louis University. We have been particularly fortunate to have had the expert assistance of Laurie Carman at every stage of writing and production of this second edition, and we are grateful.

A century ago Ralph Waldo Emerson wrote "Our chief want in life is somebody who shall make us do what we can." For the authors, this want has been capably filled by Gary W. Ostedt, Mathematics Editor at Wiley. His confidence, warm support, and creative enthusiasm turned insoluble problems into radiant opportunities. We are grateful for his help.

It is a pleasure to acknowledge the valuable assistance of Robert W. Pirtle and Claire E. Egielski of John Wiley & Sons and Richard C. Spangler of Tacoma Community College during the planning and production of this revision.

Finally, at every step of the seemingly endless sequence of writing, testing, and rewriting that makes a textbook, we have benefited from the active participation, suggestions, help, proofreading, concern, kibitzing, and curiosity of our children: Patty, Laurie, Mary, and Eric. They not only made it worth doing, they made it fun, and in the process they made it a better book than we could have produced without them.

Santa Barbara, California   Robert A. Carman
                                                             Marilyn J. Carman

CONTENTS

ABOUT THIS BOOK

Everyone knows that it is easy to find a book that they can read and read and read again, and never understand at all. This is especially true of textbooks, and mathematics textbooks are often the worst culprits of all. We have tried to make this book one you will understand. It is not algebra taught in a vacuum of theory and abstractions, with pages of symbols and strange roundabout ways of saying things that seem obvious, but algebra you can understand and use.

This book is designed so that you can:

★ start at the beginning or where you need to start,
★ work on only what you need to know,
★ move as fast or as slow as you wish,
★ skip material you already understand,
★ do as many practice problems as you need,
★ take self-tests to measure your progress.

If you are worried that algebra may be difficult, and you want a book you will understand, this book is designed for you.

This is no ordinary book. It is not designed for browsing or casual reading. You *work* your way through it. The ideas are arranged step-by-step in short portions or *frames*. Each frame contains information, careful explanations, examples, and questions to test your understanding. Read the material in each frame carefully, follow the examples, and answer the questions that lead to the next frame. Correct answers move you quickly through the book. Further explanation is provided when it is needed. You move through this book frame by frame. Because we know that every person is different and has different needs, each major section of the book starts with an optional preview test that will help you determine the parts on which you need to work.

As you move through the book you will notice that material not directly connected to the frames appears in boxes and cartoons. Read these at your leisure. They contain information that you may find useful, interesting, and even fun.

Because we know that word problems are a very difficult part of algebra for most students, we have taken special care throughout the book to explain how to do such problems. You'll be guided through many carefully worked examples and given helpful hints designed to make you an algebra user rather than an algebra memorizer.

Most students hesitate to ask questions. They would rather risk failure than look foolish by asking "dumb questions." To relieve you of worry over dumb questions (or DAQs), we'll ask and answer them for you. Thousands of students have taught us that "dumb questions" can produce smart students. Watch for DAQ.

More than thirty-six centuries ago, Ahmes, an Egyptian scribe copying an algebra text, wrote wonderingly that it contained "rules for inquiring into nature, and for knowing all that exists, every mystery, . . . every secret." No one today would make such a claim for an algebra textbook or even for all of mathematics. In this book we will show you how to do simple algebra, how to use it and understand it. Old Ahmes would be fascinated.

Now, turn to page 1 and let's begin.

BASIC ALGEBRA
A GUIDED APPROACH

Second Edition

Introduction to Algebra

Objective	Sample Problems	Where To Go for Help

Upon successful completion of this program you will be able to:

Identify each of the following as a term, coefficient, factor, expression, variable, constant, or equation.

		Page	Frame

1. Understand and use basic algebraic words such as *term, expression, factor, variable, coefficient,* and *equation.*

(a) $2x$ _____   Page 3   Frame 1

(b)   The 3 in $3x^2$ _____

(c)   The x in $2x + 3x^2$ _____

(d) $3x + 5$ _____

(e) $x + 4 = 9$ _____

2. Add, subtract, multiply, and divide signed numbers.

(a) $4 - (-7)$ = _____   15   14

(b) $8 - 14$ = _____

(c) $(-6.4) \times (-3.1)$ = _____

(d) $(-12.2) \div (-4.0)$ = _____

3. Work with numbers in exponential notation.

(a) 2^3 = _____   29   25

(b) $(-0.4)^2$ = _____

(c)   Write in exponent form $3 \cdot a \cdot a \cdot b \cdot b \cdot b$ = _____

(d)   Multiply $a^2 \cdot a^3$ = _____

(e)   Calculate $(a^3)^2$ = _____

(f)   Divide $\dfrac{a^6}{a^4}$ = _____

Simplify:

4. Perform the basic algebraic operations.

(a) $3a^2b + ab^2 - a^2b - 4ab^2 + ab$ = _____   43   37

(b) $-3(xy)(2xy^2)(x^3z)$ = _____

(c) $\dfrac{8ab^2c^4}{2ab^3c}$ = _____

(d) $-2(3x - 4)$ = _____

Date _____

Name _____

Course/Section _____

Objective	Sample Problems	Where To Go for Help

Write as an algebraic expression: 59   **48**

5. Translate English phrases and sentences to algebraic expressions and equations.

   (a)   Seven less than twice a given number.   = _____

   (b)   The product of some number and 16 equals that number plus six.   = _____

If $x = 2$, $a = 5$, and $b = 6$, find the value of   69   **56**

6. Calculate the numerical value of literal expressions.

   (a) $\dfrac{2b - x}{a}$ = _____

   (b) $x^2 + 3x$ = _____

   (c) $3abx^2$ = _____

(Answers to these problems are at the bottom of this page.)

If you are certain you can work all of these problems correctly, turn to page 79 for a self-test. If you want help with any of these objectives or if you cannot work one of the sample problems, turn to the page indicated. Super-students who want to be certain they learn all of this will turn to frame **1** and begin work there.

PREVIEW 1

Answers to Sample Problems

1. (a) Term   (b) Coefficient or factor   (c) Variable   (d) Expression   (e) Equation

2. (a) 11   (b) -6 (c) 19.84   (d) 3.05

3. (a) 8   (b) 0.16   (c) $3a^2b^3$ (d) a^5 (e) a^6 (f) a^2

4. (a) $2a^2b - 3ab^2 + ab$ (b) $-6x^5y^3z$ (c) $\dfrac{4c^3}{b}$ (d) $8 - 6x$

5. (a) $2x - 7$ (b) $16x = x + 6$

6. (a) 2   (b) 10   (c) 360

Introduction to Algebra

© 1971 United Feature Syndicate, Inc.

1-1 THE LANGUAGE
OF ALGEBRA

1   Before you decide whether or not algebra is a serious business, you should learn what it
is and why people bother to learn it. What is algebra? The simplest answer is that
algebra is arithmetic in disguise—generalized arithmetic.

Arithmetic statements and operations always involve specific numbers: $2 + 3 = 5$,
$3 \times 4 = 12$, $28 \div 4 = 7$, and so on. Algebra statements can tell us something about
many numbers, or even all numbers. For example, what do the following arithmetic
statements have in common?

$$3 - 2 = 1$$
$$4 - 2 = 2$$
$$5 - 2 = 3$$
$$10 - 2 = 8$$
$$101 - 2 = 99$$

The answer is that they are all statements of the form "some number minus 2 equals
some other number," or, in symbols,

$$\square - 2 = \triangle$$

Where the symbols $\square$ and $\triangle$ represent numbers.

This is an algebraic equation and we can use it to study and reason about *all* arithmetic statements like those listed above. A mathematician is more interested in this
general equation than in any particular example of it. Arithmetic teaches you how to
make calculations relating to special one-shot situations. With algebra you begin to
study mathematics as a tool for general abstract thought.

Ready to learn about algebra? Then turn to **2** to continue.

---

3   **1-1 The Language of Algebra**

2 Many of the ideas behind algebra started thousands of years ago, but the kind of notation you see in a modern algebra book dates from the sixteenth century when François Vieta, a French lawyer and amateur mathematician, began the systematic use of letters to represent numbers. A mathematical statement in which letters are

Literal Expression used to represent numbers is called a *literal* expression. Letters can represent single numbers or entire sets of numbers. Algebra is the arithmetic of literal expressions. It is a kind of symbolic arithmetic that enables us to find answers to problems by simple operations with letters rather than by repeated and difficult arithmetic with numbers.

Any letters will do. In mathematics, English, Greek, or even Hebrew alphabets are used, including lowercase, capital, and even script letters. The letters

a, A, **A**, $\mathcal{A}$, or $\mathscr{A}$

are all different and each represents a *different* quantity in algebra even though all are the same letter of the alphabet. If you want to represent the distance a car travels by the letter d, you should be consistent and use the symbol d and not confuse it with D, **D**, $\mathfrak{D}$, or $\mathscr{D}$.

People who use algebra often choose the symbols they use on the basis of their memory-jogging value. Time is represented by t, distance by d, cost by c, area by A, and so on. The symbol is chosen to remind you of its meaning.

For no special reason, other than habit, mathematicians usually reserve the last six letters of the alphabet, u, v, w, x, y, and z to represent unknown quantities.

What letters would you use for algebra symbols to represent the following quantities:

(a) Radius of a circle   (b) Interest on a loan
(c) Speed of motion   (d) Work done
(e) Rate of change   (f) Volume of a box

Check your answers in **3**.

---

3 (a) R or r (b) i or I
(c) s or sometimes v for   (d) W
    velocity
(e) R or r (f) V

The Numbers of
Algebra

Numbers are the basic stuff of both arithmetic and algebra, and it is important that you have a clear understanding of the various kinds of numbers that appear in mathematics. Fortunately, we have a simple, easy to understand scheme for classifying all numbers. Look at the diagram on page 5. The familiar numbers we use to

Counting or Natural count objects, 1, 2, 3, 4, and so on, are called the *counting numbers* or the *natural*
Numbers *numbers*. These are shown at the bottom of the diagram. The set of numbers known
Integers as *integers* includes the positive natural numbers, 1, 2, 3, ..., their negative counterparts, -1, -2, -3, ..., and zero.

Rational Numbers The set of *rational numbers* includes any numbers that can be written as a fraction $\dfrac{a}{b}$

where a and b can be any integers except that b cannot equal zero. All fractions, such as $\dfrac{1}{2}, \dfrac{3}{2}$, or $-\dfrac{7}{8}$, are rational numbers and of course any integer can be written as a fraction:

$$4 = \frac{4}{1}, \, 7 = \frac{7}{1}, \text{ or } -12 = \frac{-12}{1}.$$

Any ordinary decimal number can be written as a fraction and they are therefore rational numbers.

$$0.4 = \frac{4}{10} \qquad 1.5 = \frac{3}{2} \qquad 2.375 = 2\frac{3}{8}$$

We build the integers from the natural numbers and the rational numbers from the integers.

Real Numbers

Irrational Numbers

All of the numbers that appear in elementary algebra and in most scientific or technical applications of mathematics are included in the set of *real numbers*. The real numbers include the rational numbers, that is, all positive and negative integers and fractions, and the *irrational numbers*. Irrational numbers are quantities such as $\sqrt{2}$, $\sqrt[3]{10}$, or $2 + \sqrt{5}$ that are not integers and cannot be written as fractions or finite decimals.

To test your understanding of this number classification, examine the following list and (a) draw a circle around each integer, (b) place a check ✓ over each rational number, (c) place an **x** by each natural number.

$$2, \quad -4, \quad 0, \quad 1.3, \quad \frac{1}{2}, \quad \frac{-2}{3}, \quad \sqrt{3}, \quad 10, \quad -1, \quad \frac{3}{2}, \quad 1 - \sqrt{5}$$

$$4 + 1, \quad 5 - 7, \quad 7 \div 2, \quad 3 \times 1\frac{1}{3}, \quad -8, \quad -2.4$$

Careful, some numbers may require more than one mark.

Check your answers in frame **4**.

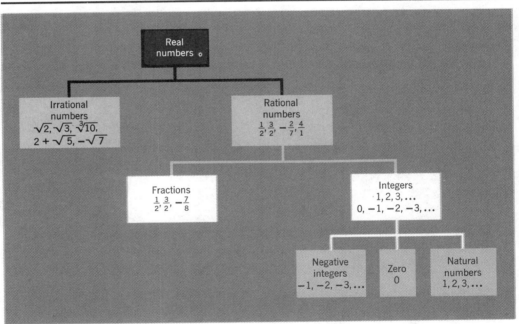

4

x✓ Ⓞ ✓ ⊝ ✓ Ⓞ ✓ 1.3 ✓ $\frac{1}{2}$ ✓ $-\frac{2}{3}$ ✓ $\sqrt{3}$ x✓ ⑩ ✓ ⊖ ✓ $\frac{3}{2}$

$1 - \sqrt{5}$ x✓ ④⊕①   ✓ ⑤⊖⑦   ✓ $7 \div 2$ x✓ ③×⑪⅓   ✓ ⊖⑧   ✓ -2.4

Algebraic Symbols

Most of the usual arithmetic symbols have the same meaning in algebra that they have in arithmetic. For example, the addition (+) and subtraction (−) signs are used in exactly the same way.

$$a + b \qquad 3 + x \qquad 1 - y \qquad Q - 5 \qquad d - e$$

Multiplication

However, the *multiplication* sign (×) of arithmetic looks like the letter x and to avoid confusion we show multiplication in algebra in other ways. The product of two algebraic quantities a and b, "a times b," may be written using

A raised dot	$a \cdot b$
Parentheses	$a(b)$ or $(a)b$ or $(a)(b)$
Or with no symbol at all	ab

Obviously this last way of showing multiplication won't do in arithmetic; we cannot write "two times four" as "24"—it looks like twenty-four. But it is a quick and easy way to indicate multiplication in algebra.

Placing two quantities side by side to show multiplication is not new and is not only an algebra gimmick; we use it every time we write 20¢ or 4 feet.

$$20¢ \quad = 20 \times 1¢$$
$$4 \text{ feet} = \ 4 \times 1 \text{ foot}$$

Write the following multiplications using no multiplication symbols:

(a)   8 times b = _____   (b) a times h = _____

(c)   2 times s times t = _____   (d)   3 times x times y = _____

(e)   4 times y times a times k = _____

Check your answers in **5**.

What does the word Algebra mean?

It comes from the title of a book written in 830 by an Arab mathematician. The title of the book was *Al·jabr W'al Muqàbala*, which means roughly "balance and simplify." *Al·jabr* becomes algebra

5   (a)   8 times $b = 8b$ (b) a times $h = ah$ (c)   2 times s times $t = 2st$

     (d)   3 times x times $y = 3xy$ (e)   4 times y times a times $k = 4yak$

We pronounce "$8b$" as "eight bee" and not "eight bees."

Parentheses

Parentheses are used in arithmetic to show that some complicated quantity is to be treated as a single number. For example,

$2 \cdot (3 + 4)$ means that the number 2 multiplies *all* of the quantity in the parentheses.

$$2 \cdot (3 + 4) = 2 \cdot 7 = 14$$

Examples:

$$5 + (11 - 7) = 5 + 4 = 9$$

$$8 - (2 + 5) = 8 - 7 = 1$$

$$3 \cdot (9 - 5) = 3 \cdot 4 = 12$$

Practice using parentheses by evaluating the following arithmetic quantities.

(a) $4 \cdot (8 - 1)$ = _____   (b) $12 - (3 - 1 + 2) =$ _____

(c) $(2 + 4 - 1) + (6 - 3 - 2) =$ _____   (d) $5 \cdot (9 - 7 + 1)$ = _____

(e) $4 \cdot \left(2 + \dfrac{1}{2}\right)$ = _____   (f) $\dfrac{2}{3} + \left(4 - 2\dfrac{1}{2}\right)$ = _____

Check your work in **6**.

---

6   (a) $4 \cdot (8 - 1) = 4 \cdot 7 = 28$

(b) $12 - (3 - 1 + 2) = 12 - 4 = 8$

(c) $(2 + 4 - 1) + (6 - 3 - 2) = 5 + 1 = 6$

(d) $5 \cdot (9 - 7 + 1) = 5 \cdot 3 = 15$

(e) $4 \cdot \left(2 + \dfrac{1}{2}\right) = 4 \cdot \dfrac{5}{2} = 10$

(f) $\dfrac{2}{3} + \left(4 - 2\dfrac{1}{2}\right) = \dfrac{2}{3} + \dfrac{3}{2} = \dfrac{13}{6} = 2\dfrac{1}{6}$

If you had any trouble working with the fractions in problems (e) and (f) you should take time out to review the arithmetic of fractions beginning on page 521.

In algebra, as in arithmetic, parentheses ( ), brackets [ ], or braces { }, indicate that whatever is enclosed in them should be treated as a single quantity. An expression such as $(3x^2 - 4ax + 2by^2)^2$ should be thought of as (something)^2. The expression $(2x + 3a - 4) - (x^2 - 2a)$ should be thought of as (first quantity) − (second quantity).

$\dfrac{3x - 4y}{7 - x}$ should be thought of as $\dfrac{\text{first quantity}}{\text{second quantity}}$.

It is a division of two quantities.

Parentheses are the punctuation marks of algebra. Like the period, comma, or semicolon in regular sentences, they tell you how to read an equation and get its correct meaning.

Write the following using algebraic notation:

(a) 8 times $(2a + b)$ = _____   (b) $(a + b)$ times $(a - b)$ = _____

(c) $(x + y)$ times $(x + y) =$ _____   (d) 3 times x times $(1 - y) =$ _____

(e) Add x to $(y - 2)$ and multiply the sum by $(x^2 + 1) =$ _____

Check your work in **7**.

---

7   (a) $8(2a + b)$ (b) $(a + b)(a - b)$
(c) $(x + y)^2$ (d) $3x(1 - y)$
(e) $[x + (y - 2)](x^2 + 1)$ or $(x + y - 2)(x^2 + 1)$

Division

In arithmetic we would write "48 divided by 2" as

$$2\overline{)48} \qquad \text{or} \qquad 48 \div 2 \qquad \text{or} \qquad 48/2$$

These notations are used very seldom in algebra. Most often division is written as a fraction.

"x divided by y" is written $\dfrac{x}{y}$

"$(2n + 1)$ divided by $(n - 1)$" is written $\dfrac{(2n + 1)}{(n - 1)}$ or $\dfrac{2n + 1}{n - 1}$

Write the following divisions using algebra notation:

(a) a divided by b = _____   (b) x divided by 2   = _____

(c) 3 divided by y = _____   (d) $(a + 2)$ divided by $(b + 3)$ = _____

(e) Multiply e by h and divide the product by $(2d + 1)$ = _____

Turn to **8** to check your work.

---

8   (a) $\dfrac{a}{b}$ (b) $\dfrac{x}{2}$ (c) $\dfrac{3}{y}$ (d) $\dfrac{a + 2}{b + 3}$ (e) $\dfrac{eh}{2d + 1}$

In any language letters are grouped to form words, and punctuation marks such as commas, periods, or semicolons are used to group words into phrases or sentences. In the language of algebra, letters and numerals are combined to form literal expressions or algebraic expressions. An *expression* is a general name for any collection of numerals and letters connected by arithmetic operation signs, parentheses, or other grouping symbols. For example,

Expression

$2x^2 + 4$

is an algebraic expression.

$(b - 1)$, $(x + y)$, x, 5, and $2(x^2 - 3ab)$ are all algebraic expressions.

An algebraic expression may be a product or multiplication of several factors or a sum of two or more terms. When two variables, constants or expressions are multiplied together, each multiplier is called a *factor* of the product. For example,

Factor

a and b are factors of the product ab.

$2x$ and $(x + 1)$ are factors of the product $2x(x + 1)$.

The expression $(k - 1)(2k + 1)$ is the product of factors $(k - 1)$ and $(2k + 1)$.

A *constant* is any numerical quantity whose value does not change.

2, $\frac{1}{4}$, 4.3, and π are all constants.

Term

A *term* is a general name for a numeral, literal number, or the product or quotient of numerals and literal numbers. For example,

2, x, and $2x$ are terms.

x^2, $3x^2$, $5ab$, $2\pi r$, and $\dfrac{3}{c}$ are all terms.

Notice that these terms do not contain addition or subtraction signs. The sum of two or more terms is an expression.

$x + 2$ is an expression, the sum of terms x and 2.
$3x^2 + 2x + 1$ is an expression whose first term is $3x^2$.
$5ab + c^2 + 3a$ is an expression, the sum of three terms.

Fill in the blank spaces in the following sentences with the correct words:

(a) In the _____ $3a^2 + 2a - 4$, $2a$ is the second _____ .

(b) The _____ of $2x$ are 2 and x.

(c) The first term of the _____ $2y - 1$ is _____ , the _____ of 2 and y.

(d) In the _____ $3xy^2 - 1$, $3xy^2$ and 1 are both _____ .

(e) In the _____ $a + b$, a and b are _____ .

Hop to **9** to check your answers.

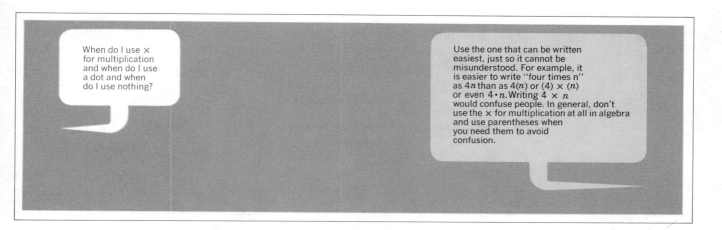

9   (a) In the *expression* $3a^2 + 2a - 4$, $2a$ is the second *term*.
       (b) The *factors* of $2x$ are 2 and x.
       (c) The first term of the *expression* $2y - 1$ is $2y$, the *product* of 2 and y.
       (d) In the *expression* $3xy^2 - 1$, $3xy^2$ and 1 are both *terms*.
       (e) In the *expression* $a + b$, a and b are *terms*.

Algebraic Equations

In the sentence

"The boy jumped over the fence."

The word *boy* stands for a person or a number of people. Many more specific words may be substituted to give true statements:

John jumped over the fence.
Bill jumped over the fence.
Dave jumped over the fence.

or, in general,

☐ jumped over the fence.

Variables

The word *boy* or the box ☐ are *variables*. They stand for a particular person or a set of persons. Variables in English sentences are pronouns or impersonal nouns that are non-specific but can be given exact meanings.

Some people are happy when it rains.
She enjoys skiing.
He is a good student.

Variables are important in algebra sentences, too. The equation

               Copyright © 1982 by John Wiley & Sons, Inc.

$\square + 3 = 7$

is a statement that some number $\square$ added to 3 is equal to 7. The symbol $\square$ is a variable representing a particular number.

What is the value of $\square$ in the equation above that makes it a true statement?

Try it. Guess if you must, and then check your guess. Look in **10** for the answer.

---

10   If $\square + 3 = 7$, then $\square = 4$ of course.

Congratulations, you have just solved your first algebraic equation.

Normally this equation would be written

$x + 3 = 7$

We use a letter for the variable instead of a box because letters are easier to write and because we may have several variables and will need a different symbol for each. The equals sign ($=$) is used in algebra and has the same meaning there that it has in arithmetic. The arithmetic equation

$4 + 1 = 3 + 2$

means that the quantity named on the left of the equals sign $(4 + 1)$ has the same value or names the same number as the quantity on the right $(3 + 2)$.

In algebra the equality sign is used to compare algebraic expressions. The *algebraic equation*

$x + 3 = 7$

*Left and Right
Member of
an Equation*

means that the expression on the left $(x + 3)$, the *left member* of the equation, has the same value as the quantity on the right (7), the *right member* of the equation.

In the language of algebra, an equation is equivalent to a statement or a sentence containing numerals and literal numbers. Some equations are always true, some false, and some neither true nor false. The equations

$a = a$
$a + a = 2a$
$a + b = b + a$

or

$x + 0 = x$

are *always* true for real numbers, no matter what the value of the variables in the equation.

The equations

$x + 1 = x$

or $a = 2 + a$

are never true, no matter what value the variable is given.

Division by Zero

Some equations are not simply false, they are *meaningless*. For example,

$\dfrac{x}{0} = y$

has no meaning because division by zero is not defined for real numbers.

Equations such as

$x + 3 = 7$
$6x - 5 = 13$
$x^2 - 3x + 2 = 0$

are true only for some certain value or values of the variable.

$x + 3 = 7$ is true only for $x = 4$.
$6x - 5 = 13$ is true only for $x = 3$.
$x^2 - 3x + 2 = 0$ is true for $x = 1$ or $x = 2$.

Conditional Equations

Equations of this sort whose truth or falsity depends on the value given to the variable are called *conditional equations* or *open sentences*. They are by far the most interesting kind of equation in algebra and we will study them in detail in this book.

Label each of the following algebraic equations as either T, always true; F, always false; or C, conditional.

(a) $x - 2 = 0$ ⎯⎯⎯⎯⎯⎯ (b) $x + 2 = 2 + x$ ⎯⎯⎯⎯⎯⎯

(c) $2x = x + 5$ ⎯⎯⎯⎯⎯⎯ (d) $x = x - 4$ ⎯⎯⎯⎯⎯⎯

(e) $(x + y) + z = x + (y + z)$ ⎯⎯⎯⎯⎯⎯ (f) $3y = 1 - y$ ⎯⎯⎯⎯⎯⎯

Check your answers in **11**.

---

11 (a) C (b) T (c) C (d) F (e) T (f) C

Inequality

In addition to the study of equations and literal expressions, algebra often includes the study of inequalities. An *inequality* is a statement that a given quantity is greater than or less than some other quantity. For example,

$2 < 3$ is read "2 is less than 3,"

and

$3x > 5$ is read "3x is greater than 5."

The arrowhead-shaped symbol for inequality always points toward the smaller quantity.

The first inequality above is clearly true. The second inequality is true for certain values of the variable x and it is false for other values of x.

The inequality

$2 \leq x + 1$ is read "2 is less than or equal to $x + 1$,"

and the inequality

$8 \geq 4x$ is read "8 is greater than or equal to $4x$."

The first inequality is true when $2 = x + 1$ or when $2 < x + 1$. The second inequality is true when $8 = 4x$ or when $8 > 4x$.

You will learn how to solve algebraic inequalities later in this book. Try these problems for practice in using the inequality signs correctly.

Insert the proper symbol $<$, $>$, $\leq$, or $\geq$ between each pair of numerals:

(a) $7 \; > \; 3$ (b) $\dfrac{1}{2} \; > \; \dfrac{1}{4}$ (c) $1 \; < \; 10$ (d) $\dfrac{1}{2} \; > \; 1$

(e) $0.5 \; > \; 0.01$ (f) $0.7 \; < \; \dfrac{3}{4}$ (g) $\dfrac{1}{10} \; < \; 0.2$ (h) $0.4 \; > \; \dfrac{2}{5}$

Look on page 12 to check your answers.

12   (a) $7 > 3$ (b) $\frac{1}{2} > \frac{1}{4}$ (c) $1 < 10$ (d) $\frac{1}{2} < 1$

      (e) $0.5 > 0.01$ (f) $0.7 < \frac{3}{4}$ (g) $\frac{1}{10} < 0.2$ (h) $0.4 \leq \frac{2}{5}$

                                                                    or $0.4 \geq \frac{2}{5}$

Now turn to **13** for a set of practice problems on what you have learned so far in this unit.

Unit I
Introduction
to Algebra

The answers to all problems are on page 535.

A. **Complete each of the following statements using the words** *variable, expression, factor, equation, term, inequality, conditional equation, right member, left member.*

1. In the _____ $2x^3 - x$, $2x^3$ is the first _____, and x is a _____.

2. The _____ $3a^4b + 2ab + 4$ has three _____.

3. x and y are the _____ of xy.

4. $x + 2 = 7$ is a _____ whose _____ is the _____ $x + 2$.

5. $(y - 1)$ is a _____ of the product $(x + 2)(y - 1)$.

6. $x < 2$ is an _____ and x is a _____.

7. In the _____ $x^2 - 2x = 13$, the _____ is $x^2 - 2x$, the _____ is 13, and the _____ is x.

8. In the _____ $3r^2h + rh$, rh is a _____.

9. In the _____ $2m^2 - m + 3 = 0$, the _____ is $2m^2$.

10. The _____ of the _____ $a = 2$ is 2.

B. **Write the following in algebraic notation.**

1. x divided by y = _____

2. A minus 3 = _____

3. 6 times x times y = _____

4. a plus b^2 = _____

5. 3 plus y = _____

6. x^2 times y^2 = _____

7. 2 times $(2a + b)$ = _____

8. $(a + 1)$ divided by b = _____

9. x minus y = _____

10. 3 times q times s times t = _____

11. $(a + b)$ times $(a - b)$ = _____

12. A plus B plus C = _____

13. 3 times a = _____

14. a divided by 2 = _____

15. 1 minus x = _____

16. t minus q plus 1 = _____

17. 2 plus y minus t = _____

18. 4 times $(x - 2y)$ = _____

19. $(3p + t)$ divided by $2x$ = _____

20. $(x - 3)$ divided by 2 = _____

Date

Name

Course/Section

21. $5A$ times $(A + 2) =$ _____   22. x plus $2y$ minus $3z =$ _____

23. $(x - y + 1)$ times $(3x - 1) =$ _____   24. $(1 + 2a)$ times $(1 - 2b) =$ _____

C. Complete each statement by inserting the proper symbol, $<$, $>$, or $=$.

1. 2   10

2. $\dfrac{1}{3}$ $\dfrac{1}{4}$

3. $\dfrac{1}{4}$ $\dfrac{1}{3}$

4. 3   3

5. x $x + 2$

6. $2 + 3$ $8 - 3$

7. 0   1

8. 1   100

9. 1   0.01

10. 3.4   3.04

11. $2\dfrac{1}{2}$ 2.5

12. $a + 1$ $1 + a$

13. $2 + 3 + 5$ $5 + 2 + 3$

14. $\dfrac{2}{3}$ $\dfrac{7}{8}$

15. 4   2

16. $3\dfrac{1}{4}$ $\dfrac{7}{2}$

17. 0.8 $\dfrac{4}{5}$

18. $\dfrac{21}{5}$ $\dfrac{13}{3}$

19. $7 \cdot 8$ $9 \cdot 6$

20. $5 \div 3$ $6 \div 4$

When you have completed this problem set either return to the preview test on page 1 or continue in **14** with the study of signed numbers.

---

14 What kind of number would you use to name each of the following: a golf score two strokes below par, a loss of $2 in a poker game, a debt of $2, a loss of two yards on a football play, a temperature two degrees below zero, or the year 2 B.C.? The answer is that they are all *negative numbers,* all equal to −2.

Mathematicians of a thousand years ago believed that an arithmetic expression like 3 − 5 had no meaning. Of course merchants used negative numbers to represent debts, but no mathematician would accept a negative number as the solution to an algebra problem. Even the word "negative" comes from the Latin word *negare,* meaning "to deny"—negative numbers, or "false numbers" as they were called, were denied a place in mathematics. It was not until the sixteenth century that negative numbers became an accepted part of mathematics.

Number Line

The visual invention that helped mathematicians make sense of the negative numbers is called a *number line.* To construct a number line:

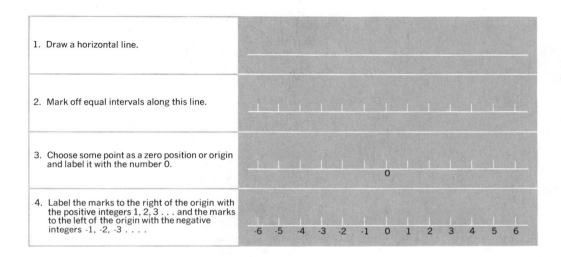

Use the number line above to locate the numbers 3, −2, 1.5, and −3. Mark these numbers by placing dots on the line.

Check your answers in **15**.

___

15

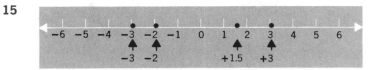

The number line can be drawn to any length (it should be infinitely long, of course, but we can draw only part of it), and we may choose any point on the line as the zero position and any spacing between integers. We call all numbers to the right of the origin *positive* numbers and all numbers to the left of the origin *negative* numbers. Only the positive and negative integers are marked and labeled, but *any* real number can be located there. In fact, the number line is important because there is exactly one point on the line associated with each real number, whether it be a natural number, decimal, fraction or square root, positive or negative.

Draw a number line and locate the following numbers on it.

(a) 5   (b) −5   (c) $2\frac{1}{4}$ (d) −1.2   (e) +1.3

(f)   0.4   (g) $-\dfrac{8}{3}$ (h) -0.5 (i) -4.1 (j)   3

Compare your drawing with ours in **16**.

16

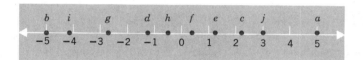

Signed numbers are a bookkeeping concept. They were used in bookkeeping by the ancient Greeks, Chinese, and Hindus more than 2000 years ago. Chinese merchants wrote positive numbers in black and negative numbers in red in their account books. The Hindus used a dot above the number or a circle around it to show that a number was negative. (Being "in the red" or "in the hole" still means to be in debt or to have a negative income.) We use a minus sign $(-)$ to show that a number is negative, and because this symbol is also used for subtraction, some confusion results.

The $+$ and $-$ signs have a dual role in mathematics. In simple arithmetic $+$ means "add" and $-$ means "subtract." But in algebra these symbols are also *direction* signs for the number line. The number $+3$ names a point on the number line three units to the *right* of the zero, in the *positive* direction. The number -3 names a point on the number line three units to the *left* of the zero in the *negative* direction.

The direction sign of a negative number is always shown. The direction sign of a positive number is always understood to be there, but it is usually not written out. We write 2 instead of $+2$.

 Be careful to read -3 as "negative 3" rather than "minus 3" and $+4$ as "positive 4" rather than "plus 4" to avoid confusion with addition and subtraction.

Comparing Signed Numbers

Comparing numbers is easy with the number line. We can see at a glance whether one number is less than or greater than another number. For any pair of numbers, the number to the *left* on the number line is *less* than the number to the right.

Examples:

$2 < 4$ $+2$ is to the *left* of $+4$ on the number line; therefore $+2$ is *less than* $+4$.

$-3 < -1$ -3 is to the left of -1 on the number line; therefore -3 is *less than* -1.

         We could also write this inequality as $-1 > -3$. The arrowhead-shaped symbol points to the smaller number, the number on the left on the number line.

Now, try these problems for practice in comparing numbers.

1.   Label each of the following inequality statements as either *true* or *false*:

    (a) $2 < 1$ (b) $-2 < -1$ (c) $0 < 4$ (d) $-1 > -4$

    (e) $\dfrac{1}{2} < -2$ (f) $-0.4 > 0.4$ (g) $-4 > 1$ (h) $4 > -1$

    (i) $-2 < 0.1$ (j) $2 < -0.1$ (k) $0.2 < 0.3$ (l) $\dfrac{1}{3} > \dfrac{1}{4}$

2.   Insert either $<$ or $>$ to make these true statements.

    (a)   3   2   (b) -1 -3 (c) $\dfrac{1}{2}$ $-\dfrac{1}{4}$ (d)   0 -1

    (e) -0.4 0.2   (f) -0.5 -0.7 (g) -4 4   (h) -1 0.1

(i) $-1\frac{1}{2}$ $-\frac{1}{2}$ (j) $-\frac{1}{2}$ 1   (k) -0.01 -0.1 (l) -3 0

Draw a number line first, and use it to locate and compare these numbers. When you have completed these check your answers in **17**.

17
1. (a) F   (b) T   (c) T   (d) T
   (e) F   (f) F   (g) F   (h) T
   (i) T   (j) F   (k) T   (l) T

2. (a) >   (b) >   (c) >   (d) >
   (e) <   (f) >   (g) <   (h) <
   (i) <   (j) <   (k) >   (l) <

Arithmetic with Signed Numbers

Addition and subtraction of signed numbers may be pictured using the number line. First, write every problem as a sum of numbers with direction signs shown:

Examples: $3 + 4$ becomes $(+3) + (+4)$
               $5 - 2$ becomes $(+5) + (-2)$

Second, imagine every positive number as a hop to the right and every negative number as a hop to the left.

Example: $+3$ is

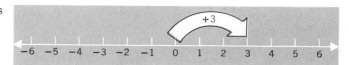

Example: -2 is

Example:
The sum $3 + 2$ is
                $(+3) + (+2)$ or

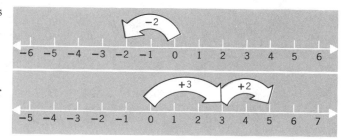

Starting at the origin, hop three spaces to the right ($+3$) and then hop two spaces to the right ($+2$). The trip along the number line ends at the point $+5$; therefore, $3 + 2 = 5$.

Example:
The sum of $6 - 4$ is
                $(+6) + (-4)$ or

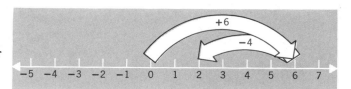

Starting at the origin, hop six spaces to the right ($+6$), followed by a hop of four spaces to the left (-4). The trip ends at the point $+2$; therefore, $6 - 4 = 2$.

Example:
The sum $4 - 7$ is
                $(+4) + (-7)$ or

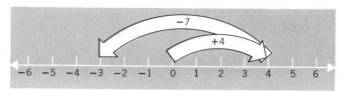

Again, starting at the origin, hop four spaces to the right $(+4)$, followed by a hop of seven spaces to the left (-7). The trip ends at the point -3; therefore, $4 - 7 = -3$.

The sign, $+$ or $-$, gives the direction of the hop and the number to which the sign is attached gives the size of the hop.

Use this visual method to perform the following additions and subtractions:

(a) $4 + 2$ (b) $7 - 3$ (c) $4 - 6$ (d) $-2 - 3$

(e) $-3 + 5$ (f) $2 - 6$ (g) $-4 + 2$ (h) $-6 - 2$

When you have finished hopping around the number line, hop to **18** and check your work.

---

18   (a) $4 + 2 = (+4) + (+2) = 6$

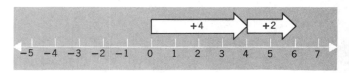

(b) $7 - 3 = (+7) + (-3) = 4$

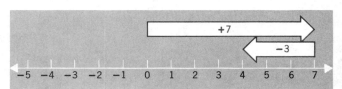

(c) $4 - 6 = (+4) + (-6) = -2$

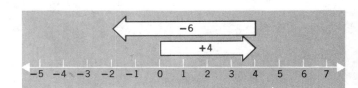

(d) $-2 - 3 = (-2) + (-3) = -5$

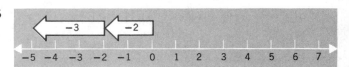

(e) $-3 + 5 = (-3) + (+5) = 2$

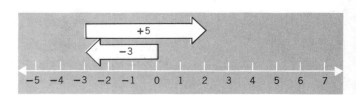

(f) $2 - 6 = (+2) + (-6) = -4$

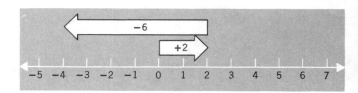

(g) $-4 + 2 = (-4) + (+2) = -2$

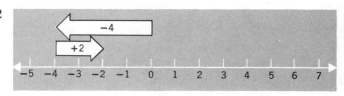

(h) $-6 - 2 = (-6) + (-2) = -8$

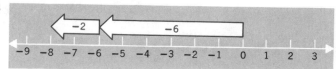

The number line enables us to visualize the addition or subtraction of signed numbers, but in order to actually use these numbers in algebra we need a rule that involves no pictures. This rule uses the new concept of absolute value.

Absolute Value

The *absolute value* $|n|$ of any number n is defined as the distance on the number line from the origin to the point associated with that number. We indicate absolute value by placing vertical bars on each side of the number.

Examples: $|2| = 2$

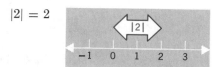

Read $|2|$ as "absolute value of 2."

$|-3| = 3$

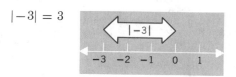

Read $|-3|$ as "absolute value of negative 3."

$$\left|-\frac{2}{3}\right| = \frac{2}{3} \qquad |-0.7| = 0.7$$

$$|0| = 0$$

For any negative real number the absolute value is the positive number you obtain when you drop the sign on the number.

Find the following absolute values.

(a) $|8| = \underline{\quad 8 \quad}$ (b) $\left|-\frac{3}{8}\right| = \underline{\quad \frac{3}{8} \quad}$ (c) $|0.6| = \underline{\quad 0.6 \quad}$

(d) $|-12| = \underline{\quad 12 \quad}$ (e) $|-0.04| = \underline{\quad 0.04 \quad}$ (f) $|-324| = \underline{\quad 324 \quad}$

The correct answers are in **19**.

---

19   (a) $|8| = 8$ (b) $\left|-\frac{3}{8}\right| = \frac{3}{8}$ (c) $|0.6| = 0.6$

(d) $|-12| = 12$ (e) $|-0.04| = 0.04$ (f) $|-324| = 324$

Adding and Subtracting Signed Numbers

To add or subtract two signed numbers follow this procedure.

★ **First,** if the two numbers have the *same* sign, rewrite the problem as a sum. Then add the absolute values of the numbers and give this sum the sign of the original numbers.

Example: $9 + 3 = +12$

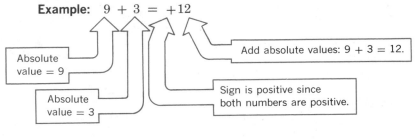

Absolute value = 9

Absolute value = 3

Add absolute values: $9 + 3 = 12$.

Sign is positive since both numbers are positive.

Example: $-16 - 11 = (-16) + (-11) = -27$

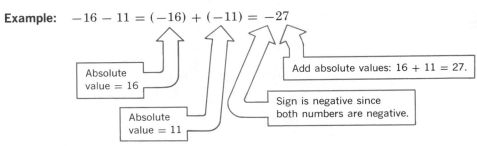

Absolute value = 16

Absolute value = 11

Add absolute values: $16 + 11 = 27$.

Sign is negative since both numbers are negative.

★ **Second,** if the two numbers have the *opposite* sign, subtract their absolute values and give this difference the sign of the number with the larger absolute value.

Example: $12 - 7 = 12 + (-7) = +5$

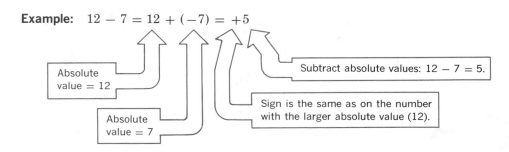

Absolute value = 12

Absolute value = 7

Subtract absolute values: $12 - 7 = 5$.

Sign is the same as on the number with the larger absolute value (12).

Example: $-14 + 5 = -9$

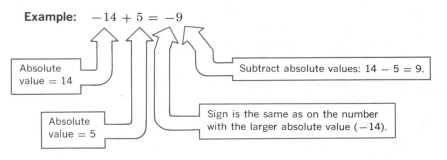

Absolute value = 14

Absolute value = 5

Subtract absolute values: $14 - 5 = 9$.

Sign is the same as on the number with the larger absolute value (-14).

★ **Third,** replace any expression of the form $-(-a)$ with $+a$, then continue as before.

Example: $8 - (-4) = 8 + 4 = 12$

$-(-4) = 4$

Example: $-9 - (-2) = -9 + 2 = -7$

$-(-2) = +2$

More examples:

$8 + 2 = +10$
$\qquad$ $+8$ and $+2$ have the same sign. Add and give the sum (10) the sign of the original numbers ($+$).

$$8 - 2 = 8 + (-2) = +6$$

$+8$ and -2 have opposite signs. Take the difference (6) of the absolute values and give it the sign $(+)$ of the number with the larger absolute value (8).

$$-7 + 3 = -4$$

-7 and $+3$ have opposite signs. Take the difference in absolute values $(7 - 3 = 4)$ and give it the sign $(-)$ of the number with the larger absolute value (-7).

$$7 - (-3) = 7 + 3 = 10$$

Replace $-(-3)$ by $+3$. Add.

$$-8 - (-5) = -8 + 5$$
$$= -3$$

Replace $-(-5)$ by $+5$. -8 and $+5$ have opposite signs. Take the difference in absolute values $(8 - 5 = 3)$ and give it the sign $(-)$ of the number with the larger absolute value (-8).

$$-7 - 12 = -7 + (-12)$$
$$= -19$$

-7 and -12 have the same signs. Add the absolute values $(7 + 12)$ and give the sum the sign $(-)$ of the original numbers.

Remember: When no sign is given the $+$ sign is implied.

Beware! Never use two signs together without parentheses. The phrase "$+4$ added to -2" is written $4 + (-2)$ but *never* $4 + -2$.

Use this procedure to work the following problems.

(a) $11 - 14$ $=$ _____

(b) $-6 - 5$ $=$ _____

(c) $-10 - (-7) =$ _____

(d) $13 - (-6) =$ _____

Check your work in **20**.

---

20   (a) $11 - 14 = 11 + (-14)$
$$= -3$$

The signs are different $(+11$ and $-14)$. Subtract absolute values $(14 - 11 = 3)$ and attach the sign $(-)$ of the number (-14) with the larger absolute value.

(b) $-6 - 5 = -6 + (-5)$
$$= -11$$

The signs are the same $(-6$ and $-5)$. Add $(6 + 5 = 11)$ and attach the sign $(-)$.

(c) $-10 - (-7) = -10 + 7 = -3$

Replace $-(-7)$ by $+7$. The signs are now different $(-10$ and $+7)$. Subtract absolute values $(10 - 7 = 3)$ and attach the sign $(-)$ of the number (-10) with the larger absolute value.

(d) $13 - (-6) = 13 + 6 = 19$

Replace $-(-6)$ by $+6$. The signs are the same $(+13$ and $+6)$. Add $(13 + 6 = 19)$ and attach the sign $(+)$.

Adding a List of Signed Numbers

To add a list of signed numbers, first add the positive numbers, then add the negative numbers, and finally combine these two partial sums.

For example:

$$24 - 5 + 7 - 11 - 3 - 9 + 6 = \underline{\hspace{2cm}}$$

Add positive numbers: $24 + 7 + 6 = 37$

Add negative numbers: $-5 - 11 - 3 - 9 = -28$

Combine these partial sums: $37 - 28 = +9$

Example: $4 - 7 - 15 + 1 - 5 + 10 - 12 + 2 = $ _____

$$= (4 + 1 + 10 + 2) + (-7 - 15 - 5 - 12)$$
$$= 17 + (-39)$$
$$= 17 - 39$$
$$= -22$$

Now here are a few problems for practice:

(a) $3 - 7$

(b) $-13 + 6$

(c) $14 - (-18)$

(d) $3\frac{1}{2} - \left(-1\frac{1}{2}\right)$

(e) $-4 - (-12)$

(f) $\frac{1}{2} - 2\frac{1}{2}$

(g) $-9 - 4\frac{1}{2}$

(h) $17 - (+6)$

(i) $-13 - 3\frac{1}{4}$

(j) $2.4 - (-1.3)$

(k) $1.2 - 6.5$

(l) $0 - (-4)$

Work carefully. The answers are in **21**.

---

21   (a) -4

(b) -7

(c) 32

(d) $+5$

(e) $+8$

(f) -2

(g) $-13\frac{1}{2}$

(h) 11

(i) $-16\frac{1}{4}$

(j) 3.7

(k) -5.3

(l) $+4$

Multiplying Signed Numbers

The *product* of two positive numbers is another positive number, something you learned back in arithmetic. $3 \times 4 = (+3) \times (+4) = +12$.

The product of numbers with opposite signs can be understood in terms of repeated addition. For example,

$$3 \times (-4) = (+3) \times (-4) = -12$$

because

$$3 \times (-4) = (-4) + (-4) + (-4) = -12$$

In the same way,

$$(-3) \times (+4) = -12$$

because

$$4 \times (-3) = (-3) + (-3) + (-3) + (-3) = -12$$

Finally, the product of two negative numbers is always a positive number. For example,

$$(-3) \times (-4) = +12$$

Do the following multiplications.

(a) $(+5) \times (-7) = $ _____

(b) $(-2) \times (-8) = $ _____

(c) $(-6) \times (+5) = $ _____

(d) $(-3) \times (-5) = $ _____

Look in **22** for the answers.

---

22   (a) -35 (b) $+16$ (c) -30 (d) $+15$

For these examples we can state the following general rule for the multiplication and division of two signed numbers:

Rule for Multiplying or Dividing Signed Numbers

★   If both numbers have the *same* sign, the result of the multiplication or division will be a *positive* number.

★   If the numbers have *opposite* signs, the result of the multiplication or division will be a *negative* number.

FACTORS or DIVISORS	PRODUCT or QUOTIENT
Same sign $(+) \times (+)$ $(-) \times (-)$	$+$
Opposite sign $(+) \times (-)$ $(-) \times (+)$	$-$

For multiplication: Same sign $\begin{cases} 4 \times 3 = (+4) \times (+3) = +12 \\ (-4) \times (-3) = +12 \end{cases}$

                        Opposite sign $\begin{cases} (-4) \times (+3) = -12 \\ (+4) \times (-3) = -12 \end{cases}$

For division: Same sign $\begin{cases} (+8) \div (+2) = \dfrac{+8}{+2} = +4 \\[2mm] (-8) \div (-2) = \dfrac{-8}{-2} = +4 \end{cases}$

                   Opposite sign $\begin{cases} (-8) \div (+2) = \dfrac{-8}{+2} = -4 \\[2mm] (+8) \div (-2) = \dfrac{+8}{-2} = -4 \end{cases}$

This is a very important rule in algebra. Work the following set of problems to be certain you understand it.

(a) $(-1.2) \times (-3)$ (b) $(0.02) \times (-1.5)$ (c) $(-2.4) \times (3.5)$

(d) $(-1.5) \times (-2)$ (e) $(-3) \times (-3)$ (f) $(-2) \times (-2) \times (-2)$

(g) $\left(\dfrac{1}{2}\right) \times \left(-\dfrac{2}{3}\right)$ (h) $\left(-\dfrac{2}{5}\right) \times \left(-\dfrac{1}{4}\right)$ (i) $\left(-\dfrac{3}{2}\right) \times \left(\dfrac{1}{6}\right)$

(j) $(-4.5) \div (-5)$ (k) $(20) \div (-4)$ (l) $(-3.6) \div (2)$

(m) $\dfrac{-7.2}{+6}$ (n) $\dfrac{-27}{-3}$ (o) $\dfrac{+4.2}{-0.7}$ (p) $(-4) \times 0$

(q) $(-1) \times 4$ (r) $(-1) \times (-4)$ (s) $(-4) \div (-1)$

Check your answers in **23**.

---

23   (a) $+3.6$ (b) -0.03 (c) -8.4 (d) $+3$ (e) $+9$

       (f) -8 (g) $-\dfrac{1}{3}$ (h) $+\dfrac{1}{10}$ (i) $-\dfrac{1}{4}$ (j) $+0.9$

       (k) -5 (l) -1.8 (m) -1.2 (n) $+9$ (o) -6

       (p) 0 (q) -4 (r) $+4$ (s) $+4$

Notice on the last three problems that multiplying or dividing by (-1) simply changes the sign of the number being multiplied or divided.

$$4 \times (-1) = -4 \qquad\qquad 4 \div (-1) = -4$$
$$(-4) \times (-1) = +4 \qquad\qquad (-4) \div (-1) = +4$$

Now, for more practice in the arithmetic of signed numbers turn to **24** for a set of problems.

Unit I
Introduction
to Algebra

The answers to all problems are on page 535.

A. Add or subtract as indicated.

1. $4 - 7 =$ _____

2. $-19 + 5 =$ _____

3. $-12 - 7 =$ _____

4. $15 + (-9) =$ _____

5. $-23 + (-9) =$ _____

6. $-18 - (-12) =$ _____

7. $13 + (-7) =$ _____

8. $21 - (-19) =$ _____

9. $-41 + 27 =$ _____

10. $101 - (-23) =$ _____

11. $-32 - 49 =$ _____

12. $-9 + (-16) - (+17) =$ _____

13. $-203 + (-99) =$ _____

14. $-78 - (-78) =$ _____

15. $-78 + (-78) =$ _____

16. $84 - (-28) - (-36) =$ _____

17. $2.3 - 4.7 =$ _____

18. $-4.0 - (-1.7) =$ _____

19. $8.1 + (-7) =$ _____

20. $-.05 - 3 =$ _____

21. $2.08 - 16 =$ _____

22. $-12.5 - (-18.4) =$ _____

23. $32 + (-2.04) =$ _____

24. $-13.7 - 14.3 =$ _____

25. $-\frac{1}{4} + \left(-\frac{1}{4}\right) =$ _____

26. $\frac{2}{3} - \left(-1\frac{1}{3}\right) =$ _____

27. $-7 + \frac{3}{4} =$ _____

28. $-\frac{3}{5} - \frac{4}{5} =$ _____

29. $-2\frac{3}{4} + \left(-1\frac{1}{3}\right) =$ _____

30. $6\frac{1}{3} - \left(-\frac{2}{5}\right) =$ _____

31. $-18\frac{2}{3} - \left(-5\frac{1}{6}\right) =$ _____

32. $\left(-\frac{1}{4}\right) + \left(-\frac{1}{3}\right) + \frac{1}{2} =$ _____

B. Multiply or divide as indicated.

1. $(-3) \times (-4) =$ _____

2. $(-7) \times 8 =$ _____

3. $5 \times (-9) =$ _____

4. $(-1) \times (-9) =$ _____

5. $6 \times (-7) =$ _____

6. $(-9) \times (-9) =$ _____

7. $(-4) \times 0 =$ _____

8. $(-1) \times 1 =$ _____

9. $(7) \times (-4) =$ _____

10. $(-3) \times 12 =$ _____

11. $(-21) \times (-2) =$ _____

12. $4 \times (-15) =$ _____

Date

Name

Course/Section

13. $\dfrac{-9}{3} = $ _____

14. $\dfrac{-56}{-8} = $ _____

15. $\dfrac{-27}{3} = $ _____

16. $\dfrac{36}{9} = $ _____

17. $64 \div (-8) = $ _____

18. $-81 \div 9 = $ _____

19. $-36 \div (-4) = $ _____

20. $51 \div (-3) = $ _____

21. $(-0.2) \times (-0.4) = $ _____

22. $(-1) \times 0.72 = $ _____

23. $0.9 \times (-10) = $ _____

24. $(-0.5) \times (-8) = $ _____

25. $\dfrac{-1.2}{0.3} = $ _____

26. $\dfrac{-4}{-0.5} = $ _____

27. $\dfrac{3.6}{-9} = $ _____

28. $\dfrac{-0.555}{15} = $ _____

29. $\left(-\dfrac{2}{3}\right) \times \left(-\dfrac{1}{2}\right) = $ _____

30. $8 \times \left(-\dfrac{3}{4}\right) = $ _____

31. $-\dfrac{4}{5} \times 10 = $ _____

32. $-1\dfrac{2}{3} \times \left(-\dfrac{3}{5}\right) = $ _____

C. **For each of the following insert $<$ or $>$ to make true statements:**

1. 5 _____ -3

2. -4 _____ 0

3. 9 _____ -10

4. 1 _____ -1

5. $\dfrac{3}{4}$ _____ $-1\dfrac{1}{4}$

6. 0 _____ $\dfrac{1}{2}$

7. -0.9 _____ 0.01

8. 0.2 _____ -6

9. -1.4 _____ -1.3

10. -0.5 _____ -0.05

11. 0.5 _____ -95

12. -0.001 _____ -0.9

13. $-2 + (-3)$ _____ $-2 - (-3)$

14. $5 \times (-1)$ _____ $(-2) \times (-2)$

15. $\dfrac{-16}{2}$ _____ $(-4) \times (-2)$

16. $\dfrac{-45}{9}$ _____ $(-2) \times 3$

17. $25 \times (-2)$ _____ $1 + (-1)$

18. -19 _____ $(-3) \times 6$

19. $-\dfrac{3}{4}$ _____ -0.7

20. -1.25 _____ $-1\dfrac{1}{5}$

21. $-\dfrac{1}{3}$ _____ $-\dfrac{1}{4}$

D. **Perform the indicated operations:**

1. $(-9) \times (-2) \times (-5) = $ _____

2. $\dfrac{-4 - 6}{-10} = $ _____

3. $\dfrac{-7-(-2)}{15} =$ _____

4. $\dfrac{-104+(-32)}{-2} =$ _____

5. $\dfrac{-5-(-5)}{5} =$ _____

6. $-14+(-62)-(-81) =$ _____

7. $\dfrac{-294}{7} =$ _____

8. $\dfrac{-396}{-3} =$ _____

9. $\dfrac{785}{-157} =$ _____

10. $-0.04+(-1.9)-(-7.3) =$ _____

11. $2.6-8.06+4.09 =$ _____

12. $-3.85-4.7-15.9 =$ _____

13. $(-1.93)\times(-0.14) =$ _____

14. $\dfrac{-1.8+(-1.8)}{-1.8} =$ _____

15. $\dfrac{0.0267}{-0.003} =$ _____

16. $28.357\div(-1) =$ _____

17. $-0.0001\div(-1) =$ _____

18. $-2006\div0.5 =$ _____

19. $-1\dfrac{2}{3}+\left(-3\dfrac{1}{6}\right)+\left(-2\dfrac{1}{4}\right)+\left(-\dfrac{5}{6}\right) =$ _____

20. $\dfrac{-5}{9}+\left(-\dfrac{2}{3}\right)-1\dfrac{1}{6}-\left(-2\dfrac{1}{2}\right) =$ _____

21. $\left(-\dfrac{1}{2}\right)\times\left(\dfrac{1}{3}\right)\times\left(-\dfrac{1}{4}\right)\times\left(\dfrac{1}{5}\right) =$ _____

22. $\dfrac{-7\frac{2}{5}+(-2\frac{3}{5})}{(-2)\times(-5)} =$ _____

23. $\dfrac{3\frac{1}{3}\times(-\frac{3}{10})}{(-0.25)\times(-4)} =$ _____

24. $\dfrac{-0.2+\frac{1}{5}}{-7-8.9-(-14.9)} =$ _____

25. $\dfrac{(-1.5)\times(-7)\times(-\frac{2}{3})}{-4+(-6)-(-9)+(-13)} =$ _____

26. $\dfrac{-45+(-1.8)+42}{(-0.3)\times(-0.4)} =$ _____

E. Word Problems

1. An elevator started on the fourth floor and rose 23 floors. It then came down 18 floors. At which floor was it on at that time? _____

2. Margaret owed $38.29, $17.95, $8.43, and $23.75. She collected her $62.48 paycheck and $15.00 owed her by a friend. What was Margaret's net financial status? _____

3. The East Pittsburgh Clinkers football team gained 16 yards on the first play, lost 9 yards on the second play, and lost 23 yards on the third. What was the net result of the three plays? _____

4. One winter day in Bemediji, Minnesota the thermometer at 10 A.M. read 10° F. The temperature dropped $2\frac{1}{2}$° per hour for three hours then rose one-half degree during the next hour. In the next four hours the temperature dropped 5 more degrees. What was the temperature at that time? _____

Date _____

Name _____

Course/Section _____

5. A plane was cruising at 22,000 feet, rose 4,000 feet, then descended 8,000 feet and leveled off. What was the new cruising altitude? _____

6. The value of a share of TNT stock was $3\frac{3}{4}$ on the day of purchase. The next day it dropped $\frac{3}{8}$; the next day it dropped $\frac{1}{4}$. The following day it rose $1\frac{1}{2}$ and another $\frac{3}{4}$ the next day. What was the share value at that point? _____

F. Brain Boosters (Brain Booster problems are more difficult and more fun than the regular problems. You will find them challenging, even tricky, so don't expect to be as successful with them as you are with the others.)

1. Find the numerical value of each of the following.

   (a) $2 - 3$ (b) $4 - (2 - 3)$
   (c) $-(4 - (2 - 3))$ (d) $-1(-1 - (-1) - (-(-1)))$
   (e) $1 - (2 - (3 - (4 - \ldots (9 - (10 - 11)))\ldots)$

2. What arithmetic symbol can we place between 2 and 3 to make a number that is greater than 2 but less then 3? _____

3. (a) Insert the following numbers, -5, -4, -3, -2, -1, 0, 1, 2, and 3, into the circles so that each side of the triangle sums up to -4. There are two possible ways to do it.

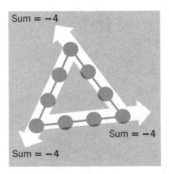

Sum = -4

Sum = -4

Sum = -4

   (b) Insert the same numbers to get a sum of -3.

4. A phonograph record has an outside diameter of 12 in., with a $\frac{1}{2}$ in. margin on the outer rim and a smooth center of $2\frac{1}{2}$ in. diameter. If there are 50 grooves to the in., how far does the needle move when the record is played? _____

When you have had the practice you need, either return to the preview test on page 1 or continue in **25** with the study of exponents.

25 When the same number appears many times as a factor, writing the product may become monotonous, tiring, and even inaccurate. It is easy, for example, to miscount the twos in

$$16{,}384 = 2 \times 2 \times 2 \times 2 \times 2 \times 2 \times 2 \times 2 \times 2 \times 2 \times 2 \times 2 \times 2 \times 2$$

or the tens in

$$100{,}000{,}000{,}000 = 10 \times 10 \times 10 \times 10 \times 10 \times 10 \times 10 \times 10 \times 10 \times 10 \times 10$$

Products of this kind are usually written in a shorthand form as 2^{14} and 10^{11}. In this exponential notation the raised number 14 indicates the number of times 2 is to be used as a factor. For example,

$$2 \times 2 = 2^2 \qquad \text{product of } \underline{\text{two}} \text{ factors of } 2$$
$$2 \times 2 \times 2 = 2^3 \qquad \text{product of } \underline{\text{three}} \text{ factors of } 2$$
$$\boxed{2 \times 2 \times 2 \times 2} = 2^4 \qquad \text{product of } \underline{\text{four}} \text{ factors of } 2$$

Four 2s

Write $3 \times 3 \times 3 \times 3 \times 3$ in exponential form.

$$3 \times 3 \times 3 \times 3 \times 3 = \underline{\hspace{2cm}}$$

Check your answer in **26**.

---

26 $\underbrace{3 \times 3 \times 3 \times 3 \times 3}_{\text{Five factors of 3}} = 3^5$ 　　　　　Read it as "three to the fifth power."

Base
Exponent

In this expression 3 is called the *base* and 5 is called the *exponent*.

$$3^5 = 3 \times 3 \times 3 \times 3 \times 3 = 243$$

Exponent — ; Base —

The exponent 5 tells you how many times the base 3 must be used as a factor in the product, and 3^5 or 243 is called the fifth *power* of 3.

Write each of the following products in exponential form and identify the base and exponent:

(a)　$5 \times 5 \times 5 \times 5 = \underline{\hspace{1.5cm}}$　　　　base $= \underline{\hspace{1.5cm}}$　　exponent $= \underline{\hspace{1.5cm}}$

(b)　$7 \times 7 = \underline{\hspace{1.5cm}}$　　　　　　　base $= \underline{\hspace{1.5cm}}$　　exponent $= \underline{\hspace{1.5cm}}$

(c)　$10 \times 10 \times 10 \times 10 \times 10 = \underline{\hspace{1.5cm}}$　　base $= \underline{\hspace{1.5cm}}$　　exponent $= \underline{\hspace{1.5cm}}$

(d)　$1 \times 1 \times 1 = \underline{\hspace{1.5cm}}$　　　　　base $= \underline{\hspace{1.5cm}}$　　exponent $= \underline{\hspace{1.5cm}}$

(e)　$3 \times 3 \times 3 \times 3 \times 3 \times 3 = \underline{\hspace{1.5cm}}$　　base $= \underline{\hspace{1.5cm}}$　　exponent $= \underline{\hspace{1.5cm}}$

The answers are in **27**. Go there when you finish these.

---

27 (a) $5 \times 5 \times 5 \times 5$ $= 5^4$ base = 5 exponent = 4

(b) 7×7 $= 7^2$ base = 7 exponent = 2

(c) $10 \times 10 \times 10 \times 10 \times 10 = 10^5$ base = 10 exponent = 5

(d) $1 \times 1 \times 1$ $= 1^3$ base = 1 exponent = 3

(e) $3 \times 3 \times 3 \times 3 \times 3 \times 3 = 3^6$ base = 3 exponent = 6

The following properties are very important.

★ Any power of 1 is equal to 1.

$1^n = 1$

$1^2 = 1 \times 1 = 1$
$1^3 = 1 \times 1 \times 1 = 1$
$1^4 = 1 \times 1 \times 1 \times 1 = 1$ and so on.

★ Zero to any positive power is equal to zero.

$0^n = 0$

★ When the base is 10, the exponent number is equal to the number of zeros in the final product.

$10^2 = 100$
$10^3 = 1000$
$10^4 = 10000$ and so on

4 zeros

★ For any real number, n,

$n^1 = n$

and for any real number n except $n = 0$,

$n^0 = 1$

For example, $2^1 = 2$, $3^1 = 3$, $4^1 = 4$, and so on.
$2^0 = 1$, $3^0 = 1$, $4^0 = 1$, and so on.
0^0 is not defined

To see why $2^0 = 1$, look at the following sequence of multiplications:

$2^3 = 2 \cdot 2 \cdot 2$
$\quad = 1 \cdot 2 \cdot 2 \cdot 2$ The value does not change if we multiply by 1.
$2^2 = 1 \cdot 2 \cdot 2$
$2^1 = 1 \cdot 2$
$2^0 = 1$

Now try the following set of problems on exponent arithmetic.

1. Evaluate:

(a) 2^3 (b) 3^2 (c) 4^2 (d) 5^3

(e) 2^5 (f) 5^1 (g) 4^0 (h) 1^7

(i) 10^3 (j) 4^3 (k) $(-2)^2$ (l) $(-3)^3$

(m) $(-1)^5$ (n) $\left(\dfrac{1}{2}\right)^2$ (o) $\left(\dfrac{1}{4}\right)^3$ (p) $(1.2)^3$

(q) $(0.5)^2$ (r) $(-2.5)^3$ (s) $(0.06)^2$ (t) $(-0.1)^4$

2. Write in exponential form:

(a) $2 \cdot 2 \cdot 2 \cdot 2$ (b) $3 \cdot 3$ (c) 10,000

(d) $8 \cdot 8 \cdot 8 \cdot 8 \cdot 8 \cdot 8$ (e) $2 \cdot 3 \cdot 3$ (f) $4 \cdot 5 \cdot 5 \cdot 5$

(g) $2 \cdot 2 \cdot 2 \cdot 3 \cdot 3$ (h) $1 \cdot 1 \cdot 1 \cdot 1 \cdot 1$ (i) $6 \cdot 6 \cdot 6 \cdot 6 \cdot 6$

(j) $2 \cdot 2 \cdot 3 \cdot 3 \cdot 3 \cdot 5$ (k) $10 \cdot 10 \cdot 10 \cdot 10$ (l) $7 \cdot 7 \cdot 7 \cdot 9 \cdot 9 \cdot 12$

Check your answers in **28**.

---

28   1.   (a)   8   (b)   9   (c)   16   (d)   125   (e)   32

(f)   5   (g)   1   (h)   1   (i)   1000   (j)   64

(k)   4   (l) -27 (m) -1 (n) $\dfrac{1}{4}$ (o) $\dfrac{1}{64}$

(p)   1.728   (q)   0.25   (r) -15.625 (s)   0.0036   (t)   0.0001

2.   (a) 2^4 (b) 3^2 (c) 10^4 (d) 8^6

(e) $2^1 \cdot 3^2$ (f) $4^1 \cdot 5^3$ (g) $2^3 \cdot 3^2$ (h) 1^5

(i) 6^5 (j) $2^2 \cdot 3^3 \cdot 5^1$ (k) 10^4 (l) $7^3 \cdot 9^2 \cdot 12^1$

Of course, this arithmetic shorthand notation may be applied to algebra and used with literal numbers.

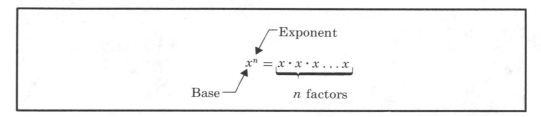

$x^0 = 1$ for $x \neq 0$
$x^1 = x$
$x^2 = x \cdot x$
$x^3 = x \cdot x \cdot x$ and so on.

x^2 is read "x squared" or "x to the second power."

x^3 is read "x cubed" or "x to the third power."

x^4 is read "x to the fourth power."

and so on.

If $a = 4$, find: (a) a^3 (b) $3a$

Be careful.

Compare your answers with ours in **29**.

---

29   For $a = 4$ (a) $a^3 = 4^3$: $4 \times 4 \times 4 = 16 \times 4 = 64$
(b) $3a = 3 \times 4 = 12$

Coefficients

In an expression such as $3a$, the number 3 is called the *coefficient* of a. A coefficient is simply a multiplier or factor.

In the literal expression $2ax$, $2a$ is the coefficient of x, $2x$ is the coefficient of a, and 2 is the coefficient of ax. We usually refer to the numerical part of a term as its *numerical coefficient*. The numerical coefficient in $5a$ is 5. The numerical coefficient in the term $3xy^2$ is 3. The coefficient of x in the expression $x + 2$ is 1.

Write each of the following products in exponential form and underline the numerical coefficient in each if there is one.

(a) $b \cdot b \cdot b \cdot b$ (b) x cubed

(c) $z \cdot z \cdot z \cdot z \cdot z$ (d) $3 \cdot x \cdot x \cdot y$

(e) $16 \cdot a \cdot a \cdot a \cdot d \cdot d$

(f) $8 \cdot r \cdot s \cdot t \cdot t \cdot v \cdot v \cdot v$

(g) $(x + 1) \cdot (x + 1)$

(h) $(a + b) \cdot (a + b) \cdot (a + b)$

(i) $3 \cdot x \cdot x \cdot (x - 1) \cdot (x - 1)$

(j)   The cube of $(a - d)$

(k)   The fourth power of $(q + t)$

(l)   The product of 5 and the square of $(p + 2)$

(m) $7 \cdot a \cdot b \cdot b \cdot c \cdot (d + e) \cdot (d + e)$

(n)   The third power of $(z - 1)$

(o)   The cube of ab

Look in **30** to check your answers.

30

(a) b^4

(b) x^3

(c) z^5

(d) $\underline{3}x^2y$

(e) $\underline{16}a^3d^2$

(f) $\underline{8}rst^2v^3$

(g) $(x + 1)^2$

(h) $(a + b)^3$

(i) $\underline{3}x^2(x - 1)^2$

(j) $(a - d)^3$

(k) $(q + t)^4$

(l) $\underline{5}(p + 2)^2$

(m) $\underline{7}ab^2c(d + e)^2$

(n) $(z - 1)^3$

(o) $(ab)^3$ or a^3b^3

If no numerical coefficient is given, the coefficient is understood to be 1.

Negative Exponents

Finding exponential products when the base is a negative number is difficult for some students. It will help if you keep in mind that:

$$(-a)^2 = (-a) \cdot (-a) = +a^2$$

$$(-a)^3 = (-a) \cdot (-a) \cdot (-a) = (+a^2) \cdot (-a) = -a^3$$

$$(-a)^4 = (-a) \cdot (-a) \cdot (-a) \cdot (-a) = (+a^2) \cdot (+a^2) = +a^4$$

or, in general,

$(-a)^n = +a^n$ when the exponent n is an even number, $n = 2, 4, 6, 8 \ldots$

and

$(-a)^n = -a^n$ when the exponent n is an odd number, $n = 1, 3, 5, 7 \ldots$

For example, we know that

$(-x)^5 = -x^5$ since the exponent is 5, an odd number

and

$(-x)^8 = +x^8$ since the exponent is 8, an even number.

Do these few problems to get the idea firmly established:

(a) $(-y)^3$ (b) $(-H)^4$ (c) $(-q)^7$ (d) $(-z)^5$

(e) $(-2x)^3$ (f) $\left(-\dfrac{1}{2}A\right)^2$ (g) $(-3t)^4$ (h) $(-5q)^1$

(i) $(-1.5B)^2$ (j) $[-(x+1)]^3$ (k) $(-2s)^6$ (l) $(-2x)^5$

Turn to **31** to check your answers.

---

31 (a) $-y^3$ (b) $+H^4$ (c) $-q^7$ (d) $-z^5$

(e) $-8x^3$ (f) $+\dfrac{1}{4}A^2$ (g) $+81t^4$ (h) $-5q$

(i) $+2.25B^2$ (j) $-(x+1)^3$ (k) $+64s^6$ (l) $-32x^5$

Multiplying Numbers in Exponential Form

The multiplication of powers of numbers is very simple and quick. The product $x^2 \cdot x^3$ can be calculated as

$$x^2 \cdot x^3 = (x \cdot x) \cdot (x \cdot x \cdot x)$$
$$= (x \cdot x \cdot x \cdot x \cdot x)$$
$$= x^5$$

The x^2 part contributes x as a factor twice, and the x^3 part contributes x as a factor three times. The product of x^2 and x^3 therefore has x as a factor five times.

> To multiply exponent numbers having the same base add the exponents:
>
> $$x^m \cdot x^n = x^{m+n}$$

For example,

$$2^3 \cdot 2^4 = 2^{3+4} = 2^7$$
$$3 + 4 = 7$$

You should notice in particular that

$$x \cdot x^3 = x^1 \cdot x^3 = x^{1+3} = x^4$$

the exponent is understood to be 1.

✳ Careful

Only powers of the *same* base may be multiplied in this way. Products such as $x^2 \cdot y^3$ *cannot* be multiplied in this way.

For example, $3^2 \cdot 4^3$ *cannot* be multiplied by adding exponents.

Multiply each of the following using this rule.

(a) $m^3 \cdot m^6$ m^9 (b) $q \cdot q^5$ q^6 (c) $r^2 \cdot r^2 \cdot r^3$ R^7

(d) $y^4 \cdot y \cdot y^3$ y^8 (e) $p^2(p^3)$ p^5 (f) $a(a^2)$ a^3

(g) $z^i \cdot z^j$ z^{i+j} (h) $t \cdot t^n$ t^{1n} (i) $p^5 \cdot q^3$ $p^5 \cdot q^3$

Check your answers in **32**.

---

32 (a) m^9 (b) q^6 (c) r^7 (d) y^8

(e) p^5 (f) a^3 (g) z^{i+j} (h) t^{n+1} (i) p^5q^3

It is easy to see from the rule for the multiplication of powers that

$$x^2 \cdot x^2 = x^4$$
$$x^3 \cdot x^3 = x^6$$
and $$x^4 \cdot x^4 = x^8$$

Or, in general, $\quad (x^n)^2 = x^n \cdot x^n = x^{n+n} = x^{2n}$
$$(x^n)^3 = x^n \cdot x^n \cdot x^n = x^{n+n+n} = x^{3n}$$
$$(x^n)^4 = x^n \cdot x^n \cdot x^n \cdot x^n = x^{n+n+n+n} = x^{4n}$$

and so on.

*Power of a Number
in Exponential Form*

To find the power of an exponent number, that is, to raise x^n to the mth power, multiply the exponents.

$$\boxed{(x^n)^m = x^{nm}}$$

Examples: $\quad (y^3)^2 = y^{3 \cdot 2} = y^6$
$$(x^4)^3 = x^{4 \cdot 3} = x^{12}$$
$$(-x^2)^4 = (-1)^4 \cdot (x^2)^4 = +1 \cdot x^{2 \cdot 4} = x^8$$

If several variables are present in the expression the procedure is very similar.

$$(x^n y^p)^m = x^{nm} y^{pm}$$

Examples: $\quad (xy)^2 = x^2 y^2$

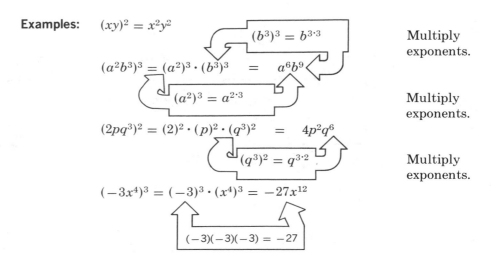

$(a^2 b^3)^3 = (a^2)^3 \cdot (b^3)^3 = a^6 b^9$ — Multiply exponents.

$(2pq^3)^2 = (2)^2 \cdot (p)^2 \cdot (q^3)^2 = 4p^2 q^6$ — Multiply exponents.

$(-3x^4)^3 = (-3)^3 \cdot (x^4)^3 = -27x^{12}$ — Multiply exponents.

$(-3)(-3)(-3) = -27$

A few problems will help. Find:

(a) $(y^2)^2$ (b) $(z^2)^3$ (c) $(w^4)^2$ (d) $(B^2)^4$

(e) $(a^4 b^5)^2$ (f) $(pq^2)^3$ (g) $(R^3)^5$ (h) $(z^a)^p$

(i) $(2x)^3$ (j) $(\frac{1}{2}a^2 b^4)^2$ (k) $(\frac{2}{3}xy^2)^3$ (l) $(-2p^2 q^5)^2$

(m) $(-2y)^4$ (n) $(-3a^2 b^5)^3$ (o) $(-5x^5 y^3)^2$ (p) $(-x^2 yz^3)^5$

You will find the answers in **33**.

---

33   (a) y^4 (b) z^6 (c) w^8 (d) B^8

(e) $a^8 b^{10}$ (f) $p^3 q^6$ (g) R^{15} (h) z^{ap}

(i) $8x^3$ (j) $\frac{1}{4}a^4 b^8$ (k) $\frac{8}{27}x^3 y^6$ (l) $4p^4 q^{10}$

(m) $16y^4$ (n) $-27a^6 b^{15}$ (o) $25x^{10} y^6$ (p) $-x^{10} y^5 z^{15}$

*Dividing Numbers
in Exponential Form*

The division of numbers in exponential form involves a rule similar to the multiplication rule. For example,

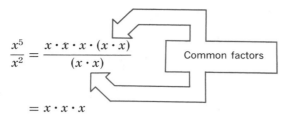

$$\frac{x^5}{x^2} = \frac{x \cdot x \cdot x \cdot (x \cdot x)}{(x \cdot x)}$$

Common factors

$$= x \cdot x \cdot x$$

$= x^3$ We have divided both numerator and denominator of the fraction by the common factor $(x \cdot x)$.

Notice that the final exponent is the difference between the exponent of the numerator and the exponent of the denominator.

$$\frac{x^5}{x^2} = x^{5-2} = x^3$$

When the base of the number is the same, we can simplify a quotient by dividing out common factors. This is equivalent to subtracting the exponent of the denominator from the exponent of the numerator.

$$\boxed{\frac{x^m}{x^n} = x^{m-n}}$$

Examples: $\dfrac{x^{10}}{x^4} = x^{10-4} = x^6$

$$\frac{x^3}{x^2} = x^{3-2} = x^1 = x$$

$$\frac{x^2}{x^5} = x^{2-5} = x^{-3}$$

$$\frac{x^4}{x} = \frac{x^4}{x^1} = x^{4-1} = x^3$$

Always subtract the exponent in the denominator from the exponent in the numerator.

Think of this operation as dividing out the common factors on top and bottom of the fraction.

 Again, be careful: only powers of the *same* base may be divided in this way.

Divide each of the following using this rule.

(a) $\dfrac{Q^5}{Q^3}$ (b) $\dfrac{A^6}{A^2}$ (c) $\dfrac{r^3}{r}$ (d) $\dfrac{t^2}{t^2}$

(e) $\dfrac{2z^4}{z^2}$ (f) $\dfrac{3y^7}{2y^4}$ (g) $\dfrac{p^4}{p^0}$ (h) $\dfrac{x^3}{x^5}$

Check your answers in **34**.

---

34 (a) Q^2 (b) A^4 (c) r^2 (d) 1

    (e) $2z^2$ (f) $\dfrac{3y^3}{2}$ (g) p^4 (h) x^{-2}

Zero Exponents Notice in problem (d) that $\dfrac{t^2}{t^2} = t^{2-2} = t^0$ and any non-zero number to the zeroth power is defined to be 1.

From arithmetic, $\dfrac{2}{2} = 1, \dfrac{7}{7} = 1, \dfrac{798}{798} = 1$, and so on.

In algebra, $\dfrac{x^3}{x^3} = 1, \dfrac{y^5}{y^5} = 1, \dfrac{z^{16}}{z^{16}} = 1$, and so on.

In problem (h) above $\dfrac{x^3}{x^5} = x^{3-5} = x^{-2}$.

Negative Exponents

To understand the meaning of the negative exponent, consider the following: divide out the common factors in $\dfrac{x^3}{x^5}$.

$$\frac{x^3}{x^5} = \frac{x \cdot x \cdot x}{x \cdot x \cdot x \cdot x \cdot x} = \frac{1}{x \cdot x} = \frac{1}{x^2}$$

But $\qquad \dfrac{x^3}{x^5} = x^{3-5} = x^{-2} \qquad$ so $\qquad x^{-2} = \dfrac{1}{x^2}$

When a negative exponent appears it always has this meaning.

$$\boxed{x^{-n} = \frac{1}{x^n}}$$

Any number to a negative power is equal to the fraction 1 divided by that same number to the positive power.

For example, $x^{-2} = \dfrac{1}{x^2}$

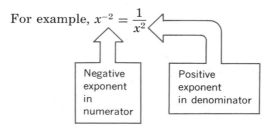

Negative exponent in numerator

Positive exponent in denominator

More examples:

$2^{-3} = \dfrac{1}{2^3} = \dfrac{1}{8}$

$3^{-2} = \dfrac{1}{3^2} = \dfrac{1}{9}$

$(-2)^{-4} = \dfrac{1}{(-2)^4} = \dfrac{1}{2^4} = \dfrac{1}{16}$

$a^{-5} = \dfrac{1}{a^5}$

$(x^2)^{-3} = x^{2 \cdot (-3)} = x^{-6} = \dfrac{1}{x^6}$

$(p^2 q)^{-4} = (p^2)^{-4}(q)^{-4} = p^{2 \cdot (-4)} \cdot q^{-4} = p^{-8}q^{-4} = \dfrac{1}{p^8 q^4}$

$-2^{-2} = -(2)^{-2} = -\dfrac{1}{2^2} = -\dfrac{1}{4}$

Rewrite each of the following with positive exponents, and simplify if possible.

(a) 4^{-2} (b) 5^{-1} (c) 3^{-3} (d) -3^{-3}

(e) x^{-6} (f) $(a^3)^{-5}$ (g) $(y^{-2})^4$ (h) $(-3)^{-2}$

(i) $(a^2 b^3)^{-2}$ (j) $(-x)^{-3}$ (k) $-t^{-4}$ (l) $(y^{-2})^{-4}$

Write without a denominator.

(m) $\dfrac{x^2}{y^3}$ (n) $\dfrac{x^{-4}}{y^{-6}}$ (o) $\dfrac{3^2 x^5}{3^4 x^{-2}}$ (p) $\dfrac{2x^{-2} y^2}{16 x^{-4} y^5}$

Check your answers in **35**.

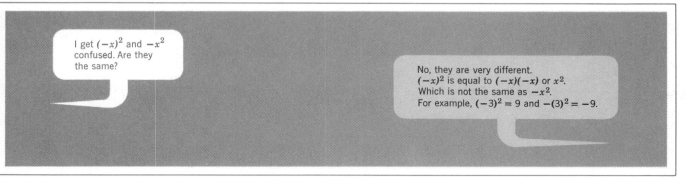

35 (a) $\dfrac{1}{4^2} = \dfrac{1}{16}$ (b) $\dfrac{1}{5^1} = \dfrac{1}{5}$ (c) $\dfrac{1}{3^3} = \dfrac{1}{27}$

(d) $-\dfrac{1}{3^3} = -\dfrac{1}{27}$ (e) $\dfrac{1}{x^6}$ (f) $\dfrac{1}{a^{15}}$

(g) $\dfrac{1}{y^8}$ (h) $\dfrac{1}{(-3)^2} = \dfrac{1}{9}$ (i) $\dfrac{1}{a^4 b^6}$

(j) $\dfrac{1}{(-x)^3} = -\dfrac{1}{x^3}$ (k) $-\dfrac{1}{t^4}$ (l) $y^{(-2)(-4)} = y^8$

(m) $x^2 y^{-3}$ (n) $x^{-4} y^6$ (o) $3^{-2} x^7$

(p) $2^{-3} x^2 y^{-3}$

We will work with negative exponents more extensively in Unit 5.

Now turn to **36** for a set of practice problems on exponential notation.

PROBLEM SET 1-3

36

Exponents

Answers are on page 536.

A.   Write each of the following products in exponential form:

1. $10 \cdot 10 \cdot 10 = $ _____

2. $5 \cdot 5 \cdot 5 \cdot 5 \cdot 5 = $ _____

3. $7 \cdot 7 = $ _____

4. $c \cdot c \cdot c \cdot c \cdot c \cdot c \cdot c = $ _____

5. $10 \cdot a \cdot a \cdot a = $ _____

6. $3 \cdot 3 \cdot x \cdot x \cdot y \cdot y = $ _____

7. $n \cdot n \cdot p \cdot p \cdot p = $ _____

8. $e \cdot e \cdot f \cdot g \cdot g \cdot g = $ _____

9. $(x + y)(x + y) = $ _____

10. $(3 - b)(3 - b)(3 - b)(3 - b) = $ _____

11. $6 \cdot 6 \cdot x \cdot x \cdot (x - 2)(x - 2) = $ _____

12. $7 \cdot 7 \cdot 7 \cdot t \cdot t \cdot t \cdot (t - r)(t - r) = $ _____

13. $(d - c)(d - c)(d - c)(d - c)(d - c) = $ _____

14. $(x + y)(x + y)(x - y)(x - y) = $ _____

15. $10 \cdot 10 \cdot 10 \cdot 10 \cdot g \cdot g \cdot (g - 2)(g - 2) = $ _____

B.   Find the value of these:

1. $2^5 = $ _____

2. $7^2 = $ _____

3. $5^0 = $ _____

4. $10^6 = $ _____

5. $1^9 = $ _____

6. $3^5 = $ _____

7. $6^3 = $ _____

8. $8^1 = $ _____

9. $10^4 = $ _____

10. $2^{10} = $ _____

11. $(-3)^2 = $ _____

12. $(-1)^3 = $ _____

13. $(-5)^3 = $ _____

14. $(-2)^6 = $ _____

15. $(-1)^7 = $ _____

16. $(-7)^0 = $ _____

17. $(\frac{1}{3})^3 = $ _____

18. $(\frac{1}{6})^2 = $ _____

19. $(\frac{3}{4})^3 = $ _____

20. $(\frac{2}{5})^4 = $ _____

21. $(1\frac{1}{2})^2 = $ _____

22. $(\frac{5}{8})^2 = $ _____

23. $(2\frac{1}{2})^2 = $ _____

24. $(\frac{3}{10})^3 = $ _____

25. $(0.9)^2 = $ _____

26. $(0.1)^3 = $ _____

27. $(3.4)^2 = $ _____

28. $(-0.4)^2 = $ _____

29. $(-0.03)^3 = $ _____

30. $(-1.5)^3 = $ _____

31. $(-0.01)^2 = $ _____

32. $(2.06)^2 = $ _____

C.   Simplify the following:

1. $(2x)^4 = $ _____

2. $(-5c)^3 = $ _____

3. $(7mn)^2 = $ _____

4. $(-3a)^3 = $ _____

Date _____

Name _____

Course/Section _____

5. $(-2y)^6 = $ _____

6. $\left(-\dfrac{1}{3}d\right)^2 = $ _____

7. $(0.1yt)^3 = $ _____

8. $(10xy)^5 = $ _____

9. $(-1.2rs)^3 = $ _____

10. $\left(-\dfrac{3}{4}a^2c^2\right)^2 = $ _____

11. $(0.2n^2)^2 = $ _____

12. $\left(-\dfrac{1}{3}b\right)^4 = $ _____

13. $y^2 \cdot y^5 = $ _____

14. $t^6 \cdot t^8 = $ _____

15. $e^7 \cdot e^3 \cdot e^9 = $ _____

16. $-x^2 \cdot x^3 = $ _____

17. $2a^3 \cdot a^2 \cdot a = $ _____

18. $\dfrac{1}{2}c^8 \cdot c^3 \cdot c^4 = $ _____

19. $3n \cdot n^5 = $ _____

20. $-2h^2 \cdot h^8 = $ _____

21. $b^4(b^2c) = $ _____

22. $3w^3 \cdot x^2 \cdot (w^2x^2) = $ _____

23. $y^n y^3 = $ _____

24. $\dfrac{e^4}{e^2} = $ _____

25. $\dfrac{t^8}{t} = $ _____

26. $\dfrac{m^{17}}{m^9} = $ _____

27. $\dfrac{k^2}{k^0} = $ _____

28. $\dfrac{15y^5}{3y} = $ _____

29. $\dfrac{48x^9}{16x^9} = $ _____

30. $\dfrac{21c^3d^2}{7c^2d^2} = $ _____

31. $\dfrac{-32a^3b^2}{-16ab} = $ _____

32. $\dfrac{3.4s^6}{-1.7s^4} = $ _____

33. $(3x^2y^2z)^4 = $ _____

34. $\left(\dfrac{1}{2}pq^3t^2\right)^4 = $ _____

35. $\left(\dfrac{2}{3}a^3bc^5\right)^2 = $ _____

36. $\left(\dfrac{x^2}{y^3}\right)^3 = $ _____

37. $\left(\dfrac{2}{x}\right)^3 = $ _____

38. $\left(\dfrac{3ab^2}{c^3}\right)^2 = $ _____

39. $x^{19} \cdot x^{24} = $ _____

40. $a^{13} \cdot a \cdot a^{11} = $ _____

41. $p^{12} \cdot p^9 \cdot p^{31} = $ _____

42. $(x^7)^9 = $ _____

43. $(x^{11})^{11} = $ _____

44. $(a^{21})^7 = $ _____

45. $(2x^5)^{10} = $ _____

46. $(3a^7)^6 = $ _____

47. $\left(\dfrac{x^5 \cdot x^{17}}{y^9}\right)^8 = $ _____

48. $t^4(t^{11} \cdot t^8) = $ _____

49. $3x^{13} \cdot x^{21} = $ _____

50. $y^{20} \cdot y^{18} = $ _____

D. **Find the value of the following:**

1. $10^8 =$ _____

2. $(-3)^3 =$ _____

3. $\left(\dfrac{2}{3}\right)^4 =$ _____

4. $(0.5)^3 =$ _____

5. $(1.2)^2 =$ _____

6. $(-2)^3 =$ _____

7. $\left(\dfrac{1}{2}\right)^3 =$ _____

8. $\left(\dfrac{3}{4}\right)^3 =$ _____

9. $(2.5)^3 =$ _____

10. $(2)^{-2} =$ _____

11. $(2)^{-5} =$ _____

12. $(0.3)^{-3} =$ _____

13. $(0.25)^{-2} =$ _____

14. $-4^2 =$ _____

15. $(-4)^2 =$ _____

16. $-4^{-2} =$ _____

17. $(-4)^{-2} =$ _____

18. $(0.6)^0 =$ _____

Write the following using only positive exponents:

19. $x^{-4} =$ _____

20. $y^{-5} =$ _____

21. $(a^2)^{-4} =$ _____

22. $(b^3)^{-6} =$ _____

23. $(x^{-3})^6 =$ _____

24. $(t^{-1})^4 =$ _____

25. $(x^3 y^5)^{-2} =$ _____

26. $(r^4 t^{-2})^3 =$ _____

27. $(2xy^3)^{-3} =$ _____

28. $(p^{-4}q^3)^{-2} =$ _____

29. $(x^5 y^{-3})^{-4} =$ _____

30. $(3np^4)^{-2} =$ _____

Write without a denominator:

31. $\dfrac{1}{x^3} =$ _____

32. $\dfrac{a^4}{x^5} =$ _____

33. $\dfrac{d^{-2}}{e^{-5}} =$ _____

34. $\dfrac{2}{8x^{-2}} =$ _____

35. $\dfrac{x^{-3}y^4}{9x^2 y^{-3}} =$ _____

36. $\dfrac{x^3 y^{-2}}{5x^4 y^{-5}} =$ _____

37. $\dfrac{x^{-6}y^2}{y^{-4}x^{-2}} =$ _____

38. $\dfrac{x^{-3}y^7}{2x^5 y^4} =$ _____

39. $\dfrac{x^2 y^{-8}}{9x^{-6}y^{-3}} =$ _____

E. **Simplify the following:**

1. $(0.3y)^4 =$ _____

2. $\left(-\dfrac{1}{2}b\right)^5 =$ _____

3. $(10cq)^6 =$ _____

4. $\left(-\dfrac{1}{7}p\right)^2 =$ _____

5. $f^{12} \cdot f^3 \cdot f^7 =$ _____

6. $8y(y^4) =$ _____

7. $(x^2 y)^2 \cdot y^3 =$ _____

8. $(tm)^3 t^3 m^4 =$ _____

9. $\dfrac{2a^3 \cdot a^4}{a^2} =$ _____

10. $\dfrac{k^3 \cdot k^{16}}{k \cdot k^3} =$ _____

11. $\dfrac{3p \cdot p^2}{3p} =$ _____

12. $\dfrac{36x^2 y^3}{9xy} =$ _____

13. $\dfrac{56b^6 d}{8b^3 d} =$ _____

14. $(0.12n)^3 \cdot n^4 =$ _____

Date _____

Name _____

Course/Section _____

41 **Problem Set 1-3**

15. $\dfrac{6z^4}{z^4} = $ _____

16. $\dfrac{(3q)^4}{q^2} = $ _____

17. $\dfrac{77x^2 \cdot x^4}{11x^4} = $ _____

18. $\dfrac{(1.5h)^2 \cdot h^3}{0.3h} = $ _____

19. $\left(\dfrac{1}{4}\,a^3\right)^2 \cdot 8a^5 = $ _____

20. $\dfrac{6d^4c^2 \cdot (d^3 \cdot c^4)}{12d^7c^6} = $ _____

21. $(a^2b^3)^{-2} = $ _____

22. $(2x^2)^{-2} = $ _____

23. $(5xy^2)^{-3} = $ _____

24. $(4pq^3t)^{-3} = $ _____

25. $x^4 \cdot x^2 \cdot x^{-3} = $ _____

26. $2a \cdot a^{-4} \cdot a^3 = $ _____

27. $3t^2 \cdot t^{-5} \cdot t = $ _____

28. $y \cdot y^3 \cdot y^{-6} = $ _____

29. $2x \cdot 3x^{-4} \cdot x^3 = $ _____

30. $2p^{-2} \cdot p^{-3} \cdot 4p^7 = $ _____

F.   Brain Boosters

1. Which is greater?   (a) $((2^2)^2)^2$ (b) $(2^2)^{(2^2)}$ (c) $(2)^{(2^2)^2}$

2. Simplify $\dfrac{5^7 \times 12^4 \times 32^2 \times (3^2 \times 4^2 \times 5)^2}{(3 \times 15 \times 2^3)^8}$ _____

3. Are these equations true? Check them.

(a) $1^1 + 2^1 = 3^1$
(b) $6^{-1} + 3^{-1} = 2^{-1}$
(c) $1 + 5 + 6 = 2 + 3 + 7$
(d) $1^2 + 5^2 + 6^2 = 2^2 + 3^2 + 7^2$
(e) $1 + 5 + 8 + 12 = 2 + 3 + 10 + 11$
(f) $1^2 + 5^2 + 8^2 + 12^2 = 2^2 + 3^2 + 10^2 + 11^2$
(g) $1^3 + 5^3 + 8^3 + 12^3 = 2^3 + 3^3 + 10^3 + 11^3$

4. Show that the following equations are true and predict the next equation.

$3^2 + 4^2 = 5^2$
$10^2 + 11^2 + 12^2 = 13^2 + 14^2$
$21^2 + 22^2 + 23^2 + 24^2 = 25^2 + 26^2 + 27^2$

_____

When you have had the practice you need either return to the preview test on page 1 or continue to **37** where you will learn to work with literal numbers.

_____

Adding quantities, whether in arithmetic or algebra, is a matter of counting.

Adding apples,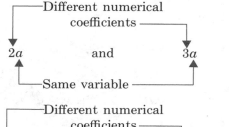

or money, $2¢ + 3¢ = 5¢$

or anything else, 2 *things* + 3 *things* = 5 *things*

means counting the number of similar objects or like quantities.

We may add similar algebraic quantities in exactly the same way.

Two terms are said to be *like terms* if they differ only in their numerical coefficients.

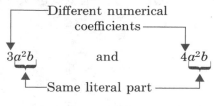

$2x$ and $3x^2$ are *not* like terms because the variable x appears to different powers.

To add like terms, add their numerical coefficients and keep the same literal part.

For example,

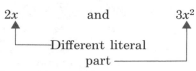

We are adding *like* quantities, whether apples, ¢, or, in this case, "a"s.

$3a^2b + 4a^2b = 7a^2b$

$3 + 4 = 7$

The literal part of each term is a^2b.

$$3 \; \boxed{a^2b} = \boxed{a^2b} + \boxed{a^2b} + \boxed{a^2b}$$

Try these problems to get the idea firmly in your "mental muscles":

Add:

(a) $3x + 5x$

(b) $3p + 6p$

(c) $q + 2q$

(d) $3y + y + 5y$

(e) $11z + 4z$

(f) $7st + 2st + 3st$

(g) $d^2 + 3d^2$

(h) $4x^2y + 5x^2y$

(i) $2wm + 8wm + 3w^2m$

(j) $5ab^2c^3 + 7ab^2c^3$

(k) $2xy^2 + 3x^2y$

(l) $7h + 2h^2$

(m) $7M^4 + 2M^4$

(n) $E + 2E + 3E$

(o) $2(a + 1) + 4(a + 1)$

Work carefully, adding only like terms.

Check your answers in **38**.

38   (a) $8x$ (b) $9p$ (c) $3q$ (d) $9y$ (e) $15z$
        (f) $12st$ (g) $4d^2$ (h) $9x^2y$ (i) $10wm + 3w^2m$ (j) $12ab^2c^3$
        (k) $2xy^2 + 3x^2y$ (l) $7h + 2h^2$ (m) $9M^4$ (n) $6E$ (o) $6(a + 1)$

You should have noticed that in problems (i), (k), and (l) we cannot combine all of the terms because they are not all like terms. In (l) $7h$ and $2h^2$ are *not* like terms because the variable or literal part is not the same in both terms.

When both like and unlike terms are included in the same sum, we may rearrange the terms and add the like terms in any order we wish. For example, we add $2a + 3a$ or $3a + 2a$ to get $5a$. The order in which the terms are added does not change the final sum. Furthermore, with three terms, we may add

$2a + 3a + 5a$ as $(2a + 3a) + 5a = 5a + 5a$
                    or $2a + (3a + 5a) = 2a + 8a$
               to get $10a$

Again, the way in which the terms are grouped for addition does not change the final sum.

Commutative Law of Addition

Two general properties of real numbers assure us that this kind of rearranging is legal. The first property is called the *commutative law of addition*:

For any real numbers a and b,

$$\boxed{a + b = b + a}$$

        $2 + 3 = 3 + 2$

Like a commuter traveling from the suburbs to the city, the terms can change positions.

Associative Law of Addition

The second property is called the *associative law of addition*:

For any real numbers a, b, and c,

$$\boxed{(a + b) + c = a + (b + c)}$$

     $(2 + 3) + 4 = 2 + (3 + 4)$

The terms may associate or join together a with b or b with c.

These two laws together tell us that we can add a list of numbers in any order we wish.

Most people take for granted this ability to rearrange numbers when they do addition, but it should be comforting for you to know that the laws hold for any of the quantities you might find in algebra—positive or negative integers, decimal numbers, fractions, numbers in exponential form, square roots, and any of the letters used to represent these numbers.

These two laws enable us to simplify sums of many terms. For example, the sum

$3a + 2b + a + 5b + 4a$

becomes

$(3a + a + 4a) + (2b + 5b)$ We rearrange and group like terms together.

or finally,

$8a + 7b$ adding like terms.

Another example:

$2x^2 + 4y + x^2 + 5y + xy$

becomes

$(2x^2 + x^2) + (4y + 5y) + xy$

or finally,

$3x^2 + 9y + xy$

Try these problems. Simplify by combining like terms:

(a) $2ab + 5ab + ab + b^2$ (b) $3p + 2qt + qt + p + tp$

(c) $s^2t + st + st^2 + st^2 + st$ (d) $2d^2 + 3d + d^2 + d + 2 + d + 1$

(e) $3xy^2 + 2xy^2 + xy^2$ (f) $4ce + 2e + 3f + 2ce + e + f$

(g) $a^2b + 4ab + 3b^2 + ab + 2a^2b$ (h) $3uv + 2uv + u^2 + uv + 2u^2$

(i) $a + b + c + 2a + ab + 3b + 2c$ (j) $x + 2y + 3z + 2x + y + z + 4$

(k) $p^2qr + pq^2r + 3p^2qr + pqr^2$ (l) $sty + s^3ty + 2s^3ty + 2sty + st^2y$

(m) $8b^2cd + 3bc^2d + 2bcd + 3b^2cd + bcd$

(n) $4xyz + 2xyz^2 + 2xyz + 3xy^2z + 4xy^2 + xy^2z$

Check your work in **39**.

---

39   (a) $8ab + b^2$ (b) $4p + 3qt + tp$

(c) $s^2t + 2st + 2st^2$ (d) $3d^2 + 5d + 3$

(e) $6xy^2$ (f) $6ce + 3e + 4f$

(g) $3a^2b + 5ab + 3b^2$ (h) $6uv + 3u^2$

(i) $3a + 4b + 3c + ab$ (j) $3x + 3y + 4z + 4$

(k) $4p^2qr + pq^2r + pqr^2$ (l) $3sty + 3s^3ty + st^2y$

(m) $11b^2cd + 3bc^2d + 3bcd$ (n) $6xyz + 2xyz^2 + 4xy^2z + 4xy^2$

Subtracting
Literal Numbers

The subtraction of literal numbers is a process exactly similar to addition. For example,

$5a - 2a = 3a$
$5 - 2 = 3$

$(\boxed{a} + \boxed{a} + \boxed{a} + \boxed{a} + \boxed{a}) - (\boxed{a} + \boxed{a}) = \boxed{a} + \boxed{a} + \boxed{a}$

The literal part of each term is a. We are subtracting "a"s.

and

$6a^2b - 4a^2b = 2a^2b$
$6 - 4 = 2$

$(\boxed{a^2b} + \boxed{a^2b} + \boxed{a^2b} + \boxed{a^2b} + \boxed{a^2b} + \boxed{a^2b})$
$- (\boxed{a^2b} + \boxed{a^2b} + \boxed{a^2b} + \boxed{a^2b}) = \boxed{a^2b} + \boxed{a^2b}$

The literal part of each term is a^2b.

To simplify a series of terms being added or subtracted, group together like terms first, then add or subtract.

Example:

$3a + 4b - a + 2b + 2a - 3b$ becomes

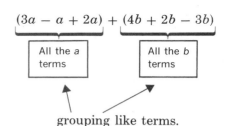

$\underbrace{(3a - a + 2a)}_{\text{All the } a \text{ terms}} + \underbrace{(4b + 2b - 3b)}_{\text{All the } b \text{ terms}}$

grouping like terms.

or $4a + 3b$

after adding and subtracting like terms.

Ready for some problems? Simplify by combining like terms.

(a) $5d^3 - 2d^3$ (b) $7e - 8e + 2e$ (c) $4x^2 - x^2 + 2x^2$

(d) $16y^2 - 8y^2$ (e) $5c - 8c + 4c + c$ (f) $5ab - 3ab$

(g) $t - 3t^2 + 5t^2$ (h) $2r - 5r$ (i) $3p^2q - 4p^2q + 5p^2q$

(j) $5x + 4xy - 6x - 2xy$

(k) $3s + 2t + s - t - 2s - 4$

(l) $3ab^2 + a^2b - ab^2 + 2a^2b - a^2b$

(m) $5q - 6m - 2m - q - 6q + m$

(n) $-2xy^2 + 3x^2y - xy + 5xy^2 - x^2y$

(o) $x + 2y - 3z - 2x - y + 5z - x + 2y - z$

(p) $8b^2cd + 3bc^2d - 2bcd - 3b^2cd + bcd$

(q) $17ef + 9eh - 6ef + eh - 6eh - ef$

Check your answers in **40**.

40

(a) $3d^3$	(b) e	(c) $5x^2$	(d) $8y^2$
(e) $2c$	(f) $2ab$	(g) $t + 2t^2$	(h) $-3r$
(i) $4p^2q$	(j) $-x + 2xy$	(k) $2s + t - 4$	(l) $2ab^2 + 2a^2b$
(m) $-2q - 7m$	(n) $3xy^2 + 2x^2y - xy$	(o) $-2x + 3y + z$	
(p) $5b^2cd + 3bc^2d - bcd$		(q) $10ef + 4eh$	

Multiplying Literal Numbers

In order to *multiply* algebraic expressions you must first be able to multiply literal numbers expressed in exponential form. If you need practice in working with exponents return to page 29 for some help.

As we have already seen, multiplication in algebra is shown by placing the factors side by side, xy

or using parentheses, $(x)(y)$

or using exponents when the base numbers are the same, $xx = x^2$

or sometimes all of these at once, $x(yz^2)$.

Furthermore, as in ordinary arithmetic, the factors may be multiplied in any order we wish. For example, the multiplication "x times y" can be performed as xy or as yx. In either case, the order of the factors does not change the final product.

Three factors may be rearranged and multiplied *in any order we wish*. The multiplication "x times y times z" can be performed as (xy) times z or as x times (yz). In either case the final product is the same.

Commutative Law of Multiplication

These "rearrangement" rules result from two general properties that are true for all real numbers. The first property is called the *commutative law of multiplication*:

For any real numbers a and b,

$$\boxed{ab = ba}$$ The commutative law of multiplication.

$2 \cdot 3 = 3 \cdot 2$

Associative Law of Multiplication

The second property is called the *associative law of multiplication*:

For any real numbers a, b, and c,

$$\boxed{(ab)c = a(bc)}$$ The factors can associate or join together a with b or b with c.

$(2 \cdot 3) \cdot 4 = 2 \cdot (3 \cdot 4)$

These two laws together tell us that we can multiply factors in any order we wish.

To multiply simple algebraic terms:

First, rearrange the factors so that similar factors are grouped together.

In other words, group all numerical coefficients together, then group all powers with the same base together—all powers of x together, all powers of y together, all powers of a together, and so on. With experience you will learn to do this mentally.

Second, multiply any numerical coefficients.

Third, multiply literal factors using the rules of exponent numbers.

For example,

$(2a)(3a^2b)$

becomes

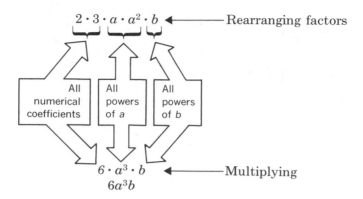

Try these problems. Multiply and simplify:

(a) $2x \cdot x^3$ (b) $3(2d^2)$ (c) $(a^2)(4a^3)$
(d) $y \cdot y^2$ (e) $z \cdot z^3$ (f) $(4st^2) \cdot (t^3)$
(g) $(2p^2q)(5p^3q)$ (h) $(2gh)(4g^2h)(gh^2)$ (i) $(2z)(6x^2z^3)$
(j) $(2a^2)(3a^3)$ (k) $(3de^2)(5d^2e)$ (l) $(3a^2b^2)(3ac)(a^3b)$
(m) $(3a)(2a)(5a)$ (n) $(x)(xy)(x^2y^2)$ (o) $(pq)(pq)(pq)$
(p) $(3de)(4ef)(2df)$ (q) $(3c^2)(3b)(2bc)$ (r) $2(xy)(x^2)(3x^2yz)$

Work carefully, then check your answers in **41**.

---

41   (a) $2x^4$ (b) $6d^2$ (c) $4a^5$ (d) y^3
(e) z^4 (f) $4st^5$ (g) $10p^5q^2$ (h) $8g^4h^4$
(i) $12x^2z^4$ (j) $6a^5$ (k) $15d^3e^3$ (l) $9a^6b^3c$
(m) $30a^3$ (n) x^4y^3 (o) p^3q^3 (p) $24d^2e^2f^2$
(q) $18b^2c^3$ (r) $6x^5y^2z$

Multiplying
Negative Factors

To multiply negative factors follow exactly the same procedure, but remember the rules for multiplying signed numbers.

$$(-x)(y) = -xy$$
$$(x)(-y) = -xy$$
$$(-x)(-y) = +xy$$

Example:

$$(-2xy)(3x^2y) = (-2) \cdot 3 \cdot x \cdot x^2 \cdot y \cdot y$$
$$= -6x^3y^2$$

Example:

$$(-4a^2b)(-2ab^3) = (-4) \cdot (-2) \cdot a^2 \cdot a \cdot b \cdot b^3$$
$$= 8a^3b^4$$

Try these. Multiply as shown:

(a) $(-1)(a)$ (b) $-x(2y)$ (c) $-a(a^3)$
(d) $(-a^2)(-ab^2)$ (e) $-2(3pq)$ (f) $4(-3de^2)$
(g) $-3st^2(2s^3t)$ (h) $-x(xy)$ (i) $-2(mn^2p)(-3m^2n)$
(j) $-3z(4x^3y^2)(-2z^2)(0)$ (k) $(-a)(-a)(-a)$ (l) $(-2d)(-2d^2)(-2d^3)$

Turn to **42** to check your answers.

---

42 (a) $-a$ (b) $-2xy$ (c) $-a^4$ (d) a^3b^2
      (e) $-6pq$ (f) $-12de^2$ (g) $-6s^4t^3$ (h) $-x^2y$
      (i) $6m^3n^3p$ (j) 0 (k) $-a^3$ (l) $-8d^6$

On problem (j) zero multiplied by any number equals zero, of course.

Using Parentheses

Mathematical expressions often require some form of punctuation to make clear their meaning. The symbols most often used for this kind of punctuation are parentheses ( ), brackets [ ], and braces { }. As you have already seen, we can use parentheses to isolate a negative number as in

$$4 + (-5) = -1$$

or to show a product as in

$$(-3)(-4) = 12 \quad \text{or} \quad (-2)^3 = (-2)(-2)(-2) = -8$$

Grouping Symbols

Parentheses, brackets, and braces are most useful in algebra as *grouping symbols*, where they allow us to show that an algebraic expression is to be considered in its entirety. For example, in the expression

$a + (x + 1)$ the parentheses tell us to add the variable a to the quantity $x + 1$.

In the product

$2(a + b)$ the parentheses tell us to multiply 2 and the entire quantity $a + b$.

The rule that allows us to rewrite an algebraic expression with parentheses as an equivalent algebraic expression without parentheses is called the *distributive law*.

Distributive Law

$$\boxed{a(b + c) = ab + ac}$$ This rule says that we can distribute or spread out the effect of the multiplier a to each term in the expression in parentheses.

The distributive law is a basic property of the real numbers, and we can use it to rewrite, simplify, or otherwise change the form of an algebraic expression written with parentheses.

Examples:

$$\boxed{4}\ (2 + 3) = \boxed{4}\ \cdot 2 + \boxed{4}\ \cdot 3 = 8 + 12 = 20$$

Multiply each term inside the parentheses by 4.

$$\boxed{3}\ (4 - a) = \boxed{3}\ \cdot 4 + \boxed{3}\ (-a) = 12 - 3a$$
$$\boxed{-2}\ (a + x) = (\ \boxed{-2}\ ) \cdot a + (\ \boxed{-2}\ )x = -2a - 2x$$
$$\boxed{-3}\ (3x - y) = (\ \boxed{-3}\ )(3x) + (\ \boxed{-3}\ )(-y) = -9x + 3y$$
$$-(a - 2b) = (\ \boxed{-1}\ )(a - 2b) = (\ \boxed{-1}\ )a + (\ \boxed{-1}\ )(-2b) = -a + 2b$$
$$4 + x(x + 1) = 4 + \boxed{x}\ \cdot x + \boxed{x}\ \cdot 1 = 4 + x^2 + x$$
$$5 - (2a - 3b) = 5 + (\ \boxed{-1}\ )(2a - 3b) = 5 + (\ \boxed{-1}\ )(2a) + (\ \boxed{-1}\ )(-3b) = 5 - 2a -$$

Of course the parentheses can contain any number of terms.

Example: $3(a - 2b + 3c - d) = 3 \cdot a + 3(-2b) + 3(3c) + 3(-d)$
$$= 3a - 6b + 9c - 3d$$

Use the distributive law to rewrite the following expressions without parentheses.

(a) $4(x - y)$

(b) $5(2p - q)$

(c) $-(1 - x)$

(d) $-(2x + 1)$

(e) $2(-2a - 3b)$

(f) $3(-x - 4)$

(g) $5 + x(x - 2)$

(h) $4 + y(3 - 2y)$

(i) $1 - (3p - 4q)$

(j) $3a - (2a - b)$

(k) $2y - (x + 4 - y)$

(l) $a - (1 - 3a - b)$

Check your answers in **43**.

_____

43

(a) $4x - 4y$

(b) $10p - 5q$

(c) $-1 + x$

(d) $-2x - 1$

(e) $-4a - 6b$

(f) $-3x - 12$

(g) $5 + x^2 - 2x$

(h) $4 + 3y - 2y^2$

(i) $1 - 3p + 4q$

(j) $a + b$

(k) $3y - x - 4$

(l) $4a - 1 + b$

To simplify an expression containing several sets of nested grouping symbols, use the distributive law to remove the innermost set of symbols first, then combine like terms and continue.

Example:

$$2[a + 2(a - b)] = 2[a + 2a - 2b] \quad \text{Using the distributive law}$$

Work with the innermost parentheses first.

$$= 2[3a - 2b] \quad \text{Combining like terms}$$
$$= 2(3a) + 2(-2b) \quad \text{Using the distributive law again}$$
$$= 6a - 4b$$

More examples:

$$-4[2x - (1 - x)] = -4[2x - 1 + x]$$
$$= -4[3x - 1]$$
$$= -12x + 4$$

$$1 - 3[1 - 4(2 - x)] = 1 - 3[1 - 8 + 4x]$$
$$= 1 - 3[-7 + 4x]$$
$$= 1 + 21 - 12x$$
$$= 22 - 12x$$

Use the distributive law to rewrite the following expressions without parentheses or brackets.

(a) $2[a + 2(a - b)]$

(b) $-[2y - (4 + 3y)]$

(c) $4 - [2x - (-x - 2)]$

(d) $a - 4[a - 4(a - 4)]$

(e) $2 - x[2 - x(2 - x)]$

(f) $2y - [y^2 - 2(3y - 1)]$

Check your answers in **44**.

_____

44

(a) $6a - 4b$

(b) $y + 4$

(c) $2 - 3x$

(d) $13a - 64$

(e) $2 - 2x + 2x^2 - x^3$

(f) $8y - y^2 - 2$

Division of Algebraic Expressions

Division of simple algebraic expressions is no more difficult than simplifying fractions in arithmetic. We can simplify the fraction $\frac{16}{24}$ by writing both numerator and denominator as products of primes,

$$\frac{16}{24} = \frac{2 \cdot 2 \cdot 2 \cdot 2}{2 \cdot 2 \cdot 2 \cdot 3}$$

and dividing out, or cancelling, equal factors.

$$\frac{16}{24} = \frac{(2 \cdot 2 \cdot 2) \cdot 2}{(2 \cdot 2 \cdot 2) \cdot 3}$$
Divide top and bottom of the fraction by $(2 \cdot 2 \cdot 2)$.

$$= \frac{2}{3}$$

The same process is followed in algebra:

Example:

$$\frac{15a^3}{3a} = \frac{3 \cdot 5 \cdot a \cdot a \cdot a}{3 \cdot a}$$

$$= \frac{(3 \cdot a) \cdot 5 \cdot a \cdot a}{(3 \cdot a)}$$
Divide top and bottom of the fraction by $(3 \cdot a)$.

$$= 5 \cdot a \cdot a$$

$$= 5a^2$$

Example:

$$\frac{12x^2yz^3}{4xyz} = \frac{3 \cdot 2 \cdot 2 \cdot x \cdot x \cdot y \cdot z \cdot z \cdot z}{2 \cdot 2 \cdot x \cdot y \cdot z}$$

$$= \frac{(2 \cdot 2 \cdot x \cdot y \cdot z) \cdot 3 \cdot x \cdot z \cdot z}{(2 \cdot 2 \cdot x \cdot y \cdot z)}$$
Divide top and bottom of the fraction by $(2 \cdot 2 \cdot x \cdot y \cdot z)$.

$$= 3 \cdot x \cdot z \cdot z$$

$$= 3xz^2$$

You should remember from your study of exponential numbers that

$$\frac{x^m}{x^n} = x^{m-n}$$

so the process of simplifying algebra fractions is even simpler and quicker.

Example:

$$\frac{x^3y^4}{xy^2} = \frac{x^3}{x^1} \cdot \frac{y^4}{y^2} = x^{3-1} \cdot y^{4-2} = x^2y^2$$

Example:

$$\frac{2ab^2c^4}{ab^3c} = 2 \cdot \frac{a^1}{a^1} \cdot \frac{b^2}{b^3} \cdot \frac{c^4}{c^1} = 2 \cdot a^{1-1} \cdot b^{2-3} \cdot c^{4-1} = 2b^{-1}c^3 = \frac{2c^3}{b}$$

  Remember that $a^{1-1} = a^0 = 1$

$$\text{and } \ b^{-1} = \frac{1}{b}$$

If you need to review any of these operations with exponential numbers, return to page 29. Otherwise, continue by working the following set of problems.

Simplify:

(a) $\dfrac{x^3}{x}$
(b) $\dfrac{y^5}{y^2}$
(c) $\dfrac{y^4}{y^5}$
(d) $\dfrac{9a^3b}{3a}$

(e) $\dfrac{10ge^3}{2e}$
(f) $\dfrac{6p^4}{2p^2}$
(g) $\dfrac{s^4t^7}{st^7}$
(h) $\dfrac{m^2n^6}{m^3n^2}$

(i) $\dfrac{16h^3}{4h^3}$ (j) $\dfrac{7a^2b^3cd^2}{b}$ (k) $\dfrac{2xyz}{xy}$ (l) $\dfrac{cde^3}{cde}$

(m) $\dfrac{5e^3f}{5e^2}$ (n) $\dfrac{0}{2x}$ (o) $\dfrac{8d^5g^3}{2d^4g^2}$ (p) $\dfrac{65x^3y^2z}{13xy^2z}$

(q) $\dfrac{48p^2q^4r^3}{16p^3qr}$ (r) $\dfrac{abcd}{ab^2c}$ (s) $\dfrac{2s^2t}{0}$ (t) $\dfrac{3mp^2}{-mp^2}$

The correct answers are in **45**.

---

45   (a) x^2 (b) y^3 (c) $\dfrac{1}{y}$ (d) $3a^2b$ (e) $5ge^2$

(f) $3p^2$ (g) s^3 (h) $\dfrac{n^4}{m}$ (i) 4 (j) $7a^2b^2cd^2$

(k) $2z$ (l) e^2 (m) ef (n) 0 (o) $4dg$

(p) $5x^2$ (q) $\dfrac{3q^3r^2}{p}$ (r) $\dfrac{d}{b}$ (s) Not defined   (t) -3

In problem (n), a zero numerator means that the value of the fraction is zero. In problem (s), the value of the fraction is not defined since division by zero is not allowed. In problem (t), dividing out the literal part leaves $\dfrac{3}{-1}$ and this equals -3.

(If you had forgotten this, you may want to review the section on signed numbers starting on page 15.)

Combined Operations on Algebraic Expressions

Many of the algebraic calculations you will do will involve a combination of the four operations: addition, subtraction, multiplication, and division.

Example: to simplify

$3a(2a^2b) - a^3b + 2b(a)^2 - a^2b$

First, multiply

$6a^3b - a^3b + 2a^2b - a^2b$

Second, combine like terms:

$5a^3b + a^2b$

At first you should take problems like these one step at a time, writing out each step as we have done above.

Another example: simplify

$$c^2d - \dfrac{3c^3d^2}{cd} + 4c(cd) - c \quad\longrightarrow\quad \dfrac{3c^3d^2}{cd} = \dfrac{3c^2d(cd)}{(cd)} = 3c^2d$$

First, multiply or divide. $4c(cd) = 4c^2d$

$c^2d - 3c^2d + 4c^2d - c$

Second, combine like terms.

$2c^2d - c$

Ready for a few of these more complex problems?
If so, simplify these:

1.   (a) $2t(t^2) - t^3$ (b) $3m^2(4m) - 5m^3$ (c) $3g(h^2) + g$

51   **1-4 Working with Literal Numbers**   Copyright © 1982 by John Wiley & Sons, Inc.

(d) $q + q(qt) + q^2t$

(e) $2q(t^2) - t(qt) - t^2$

(f) $2x(3y) - 2y(2x) - xy$

(g) $4c(cd^2) + 2cd(cd) - 5c^2d^2$

(h) $x - 2x(xy) + y(x^2)$

(i) $z^4 - z^3(2z)$

2. (a) $\dfrac{4a^2 + 3a^2}{a}$

(b) $\dfrac{4d^2}{d} - 3d$

(c) $\dfrac{q^2 + 3q^2}{2q} + 2q$

(d) $s^2 + \dfrac{s^2 + 3s^2}{2}$

(e) $t - \dfrac{2t^3 + 4t^3}{3t}$

(f) $3r^2 + \dfrac{r^2}{r}$

(g) $\dfrac{3r^2 + r^2}{r}$

(h) $\dfrac{2p^3 - p^3}{p} + \dfrac{3p^3 - p^3}{p^2}$

(i) $\dfrac{k + 2k}{k} - \dfrac{5k^2 + k^2}{3}$

Check your answers in **46**.

46 1. (a) t^3 (b) $7m^3$ (c) $3gh^2 + g$ (d) $q + 2q^2t$
(e) $qt^2 - t^2$ (f) xy (g) c^2d^2 (h) $x - x^2y$
(i) $-z^4$

2. (a) $7a$ (b) d (c) $4q$ (d) $3s^2$
(e) $t - 2t^2$ (f) $3r^2 + r$ (g) $4r$ (h) $p^2 + 2p$
(i) $3 - 2k^2$

If you had trouble with any of these you may find it helpful to return to page 43 and review this section before you continue.

In order to become expert and accurate at doing problems like these, you must work slowly and carefully at first. Develop the habit of being very fussy about doing each step correctly. Even the best mathematician knows that he or she must be careful not to lose terms or misread x^2 for x^3 or make other simple mistakes.

Quickness at algebra will come with practice. If you are confident that you understand this section on working with literal numbers, turn to **47** for a set of practice problems.

SUBSCRIPTS

Mathematicians, scientists, and others who use algebra in their work often use subscripts to help name variables and constants. You may find these strange looking and tricky to pronounce. Here is a bit of help.

Suppose you measured the air temperature every morning at a certain time. You might label the temperature variable as T and call the temperature on the first day T_1 (say "tee-sub-one"), the temperature on the second day T_2, the temperature on the third day T_3, and so on. The little numbers slightly below the letters are called subscripts. Any sort of subscripts can be used and, usually we choose subscripts that have some memory jogging value. For example, if we measured all the weights of the boys in a sixth grade class, we could call Al's weight W_A (say "W-sub-A") or W_{Al}, Bill's weight might be W_B or W_{Bill}, Sam's weight would be W_S or W_{Sam}, and so on.

Subscripts can be handy reminders of what the variable letter represents.

Unit I
Introduction
to Algebra

The answers are on page 537.

A. Simplify by combining like terms (addition):

1. $12c + (-4c) =$ _____

2. $9b + 8b =$ _____

3. $7x + 11x =$ _____

4. $(-15r) + 7r =$ _____

5. $5n + 9n =$ _____

6. $-7t + (-2t) =$ _____

7. $3h^2 + (-7h^2) =$ _____

8. $-16a^2b + 4a^2b =$ _____

9. $y + (-5y^2) =$ _____

10. $2c^3d + 4c^3d =$ _____

11. $-5abc + abc =$ _____

12. $-6z + (-9z) =$ _____

13. $10ab + (-5ab) + 17ab =$ _____

14. $-2cd + 4cd + (-12cd) =$ _____

15. $15x^2 + 9x^2 + x^2 + 3x^2 =$ _____

16. $-9vt + 7vt + (-8vt) =$ _____

17. $8c^3de + (-2c^3de) + (-17c^3de) =$ _____

18. $4j^2k + 7jk + 2jk^2 =$ _____

19. $-3y + 4y + 8y =$ _____

20. $15b^2c + (-2b^2c) + (-9b^2c) =$ _____

B. Simplify by combining like terms (subtraction):

1. $9b - (2b) =$ _____

2. $14a - (-6a) =$ _____

3. $-7p - (-9p) =$ _____

4. $(-15r) - (12r) =$ _____

5. $-13y - (-13y) =$ _____

6. $0 - (-25x) =$ _____

7. $-9a - (3a) =$ _____

8. $7d - d =$ _____

9. $-4.2t + 4.2t =$ _____

10. $7r - 7r =$ _____

11. $-3x^2 - 5x^2 =$ _____

12. $0.6cd - 0.9cd =$ _____

13. $3y^2z^2 - 2y^2z^2 =$ _____

14. $-5a^2b^2 - a^2b^2 =$ _____

15. $0.1x^2y^2 - 0.7x^2y^2 =$ _____

16. $3(a + b) - 7(a + b) =$ _____

17. $8d - 0 =$ _____

18. $4a^2 - 15a^2 =$ _____

19. $(-7rs^2) - (2rs^2) =$ _____

20. $7x^2y - 9x^2y =$ _____

21. $-11xy - 3xy =$ _____

C. Multiply:

1. $(5)(4a) =$ _____

2. $(6c)(4c) =$ _____

3. $(m^2)(m^3) =$ _____

4. $(3y^2)(y) =$ _____

Date _____

Name _____

Course/Section _____

5. $-8r(-2s) = \underline{\qquad}$

6. $(3z^3)(5z) = \underline{\qquad}$

7. $(-ab)(abc) = \underline{\qquad}$

8. $(9x^3y^5)(-2xy^3) = \underline{\qquad}$

9. $(7d)^2 = \underline{\qquad}$

10. $(21st^2)(-\frac{1}{3}s^2t) = \underline{\qquad}$

11. $(-5d^2e)(4d^2) = \underline{\qquad}$

12. $(-3z)(6z^3)(z^2) = \underline{\qquad}$

13. $(\frac{2}{3}y^3)(-12y) = \underline{\qquad}$

14. $(-4b)^3 = \underline{\qquad}$

15. $-2(r^2s)(-4r^3) = \underline{\qquad}$

16. $(3c)^2(-2c)^4 = \underline{\qquad}$

17. $5(-4n)^2 = \underline{\qquad}$

18. $4a^2(2a)(-a^3) = \underline{\qquad}$

19. $-10(3x^2y)(2y^2z)(z^3) = \underline{\qquad}$

20. $2(bc^2)(3b)^2 = \underline{\qquad}$

D. Divide:

1. $\dfrac{16c}{2} = \underline{\qquad}$

2. $\dfrac{32d}{-4} = \underline{-8d}$

3. $\dfrac{35y^2}{-7y} = \underline{\qquad}$

4. $\dfrac{-15ab}{3b} = \underline{-5a}$

5. $\dfrac{-36p^3}{-9p^2} = \underline{\qquad}$

6. $\dfrac{-7v^3}{v^3} = \underline{-7}$

7. $\dfrac{-24x^5}{3x} = \underline{\qquad}$

8. $\dfrac{40t^3}{-4t} = \underline{-10t^2}$

9. $\dfrac{22m^3n}{2m^4n} = \underline{\qquad}$

10. $\dfrac{49r^5s}{-7r^2} = \underline{-7r^3s}$

11. $\dfrac{1.2b^2c}{0.4c^3} = \underline{\qquad}$

12. $\dfrac{-50d^2e}{-5de} = \underline{10d}$

13. $\dfrac{3.6p^4r}{0.9p^3r^4} = \underline{\qquad}$

14. $\dfrac{(-2a)^3b}{8a^2b} = \underline{-a}$

15. $\dfrac{-5m^2n}{(mnp)^3} = \underline{\qquad}$

16. $\dfrac{-5d^4z}{45d} = \underline{-\dfrac{d^3z}{9}}$

17. $\dfrac{15(x+y)}{3(x+y)} = \underline{\qquad}$

18. $\dfrac{-63x^8y^2z}{9x^3y} = \underline{-7x^5yz}$

19. $\dfrac{57a^{10}b^7}{3a^6b} = \underline{\qquad}$

20. $\dfrac{(-5y)^3}{-y^2} = \underline{125y^5}$

21. $\dfrac{-96w^2x^5y^4}{8wy^6} = \underline{\qquad}$

E. Perform the indicated operations:

1. $4c + 7d + 6c + 9d = \underline{\qquad}$

2. $-4y + 5x - 7y - 2x = \underline{-11y+3x}$

3. $6ab - 5bc + 7ab - bc = \underline{\qquad}$

4. $7m - 3m^2 + 2m^2 = \underline{7m-m^2}$

5. $\dfrac{5e + 3e}{e} =$ _____

6. $\dfrac{(5a^2)(2a^3)}{a^4} = 10a$

7. $\dfrac{(3y)^3}{y^2} =$ _____

$\dfrac{2}{a^3c^2}$ **8.** $\dfrac{6ac - 2ac + 12ac}{(2ac)^3} = \dfrac{18c}{a^3}$

9. $\dfrac{32d^3}{-2d} - 5d^2 =$ _____

10. $6a^2y - 3a^2y + 7ay^2 - 7ay^2 = 3a^2y$

11. $15 - b^2 + 2b + 4 + 3b^2 =$ _____

12. $\dfrac{4k^2 + 8k^2}{3k^2} = 4$

13. $3f - \dfrac{6f^6}{3f^6} =$ _____

14. $\dfrac{7abc^2}{abc} + 2c = 9c$

15. $\dfrac{(0.9xy)(-3x^2y^3)}{xy} =$ _____

16. $\dfrac{(\frac{2}{3}b^2z^2)(-9b^5z^{10})}{-6b^5z^6} = b^2z^6$

17. $(-1.2d^3)(5dc^3) =$ _____

18. $-9ab + 6a - 2ab + 4b - 6ab + 10 - 6a = -17ab + 4b + 10$

19. $-0.04am + 1.2am + 0.24am - 5.62am =$ _____

20. $(4y)(-2y) - 7(2y^2) = 112y^4 - 22y^2$

21. $-3d^2 + \dfrac{(4d)(-2d^3) + 3d^4}{d^2} =$ _____

22. $\left(\dfrac{1}{4}q^2r\right)(24qs)\left(\dfrac{1}{2}qr^2s\right) = 3q^4r^3s^2$

23. $\dfrac{18x^3y^{10}}{2xy^5} + \dfrac{7x^8y^6}{21x^6y} =$ _____

$1.3b^2$

24. $0.3b^2 + \dfrac{1.5b^3 - 1.2b^3}{0.3b} = 1.6b^2$

25. $5m^2n + 2m^2 - 6m^2n + n^2 =$ _____

26. $-7z^3 + 2z(4z^2) - 8z^2(z) =$ _____

27. $18gh + 9g^2h + 12gh^2 =$ _____

$.3b^2 + \dfrac{.3b^3}{.3b}$

28. $\dfrac{(36x)(x^2y) + (11y)(-2x^3)}{7xy} = 2x^2$

29. $\dfrac{(21b^3c)(-2b)(-3c)}{18b^2c^2} =$ _____

$.3b^2 + .3b^2$

$.3b^2 + b^2$

$.3b^2$

$1.3b^2$

30. $\dfrac{(15e)(-3f^2)}{5f} + \dfrac{(7e^2f)(3ef^2)}{e^2f^2} = 12ef$

F. Simplify using the distributive law:

1. $5(a - 2b)$

2. $7(2x - z)$ $14x - 7z$

3. $-(x - 3y)$

4. $-(4a - t)$ $-4a + t$

5. $2(-p - 4q)$

6. $6(-2x - y^2)$ $-12x - 6y^2$

7. $-4(x - 4)$

8. $-5(x^2 - 1)$ $-5x^2 + 5$

9. $1 - (4x - 1)$ $2 - 5p - 2q$ **10.** $2 - (5p - 2q)$ $-10p + 4q$

11. $3 - (m^2 - 2m)$ $8 - 3x - 4y$ **12.** $8 - (3x - 4y)$ $-24x + 32y$

13. $4x - 2(x - 2)$

14. $9m - 6(2 - m)$ $24m - 9m^2 - 12$

15. $p - (2 + 4p - q^2)$

16. $t - (1 - 5t - t^3)$ $-t + 5t + t^4$

17. $3x - x(1 - x)$

18. $5a - a(7 - a)$ $a^2 - 2a$

19. $5y - 2y(y - 2)$

20. $6x - 3x(x - 1)$ $9x - 3x^2$

Date _____

Name _____

Course/Section _____

21. $3[x + 3(x - y)]$

22. $4[a - 3(a - b)]$ $12b-8a$

23. $-[x - (x - 4)]$

24. $-[2b - (2 - 3b)]$ $2-5b$

25. $1 - [1 - (1 - x)]$

26. $6 - [4 - (2 - y)]$ $4-y$

27. $4 - 2[x - 3(1 - x)]$

28. $8 - 3[p - 2(p - 1)]$ $2+3p$

29. $3 - a[1 - a(1 - a)]$

30. $-4 - x[2 - x(x - 1)]$ x^3-x^2-2x-4

G. Brain Boosters

1. If you have x \$5 bills and y \$2 bills, how many 25¢ pieces can you get for these bills? _____

2. The following nursery rhyme dates back to the eighteenth century and a version of it appeared in a mathematics book published in 1202.

> As I was going to St. Ives,
> I met a man with seven wives;
> Every wife had seven sacks;
> Every sack had seven cats;
> Every cat had seven kits.
> Kits, cats, sacks, and wives,
> How many were going to St. Ives?

Answer the question in the last line. _____

3. Simplify $\dfrac{63a^4b^3c^7}{160a^2b^7c} \times \dfrac{80a^3b^4c}{231ab^3c^3} \times \dfrac{\frac{1}{3}c^2}{\frac{1}{2}ab}$ _____

When you have had the practice you need either return to the preview test on page 1 or continue in **48** with translating English to algebra.

48

The mathematics you use every day in your work or at play usually appears wrapped in words and hidden in sentences. Neat little sets of directions seldom are attached; no "Divide and reduce to lowest terms" or "multiply and simplify your answer." The difficulty with real problems, or "word problems," is that they must be translated from words to mathematical symbols. You need to learn to *talk* algebra and *think* algebra, not simply juggle numbers and letters.

Algebra is most useful as a tool for applying mathematical logic to real situations and solving real problems. However, in order to use algebra for practical problem solving, you must be able to translate simple English sentences and phrases into mathematical equations or expressions. Fortunately, certain words and phrases appear again and again in word problems. They are signals alerting you to the mathematical operations to be used. Here is a list of such *signal words*:

Signal Words

SIGNAL WORDS

English Term	Math Translation	Example
Equals Is, is equal to, was, are, were The same as . . . What is left is . . . The result is . . . Gives, makes Leaving, leaves	$=$	$a = b$
Plus Sum of Increased by More than	$+$	$a + b$
Minus b, subtract b Less b Decreased by b Take away b Reduced by b Diminished by b b subtracted from a	$-$	$a - b$
Times, multiply, of Multiplied by Product of OF	$\times$	$a \times b$ or ab
Divide Divided by b Quotient of	$\div$	$a \div b$ or $\dfrac{a}{b}$
Twice, twice as much, double	$\times 2$	$2a$
Is less than Is more than	$<$ $>$	$a < b$ $b > a$

Translate the phrase "three plus some number" to an algebra expression. Try it, then turn to **49** to check your answer.

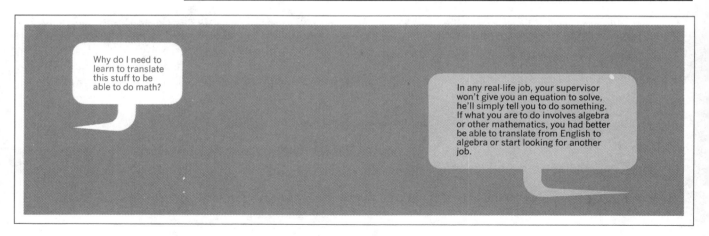

49 First, make a word equation by using parentheses:

(Three) (plus) (some number)

Second, substitute mathematical symbols:

(Three) (plus) (some number)

$$3 \qquad + \qquad n$$

or $3 + n$

Signal words are translated directly to math symbols. Unknown quantities are represented by letters of the alphabet—any letter will do.

Easy?

Translate the following phrases into mathematical expressions:

(a) "A number divided by seven" _____

(b) "Three more than some number" _____

(c) "One-half of some quantity" _____

(d) "The sum of four and some unknown number" _____

(e) "A number diminished by eleven" _____

(f) "Five multiplied by a number" _____

(g) "Seven more than twice a given number" _____

You will find our step-by-step translations in **50**.

50 (a) (A number) (divided by) (seven)

$$N \qquad \div \qquad 7 \qquad \text{or} \quad N \div 7 \text{ or } \frac{N}{7}$$

(b) (Three) (more than) (some number)

$$3 \qquad + \qquad x \qquad \text{or} \quad 3 + x$$

(c)   (One-half) (of) (some quantity)

$$\frac{1}{2} \qquad \times \qquad Q$$

or $\frac{1}{2}Q$ or $\frac{Q}{2}$

(d)   The sum of (four) (and) (some unknown number)

$$4 \qquad + \qquad x$$

or $4 + x$

(e)   (A number) (diminished by) (eleven)

$$A \qquad - \qquad 11$$

or $A - 11$

(f)   (Five) (multiplied by) (a number)

$$5 \qquad \times \qquad b$$

or $5b$

(g)   (Seven) (more than) (twice a given number)

$$7 \qquad + \qquad 2N$$

or $7 + 2N$

Of course, any letter can replace those used in the expressions above. In problem (g) you could write $7 + 2a$, $7 + 2x$, $2p + 7$, or many other equivalent expressions.

Here is some more practice. Translate these into algebra expressions:

(a) d more than k _____

(b) g more than m _____

(c) q subtracted from p _____

(d)   The product of a and b _____

(e) x divided by y _____

(f)   14 increased by z _____

(g)   Twice a number plus 4 _____

(h)   The product of two numbers minus 5 _____

(i)   Six less than a number _____

(j)   Six less a number _____

(k)   Six more than a number _____

(l)   A number less six _____

(m)   Twice the square of a number _____

(n)   Three times the product of two numbers _____

(o)   The quotient of two numbers _____

(p)   Three times a number divided by twice another number _____

Look for our answers in **51**.

51

(a) $d + k$ (b) $m + g$ (c) $p - q$ (d) ab

(e) $\frac{x}{y}$ (f) $14 + z$ (g) $2N + 4$ (h) $xy - 5$

(i) $a - 6$ (j) $6 - a$ (k) $a + 6$ (l) $a - 6$

(m) $2x^2$ (n) $3xy$ (o) $\frac{x}{y}$ (p) $\frac{3a}{2b}$

Translating Sentences

So far we have translated only phrases, but complete sentences can also be translated into mathematical symbols. An English phrase translates into an algebra expression and an English sentence translates into an algebra equation. For example, the sentence

"John's age is equal to Kevin's age."   translates to

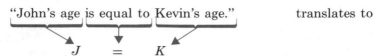

$$J \quad = \quad K$$

Each word or phrase in the sentence is translated directly into a mathematical term, variable, number, expression, or arithmetic operation sign.

Example:

"Marie's score was twice as high as Dan's score."

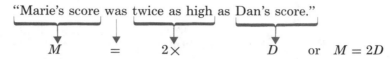

$$M \quad = \quad 2\times \quad\quad D \quad \text{or} \quad M = 2D$$

Example:

"The regular price is $4 more than the sale price."

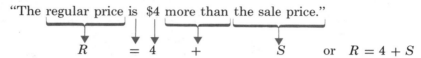

$$R \quad = 4 \quad + \quad\quad S \quad \text{or} \quad R = 4 + S$$

Translate this sentence into mathematical form as we did above.
"Four times Tom's age is 72."
Do it here . . .

. . . then check your translation in **52**.

---

52   Four times Tom's age is 72.

$$4 \quad \times \quad T \quad = 72$$

$$4T = 72$$

Follow these steps when you must translate English sentences into algebra equations:

Example: "We noticed that Bill's age plus 8 years is exactly equal to three times Ron's age."

Step 1.   Cross out all unnecessary words.

~~We noticed that~~ Bill's age plus 8 ~~years~~ is ~~exactly~~ equal to three times Ron's age."

Step 2.   Make a word equation using parentheses.

(Bill's age) (plus) (8) (is equal to) (three) (times) (Ron's age)

Step 3.   Substitute a letter or an arithmetic symbol for each parentheses.

$$B \quad + \quad 8 \quad = \quad 3 \quad \times \quad R$$

Step 4.   Combine and simplify. $B + 8 = 3R$

Translating English sentences into the language of algebra requires that you read the English sentences—the word problems—very differently from the way you read stories or newspaper articles. Very few students are able to write out the mathematical equation after reading the problem only once. You should expect to need to read it several times and you'll want to read it slowly. No speed reading here! The ideas in math problems are usually concentrated in a few key words and you must find them. If you find a word that is not in your vocabulary, stop reading and look it up in a dictionary. It may be important. We will discuss word problems again in Unit 2 and they will appear in every section of this book.

Here is another example of translating a sentence into an algebra equation:

"It is an interesting coincidence that the cost of the desk is just $8 more than double the cost of the chair."

Step 1. "~~It is an interesting coincidence that the~~ cost of the desk is ~~just~~ $8 more than double ~~the~~ cost of the chair."

Step 2. (Desk cost) (is) (8) (more than) (double) (chair cost)

Step 3. $\quad\quad\quad D \quad = 8 \quad + \quad\quad 2 \quad \times \quad C$

Step 4. $D = 8 + 2C$

Another example:

"If Henry's year-end bonus is added to his yearly salary, his total income is exactly three times the minimum starting salary."

Step 1. "~~If Henry's year-end~~ bonus ~~is~~ added to ~~his yearly~~ salary, ~~his total income~~ is ~~exactly~~ three times ~~the~~ minimum ~~starting salary.~~"

Step 2. (bonus) (added to) (salary) (is) (three) (times) (minimum)

Step 3. $\quad\quad B \quad + \quad\quad S \quad = \quad 3 \quad \times \quad\quad M$

Step 4. $B + S = 3M$

The more sentences you translate, the easier it gets.

Translate the following sentence to an algebra equation using the four-step procedure shown above.

"The surprising thing about it is that this year's bill is only $3 more than twice last year's bill."

Check your work in **53**.

---

53   "~~The surprising thing about it is that~~ this ~~year's bill~~ is only $3 more than twice last ~~year's bill.~~"

(This) (is) (3) (more than) (twice) (last)

$\quad T \quad = 3 \quad + \quad\quad 2\times \quad L$

$T = 3 + 2L$

Translate each of these sentences into an equivalent algebra equation.

(a)   The difference between two numbers is seven.
(b)   The bus driver is exactly twice as old as her son.
(c)   The larger of two numbers is five more than the smaller number.
(d)   The sum of the two purchases is $18.
(e)   My math teacher said that the first score is five more than twice the second score.
(f)   Dave's age is three times what Eric's age was six years ago.
(g)   The height of the building is only 1.6 times its width.
(h)   By weighing we learned that three-fourths of a sack of cement equals one-half of a sack of cement plus ten pounds.

(i)   Three-eighths of what number is six?
(j)   Sixteen divided by some unknown number equals that number less six.
(k)   The product of two numbers is twice their sum.
(l)   The sum of two numbers divided by their difference is equal to 3.
(m)   The cost of the chair minus the discount was only $6.
(n)   Twice the square of a certain number plus a second number is equal to 23.
(o)   The cost of two-thirds of a load of wood is $20.
(p)   The top score on the exam minus the lowest score was four times the average score.

Turn to **54** to check your answers.

54

(a) $p - q = 7$ (b) $B = 2s$

(c) $L = 5 + s$ (d) $x + y = 18$

(e) $F = 5 + 2s$ (f) $D = 3(E - 6)$

(g) $H = 1.6W$ (h) $\dfrac{3}{4}C = \dfrac{1}{2}C + 10$

(i) $\dfrac{3}{8}N = 6$ (j) $\dfrac{16}{x} = x - 6$

(k) $AB = 2(A + B)$ (l) $\dfrac{A + B}{A - B} = 3$

(m) $C - D = 6$ (n) $2x^2 + y = 23$

(o) $\dfrac{2}{3}L = 20$ (p) $T - L = 4A$

Now, for more practice in translating English to algebra, turn to **55** for a set of practice problems.

I saw a math book where they wrote negative 2 as ⁻2, with a little raised negative sign. Why do that?

Some elementary textbooks use that notation in hopes of avoiding confusion between "negative 2," the number, and "minus 2," the operation. We will not use it in this book.

Unit I
Introduction
to Algebra

The answers are on page 539.

A. Translate each of the following phrases to a mathematical operation symbol:

1. Add to ___+___   2. Increased by _____

3. Four less ___4-___   4. The cube of x _____

5. Multiplied by ___×___   6. Times _____

7. Two less than a ___$a-2$___ $2<A$ 8. As many as _____

9. The sum of ___+___   10. Three pounds extra _____

11. Is ___=___   12. Diminished by _____

13. Divided by ___÷___   14. Of _____

15. Two-thirds of ___$\frac{2}{3}x$___   16. Is less than _____

17. Twice ___×2___   18. Equals _____

19. Is greater than ___>___   20. Five inches more _____

21. Four more than ___4+___   22. a subtracted from b _____

23. The square of a ___a^b___   24. Reduced by two _____

25. The same as ___≥___   26. Were _____

27. Quotient of ___÷___   28. Equals _____

29. Reduced by 5 _____ 30. Are _____

B. Translate each of the following phrases to an algebraic expression:

1. a divided by b ___$\frac{a}{b}$___   2. The sum of x and y ___$x+y$___

3. p increased by 4 ___$p+4$___   4. 4 times t ___$4t$___

5. y divided by 4 ___$y/4$___   6. d less e ___$d-e$___

7. The product of r and m ___$r×m$___   8. K plus 10 ___$K+10$___

9. z reduced by 5 ___$z÷5$___   10. B multiplied by 9 ___$9B$___

11. The difference of h and g ___$h-g$___   12. Two less than M ___$M-2$___

13. Five more than u ___$5+u$___   14. The sum of x and half of y ___x___

15. The quotient of $3a$ and $5b$ ___$3a/5b$___   16. Twice y divided by two _____

17. Triple a number ___x^3___   18. 8 less g

Date _____

Name _____

Course/Section _____

19. The cube of a number _____
20. The square of W _____

21. Four times the square of a number, minus 7 _____
22. The product of the square of a and the square of b _____

23. The sum of the square of a and the square of b _____
24. The product of q and t divided by the product of r and s _____

25. The product of q and t divided by the sum of q and t _____
26. The square of M minus the product of P and Q _____

27. Five times the product of A and B _____
28. Four times the sum of D and the square of E _____

29. Twice the product of the square of x and the square of t divided by the sum of x and t _____
30. The quotient of the quantity x plus one and the square of the quantity x minus one _____

31. The product of mass and velocity _____
32. One-half of g times the square of the time _____

33. Ten less than four times a number _____
34. The larger number minus six times the smaller _____

35. Twice the distance, plus three miles _____
36. The sum of the airplane's ground speed and the wind speed _____

37. Fifteen dollars more than the selling price minus the discount _____
38. One-half of the mass times the square of the velocity _____

39. Twice the difference of two numbers _____
40. The product of a number and the square of a second number _____

C. Write each of these sentences as an algebraic equation:

1. Three times a number, plus four is equal to six. _____

2. The sum of two numbers is 31. _____

3. Sally's age is five years less than Charlie's age. _____

4. The product of Al's age and Bill's age is 10 times Cathy's age. _____

5. Five times a number, decreased by 7 is 88. _____

6. The difference between the cost of the shirt and the jacket is twice the cost of the dress. _____

7. Four times a number less three gives twice another number. _____

8. Jill's age four years ago was the same as twice Nan's age three years from now. _____

9. The sum of six times a number and its square is 27. _____

10. The area of a circle is equal to π times the radius squared. _____

11. The volume of a cylinder is equal to $\frac{1}{4}$ of its height times π times its diameter squared. _____

12. The book costs \$3 more than the sum of the costs of the paper and the notebook. _____

13. Some number increased by 10 is equal to 16. _____

14. The perimeter of a rectangle is equal to twice its length plus twice its width. _____

15. The volume of a cube is equal to the cube of its edge. _____

16. The sum of two numbers decreased by four is equal to twice their product. _____

17. The Fahrenheit temperature is equal to 32 plus the product of 1.8 times the Celsius temperature. _____

18. The product of one number and the square of another is six. _____

19. One-half the sum of two numbers equals 7. _____

20. The sum of the squares of two numbers is four more than their sum. _____

21. Ten years from now the father's age will be three times what his daughter's age was six years ago. _____

22. The 160-lb package was divided into three parts, so that the largest part weighed twice the smallest. _____

23. The volume of a sphere is π times one-eighth of its diameter cubed. _____

24. The interest on the loan is equal to the principle times the interest rate times the time of the loan. _____

25. The IQ is found by dividing the mental age by the chronological age and multiplying by 100. _____

26. The distance is equal to the velocity times the time. _____

27. The area of a circle is one-fourth times π times the diameter squared. _____

28. The larger number is six more than twice the smaller number. _____

29. The length of the lot is 20 ft longer than twice the width. _____

30. A certain number increased by 9 is the same as four times that number. _____

31. The sum of the two distances is one-half of the total trip. _____

32. The yearly cost of gasoline plus the cost of servicing is equal to one-third the original price of the car. _____

Date _____

Name _____

Course/Section _____

D. Brain Boosters

Here is a tricky algebraic translation problem devised by David L. Silverman of Los Angeles, California and published in the June 1977 issue of the mathematics journal EUREKA.

A dozen, a gross, and a score,
Plus three times the square root of four,
   Divided by seven,
   Plus five times eleven,
Is nine squared and not a bit more.

Translate this poem into a numerical equation.

(One dozen = 12, one gross = 144, one score = 20.)

When you have had the practice you need either return to the preview test on page 1 or continue in **56** with the study of literal expressions.

Formula

56   In scientific and technical work, in business and mathematics, and in many other practical situations, the language of algebra is used to express relationships between quantities. A *formula* is a rule for finding the numerical value of one quantity from the values of other quantities. The formula or rule is written in mathematical language because that gives a brief, convenient to use, and easy to remember form. Consider the following examples.

Example 1

English language rule: The volume of a sphere can be calculated as π times the cube of the diameter of the sphere divided by six.

Algebraic formula: $V = \dfrac{\pi D^3}{6}$

Example 2

English language rule: The gravitational attraction force between two masses M and m is the product of the masses times the constant G divided by the square of the distance between the masses.

Algebraic formula: $F = \dfrac{GMm}{d^2}$

It is easy to see why many people prefer to write such information in an algebraic shorthand instead of writing it out in words.

Notice that the letters used in the formulas usually are chosen to remind you of what they represent: D for diameter, V for volume, F for force, and so on.

It is important that you be able to evaluate algebraic expressions by substituting numbers in them. For example, in retail stores the following formula is used:

$M = R - C$ where M is *markup* on an item,
  R is the *retail selling price,*
  and C is the *original cost.*

Find M if $R = \$25$ and $C = \$21$.

$M = $ _____ $-$ _____ $=$

Check your work in **57**.

---

57 $M = \$25 - \$21 = \$4$

Evaluating Formulas

Easy? Of course. A formula is a recipe for a calculation. Simply substitute the given numbers for the corresponding letters and then do the arithmetic. You will find that you need to work with formulas in science and technical courses, and in many other areas. Evaluating formulas represents the most used part of algebra for most people.

Let's look at an example:

Evaluate the expression $5 + x$ for $x = 2$.

Step 1.   Put the given numerical value in parentheses (2) and substitute for its variable in the algebra expression.
  $5 + (2)$

Step 2.   Do the arithmetic indicated.
  $5 + 2 = 7$

Looks easy. Try this one:

Evaluate the expression $2x^2$ for $x = 3$.

Check your work in **58**.

---

58   Do it this way: $2x^2$ for $x = 3$

Step 1. $2(3)^2$

Step 2. $2 \cdot 3^2 = 2 \cdot 9 = 18$

Notice in Step 2 that before we do the arithmetic we may need to write in the arithmetic signs that are omitted in algebra, especially the multiplication sign.

Using Parentheses ⇨   Also notice that when we substitute a numerical value into the formula we keep the number in parentheses. That may seem like extra writing, but it is worth it; you will make fewer mistakes.

Evaluate $4x + 15$ for $x = -3$

Choose an answer:

(a)   3   go to   **59**.
(b)   16   go to   **60**.

I plugged $x = -3$ into the equation $y = 2x + 10$ and got $y = 2 - 3 + 10$ or 9. OK?

No. Always use parentheses around the number you substitute into the equation: $y = 2(-3) + 10 = -6 + 10 = 4$.

---

59   Right you are.

$4x + 15$ for $x = -3$

Step 1. $4(-3) + 15$

Step 2. $4 \cdot (-3) + 15 = -12 + 15 = 3$

Look at the mess you might get caught in if you had *not* used parentheses:

Find $4x + 15$ for $x = -3$

substituting $4 - 3 + 15 = 19 - 3 = 16$, which is *incorrect!*

⇨   Always use parentheses when substituting numerical values in an algebra formula, especially with negative numbers.

Find the value of the following literal expressions:

(a) $2x^2 + 4x$ for $x = 3$
(b) $2(8 + x)$ for $x = -5$
(c) $(3x)^2$ for $x = 2$

Our step-by-step explanations are in **61**.

---

60 No. Apparently you attempted to substitute -3 for x in the formula without putting the -3 in parentheses first, and your calculation probably looked like this

$$4 - 3 + 15 = 19 - 3 = 16$$

Which is incorrect. Look at the formula. Do you see that 4 and x are multiplied together?

Return to **58** and try again.

---

61 (a) $2x^2 + 4x$ for $x = 3$
            $= 2(3)^2 + 4(3)$ Substitute (3) for x
            $= 2 \cdot 3^2 + 4 \times 3$ Insert arithmetic signs as needed
            $= 2 \cdot 9 + 4 \times 3$ Do the arithmetic
            $= 18 + 12$
            $= 30$

    (b) $2(8 + x)$ for $x = -5$
            $= 2(8 + (-5))$ Substitute (-5) for x.
            $= 2 \cdot (8 - 5)$ Insert arithmetic signs as needed.
            $= 2 \cdot 3$ Do the arithmetic.
            $= 6$

Notice that the careless student who substituted -5 for x without using parentheses would be faced with $2(8 + -5)$ in the second line. Having two arithmetic operation signs side by side like this is terribly confusing.

    (c) $(3x)^2$ for $x = 2$
           $= (3(2))^2$ Substitute (2) for x.
           $= (3 \cdot 2)^2$ Insert arithmetic signs as needed.
           $= 6^2$ Do the arithmetic.
           $= 36$

Order of Operations

In each of these problems there may be confusion about the order in which the arithmetic should be done. For example, in the arithmetic expression,

$$5 + 6 \times 2$$

Should you multiply first: $5 + (6 \times 2) = 5 + 12 = 17$
Or should you add first: $(5 + 6) \times 2 = 11 \times 2 = 22?$

 The answer is that you should always do your calculations in the following *order of operations.*

★ First, perform all operations inside a parentheses or other grouping symbol such as { } or [ ] before doing any of the operations outside the parentheses.

     If several sets of parentheses are nested, start with the inner set first.

★ Second, perform all calculations above or below a fraction bar.

★ Third, calculate all powers of numbers.

★ Fourth, perform all multiplications and divisions, working from left to right.

★ Fifth, perform additions and subtractions last, working from left to right.

If you have any doubt about the order in which calculations are to be done, refer back to this list. Ready for a bit of practice?

1. Evaluate each of the following literal expressions for $x = 3$.

    (a) $3x^2 + 4$ (b) $(2x + 1)^2$ (c) $\dfrac{x - 1}{2}$

    (d) $2x^3$ (e) $x^2 - 5x + 10$ (f) $3(x^2 - 2)$

    (g) $\dfrac{(2x - 1)^2}{5}$ (h) $3(2x^2 - x)$ (i) $2(x - 1)(x + 1)$

2. Evaluate each of the following for $a = 2$ and $b = 3$.

(a) $a(b + 2)$ (b) $2a^2b$ (c) $ab^2 - b$
(d) $3(a + b)$ (e) $5(b - a)$ (f) $3a - 2b$
(g) $a^2 + b^2$ (h) $\dfrac{a + b}{2}$ (i) $ab^2 - a^2b$
(j) $b + \dfrac{1}{a}$ (k) $\dfrac{b^2}{a}$ (l) $(a + b)(a - b)$

3. Evaluate each of the following for $p = 1$, $q = 3$, $r = 4$, and $t = -2$.

(a) pqr (b) $qt^2 + p$ (c) pq^2r
(d) $qr - t^2$ (e) $p + (r - 3t)$ (f) $p(r - t)$
(g) $r(q + t)^2$ (h) $\dfrac{r + t}{q}$ (i) $r^2 - q^2 + p^2 - 3rqp$

Take your time and work carefully.
Check your answers in **62** when you finish.

62
1. (a) 31   (b) 49   (c) 1
   (d) 54   (e) 4   (f) 21
   (g) 5   (h) 45   (i) 16

2. (a) 10   (b) 24   (c) 15
   (d) 15   (e) 5   (f) 0
   (g) 13   (h) $2\dfrac{1}{2}$ (i) 6
   (j) $3\dfrac{1}{2}$ (k) $4\dfrac{1}{2}$ (l) -5

3. (a) 12   (b) 13   (c) 36
   (d) 8   (e) 11   (f) 6
   (g) 4   (h) $\dfrac{2}{3}$ (i) -28

In most practical or applied problems, the quantities in the formulas and numbers being substituted for them will have units associated with them. For example, the area of a rectangle is given by the formula

$A = LW$ where A is the area, L is the length of one side, and W is the width.

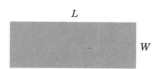

If the length and width are expressed in inches, the area calculated from the formula will be in square inches, abbreviated sq in. or in.^{2} If L and W are given in feet, the area will be in square feet, abbreviated sq ft or ft^2. Usually the problem will make it clear what units the formula variables should have. If it does not, carry the units along when you do the arithmetic. For example, evaluate

$$v = d/t \quad \text{for} \quad d = 50\,\text{ft} \quad \text{and} \quad t = 2\,\text{sec}$$

Try it. Your answer should have both a numerical size and units.
Check your work in **63**.

---

63 $v = \dfrac{d}{t}$ for $d = 50$ ft and $t = 2$ sec

$v = \dfrac{50 \text{ ft}}{2 \text{ sec}}$ $= 25 \text{ ft/sec}$

Carry the units along

First, do the arithmetic, $\dfrac{50}{2} = 25$.

Second, keep the units in the answer in the same algebraic form they have in the problem.

Now, turn to **64** for a set of practice problems involving actual formulas taken from many areas outside of mathematics.

---

64 Solve the following problems. In each problem evaluate the formula given and perform the required arithmetic.

1. The perimeter of a rectangle is given by the formula $P = 2L + 2W$, where L is the length of the rectangle and W is its width. Find P when W is $8\frac{1}{2}$ in. and L is 11 in.

2. The current in a simple electrical circuit is given by the formula $i = \dfrac{V}{R}$, where V is the voltage and R is the resistance of the circuit. Find the current in a circuit whose resistance is 10 ohms and which is connected across a 120-volt power source.

3. Find the power used in an electric light bulb, $P = i^2R$, when the current $i = 0.8$ amperes and the resistance $R = 150$ ohms. P will be in watts.

4. Find the surface area of a sphere, $A = 4\pi R^2$, when π equals roughly 3.14 and $R = 10$ cm.

5. The Fahrenheit temperature F is related to the Celsius temperature C by the formula $F = \dfrac{9C}{5} + 32$. Find the Fahrenheit temperature when $C = 40°$.

6. The volume of a round steel bar depends on its length L and diameter D according to the formula $V = \dfrac{\pi D^2 L}{4}$. Find the volume of a bar 20 inches long and 3 inches in diameter. Use $\pi = 3.14$.

7. If D dollars is invested at p percent interest for t years, the amount A of the investment is $A = D\left(1 + \dfrac{pt}{100}\right)$. Find A if $D = \$1000$, $p = 4\%$, and $t = 5$ years.

8. The total resistance R of two resistances a and b connected in parallel, is $R = \dfrac{ab}{a + b}$. What is the total resistance if $a = 200$ ohms and $b = 300$ ohms?

9. If an object is dropped from the roof of a building, it falls a distance D feet in t seconds, where $D = \frac{1}{2}gt^2$ and $g = 32$ ft/sec^2. How far will it fall in 5 seconds? (This formula ignores air friction.)

10. The energy of motion E of a ball with weight W moving at speed v is given by the formula $E = \dfrac{Wv^2}{2g}$ where $g = 32$ ft/sec^2, $W = 2$ lb, and $v = 16$ ft/sec.

Check your work in **65**.

---

1. $P = 2L + 2W = 2\left(8\frac{1}{2}\right) + 2(11)$

$$= 17 + 22 = 39 \text{ in.}$$

2. $i = \dfrac{V}{R} = \dfrac{120}{10} = 12 \text{ amps}$

3. $P = i^2R = (0.8)^2 150$
$$= (0.8)^2 \times 150$$
$$= 0.64 \times 150 = 96 \text{ watts}$$

4. $A = 4\pi R^2 = 4(3.14)(10)^2$
$$= 4 \times 3.14 \times 10^2 = 12.56 \times 100 = 1256 \text{ cm}^2$$

5. $F = \dfrac{9C}{5} + 32 = \dfrac{9(40)}{5} + 32$

$$= \dfrac{9 \times 40}{5} + 32 = \dfrac{360}{5} + 32$$

$$= 72 + 32 = 104° \text{ F}$$

6. $V = \dfrac{\pi D^2 L}{4} = \dfrac{(3.14)(3)^2(20)}{4} = \dfrac{3.14 \times 9 \times 20}{4} = 141.3 \text{ in.}^3$

7. $A = D\left(1 + \dfrac{pt}{100}\right) = (1000)\left(1 + \dfrac{(4)(5)}{100}\right)$

$$= 1000 \times \left(1 + \dfrac{20}{100}\right) = 1000 \times (1 + 0.2)$$

$$= 1000 \times 1.2 = \$1200$$

8. $R = \dfrac{ab}{a + b} = \dfrac{(200)(300)}{(200) + (300)} = \dfrac{200 \times 300}{200 + 300}$

$$= \dfrac{60000}{500} = 120 \text{ ohms}$$

9. $D = \dfrac{1}{2}gt^2 = \dfrac{1}{2}(32)(5)^2$

$$= \dfrac{1}{2} \times 32 \times 5^2 = \dfrac{1}{2} \times 32 \times 25$$

$$= 16 \times 25 = 400 \text{ ft}$$

10. $E = \dfrac{Wv^2}{2g} = \dfrac{(2)(16)^2}{2(32)}$

$$= \dfrac{2 \times 16^2}{2 \times 32} = \dfrac{\cancel{2} \times 256}{\cancel{2} \times 32} = \dfrac{256}{32} = 8 \text{ ft lb}$$

Now turn to **66** for a set of practice problems on evaluating literal expressions.

Unit I
Introduction
to Algebra

The answers are on page 540.

A. **Evaluate each of these expressions using** $x = 2$, $y = 3$, $z = 5$, $v = -2$, and $w = 6$.

1. $3x$ _____

2. y^2 _____

3. wx _____

4. $2yz$ _____

5. $z - y$ _____

6. $x + w$ _____

7. $2x + y$ _____

8. $y^2 - x^2$ _____

9. $3w - 7$ _____

10. $2w - 3y$ _____

11. $3xy^2 - 2w + 4$ _____

12. $\dfrac{1}{x} + \dfrac{y}{2} + 1$ _____

13. $y^2x - x^2y$ _____

14. $2y - 1$ _____

15. $2(y - 1)$ _____

16. $2x + y - z + 1$ _____

17. $\dfrac{yz}{x} + w$ _____

18. $2x + 3y - 1$ _____

19. $2(x + y) + 3(w - z)$ _____

20. $3x^2y - 2xz$ _____

21. $x(y + v)$ _____

22. $(x^2 + v^2)(1 - v)$ _____

23. $v(v + x)(z^5w)$ _____

24. $\dfrac{w^2}{v} - yz$ _____

25. $(1 - v)^2$ _____

26. $1 - (x - v)$ _____

27. $2(v - z)^2$ _____

28. $(x - w)^2 + v^2$ _____

29. $x^2 + y^2 - v^2$ _____

30. $\dfrac{1}{2}(x - v)(y + v)$ _____

B. **Find the value of each of these literal expressions for** $x = 4$, $y = 10$, $z = 7$, and $w = -3$.

1. $2x^2 - y$ _____

2. $3(x + y) - 2z$ _____

3. $2x + 3y - z - 5w$ _____

4. $x^2 + y^2$ _____

5. $(x + y)^2 - (z - w)^2$ _____

6. $(x + 2(y + 2)) + 2$ _____

7. $((2x + y) + 3w) - 5$ _____

8. $(x + y)(2w + 1)$ _____

9. $((x + 1)^2(y - 6)^2) + 5$ _____

10. $\dfrac{2(y - x) + 3w}{z}$ _____

11. $2xyz - 10$ _____

12. $((x + 1) + 2(y + 1))$ _____

13. $(x + x(x + x(x + 1)))$ _____

14. $3(2x^2 + 1)(y + 2)$ _____

15. $(2z - w) + (2y - 5)^2$ _____

16. $x^2 + w^2 + (x + w)^2$ _____

17. $3x^2y - w$ _____

18. $(x + 1)(x - 1)$ _____

19. xyz _____

20. $\dfrac{y + x}{y - w}$ _____

21. $x^2 + y^2 - z^2$ _____

22. $(x - 1)(y - 4)(w - 5)$ _____

23. $(2x - y)(z - y)$ _____

24. $(x + 2)^2(x - 1)$ _____

25. $3[2(w^2 - 1) + y]$ _____

26. $4[w - 2(1 - x^2)]$ _____

27. $1 - [1 + z(2 + y)]$ _____

28. $3 - 4[2 - y(w + 2)]$ _____

29. $\dfrac{1}{2}[x - y(z - w^2)]$ _____

30. $\dfrac{2}{3}[5 - (w^2 - y)]$ _____

C. Evaluate as indicated.

1. Evaluate each of these expressions for $A = 5$, $B = 2$:

   (a) $3A - B + 1 =$ _____

   (b) $3(A - B) + 1 =$ _____

   (c) $3(A - B + 1) =$ _____

   (d) $3A - (B + 1) =$ _____

   (e) $3(A - (B + 1)) =$ _____

   (f) $3(A + (B - (A - (B + 3)))) =$ _____

2. The volume of a cylinder is given by the equation $V = \dfrac{\pi D^2 H}{4}$. Calculate the volume of a cylindrical tank with height, $H = 12$ ft, diameter, $D = 6$ ft. Use $\pi \cong 3.14$.

   $V =$ _____

3. Evaluate $(W + (H(A + A)T)T)$, where $W = 4$, $H = 2$, $A = 3$, $T = 1$.

   _____

4. The distance a free-falling object drops in T seconds is $D = \frac{1}{2}gT^2$ meters where g is roughly 10. How far would a stone fall in 10 seconds? Use $g \cong 9.8$.

   $D =$ _____

5. The Celsius and Fahrenheit temperature scales are related by the equation

   $$C = \frac{5(F - 32)}{9}$$

   What is the Celsius temperature corresponding to $F = 140°$?

   $C =$ _____

6. The area of a trapezoid is $T = \dfrac{(A + B)H}{2}$, where A and B are the lengths of its parallel sides and H is its height. Find the area of the trapezoid shown.

$T = $ _____

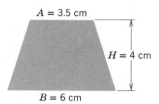

A = 3.5 cm

H = 4 cm

B = 6 cm

7. The following equation appeared on a chemistry exam:

$$P = \frac{nR(T + 273)}{V}$$

Find the value of P when $n = 5$, $R = 0.08$, $T = 27$, and $V = 3$.

$P = $ _____

8. The interest I paid on a loan can be calculated from the equation $I = P \cdot R \cdot T$, where P is the principal or amount of the loan, R is the rate of interest per year, and T is the time in years. What interest would you pay on a $2000 loan ($P$) at 8% ($R$) interest for 3 years ($T$)?

$I = $ _____

9. The volume of a sphere can be calculated from the equation $V = \left(\frac{4}{3}\right)\pi R^3$. Calculate the volume of a sphere for which the radius $R = 6$ in. (Use $\pi \cong 3.14$.)

$V = $ _____

10. The density of a rectangular plate can be calculated from the formula $D = \dfrac{M}{ABC}$. Find D if M is 2.4 kg, A is 5 m, B is 6 m, and C is 0.1 m.

$D = $ _____

11. The nth triangular number is given by the equation $T = \frac{1}{2}n(n + 1)$. Find the first four triangular numbers ($n = 1, 2, 3, 4$). Find the 10th triangular number.

$T_1 = $ _____ $T_2 = $ _____ $T_3 = $ _____ $T_4 = $ _____ $T_{10} = $ _____

12. The weight of a metal cylinder can be found from the equation $W = D(ab - \pi r^2)h$, where $a = 8$ cm, $b = 6$ cm, $r = 2$ cm, $h = 10$ cm, $D = 8\,g/cm^3$, and $\pi \cong 3.14$.

$W = $ _____

D. Brain Boosters

1. A bowler's handicap is calculated from the formula

$H = 0.8(200 - A)$ where H is the handicap and A is the bowler's three game average. Of course this formula is only useful for A less than 200. The bowler's final score is equal to her actual score plus her handicap.

Suppose you roll 140, 156, and 163 for a three-game series.

(a) What would be your three-game average A? _____

(b) What is your handicap? _____

(c) What would be your final handicap score if you rolled a 145 game?

_____

2. A golfer's handicap is calculated from the formula

$H = 0.85(A - 72)$ where H is the handicap and A is the golfer's three round average. If A is less than 72, the handicap is taken to be zero. The golfer's net score is equal to her actual score minus her handicap.

Suppose on your last three games of golf you shot 95, 87, and 82.

(a) What would be your three-round average? _____

(b) What is your handicap? _____

(c) What would be your handicap score or net score when you shoot a 90?

_____

3. The Special Theory of Relativity, developed by Albert Einstein in 1907, predicts that when a nuclear reaction takes place the net mass of the nuclei may decrease and energy will be created. The amount of energy produced is given by the famous equation $E = Mc^2$ where M is the mass decrease and c is the speed of light in a vacuum.

Find E if M is 0.000001 kg (equivalent to the mass of a drop of water) and c is 300,000,000 meters per second. (E will be in Joules.)

$E =$ _____

When you have had the practice you need, turn to 67 for a self-test covering the work of Unit 1.

Unit I
Introduction
to Algebra

Self-Test 1

67

1. $-11 - (-17)$ = _____

2. $(-3.5) \times (-2.4)$ = _____

3. $(-13.44) \div (-3.2)$ = _____

4. $(-0.02)^3$ = _____

5. Simplify $2 \cdot x \cdot x \cdot x \cdot y \cdot y \cdot z \cdot z$ = _____

6. $x^3 \cdot x^4$ = _____

7. Simplify $5xy - x^2 + 2xy + 3x^2 - xy$ = _____

8. Simplify $12p^2q + (-2p^2q) - 4p^2q$ = _____

9. Simplify $(3x)(-6x^2)(x^3)$ = _____

10. Simplify $-2(4b^2)(2b)(-ab)$ = _____

11. Simplify $(3c)^2(-2c)^3$ = _____

12. Simplify $\dfrac{24m^3n^2}{15m^4n}$ = _____

13. Simplify $\dfrac{-5x^2y}{(xy)^2}$ = _____

14. Simplify $\dfrac{16b^3}{2b} - 5b^2$ = _____

15. Simplify $-3x^3 + 2x(4x^2) - 8x^2(x)$ = _____

16. Simplify $2st + 3s(t^2) - 5st - 5t(st)$ = _____

17. Simplify $2(1 - x)$ = _____

18. Simplify $1 - 3(2y - 4)$ = _____

19. Simplify $2a - (1 - 3a)$ = _____

20. $4 - [x - 3(x - 2)]$ = _____

21. Write as a literal expression:
"The square of x minus the product of p and q." _____

Date _____

Name _____

22. Write as an algebra equation:
"The sum of five times a number and its square is 24." _____

Course/Section _____

79   Self-Test 1

23. Write as an algebra equation:
    "Six times a number less five gives
    twice another number."

    _____

24. Find $3x^2y^3 - 2xy$ for $x = 2$, $y = -2$.

    _____

25. Find $K = (A + B)(A - B)$ for $A = 4$, $B = -2$.

    _____

26. Find $V = \frac{1}{4}\pi D^2 H$ for $\pi \cong 3.14$, $D = 3$, $H = 4$.

    _____

27. Find $C = \dfrac{5(F - 32)}{9}$ for $F = 104$.

    _____

28. Find $2(s + t(s - 3))$ for $s = 6$, $t = 2$.

    _____

29. Find $E = \frac{1}{2}mv^2$ for $m = 10$, $v = 20$.

    _____

Answers are on page 540.

Unit I
Introduction
to Algebra

The Language of
Algebra

A. **Complete each of the following statements using the words** *variable, expression, factor, equation, term, inequality, conditional equation, right member,* and *left member.*

1. $(x + 7)$ is a ___factor___ of the product $(x + 7)(x - 4)$.

2. In the ___equation___ $2y^2 - 3y + 5 = 0$, the ___left___ is $2y^2 - 3y + 5$ and the ___right___ is 0. The ___variable___ is y.

3. In the ___ $5a^2b + ab$, ab is a ___term___.

4. $B > -4$ is an ___ and B is a ___.

5. In the ___ $3c^2 - 4c$, $3c^2$ is the first ___ and c is a ___.

6. $4m^3n - m^2n + 3$ is an ___ which has three ___.

7. s and t are the ___ of st.

8. The ___ of the ___ $3x = -9$ is -9.

9. In the ___ $3f^2 - 3f + 4 = 7$, the ___ is $3f^2$.

10. In the ___ $4x - 7 > 2$, x is the ___.

B. **Write the following in algebraic notation.**

1. y plus b^2 ___

2. $4y$ divided by $3x$ ___

3. $(x - y)$ times $(x + y)$ ___

4. 7 times $(5c + d)$ ___

5. 5 times m times p ___

6. t minus 15 ___

7. x plus y plus z ___

8. 5 times b times b times c ___

9. the product of h and k ___

10. $(4 + d)$ divided by e ___

11. $(x - 2)$ divided by $3x$ ___

12. $(2d - 5)$ multiplied by $3d$ ___

13. $(x - 4)$ multiplied by x^2 ___

14. $7z$ multiplied by $(2x - 1)$ ___

15. $5y$ multiplied by $(y - 2)$ ___

16. $2x$ divided by $(x - 1)$ ___

17. $10q$ divided by $(p - 6q)$ ___

18. $(3x - y)$ divided by $3y$ ___

19. $(x - 3)$ divided by $5x$ ___

20. $(3s + t)$ times t^2 ___

21. the product of $(x^2 - 1)$ and $4x$ ___

Date _____

Name _____

Course/Section _____

22. $(x - y)$ times $(2x + y)$ _____

23. $(2a + b)$ times $(a - b)$ _____

24. x multiplied by $(1 - x + x^2)$ _____

25. $2y$ multiplied by $(3 + y - 2y^2)$ _____

26. $(a - 2b + c)$ times $(a + 1)$ _____

27. the product of $(x - 3y - z)$ and $(z + x)$ _____

28. $2x$ times $(x - 2)$ times $(x^2 - 1)$ _____

29. $5y$ times $(3x - 1)$ times $(y + 1)$ _____

30. $5x$ times $(3x - 5)$ times $2y$ _____

C. **Complete each statement by inserting the proper symbol:** $<$, $>$, or $=$.

1. 11   7

2. $\dfrac{3}{4}$ $\dfrac{17}{20}$

3. $\dfrac{1}{8}$ $\dfrac{1}{7}$

4. 16 $15 + 1$

5. y $y - 1$

6. $13 - 7$ 2×3

7. x $2x$

8. 2   2.1

9. 4.5 $4\frac{1}{2}$

10. $1\dfrac{1}{10}$ 1.1

11. $\dfrac{3}{8}$ $\dfrac{8}{3}$

12. $3 + y$ $y + 3$

D. **Evaluate each of the following.**

1. $5 + (6 - 2)$ _____

2. $8 + (4 - 1)$ _____

3. $9 - (7 - 5)$ _____

4. $11 - (14 - 9)$ _____

5. $6 + (9 - 2)$ _____

6. $5 + (8 - 5)$ _____

7. $2(7 - 4)$ _____

8. $6(5 - 3)$ _____

9. $10(8 - 5)$ _____

10. $9(7 - 3)$ _____

11. $2(11 - 5)$ _____

12. $5(13 - 11)$ _____

Unit I
Introduction
to Algebra

Signed Numbers

A. Add or subtract as indicated.

1. $9 - 13$ _____

2. $-4 + 5$ _____

3. $-8 - 15$ _____

4. $-13 - (-8)$ _____

5. $13 - 8$ _____

6. $-13 - 8$ _____

7. $-34 - 17$ _____

8. $21 - (-5)$ _____

9. $-42 - (-12)$ _____

10. $78 + (-39)$ _____

11. $-7 - (-21) - 15$ _____

12. $13 + (-9) - (-10)$ _____

13. $-2.6 - (-2.6)$ _____

14. $-14 + 0.02$ _____

15. $-3\frac{1}{5} - \left(-\frac{1}{10}\right)$ _____

16. $8 + (-1.68)$ _____

17. $-7.5 + \left(-7\frac{1}{2}\right)$ _____

18. $3\frac{1}{4} - \left(-\frac{2}{3}\right) + \left(-1\frac{5}{6}\right)$ _____

19. $2\frac{1}{3} - \left(-\frac{1}{6}\right) - \left(1\frac{2}{3}\right)$ _____

20. $-6.3 + \left(-3\frac{1}{4}\right)$ _____

21. $4 - 6.3 + 0.5$ _____

22. $-5.1 - 2.7 + 3.1$ _____

23. $1.1 - 4.5 - 3.7$ _____

24. $-6.2 + 2.6 - 3.1$ _____

B. Multiply or divide as indicated.

1. $(-7) \times 3$ _____

2. $(-5) \times (-12)$ _____

3. $(-8) \times 7$ _____

4. $0 \times (-3)$ _____

5. $(-6) \times (-4)$ _____

6. $(-5) \times (-9)$ _____

7. $4 \times (-7)$ _____

8. $(-11) \times 5$ _____

9. $\frac{-21}{-7}$ _____

10. $\frac{72}{-8}$ _____

11. $\frac{-54}{9}$ _____

12. $\frac{-91}{-7}$ _____

13. $(-56) \div 8$ _____

14. $36 \div (-4)$ _____

15. $(-77) \div (-11)$ _____

16. $0 \div (-7)$ _____

17. $(-0.4) \times (-0.2)$ _____

18. $(-5) \times (-0.06)$ _____

19. $(-0.08) \times (-10)$ _____

20. $1.4 \times (-3)$ _____

21. $\left(-\frac{4}{3}\right) \times \left(-\frac{1}{2}\right)$ _____

22. $\left(-\frac{1}{2}\right) \times \left(-\frac{3}{4}\right)$ _____

23. $(2.1) \times (-5)$ _____

24. $(-0.1) \times (4.1)$ _____

25. $(0.2) \times (-3.6)$ _____

26. $(-6.4) \div (-16)$ _____

27. $(7.2) \div (-18)$ _____

28. $(-6) \div (1.5)$ _____

29. $(-7.5) \div (-0.15)$ _____

30. $(-66) \div (1.1)$ _____

Date _____

Name _____

Course/Section _____

C. **For each of the following insert $<$ or $>$ to make true statements.**

1. $-2 \quad -3$ 2. $-5 \quad -4$ 3. $0 \quad -7$

4. $-13 \quad -8$ 5. $-1\frac{1}{2} \quad -1$ 6. $-\frac{3}{4} \quad -\frac{5}{8}$

7. $-0.1 \quad 0.01$ 8. $0.02 \quad -42$

9. $-3.4 \quad -3.5$ 10. $-0.009 \quad -0.09$

11. $4 + (-2) \quad 4 - (-2)$ 12. $\frac{20}{-2} \quad (-2) \times (-5)$

13. $(-4) \times 8 \quad (-3) \times 9$ 14. $-2 \times 0 \quad -2 + 0$

15. $2 + (1 - 4) \quad (6 - 5)$ 16. $3 \times (-1) \quad (-1) \times (-3)$

17. $(-4)(-2) \quad (-4)(2)$ 18. $(-3)(4) \quad (-2)(4)$

19. $(-3)(-2) \quad (4)(-1)$ 20. $(-6)\left(\frac{1}{2}\right) \quad (12)\left(-\frac{1}{3}\right)$

D. **Perform the indicated operations.**

1. $(-1) \times (-1) \times (-1) \times (-1)$ _____ 2. $\frac{-9 \times (-3)}{3}$ _____

3. $\frac{18 - (-6)}{(-2) \times (-3)}$ _____ 4. $\frac{(-5) + (-7) - (-9)}{-2 + (-1)}$ _____

5. $-3.4 - 5.6 + 7.8$ _____ 6. $(-0.9) + (-3.5) - (-2.4)$ _____

7. $\frac{0.98}{0.02}$ _____ 8. $\frac{2.3 \times (-2.1)}{0.3}$ _____

9. $14.067 + (-9)$ _____ 10. $(-2.604) \times (-3.6)$ _____

11. $-2\frac{1}{3} + 4\frac{5}{8}$ _____ 12. $\left(-\frac{1}{4}\right) \times \left(-\frac{3}{4}\right)$ _____

13. $-8 + \left(-2\frac{1}{3}\right) - \frac{5}{6}$ _____ 14. $\left(-1\frac{2}{3}\right) \times \left(-1\frac{1}{4}\right) \times \left(-\frac{9}{10}\right)$ _____

15. $\frac{(-0.5) - (-\frac{1}{2}) - 2}{-0.5}$ _____ 16. $\frac{-12\frac{3}{8} - 11\frac{3}{4}}{-0.2}$ _____

17. $\frac{(-0.3) - (-0.6) + 1}{-0.1}$ _____ 18. $\frac{(2.1) - (-4) + (-0.1)}{-3}$ _____

19. $(-2)(-4) - (-5)$ _____ 20. $-(-3)(2) - (-5)(-2)$ _____

Unit I
Introduction
to Algebra

Exponents

A. Write each of the following in exponential form.

1. $3 \cdot 3 \cdot 3 \cdot 3$ _____

2. $y \cdot y \cdot y \cdot y \cdot y \cdot y$ _____

3. $12 \cdot 12$ _____

4. $10 \cdot 10 \cdot 10 \cdot 10$ _____

5. $2 \cdot 2 \cdot 2 \cdot b \cdot b$ _____

6. $m \cdot m \cdot p \cdot q \cdot q \cdot q \cdot q$ _____

7. $(y - 5)(y - 5)$ _____

8. $10 \cdot 10 \cdot 10 \cdot x \cdot x \cdot (2 - x)(2 - x)$ _____

9. $a \cdot a \cdot a \cdot b \cdot b \cdot (a + b)$ _____

10. $7 \cdot 7 \cdot 7 \cdot (c - 3)(c - 3)(c - 3)$ _____

11. $(-2)(-2)(-2)$ _____

12. $(a - b)(a - b)(a - b)(a - b)$ _____

13. $3 \cdot 3 \cdot x \cdot x \cdot x \cdot y$ _____

14. $2 \cdot 4 \cdot p \cdot p \cdot q \cdot q \cdot q \cdot q$ _____

15. $5 \cdot 5 \cdot 5 \cdot 5 \cdot t \cdot t$ _____

16. $2 \cdot 2 \cdot y \cdot y \cdot y \cdot y \cdot y$ _____

B. Find the value of the following.

1. 3^3 _____

2. 5^2 _____

3. 10^4 _____

4. 1^7 _____

5. 2^6 _____

6. 9^3 _____

7. 10^8 _____

8. $(-2)^3$ _____

9. $(-5)^2$ _____

10. 7^0 _____

11. $\left(\dfrac{1}{2}\right)^4$ _____

12. $\left(\dfrac{2}{3}\right)^3$ _____

13. $\left(5\dfrac{1}{2}\right)^2$ _____

14. $\left(-\dfrac{1}{3}\right)^3$ _____

15. $\left(\dfrac{5}{9}\right)^2$ _____

16. $(-4)^3$ _____

17. $(-1)^5$ _____

18. 0^6 _____

19. 1^{24} _____

20. $(2.1)^2$ _____

21. $(3.5)^2$ _____

22. $(-1.3)^2$ _____

23. 2.5^3 _____

24. 1.05^2 _____

25. $(4.5)^2$ _____

26. $(6.1)^2$ _____

27. $(-1.01)^2$ _____

28. $(2.03)^2$ _____

29. $(-1.1)^2$ _____

30. $(-3.2)^3$ _____

C. Simplify the following.

1. $(3y)^2$ _____

2. $(-4d)^3$ _____

3. $(8pq)^2$ _____

4. $(-5b)^3$ _____

5. $(-10x)^4$ _____

6. $(-2ef)^2$ _____

7. $\left(\dfrac{1}{4}g\right)^3$ _____

8. $(0.1t)^5$ _____

9. $\left(-\dfrac{1}{2}ad\right)^4$ _____

10. $(-5n^2)^3$ _____

11. $\left(-\dfrac{2}{3}s^2\right)^3$ _____

12. $(2a^2c^3)^4$ _____

13. $z^4 \cdot z^5$ _____

14. $5b^3 \cdot b^2$ _____

15. $e^3(e^5 \cdot g^3)$ _____

16. $\dfrac{x^7}{x^4}$ _____

17. $\dfrac{R^8}{R}$ _____

18. $\dfrac{k^{12}}{k^3}$ _____

19. $\dfrac{12b^5}{4b^5}$ _____

20. $\dfrac{51c^2d^3}{3cd}$ _____

21. $\dfrac{-27x^5y^3z}{9x^3y^6z^2}$ _____

22. $\dfrac{42m^5y}{4.2my}$ _____

23. $\dfrac{17y^5}{-17y^5}$ _____

24. $\dfrac{-56h^9i^5}{-8h^8i^4}$ _____

25. $\left(\dfrac{x}{2}\right)^5$ _____

26. $\left(\dfrac{x^3}{2x}\right)^6$ _____

27. $\left(\dfrac{2p^2q}{t^3}\right)^3$ _____

28. $(x^{12})^{10}$ _____

29. $x^{22} \cdot x^7$ _____

30. $p^2 \cdot p^{14}$ _____

31. $(2t^8)^8$ _____

32. $3d^5 \cdot d^{16}$ _____

33. $(-3)^{-2}$ _____

34. 4^{-3} _____

35. $(0.1)^{-2}$ _____

36. 0^0 _____

37. $\left(\dfrac{1}{4}\right)^3$ _____

38. $\left(\dfrac{1}{4}\right)^{-3}$ _____

39. $2(2x)^{-2}$ _____

40. $x^2 \cdot x^{-7} \cdot x^9$ _____

41. $y^3 \cdot y^5 \cdot y^{-1}$ _____

42. $a^{-1} \cdot a^{-4} \cdot a^3$ _____

43. $p^{-5} \cdot p^3 \cdot p^{-2}$ _____

44. $x^4 \cdot x^{-6} \cdot x^5$ _____

Write with positive exponents.

45. y^{-3} _____

46. x^{-4} _____

47. $3a^{-5}$ _____

48. $2p^{-6}$ _____

49. x^2y^{-2} _____

50. p^3q^{-4} _____

51. $x^{-2}y^{-6}$ _____

52. $b^{-3}c^{-4}$ _____

53. $5a^{-3}b^{-2}$ _____

54. $7x^{-4}y^{-5}$ _____

55. $2m^2n^{-4}$ _____

56. $5p^{-6}q^4$ _____

57. $(x^2)^{-3}$ _____

58. $(y^3)^{-4}$ _____

59. $(a^{-2})^{-3}$ _____

60. $(b^{-5})^{-3}$ _____

61. $\dfrac{1}{x^{-2}}$ _____

62. $\dfrac{2}{y^{-3}}$ _____

63. $\dfrac{b^2}{a^{-5}}$ _____

64. $\dfrac{c^3}{x^{-4}}$ _____

65. $\dfrac{x^{-2}}{y^{-3}}$ _____

66. $\dfrac{p^{-10}}{t^{-8}}$ _____

67. $\dfrac{4x^{-2}}{y^{-5}}$ _____

68. $\dfrac{6a^{-7}}{b^{-6}}$ _____

Write without a denominator.

69. $\dfrac{x^{-6}}{y^4}$ _____

70. $\dfrac{y^5}{x^7}$ _____

71. $\dfrac{a^3}{x^{-5}}$ _____

72. $\dfrac{p^{-2}}{y^{-6}}$ _____

73. $\dfrac{ax^2}{a^{-1}x^3}$ _____

74. $\dfrac{p^2q^{-4}}{q^{-2}p^5}$ _____

75. $\dfrac{x^4y}{2x^{-1}y^{-4}}$ _____

76. $\dfrac{m^{-2}n^2}{3m^{-6}n^{-7}}$ _____

77. $\dfrac{x^7y^{-8}}{4x^4y^5}$ _____

78. $\dfrac{x^{-1}y}{8x^4y^{-5}}$ _____

79. $\dfrac{x^2y^{-4}z^{-1}}{x^{-1}y^9z^{-3}}$ _____

80. $\dfrac{a^{-7}b^4c^2}{a^{-2}b^5c^{-1}}$ _____

Date _____

Name _____

Course/Section _____

Unit I
Introduction
to Algebra

Working with Literal Numbers

A. Simplify by combining like terms:

1. $4x + 3x$ _____

2. $5y + (-2y)$ _____

3. $16a + 3a$ _____

4. $(-6t) + (4t)$ _____

5. $(6p) + (-5p)$ _____

6. $(-8q) + (10q)$ _____

7. $(-7x) + (12x)$ _____

8. $(-9k) + (-13k)$ _____

9. $ab + 5ab$ _____

10. $(-3b) + (-2b)$ _____

11. $2xy + 3xy$ _____

12. $-4mp + (-3pm)$ _____

13. $-ac^2 + (3ac^2)$ _____

14. $14abc + (-3abc)$ _____

15. $(-7cd^2) + (-cd^2)$ _____

16. $12jk^2 + (-5jk^2)$ _____

17. $3y + 4y + (-8y) + y$ _____

18. $-2x^2 + 4x^2 + 5x^2$ _____

19. $(-2p) + 15p + 3p^2$ _____

20. $2pq^2t + (-3pq^2t) + (6pq^2t)$ _____

21. $8a - 6a$ _____

22. $17q - (-11q)$ _____

23. $16ab - (-19ab)$ _____

24. $0 - (-3x^2)$ _____

25. $11t - t$ _____

26. $3m - (-m) + 4m$ _____

27. $2c^2d^2 - 9c^2d^2$ _____

28. $1 - (-2x)$ _____

29. $8.1x - 3.5x$ _____

30. $0.2ax - 0.6ax$ _____

31. $2ay - 0$ _____

32. $(-18xt) - (-8xt)$ _____

33. $(-2q) - (-7q) + q$ _____

34. $-q - 2q - (-2q) - q$ _____

35. $-3pq^2r - 5pq^2r$ _____

36. $8ax^2 - 6ax$ _____

37. $2ax^2 - 5ax^2 + ax$ _____

38. $3x^2 - 7x^2 + 2x$ _____

39. $xy - xy^2 + 3xy$ _____

40. $12pt - 7pt + p$ _____

B. Multiply:

1. $4(6y)$ _____

2. $x(3x^2)$ _____

3. $2ab(a^2)$ _____

4. $2x(7x)$ _____

5. $6p(p)$ _____

6. $(y^2)(y^4)$ _____

7. $(4t^2)(t^3)$ _____

8. $(8a^5)(5a)$ _____

9. $(a)(b)(c)$ _____

10. $(a^2)(b)(c^2)$ _____

11. $(a^2)(ab^2)(ac^2)$ _____

12. $(a^2)(a^3)(a^4)3$ _____

13. $(6x)^2$ _____

14. $(5p)^2(4pq)^2$ _____

15. $(-pq^2)(-pqr)(p^2r)$ _____

16. $(-xyz^2)(-2x)(y^2)$ _____

17. $(x^2y)(ab)(3ax)$ _____

18. $2t(st^2)3s$ _____

19. $(-2c)^3$ _____

20. $4(-5q)^2$ _____

21. $3q(-2q^2)^2$ _____

22. $10(ac^2)(2a^2b)$ _____

23. $-5(xy^2)(-2x^2y)(-x)$ _____

24. $(ax)^2(bx^2)^2$ _____

25. $(xy)(x^2y)^2$ _____

26. $(2ab^2)^2(a^2b)$ _____

27. $(-2pt)(-3p^2t^4)t^2$ _____

28. $(-4bc)^2(-2b^2c)^3$ _____ **29.** $2p(p^2)(-ap^3)$ _____ **30.** $-(3x^2)(-3x)(-2x^2)$ _____

31. $(-4)(m^2)(-n)^3$ _____ **32.** $(-2ax)^3(-3a^2x^2)^2$ _____ **33.** $-(a^2b)(-bc^3)(ab^2c)$ _____

34. $-x(xy^3)(x^2y^2z)(-x^2z^2)$ _____ **35.** $(a^4t^5)(a^3t)(-a^5t^6)$ _____ **36.** $(-k^4m^2)(km^3)(-k^5m)$ _____

C. Divide:

1. $\dfrac{8x}{2}$ _____ **2.** $\dfrac{18y}{6}$ _____ **3.** $\dfrac{12p^2}{-2p}$ _____

4. $\dfrac{-15q}{3q}$ _____ **5.** $\dfrac{-40t^2}{-5t}$ _____ **6.** $\dfrac{-2x^3}{4x}$ _____

7. $\dfrac{32a^6}{-6a^2}$ _____ **8.** $\dfrac{25x^3y^2}{10xy}$ _____ **9.** $\dfrac{6a^2b^3c^5}{4ab^3c^2}$ _____

10. $\dfrac{(-2x)^3}{-x}$ _____ **11.** $\dfrac{(-3s^2t)^2}{-st}$ _____ **12.** $\dfrac{72c^3d^7}{16cd^2e}$ _____

13. $\dfrac{4x^2}{8x^5}$ _____ **14.** $\dfrac{-3a^2b}{-9a^3b^2}$ _____ **15.** $\dfrac{-8d^5y}{4d^2}$ _____

16. $\dfrac{5x + 2x}{x}$ _____ **17.** $\dfrac{(3ax^2)(2x)}{4a^2}$ _____ **18.** $\dfrac{(2p)^4}{p}$ _____

19. $\dfrac{7mn - (-2mn) + mn}{5m^2n}$ _____ **20.** $\dfrac{18a^2bc^3d}{2a^2bc}$ _____ **21.** $\dfrac{(1.4k^2)(3k^3)}{2k}$ _____

22. $\dfrac{9x^{10}y^8}{24x^7y^9}$ _____ **23.** $\dfrac{-2h + 7ah^3}{h^2}$ _____ **24.** $\dfrac{(9s^2r)(12qr^2)(2qs^3)}{6sqr^2}$ _____

25. $\dfrac{3x^2 - 4xy}{2x}$ _____ **26.** $\dfrac{2a^2b - 3ab^3}{a^2b^2}$ _____ **27.** $\dfrac{(-3a^2b)(-2ab)(4b)}{6a^2b^2}$ _____

28. $\dfrac{(4xy^2)(-2x^3y^4)(-2y^2)}{8x^2y^5}$ _____ **29.** $\dfrac{(-2x^2t)^2(3xt^3)^2}{6xt^4}$ _____ **30.** $\dfrac{(4a^5b^6)(-3a)^2(ab^3)}{-18a^5b^5}$ _____

Date _____

Name _____

Course/Section _____

Unit I
Introduction
to Algebra

Translating English to
Algebra

A. Translate each of the following phrases to a mathematical operation symbol.

1. Divided by _____

2. Equals _____

3. The same as _____

4. Reduced by 8 _____

5. Less than _____

6. Is less than _____

7. 10 less _____

8. Increased by 2 _____

9. One-half of _____

10. Two years more _____

11. Multiplied by _____

12. Of _____

13. Times _____

14. The square of y _____

15. As many as _____

B. Translate each of the following into an algebraic expression.

1. x minus y _____

2. 3 plus x _____

3. a less than t _____

4. The product of 7 and k _____

5. x divided by y _____

6. 14 less F _____

7. Twice p divided by 6 _____

8. A multiplied by B _____

9. The square of q _____

10. 16 times the square of a number _____

11. The product of k and m divided by 2 _____

12. The product of the cube of t and the square of p _____

13. Twice the height _____

14. Half the time _____

15. Triple the distance _____

16. Double the cost _____

17. Four times as many as _____

18. The sum of $\frac{1}{2}$ a number plus 10 _____

19. Twice a number subtracted from 3 _____

20. The sum of a number and twice another number _____

21. The sum of a number and its square _____

22. Three times a number subtracted from twelve _____

23. A number divided by two more than that number _____

24. Twice the sum of a number and half of that number _____

C. Write each of these sentences as an algebraic equation.

1. The sum of two numbers is 14. _____

2. Five times a number is equal to nine minus twice the number. _____

3. Kinetic energy is equal to one-half the mass times the square of the speed. _____

4. The perimeter of the rectangle is twice the width plus twice the length. _____

5. Fourteen less than a number is equal to the number squared. _____

6. One-third the sum of two numbers is equal to 7. _____

7. What number increased by 12 is equal to the square of that number? _____

8. The sum of the squares of two numbers is equal to 34. _____

9. Six times the sum of two numbers is 4 more than one of the numbers. _____

10. The volume of a cone is π times the square of the radius of its base times the **height divided by** 3. _____

11. The area of a triangle is one-half its base times its height. _____

12. One number is 4 times another. _____

13. The difference of two numbers is 4. _____

14. Jose is ten years older than his brother. _____

15. If 6 is added to a number, the sum is 14. _____

16. Mary weighs 8 pounds less than Lee. _____

17. When a number is doubled, the result is 16. _____

18. One-sixth of a number is equal to the square of another number. _____

19. Work equals force times distance. _____

20. The product of current and resistance is equal to the voltage. _____

21. The time of travel is equal to the distance divided by the speed. _____

22. The increase in potential energy is equal to the mass times g times the vertical distance moved.
_____

23. The power dissipated in an electrical circuit is equal to the current squared times the resistance.
_____

24. Fluid flow rate in a tube is equal to the pressure difference divided by the tube diameter to the fourth power.
_____

_____
Date

_____
Name

_____
Course/Section

Unit I
Introduction
to Algebra

Evaluating Literal
Expressions

A. **Evaluate each of the following expressions using** $x = 3$, $y = 2$, $t = -2$, **and** $u = 4$.

1. $5y$ _____

2. $3x^2$ _____

3. xy _____

4. $x + y$ _____

5. $x + y - u$ _____

6. $2x - t$ _____

7. $3x - y$ _____

8. $y - 2x$ _____

9. $2t^2$ _____

10. $x^2 + y^2$ _____

11. $u^2 + t^2$ _____

12. $u^2 - t^2$ _____

13. $\dfrac{x + y}{x - y}$ _____

14. $2x - 3xy + 5u$ _____

15. $t - x$ _____

16. $\dfrac{u}{y} - 1$ _____

17. $x(y + t)$ _____

18. $(2x)(3y)(t^2)$ _____

19. $x + 2y - 3t + 1$ _____

20. $2(x - u)(y + 1)$ _____

21. $(3y - 2x)(x^5 + u^7)$ _____

22. $2(x + 2) - 3(1 + y)$ _____

23. $\dfrac{2x}{y^2}$ _____

24. $2xyt$ _____

25. $3xy - y^2$ _____

26. $(1 - x)(1 - y)(1 - u)$ _____

27. $(x - 1)^2$ _____

28. $(1 - u)^3$ _____

29. $2x^3y^2 - 3$ _____

30. $2x - 3(y + 3)$ _____

31. $x + 2(u - t)$ _____

32. $2y - 3(t - 2x)$ _____

33. $t + 4(x - 2y)$ _____

34. $x^2 - 2y^2 + u$ _____

35. $2x^2 - 3t^2$ _____

36. $(x + 2y)(t - 2u)$ _____

37. $(1 + x)(1 - u)$ _____

38. $(2 + y)(1 + t)$ _____

39. $(-3 - u)(t - x)$ _____

B. **Find the value of each of these literal expressions using** $a = 2$, $b = 5$, $c = 6$, **and** $d = -3$.

1. a^2b _____

2. $2ac$ _____

3. $a^2 + 1$ _____

4. $4a^2$ _____

5. a^3d _____

6. $a^2 + b^2 - c^2$ _____

7. $\dfrac{-(a + b)}{a - b}$ _____

8. $\dfrac{a + b}{b - a}$ _____

9. $(b - 1)(b + 1)(d - 1)$ _____

10. $\dfrac{1}{2}ab^2$ _____

11. $-abcd$ _____

12. $\dfrac{2(a - b) + 4a}{c}$ _____

Date _____

Name _____

Course/Section _____

13. a^2bc _____

14. ab^2cd _____

15. abc^2 _____

16. a^2bc^2 _____

17. ab^2c^2 _____

18. $(ab)^2(2c)$ _____

19. $(a - b)(b + 4)$ _____

20. $(b + d)(b - c)$ _____

21. $(2a + 1)(2 - 3b)$ _____

22. $(2c - b)(d + 2a)$ _____

23. $a^2 - b^2 + 2c^2$ _____

24. $d^2 - 2c^2 - 3b$ _____

25. $2 - a(b - 1)$ _____

26. $3 + d(a - b)$ _____

27. $-1 - 2a(b - 2c)$ _____

28. $-4 + b(2a - c)$ _____

29. $a - a(a - b)$ _____

30. $d - d(d - c)$ _____

C. Evaluate:

1. The volume of a sphere is given by the formula $V = \dfrac{4\pi R^3}{3}$ where R is the radius of the sphere. Find the volume of a sphere whose radius R is equal to 2 ft. Use $\pi \cong 3.14$. _____

2. The gravitational potential energy of a mass m lifted a distance d near the earth is $P = mgd$. Find the potential energy if m is equal to 70 kg, d equals 10 m, and g is approximately 9.8 m/sec². _____

3. If the Celsius and Fahrenheit temperature scales are related by the equation

   $F = \dfrac{9}{5}(C + 40)$ find F when $C = 50°$. _____

4. The air resistance in pounds acting against a moving automobile is equal to 0.0025 times the square of the speed in miles per hour times the frontal area in square feet. Find the air resistance on a car moving at 40 mph if it has a frontal area of 14 sq ft. _____

5. When two electrical resistors R_1 and R_2 are put in series with a battery supplying V volts, the current in amps through the resistors is

   $i = \dfrac{V}{R_1 + R_2}$ Find the current when V is 100 volts, R_1 is 200 ohms, and R_2 is 50 ohms. _____

6. The volume of a football is given approximately by the formula

   $V = \dfrac{\pi L W^2}{6}$ where L is its length and W is its width. Find the volume of a football 12 in. long and 7 in. wide. _____ Use $\pi = 3.14$.

Solving Algebraic Equations

Objective	Sample Problems	Where To Go for Help		
			Page	Frame

Upon successful completion of this program you will be able to:

1. Use addition and subtraction to solve simple algebraic equations.

   (a) Solve $x - 7 = 4$ _____ 95 **1**

   (b) Solve $2a + 2 - a = 10$ _____

   (c) Solve $7s + 3 - s = 4 + 5s - 6$ _____

2. Use multiplication and division to solve simple algebraic equations.

   (a) Solve $3x = 18$ _____ 109 **12**

   (b) Solve $\dfrac{y}{2} = 11$ _____

   (c) Solve $\dfrac{3A}{2} - 7 = -1$ _____

3. Solve more complex algebraic equations.

   (a) Solve $5t - 7 = t + 5$ _____ 121 **19**

   (b) Solve $2(1 - x) - 1 = 3(2x + 3)$ _____

   (c) Solve $\dfrac{y}{2} + \dfrac{2}{3} = 4$ _____

4. Solve formulas.

   (a) Solve $y = 3ax$ for x _____ 133 **27**

   (b) Solve $3(y + a) = 2(b - y)$ for y _____

   (c) Solve $Q = 3x - 8y^2$ for x _____

5. Solve simple inequalities.

   (a) $3x - 4 > 1 + 2x$. Solve. _____ 143 **35**

   (b) Solve.
   $A - 2(3A + 1) < 2(1 - A) + 2$ _____

   (c) Solve. $1 - \dfrac{y}{2} \leq 2 + y$ _____

6. Solve algebraic word problems.

   (a) Find three consecutive even integers such that the sum of the first and last is 20. _____ 153 **46**

   (b) George rides his bicycle at 18 mph and passes Kim who is cycling at 10 mph in the same direction. At what time after they pass will they be exactly 2 miles apart? _____

   (c) How many pounds of candy worth $1.50 per lb must be added to 5 lb of candy worth 80¢ per lb to make a mixture worth $1 per lb? _____

Date _____

Name _____

Course/Section _____

(d) A 30 ft length of wire is cut into two
   pieces whose lengths are in the ratio of
   3 to 7. How long is the shorter piece of
   wire?   _____

(Answers to these problems are at the bottom of this page.)

If you are certain you can work all of these problems correctly, turn to page 179 for a
self-test. If you want help with any of these objectives or if you cannot work the
sample problems, turn to the page indicated. If you are a super-student and want to be
certain that you learn all of this material, you will turn to frame **1** and begin work
there.

2 Solving Algebraic Equations

2-1 INTRODUCTION

1   An arithmetic equation such as $3 + 2 = 5$ means that the number named on the left $(3 + 2)$ is the same as the number named on the right (5). An algebraic equation $x + 3 = 7$ is a statement that the sum of some number (x) and 3 is equal to 7 if we choose the correct value for x. x is a variable and should be thought of as a symbol that holds a place for a number, a blank to be filled. Many numbers might be put in the space, but only one makes this particular equation a true statement.

Find the missing numbers in these arithmetic equations:

(a) $37 + \underline{\hspace{1cm}} = 58$
(b) $\underline{\hspace{1cm}} - 15 = 29$
(c) $4 \times \underline{\hspace{1cm}} = 52$
(d) $28 \div \underline{\hspace{1cm}} = 4$

Puzzle them out, then turn to **2**

---

2   (a) $37 + \underline{21} = 58$
(b) $\underline{44} - 15 = 29$
(c) $4 \times \underline{13} = 52$
(d) $28 \div \underline{7} = 4$

It should be obvious that we could have written these equations as

$$37 + A = 58$$
$$B - 15 = 29$$
$$4C = 52$$
$$\frac{28}{D} = 4$$

(Of course, any letters would do in place of A, B, C, and D.)

How did you solve these equations? You probably "eye-balled" them—mentally juggled the other information in the equation until you found a number that made the equation true. Solving an algebraic equation is very similar except that we can't "eye-ball" it entirely. We need certain and systematic ways of solving the equation that will produce the correct answer quickly every time.

In this unit you will learn what a solution to an algebraic equation is and how to solve simple linear equations. First, you need to learn some of the language used in the mathematics of solving equations.

Turn to **3** to begin.

Root or Solution
of an Equation

3   The *solution* of an equation, sometimes called a *root* of the equation, is a number which, when substituted for the variable in the equation, makes the equation a true statement. For example, the number 2 is a solution of the equation

$x + 5 = 7$

because substituting 2 for x in the equation gives $2 + 5 = 7$, which is a true statement. We say that the solution number *satisfies* the equation.

How do you know that a given number is a solution of the equation? Easy. Substitute it in the equation, do the arithmetic, and see if the equation is true. For example, does setting 5 equal to x satisfy the equation $x + 7 = 12$?

To answer this question, first substitute the number 5 into the equation:
$(5) + 7 = 12$

Do the arithmetic: $12 = 12$

This is a true statement; therefore the number 5 satisfies the equation $x + 7 = 12$ and 5 is a solution of this equation.

Answer this question:

Does substituting 2 for x satisfy the equation $3x + 3 = 6x - 3$?

Go to **4** to check your answer.

---

4   Yes. Substituting 2 for x in $3x + 3 = 6x - 3$, we have
$$3(2) + 3 = 6(2) - 3$$
$$9 = 9$$

Since 2 satisfies the equation we say that 2 is the solution or the root of the equation.

Answer these questions:

(a)   Does the number 2 satisfy the equation $3x + 1 = 5x - 4$?
(b)   Does the number -3 satisfy the equation $2x^2 + 6x + 2 = 3$?
(c)   Does the number 3 satisfy the equation $(x + 1)(x - 2) = 4$?
(d)   Does the number -2 satisfy the equation $(x + 1)(x - 2) = 4$?

Check your answers in **5**.

What is a solution? How do I know when I have the right solution?

The solution set of any equation is the set of values of the variable for which the equation is true. In other words, plug your solution into the equation in place of the variable, do the arithmetic, and see if the equation is true.

5 (a) No. Substituting 2 for x in the equation $3x + 1 = 5x - 4$, we have
$$3(2) + 1 \overset{?}{=} 5(2) - 4$$
$$6 + 1 \overset{?}{=} 10 - 4$$
$$7 \neq 6 \qquad \text{The number 2 does not satisfy the equation.}$$

(b) No. Substituting -3 for x in the equation $2x^2 + 6x + 2 = 3$, we have
$$2(-3)^2 + 6(-3) + 2 \overset{?}{=} 3$$
$$18 - 18 + 2 \overset{?}{=} 3$$
$$2 \neq 3 \qquad \text{The number } -3 \text{ does not satisfy the equation.}$$

(c) Yes. Substituting 3 for x in the equation $(x + 1)(x - 2) = 4$ we have
$$(3 + 1)(3 - 2) \overset{?}{=} 4$$
$$4 \cdot 1 \overset{?}{=} 4$$
$$4 = 4 \qquad \text{The number 3 satisfies the equation.}$$

(d) Yes. Substituting -2 for x in the equation $(x + 1)(x - 2) = 4$ we have
$$(-2 + 1)(-2 - 2) \overset{?}{=} 4$$
$$(-1)(-4) \overset{?}{=} 4$$
$$4 = 4 \qquad \text{The number } -2 \text{ satisfies the equation.}$$

Solution Set

The *solution set* of an equation is the set of all numbers that satisfy the equation. To *solve* an equation means to find its solution set. The solution of the equation $x + 7 = 12$ is 5. This equation has the single number 5 as its solution. The equation $(x + 1)(x - 2) = 4$ is satisfied by the two numbers 3 and -2. Substituting 3 for x or -2 for x, but no other numbers, makes this equation a true statement.

Most of the equations you will encounter in this book will have as their solution a single number or a pair of numbers. But the solution of more complex equations may involve many numbers or none at all. If there are no numbers that will make the equation a true statement, then the equation has no solution. For example, the equation

$$x + 2 = x + 3$$

has no solution. There is no number that can be substituted for x that will make this equation a true statement.

Identity

An *identity* is an equation that is true for every possible value of the variable in the equation. For example, the equation

$$2x + x = 3x$$

is a true statement for any value of x. The solution of this equation includes every possible real number.

How can we find the solution set for any given equation? One way is to substitute all possible numbers into the equation, one by one, to see which ones, if any, make it true. That is, replace the variable in the equation by 1, then by 2, 3, 4, and so on through the integers, then through all fractions, decimal numbers, and so on. The *replacement set* of the equation is the set of all possible numbers that might be used to replace the variable. Unless stated otherwise, the replacement set in ordinary algebra is the set of real numbers. It is obviously not possible in a finite time, even with a computer, to test all of the real numbers. We must look for a better way to find solutions.

Replacement Set

The best way to solve an equation is to find another equivalent equation whose solution can be seen at a glance. Two equations that have the same solution set are called *equivalent equations*.

Equivalent Equations

The equation $5x = 3x + 6$ is equivalent to the equation $x = 3$. Both have the solution 3. The equation

$x - 3(2x + 1) = 5 - 4(x + 3)$ is equivalent to the equation $x = 4$. The number 4 is the solution to both equations.

Solving an Equation

An equation in x is said to be solved if it can be put into the form $x = \square$ where $\square$ is some real number.

Equations as simple as some of the ones above can be solved by guessing or by trial and error, but guessing blindly is not a very good way to do mathematics. In this unit we are going to show you how to use the properties of real numbers to transform a complicated equation such as

$x - 3(2x + 1) = 5 - 4(x + 3)$, to a simple but equivalent equation such as $x = 4$, whose solution is obvious.

Can you show that the following are equivalent equations?

$$5x - 6 = 18 + x \qquad \text{and} \qquad x = 6$$

Try it, then go to **6** to check your work.

---

6   To show that the equations $5x - 6 = 18 + x$ and $x = 6$ are equivalent, substitute the number 6 from the second equation into the first equation.

$$5(6) - 6 \overset{?}{=} 18 + (6)$$
$$30 - 6 \overset{?}{=} 18 + 6$$
$$24 = 24 \qquad \text{The number 6 satisfies both equations, and 6 is the } only \text{ number that satisfies them. Therefore the equations are equivalent.}$$

Which of the following are equivalent equations?

(a) $3a + 5 = 17 - a$ and $a = 3$
(b) $4 - y = 3 + 2(y - 1)$ and $y = 1$
(c) $2x - 3 = x - 1$ and $x = -1$
(d) $\frac{1}{2}A + 5 = 2A + 2$ and $A = 1$

Check your answers in **7**.

How do you know if two equations are equivalent?

First of all, if two equations are equivalent, they will have the same solution set. Second, you should be able to transform an equation to its equivalent by a few logical steps such as adding some number to both sides or multiplying both sides by the same number.

7   (a)   Yes. The number 3 is the solution of both equations.
    (b)   Yes. The number 1 is the solution of both equations.
    (c)   No. The solution of the first equation is 2.
    (d)   No. The solution of the first equation is 2.

Linear or First-Degree Equations

In this unit we will be working with *linear* or *first-degree* equations only. A first-degree equation contains only numerals and first powers of the variable. If x is the variable, a first-degree equation may contain terms with x, $2x$, $3x$, $\frac{1}{2}x$, and so on, but never x^2, x^3, or any higher power of x. For example,

$$x + 3 = 4$$
$$3p = 17$$
$$2 - a = 7$$
and
$$y + 2 = 3 - 4y$$

are all first-degree or linear equations. The equation $x^2 - 1 = 3$ is *not* a first-degree equation because it contains the term x^2.

The first step in transforming any first-degree equation into a simpler equivalent equation is to rearrange and combine like terms if possible. For example, the equation

$$2 + x + 6 = 2x - 6$$
can be simplified to
$$x + (2 + 6) = 2x - 6$$
or
$$x + 8 = 2x - 6$$

The equation
$$x + 2 + 3x - 4 + 5x + 3 = 1 - 2x + 5 + 3x$$
can be simplified to
$$(x + 3x + 5x) + (2 - 4 + 3) = (1 + 5) + (-2x + 3x)$$
or
$$9x + 1 = 6 + x$$

Simplify the following equations by rearranging and combining like terms.

(a) $3 - x + 2 = 2x + 3 - x$ (b) $-2 + b + 5 - 3b = 10 + b - 4$
(c) $3a + 3 - a = 10 - 2a + 5$ (d) $3q + 5 + q + 2 + 2q + 6 = q - 1$
(e) $p + 2 + p + 2p = 4 + 2p + 5$ (f) $y + 2y - 1 + y = 3 - y - 2y + 5$
(g) $\frac{1}{2} + x - 1\frac{1}{2} - 3x + 3 = 1 + x$ (h) $5z - 2 - 2z + 1 = 2 + z + 4$

Check your simplified equations in **8**.

8   (a) $5 - x = x + 3$ (b) $3 - 2b = 6 + b$
     (c) $2a + 3 = 15 - 2a$ (d) $6q + 13 = q - 1$
     (e) $4p + 2 = 2p + 9$ (f) $4y - 1 = 8 - 3y$
     (g) $2 - 2x = 1 + x$ (h) $3z - 1 = 6 + z$

2-2 SOLVING EQUATIONS BY ADDING AND SUBTRACTING

The left member of an equation and the right member of an equation both name the same number and, therefore, the general principle of equation solving is to treat every equation as a balance of the two sides.

     Copyright © 1982 by John Wiley & Sons, Inc.

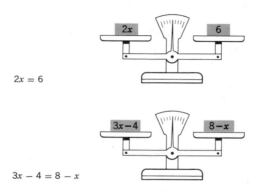

$2x = 6$

$3x - 4 = 8 - x$

Any changes made in the equation must not disturb this balance. Any operation performed on one side of the equation must also be performed on the other side.

It seems reasonable that we can add a number to each side of an equation and not change its balance. For example, the arithmetic equation

$$4 + 3 = 7$$
becomes $4 + 3 + 2 = 7 + 2$
or $\qquad 9 = 9$

when 2 is added to each side.

Imagine this process as a balancing of weights on a balance scale:

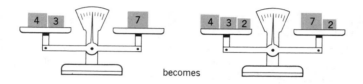

becomes

Of course we may also subtract a quantity from both sides of the equation without changing its equality.

$$4 + 3 = 7 \text{ becomes}$$
$$4 + 3 - 3 = 7 - 3 \text{ when 3 is subtracted from each side}$$
or $\qquad 4 = 4$

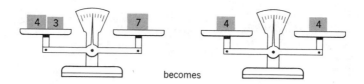

becomes

Addition and
Subtraction Rule for
Solving Equations:

This procedure is usually spelled out in a rule as follows:

> If any real number is added to or subtracted from both sides of an equation, an equivalent equation is produced.

Addition Rule

Because subtracting a number n is exactly equivalent to adding the number $-n$, this rule is often called simply the *addition rule* or the *addition axiom*. (An axiom in mathematics is a rule that is accepted without proof.)

This rule can be used directly to find the solution to some algebra equations. For example, to solve the equation

$x - 4 = 2$

add 4 to each side

$x - 4 + 4 = 2 + 4$

or

$x + (-4 + 4) = (2 + 4)$

$x = 6$

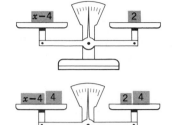

The solution is 6.

To check the solution, substitute it back into the original equation.

$(6) - 4 = 2$

or $\quad 2 = 2$

Your turn. Solve the equation $2 + x + 3 = 7 + 1$.

Compare your work with ours in **9** when you have finished.

---

9 Solve: $2 + x + 3 = 7 + 1$.

Step 1. Simplify the equation by combining like terms
$x + (2 + 3) = (7 + 1)$
$\quad x + 5 = 8$

Step 2. Add or subtract the same quantity on both sides to get an equivalent equation with only x as the left member. In this case, subtract 5 from each side.

$x + 5 \boxed{-5} = 8 \boxed{-5}$

$\quad\quad x = 3 \quad$ The solution is 3.

Step 3. Check your solution. Substitute the solution (3) into the original equation, $2 + x + 3 = 7 + 1$.
$2 + (3) + 3 = 7 + 1$
$\quad\quad 8 = 8 \quad$ The solution satisfies the original equation.

Another example: Solve $4x - 2 - 5x = 10 - 2x$.

Step 1. Simplify by combining like terms.
$(4x - 5x) - 2 = 10 - 2x$
$\quad -x - 2 = 10 - 2x$

Step 2. Add or subtract the same quantity on both sides to get an equivalent equation with only x as the left member. In this case, add $\boxed{2}$ to each side of the equation.

$-x - 2 \boxed{+2} = 10 - 2x \boxed{+2}$

$\quad\quad -x = 12 - 2x$

Now add $2x$ to each side of the equation.

$-x \boxed{+2x} = 12 - 2x \boxed{+2x}$

$\quad\quad x = 12 \quad$ The solution is 12.

Step 3. Check your solution.
$$4(12) - 2 - 5(12) = 10 - 2(12)$$
$$48 - 2 - 60 = 10 - 24$$
$$-14 = -14 \qquad \text{The solution satisfies the original equation.}$$

Remember those three steps:

1. Simplify by combining terms.
2. Add or subtract to get an equation like $x = \square$.
3. Check the solution.

Try these practice problems:

(a) $x + 6 = 11$ (b) $x - 5 = 13$
(c) $2 + y - 7 = 10 + 4$ (d) $4 = 11 + b$
(e) $8 = 7 - x$ (f) $4 - a = 0$
(g) $3x - 2 = 2x + 9$ (h) $6 + 2p = 3p - 7$
(i) $4q = 9 + 3q$ (j) $z + z + 4 - 2z = 1 - z$
(k) $-4 + r = 10$ (l) $6 + x - 5 = 8$
(m) $3x - 2 = 4x + 2$ (n) $1 + d + 2 + 2d = 5 + 2d + 4$
(o) $6w - 3 + 2w + 5 - 7w = 0$ (p) $0 = 11 - 7t + t - 2 + 5t$

Check your answers in **10**.

---

10   (a) $x + 6 = 11$

$x + 6 \boxed{-6} = 11 \boxed{-6}$ Subtract $\boxed{6}$

$x = 5$ The solution is 5.
Check: $(5) + 6 = 11$

(b) $x - 5 = 13$

$x - 5 \boxed{+5} = 13 \boxed{+5}$ Add $\boxed{5}$

$x = 18$ The solution is 18.
Check: $(18) - 5 = 13$

(c) $2 + y - 7 = 10 + 4$
$y - 5 = 14$

$y - 5 \boxed{+5} = 14 \boxed{+5}$ Add $\boxed{5}$

$y = 19$ The solution is 19.
Check: $2 + (19) - 7 = 10 + 4$
$14 = 14$

(d) $4 = 11 + b$

$4 \boxed{-11} = 11 + b \boxed{-11}$ Subtract $\boxed{11}$

$-7 = b$
$b = -7$ The solution is -7.
Check: $4 = 11 + (-7)$

(e) $8 = 7 - x$

$8 \boxed{+x} = 7 - x \boxed{+x}$ Add $\boxed{x}$

$8 + x = 7$

$8 + x \boxed{-8} = 7 \boxed{-8}$ Subtract $\boxed{8}$

$x = -1$ The solution is -1.
Check: $8 = 7 - (-1)$

(f) $\qquad 4 - a = 0$

$\qquad 4 - a \boxed{+ a} = 0 \boxed{+ a} \qquad$ Add $\boxed{a}$

$\qquad\qquad 4 = a$

$\qquad\qquad a = 4 \qquad$ The solution is 4.

Check: $\quad 4 - (4) = 0$

(g) $\qquad 3x - 2 = 2x + 9$

$\qquad 3x - 2 \boxed{+ 2} = 2x + 9 \boxed{+ 2} \qquad$ Add $\boxed{2}$

$\qquad\qquad 3x = 2x + 11$

$\qquad 3x \boxed{- 2x} = 2x + 11 \boxed{- 2x} \qquad$ Subtract $\boxed{2x}$

$\qquad\qquad x = 11 \qquad$ The solution is 11.

Check: $\quad 3(11) - 2 = 2(11) + 9$

$\qquad\qquad 31 = 31$

(h) $\qquad 6 + 2p = 3p - 7$

$\qquad 6 + 2p \boxed{+ 7} = 3p - 7 \boxed{+ 7} \qquad$ Add $\boxed{7}$

$\qquad\qquad 2p + 13 = 3p$

$\qquad 2p + 13 \boxed{- 2p} = 3p \boxed{- 2p} \qquad$ Subtract $\boxed{2p}$

$\qquad\qquad 13 = p$

$\qquad\qquad p = 13 \qquad$ The solution is 13.

Check: $\quad 6 + 2(13) = 3(13) - 7$

$\qquad\qquad 32 = 32$

(i) $\qquad 4q = 9 + 3q$

$\qquad 4q \boxed{- 3q} = 9 + 3q \boxed{- 3q} \qquad$ Subtract $\boxed{3q}$

$\qquad\qquad q = 9 \qquad$ The solution is 9.

Check: $\quad 4(9) = 9 + 3(9)$

$\qquad\qquad 36 = 36$

(j) $\quad z + z + 4 - 2z = 1 - z$

$\qquad\qquad 4 = 1 - z$

$\qquad 4 \boxed{+ z} = 1 - z \boxed{+ z} \qquad$ Add $\boxed{z}$

$\qquad\qquad 4 + z = 1$

$\qquad 4 + z \boxed{- 4} = 1 \boxed{- 4} \qquad$ Subtract $\boxed{4}$

$\qquad\qquad z = -3 \qquad$ The solution is -3.

Check: $\quad (-3) + (-3) + 4 - 2(-3) = 1 - (-3)$

$\qquad\qquad\qquad 4 = 4$

(k) $\qquad -4 + r = 10$

$\qquad -4 + r \boxed{+ 4} = 10 \boxed{+ 4} \qquad$ Add $\boxed{4}$

$\qquad\qquad r = 14 \qquad$ The solution is 14.

Check: $\quad -4 + (14) = 10$

$\qquad\qquad 10 = 10$

(l) $\qquad 6 + x - 5 = 8$

$\qquad\qquad 1 + x = 8$

$\qquad 1 + x \boxed{- 1} = 8 \boxed{- 1} \qquad$ Subtract $\boxed{1}$

$\qquad\qquad x = 7 \qquad$ The solution is 7.

Check: $6 + (7) - 5 = 8$
$$8 = 8$$

(m) $\qquad 3x - 2 = 4x + 2$

$\qquad 3x - 2 \boxed{-2} = 4x + 2 \boxed{-2} \qquad$ Subtract $\boxed{2}$

$\qquad\qquad 3x - 4 = 4x$

$\quad 3x - 4 \boxed{-3x} = 4x \boxed{-3x} \qquad$ Subtract $\boxed{3x}$

$\qquad\qquad -4 = x$

$\qquad\qquad x = -4 \qquad$ The solution is -4.

$\quad$ **Check:** $3(-4) - 2 = 4(-4) + 2$
$$-14 = -14$$

(n) $\quad 1 + d + 2 + 2d = 5 + 2d + 4$
$$3 + 3d = 9 + 2d$$

$\qquad 3 + 3d \boxed{-3} = 9 + 2d \boxed{-3} \qquad$ Subtract $\boxed{3}$

$\qquad\qquad 3d = 6 + 2d$

$\qquad 3d \boxed{-2d} = 6 + 2d \boxed{-2d} \qquad$ Subtract $\boxed{2d}$

$\qquad\qquad d = 6 \qquad$ The solution is 6.

$\quad$ **Check:** $1 + (6) + 2 + 2(6) = 5 + 2(6) + 4$
$$21 = 21$$

(o) $\quad 6w - 3 + 2w + 5 - 7w = 0$
$$w + 2 = 0$$

$\qquad w + 2 \boxed{-2} = 0 \boxed{-2} \qquad$ Subtract $\boxed{2}$

$\qquad\qquad w = -2 \qquad$ The solution is -2.

$\quad$ **Check:** $6(-2) - 3 + 2(-2) + 5 - 7(-2) = 0$
$$-12 - 3 - 4 + 5 + 14 = 0$$
$$0 = 0$$

(p) $\qquad 0 = 11 - 7t + t - 2 + 5t$
$$0 = 9 - t$$

$\quad 0 \boxed{+t} = 9 - t \boxed{+t} \qquad$ Add $\boxed{t}$

$\qquad\qquad t = 9 \qquad$ The solution is 9.

$\quad$ **Check:** $0 = 11 - 7(9) + (9) - 2 + 5(9)$
$$0 = 11 - 63 + 9 - 2 + 45$$
$$0 = 0$$

Did you notice in problem (d) that the variable was on the right?
$$4 = 11 + b$$

Subtracting 11 from each side gives

$$4 \boxed{-11} = 11 + b \boxed{-11}$$

or $\quad -7 = b$

Which of course is equivalent to $b = -7$. Interchanging right and left sides of an equation always produces an equivalent equation.

$\square = x$ is equivalent to $x = \square$.

Now, for a set of practice problems over the work of this section of Unit 1, turn to **11**.

Unit 2
Solving
Algebraic Equations

11 The answers to all problems are on page 540.

Solving Equations by
Adding and
Subtracting Terms

A. For each of the following equations, does the given number satisfy the equation?

1. $x - 7 = 8$ $x = 15$ _____

2. $m + 12 = 12$ $m = -12$ _____

3. $75 = 40 + x$ $x = 25$ _____

4. $4y + 7 = 19$ $y = 3$ _____

5. $12 = b - 15$ $b = 3.7$ _____

6. $2C + 7 = 21$ $C = 14$ _____

7. $p - 5\frac{1}{2} = 11$ $p = 5\frac{1}{2}$ _____

8. $14 - 2n = 2$ $n = 6$ _____

9. $\frac{3}{4}c - 8 = 7$ $c = 20$ _____

10. $2b - 0.5 = 3.2$ $b = 2.4$ _____

B. Solve:

1. $m + 7 = 12$ _____

2. $y - 1 = 7$ _____

3. $k + 9 = 12$ _____

4. $37 = R + 37$ _____

5. $x + 28 = 71$ _____

6. $15 + R = 102$ _____

7. $b - 24 = 0$ _____

8. $n - 450 = 5$ _____

9. $140 = z + 82$ _____

10. $y + 13 = 42$ _____

11. $a - 20 = 7$ _____

12. $K + 16 = -10$ _____

13. $w - 52 = 0$ _____

14. $t - 87 = 87$ _____

15. $54 = U + 54$ _____

16. $210 = 102 + a$ _____

17. $13 = 9 + n$ _____

18. $C - 18 = -12$ _____

19. $19 = r + 7$ _____

20. $y + 32 = 32$ _____

21. $0 = a - 7$ _____

22. $X - 17 = 16$ _____

23. $24 = r + 13$ _____

24. $4 + t = 30$ _____

C. Solve:

1. $12 = 4 - a$ _____

2. $50 = m + 15$ _____

3. $q + 39 = 52$ _____

4. $X - 22 = 15$ _____

5. $4 - C = 0$ _____

6. $17 = 12 - X$ _____

7. $12 = E - 12$ _____

8. $z - 6 = -6$ _____

9. $4.5 - d = 4.5$ _____

10. $1 + x = -9$ _____

Date _____

Name _____

Course/Section _____

11. $m - 3.5 = -4.5$ _____

12. $4\frac{1}{2} + q = -\frac{1}{2}$ _____

13. $d - 2.5 = 4.5$ _____

14. $16 - y = 9$ _____

15. $C - 1\frac{1}{4} = -2$ _____

16. $e - 0.7 = 0.8$ _____

17. $4.8 = z - 0.8$ _____

18. $32 = 7 + f$ _____

19. $0 = 12 - y$ _____

20. $1.7 + g = 0$ _____

21. $28 = 4 - G$ _____

22. $3C - 2C = 11$ _____

23. $9m + 16 = 8m$ _____

24. $7t - 4t - 6t = 0$ _____

25. $X - 3 - 2X = 0$ _____

26. $4b + 21 - b = 2b$ _____

27. $28 + 14y = 26 + 13y$ _____

28. $15 - 6f + 13 = -5f$ _____

29. $6x - 2 - 9x = 6 - 4x$ _____

30. $17e + 0.15 - 4e = 12e - 2.25$ _____

31. $x + 12 - 11x = 1 - 9x$ _____

32. $1 - z = z - 6 - 3z$ _____

33. $14 + 9x = 1 + 8x$ _____

34. $13 - 6y = 4 - 5y$ _____

35. $2 - 8p - 1 = p + 1 - 10p$ _____

36. $7t + 4 - t = 6 + 5t - 1$ _____

37. $4 - 3x + x = -x + 2 - 2x$ _____

38. $a - 8 - 4a = -6a - 3 + 2a$ _____

39. $5(2 + x) = 4x$ _____

40. $7(1 - x) + 6x = 0$ _____

41. $4(x - 1) = 5x + 1$ _____

42. $6(a + 2) = 1 + 7a$ _____

43. $2(4 - y) = -3(y - 1)$ _____

44. $4(1 - x) = -(5x + 2)$ _____

45. $1 - (x - 2) = 2(3 - x)$ _____

46. $2 - 3(y + 1) = 2(3 - 2y)$ _____

47. $3 - 2(1 - 2z) = 5(z - 1)$ _____

48. $1 - (3 - 4z) = -3(1 - z)$ _____

49. $-2 - 3(a - 4) = 2(3 - a)$ _____

50. $-5 - 4(b + 1) = -3(5 + b)$ _____

D. Translate these problems into algebraic equations and solve them:

1. A number decreased by 37 equals 17. Find the number.

   _____

2. The sum of 24 and a number is 42. What is the number?

   _____

3. After she gained 18 pounds, Mary weighed 105 pounds. What was her original weight?

   _____

4. Thirteen years ago Brian was six and one-half years old. How old is he now?

_____

5. Margaret spent $5.74 and has $2.46 left. How much money did she have originally?

_____

6. After an airplane decreased its rate of speed by 145 miles per hour, it was traveling 220 miles per hour. What was its original speed?

_____

7. An art show was attended by 974 people. This was 116 more than had attended the show in the previous year. What was last year's attendance?

_____

8. Eric wishes to buy a bicycle that costs $78. If he has already saved $29 for it, how much more must he save?

_____

9. After using a TV set for 2 years, Ed sold it for $42 less than he paid for it. If the amount Ed received was $185, how much did he pay for it originally?

_____

10. The price of a calculator increased by $12.50 so that it now sells for $37.40. What was the original price of the calculator?

_____

E. Brain Boosters

1. When George's father was 31 years old, George was 8 years old. Now the father is twice as old as George. How old is George?

_____

2. A book costs $1 plus half of its price. How much does it cost?

_____

3. Show that three times the square of an integer plus three times the integer is an even number.

_____

Date
_____

Name
_____

Course/Section

When you have had the practice you need, either return to the preview on page 93 or continue in **12** where you will learn how to solve equations using multiplication and division.

I solved an equation and checked it and the check didn't work. So then I checked the check and it was OK. Then I solved the equation again and I got the same answer as the first time.
What do I do now?

You are making some sort of error in your solution each time you solve the equation. Why not ask your instructor or tutor to go over your work with you?

12 A second rule for solving first-degree equations involves multiplying or dividing both sides of the equation by some real, non-zero number. To solve an equation, first simplify it by combining like terms. Next, use the addition rule to bring all terms containing the variable to the same side of the equation. The result of these two operations may be that you are faced with an equation that looks like this:

$$\frac{x}{2} = 7$$

How can this equation be solved?
Easy: Multiply both members of the equation by 2

$$\boxed{2}\left(\frac{x}{2}\right) = \boxed{2}(7)$$

or $\qquad x = 14 \qquad$ The solution is 14.

Check: $\quad \frac{(14)}{2} = 7 \qquad$ This is correct.

Again, think of the equation as a balance of the terms on each side of the equals sign. Multiplying both sides of the equation by the same number changes the equation into an equivalent equation, and the balance is maintained.

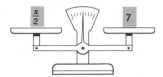
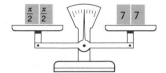

Another example:

$$\frac{x}{3} = -5$$

Multiply each member of the equation—each side—by 3 .

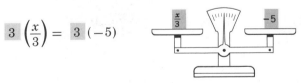
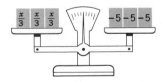

$$3\left(\frac{x}{3}\right) = \boxed{3}(-5)$$

$$x = -15 \qquad \text{The solution is } -15.$$

Check: $\quad \frac{-15}{3} = -5 \qquad$ This is correct.

You should remember from your study of arithmetic that multiplying a fraction by an integer is equivalent to multiplying the numerator by that integer. For example,

$$2 \cdot \left(\frac{3}{5}\right) = \frac{2 \cdot 3}{5} = \frac{6}{5}$$

Solve this equation in the same way:

$$\frac{y}{5} = -2$$

Our solution is in **13**.

---

13    $\dfrac{y}{5} = -2$

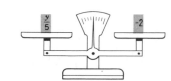 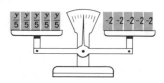

Multiply both numbers of the equation by $\boxed{5}$.

$$\boxed{5}\left(\frac{y}{5}\right) = \boxed{5}\,(-2)$$

or $y = -10$ The solution is -10.

Check: $\dfrac{(-10)}{5} = -2$ This is correct.

Division

Division is simply the reverse operation to multiplication and may be used in exactly the same way to solve equations. For example,

$3x = 42$

Dividing both members of the equation by $\boxed{3}$

$$\frac{3x}{\boxed{3}} = \frac{42}{\boxed{3}}$$

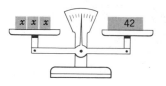 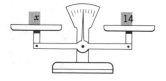

or $x = 14$ The solution is 14.

Check: $3(14) = 42$ Correct.

Of course dividing by zero is not permitted and multiplying both sides of an equation by zero transforms it into the equation $0 = 0$, which is not very helpful.

Solve this equation using division:

$4x = 76$

Check your work in **14**

14 $4x = 76$

Dividing both members of the equation by 4

$$\frac{4x}{4} = \frac{76}{4}$$

or $x = 19$ The solution is 19.

Check: $4(19) = 76$ This is correct.

We can state this procedure as a second general rule for solving equations.

Multiplication and
Division Rule for
Solving Equations:

> If both sides of an equation are either multiplied by or divided by a non-zero number, an equivalent equation is produced.

Because dividing by some non-zero number n is exactly equivalent to multiplying by the number $\frac{1}{n}$, this rule is often called simply the *multiplication rule* or the multiplication axiom.

Use this rule to solve the following equations:

(a) $2x = 15$ (b) $-3a = 27$ (c) $6b = -30$ (d) $-7y = -56$

(e) $\dfrac{d}{2} = 8$ (f) $\dfrac{t}{-3} = 6$ (g) $\dfrac{w}{6} = -2$ (h) $\dfrac{x}{-3} = -5$

Check your work in **15**.

15   (a) $2x = 15$

$$\frac{2x}{\boxed{2}} = \frac{15}{\boxed{2}} \qquad \text{Divide by } \boxed{2}.$$

$$x = 7\frac{1}{2} \qquad \text{The solution is } 7\frac{1}{2}.$$

Check: $2\left(7\frac{1}{2}\right) = 15$

(b) $-3a = 27$

$$\frac{-3a}{\boxed{-3}} = \frac{27}{\boxed{-3}} \qquad \text{Divide by } \boxed{-3}.$$

$$a = -9 \qquad \text{The solution is } -9.$$

Check: $-3(-9) = 27 \qquad$ which is correct.

(c) $6b = -30$

$$\frac{6b}{\boxed{6}} = \frac{-30}{\boxed{6}} \qquad \text{Divide by } \boxed{6}.$$

$$b = -5 \qquad \text{The solution is } -5.$$

Check: $6(-5) = -30$

(d) $-7y = -56$

$$\frac{-7y}{\boxed{-7}} = \frac{-56}{\boxed{-7}} \qquad \text{Divide by } \boxed{-7}.$$

$$y = 8 \qquad \text{The solution is } 8.$$

Check: $-7(8) = -56$

(e) $\dfrac{d}{2} = 8$

$$\boxed{2}\left(\frac{d}{2}\right) = \boxed{2}(8) \qquad \text{Multiply by } \boxed{2}.$$

$$d = 16 \qquad \text{The solution is } 16.$$

Check: $\dfrac{(16)}{2} = 8$

(f) $\dfrac{t}{-3} = 6$

$$\boxed{-3}\left(\frac{t}{-3}\right) = \boxed{-3}(6) \qquad \text{Multiply by } \boxed{-3}.$$

$$t = -18 \qquad \text{The solution is } -18.$$

Check: $\dfrac{(-18)}{-3} = 6$

(g) $\dfrac{w}{6} = -2$

$$\boxed{6}\left(\frac{w}{6}\right) = \boxed{6}(-2) \qquad \text{Multiply by } \boxed{6}.$$

$$w = -12 \qquad \text{The solution is } -12.$$

Check: $\dfrac{(-12)}{6} = -2$

(h) $\qquad\qquad \dfrac{x}{-3} = -5$

$\qquad\qquad \boxed{-3}\left(\dfrac{x}{-3}\right) = \boxed{-3}(-5) \qquad$ Multiply by $\boxed{-3}$.

$\qquad\qquad\qquad x = 15 \qquad$ The solution is 15.

Check: $\dfrac{15}{(-3)} = -5$

If the equation to be solved is more complex than these last few, it will be easier if you simplify it by combining like terms *before* you use the multiplication rule. For example,

$3x - x = 6$

should be simplified to $2x = 6$ and then solved.

The equation $\dfrac{x}{4} + 2 = 9$ should be simplified to

$\dfrac{x}{4} = 7 \qquad$ and then solved.

Equations with
Fractions

If the equation to be solved contains fractions, simplify it by multiplying both members of the equation by the smallest common denominator of the fractions. For example, in the equation

$\dfrac{x}{3} + \dfrac{x}{4} = 7 \qquad$ The smallest common denominator of 3 and 4 is 12.

$\qquad\qquad$ Multiply both members of the equation by $\boxed{12}$.

$\boxed{12}\left(\dfrac{x}{3} + \dfrac{x}{4}\right) = \boxed{12}(7)$

or, using the distributive law,

$\boxed{12}\left(\dfrac{x}{3}\right) + \boxed{12}\left(\dfrac{x}{4}\right) = \boxed{12}(7) \qquad 12 \cdot \left(\dfrac{x}{3}\right) = \dfrac{12x}{3} = 4x \qquad 12 \cdot \left(\dfrac{x}{4}\right) = \dfrac{12x}{4} = 3x$

$4x + 3x = 84 \qquad$ Now combine like terms.

$\qquad 7x = 84$
$\qquad\ x = 12 \qquad$ The solution is 12.

Check it.

Notice that we must multiply *all* of the left member of the equation, *both* terms, by 12.

Try these. Solve:

(a) $\dfrac{x}{2} + \dfrac{1}{3} = 4$ $\qquad\qquad$ (b) $\dfrac{3}{2} - \dfrac{a}{5} = 2$

(c) $\dfrac{b}{3} - \dfrac{b}{5} = 6$ $\qquad\qquad$ (d) $\dfrac{1}{x} + \dfrac{2}{3} = 1$

Work carefully and compare your answers with ours in **16**.

16 (a) $\dfrac{x}{2} + \dfrac{1}{3} = 4$ The common denominator of 2 and 3 is 6.

$$6\left(\dfrac{x}{2}\right) + 6\left(\dfrac{1}{3}\right) = 6\,(4) \qquad \text{Multiply every term by } 6.$$

$$3x + 2 = 24$$

$$3x + 2\;-2 = 24\;-2 \qquad \text{Subtract } 2.$$

$$3x = 22$$

$$\dfrac{3x}{3} = \dfrac{22}{3} \qquad \text{Divide by } 3.$$

$$x = 7\dfrac{1}{3} \qquad \text{The solution is } 7\dfrac{1}{3}.$$

Check: $\dfrac{(7\frac{1}{3})}{2} + \dfrac{1}{3} = 4 \qquad \text{or} \qquad \dfrac{22}{6} + \dfrac{2}{6} = 4$

(b) $\dfrac{3}{2} - \dfrac{a}{5} = 2$ The common denominator of 2 and 5 is 10.

$$10\left(\dfrac{3}{2}\right) - 10\left(\dfrac{a}{5}\right) = 10\,(2) \qquad \text{Multiply every term by } 10.$$

$$15 - 2a = 20$$

$$15 - 2a\;-15 = 20\;-15 \qquad \text{Subtract } 15.$$

$$-2a = 5$$

$$\dfrac{-2a}{-2} = \dfrac{5}{-2} \qquad \text{Divide by } -2.$$

$$a = -2\dfrac{1}{2} \qquad \text{The solution is } -2\dfrac{1}{2}.$$

Check: $\dfrac{3}{2} - \dfrac{(-2\frac{1}{2})}{5} = 2 \qquad \text{or} \qquad \dfrac{3}{2} + \dfrac{1}{2} = 2$

(c) $\dfrac{b}{3} - \dfrac{b}{5} = 6$ The common denominator of 3 and 5 is 15.

$$15\left(\dfrac{b}{3}\right) - 15\left(\dfrac{b}{5}\right) = 15\,(6) \qquad \text{Multiply every term by } 15.$$

$$5b - 3b = 90$$

$$2b = 90$$

$$\dfrac{2b}{2} = \dfrac{90}{2} \qquad \text{Divide by } 2.$$

$$b = 45 \qquad \text{The solution is } 45.$$

Check: $\dfrac{(45)}{3} - \dfrac{(45)}{5} = 6$

$$15 - 9 = 6$$

(d) $\dfrac{1}{x} + \dfrac{2}{3} = 1$ The common denominator of x and 3 is $3x$.

$$3x\left(\dfrac{1}{x}\right) + 3x\left(\dfrac{2}{3}\right) = 3x\,(1) \qquad \text{Multiply all terms by } 3x.$$

$$3 + 2x = 3x$$

$$3 + 2x\;-2x = 3x\;-2x \qquad \text{Subtract } 2x.$$

$$3 = x$$

or $\qquad x = 3 \qquad \text{The solution is } 3.$

Check: $\dfrac{1}{(3)} + \dfrac{2}{3} = 1$

$$\dfrac{1}{3} + \dfrac{2}{3} = 1$$

Division by Zero

In problem (d), remember that any fraction with a denominator equal to zero is meaningless. Division by zero is not defined for the real numbers. The equation $\dfrac{1}{x} + \dfrac{2}{3} = 1$ is meaningless for x equal to 0. Any number is acceptable for its solution *except* 0. Here are two examples of the troubles that can arise when you divide by zero.

Example 1: If $2 \cdot 0 = 0$ and $7 \cdot 0 = 0$

then $2 \cdot 0 = 7 \cdot 0$

If we incorrectly divide both sides of the equation by zero,

$$\dfrac{2 \cdot 0}{0} = \dfrac{7 \cdot 0}{0}$$

or $2 = 7$ which is nonsense!

Example 2: The equation $\dfrac{x}{x-1} = 3 + \dfrac{1}{x-1}$ is meaningless if the denominator $x - 1 = 0$, or $x = 1$, since this value of x gives a denominator of zero. We can go through the steps of solving the equation:

First, multiply each term by $(x - 1)$ to get $x = 3(x - 1) + 1$ or $x = 3x - 3 + 1$

then combine like terms and solve as usual to get $x = 1$.

Because the equation is meaningless for $x = 1$, we must conclude that the equation has *no* solution.

Be Careful ▷

If an equation contains a fraction with the variable or an expression containing the variable in its denominator, be very careful. It may have no solution. Always check to see what value of the variable makes the denominator zero. That number *cannot* be the solution of the equation.

For each of these equations what values of the variable are not allowed? Solve each equation

(a) $\dfrac{2}{x} + \dfrac{3}{4} = \dfrac{1}{2}$ (b) $\dfrac{1}{x-3} + \dfrac{2}{x-3} = 1$

Check your answers in **17**.

---

17 (a) The fraction $\dfrac{2}{x}$ is not defined for x equal to 0.

The equation $\dfrac{2}{x} + \dfrac{3}{4} = \dfrac{1}{2}$ can be solved by first multiplying all terms by $\boxed{4x}$:

$$\boxed{4x}\left(\dfrac{2}{x}\right) + \boxed{4x}\left(\dfrac{3}{4}\right) = \boxed{4x}\left(\dfrac{1}{2}\right)$$

$8 + 3x = 2x$

and, second, by rearranging terms:

$3x - 2x = -8$

$\qquad x = -8$ The solution is -8. Check it.

(b)   The fraction $\dfrac{1}{x-3}$ is not defined for $x-3=0$ or $x=3$. To solve the equation,

$$\frac{1}{x-3} + \frac{2}{x-3} = 1$$

first multiply each term by $(x-3)$.

$$(x-3)\left(\frac{1}{x-3}\right) + (x-3)\left(\frac{2}{x-3}\right) = (x-3)\,(1)$$

$$1 + 2 = x - 3$$

or

$$x = 6 \qquad \text{The solution is 6.} \qquad \text{Check it.}$$

For a set of practice problems that will test your ability to use the multiplication rule to solve first-degree equations, turn to **18**.

TRICKY BUSINESS

In using the multiplication rule to solve equations, you should be very careful when multiplying by the variable itself. This kind of multiplication can produce an incorrect solution. For example, suppose you are trying to solve the equation

$$3x = 6$$

The wide-awake student will divide both members of the equation by 3 to get $x = 2$

The solution is 2.

The not-so-wide-awake student might multiply both members of the equation by x to get

$$3x^2 = 6x$$

and then divide by 3 to get

$$x^2 = 2x$$

It is easy to see that substituting 2 for x satisfies this last equation, but substituting 0 for x also satisfies it. The solution set for this equation includes both 0 and 2.

Multiplying by the variable x has introduced an extra, and *incorrect*, solution that does *not* satisfy the original equation. Be careful when multiplying the members of an equation by the variable, and always check your solution to see that it does satisfy the original equation.

Unit 2
Solving
Algebraic Equations

Solving Equations by
Multiplying and
Dividing

The answers to all problems are on page 541.

A. Solve:

1. $4x = 24$ _____
2. $\dfrac{a}{3} = 3$ _____
3. $4B = 36$ _____

4. $3d = 0$ _____
5. $2y = 78$ _____
6. $5m = -45$ _____

7. $\dfrac{t}{8} = 1$ _____
8. $72 = 9L$ _____
9. $12W = 6$ _____

10. $15 = 4E$ _____
11. $2N = 0.8$ _____
12. $\dfrac{1}{4}m = 3$ _____

13. $\dfrac{1}{8}x = 7$ _____
14. $15 = 12e$ _____
15. $1.6C = 32$ _____

16. $0.9 = 0.9w$ _____
17. $\dfrac{h}{3} = -4$ _____
18. $\dfrac{x}{2} = -3$ _____

19. $\dfrac{y}{4} = 0$ _____
20. $\dfrac{d}{4} = \dfrac{3}{4}$ _____
21. $2X = -8$ _____

22. $5x = -3$ _____
23. $0.2y = 0.6$ _____
24. $0.1t = 2$ _____

25. $-6a = 8$ _____
26. $0 = 6x$ _____
27. $\dfrac{3x}{2} = 4$ _____

28. $5 = \dfrac{2x}{3}$ _____
29. $5p = \dfrac{1}{2}$ _____
30. $\dfrac{1}{3} = 4m$ _____

B. Solve:

1. $-6y = 1.2$ _____
2. $\dfrac{1}{2}t = 11$ _____
3. $\dfrac{1}{4}a = 0.25$ _____

4. $3\dfrac{1}{2}t = 7$ _____
5. $-8 = \dfrac{4}{5}C$ _____
6. $10 = \dfrac{p}{2.3}$ _____

7. $7j = 12\dfrac{1}{4}$ _____
8. $\dfrac{k}{2.7} = 0$ _____
9. $\dfrac{a}{0.25} = -1$ _____

10. $-1\dfrac{1}{2}b = 3$ _____
11. $\dfrac{5}{4} = \dfrac{x}{12}$ _____
12. $\dfrac{r}{9} = -9$ _____

13. $\dfrac{R}{17} = 7$ _____
14. $-13r = 169$ _____
15. $4.2 = \dfrac{y}{10}$ _____

16. $\dfrac{1}{2}C = 3\dfrac{1}{2}$ _____
17. $0.6 = \dfrac{m}{5}$ _____
18. $1.4r = -5.6$ _____

19. $7C = -49$ _____
20. $-12B = 144$ _____
21. $\dfrac{3}{16} = -t$ _____

22. $-5x = 12$ _____

23. $6x = -15$ _____

24. $0.8 = 4a$ _____

25. $0.2 = -5x$ _____

26. $-11y = -4.4$ _____

27. $-9z = -6.3$ _____

28. $\frac{1}{3}t = -2$ _____

29. $7x = 0$ _____

30. $5 = \frac{3}{4}y$ _____

C. Solve:

1. $\frac{1}{5}C = 0$ _____

2. $\frac{Q}{0.2} = 9$ _____

3. $-\frac{1}{9} = 3t$ _____

4. $3\frac{1}{5}M = 25$ _____

5. $35z = -7$ _____

6. $42 = \frac{3}{5}b$ _____

7. $\frac{A}{0.01} = 12$ _____

8. $7y + 3y = 45$ _____

9. $3B = 16 + 2B$ _____

10. $15N = 11N + 1$ _____

11. $3p = 8p - 40$ _____

12. $42 = 3n + 5n - n$ _____

13. $5d + 27 - 2d = 0$ _____

14. $17 - 3y + 5y = 0$ _____

15. $-121 = 4x + 7x$ _____

16. $\frac{3}{8}R + \frac{1}{4}R = 25$ _____

17. $95 = 8A - 3A + 15$ _____

18. $4\frac{2}{3}e + 5\frac{1}{3}e = -90$ _____

19. $\frac{3G}{2} = 22\frac{1}{2}$ _____

20. $\frac{a}{4} + 2a = 9$ _____

21. $x - 3x - 4 - 2x = 12$ _____

22. $x + 3x - 1 = 7$ _____

23. $y - 5y = 8$ _____

24. $2t + 6t = -3$ _____

25. $2 = 4x - x$ _____

26. $2a + 3a = 1$ _____

27. $x - 2x - 3x = -12$ _____

28. $5x - 1 + x = 7$ _____

29. $5 = 1 - 4x$ _____

30. $17 - a - 4a = 2$ _____

D. Word Problems

1. One number is four times another. If their sum is 60, what are the numbers?

   _____

2. A number is $\frac{3}{4}$ of another number. The sum of the numbers is 35. What are the numbers?

   _____

3. Dan is 3 times as old as Steve. If the sum of their ages is 52, find their ages.

   _____

4. Sue spent $\frac{2}{3}$ as much as her sister. If together the girls spent $60, how much did each girl spend?

   _____

Date _____

Name _____

5. Three plus a number is equal to twice the number minus one. What is the number?

Course/Section _____

   _____

6. Joyce and Janice sold 182 candy bars to raise track team money. If Janice sold $\frac{5}{8}$ as many candy bars as Joyce, how many did each sell?

   _____

7. A number is multiplied by 17. When the product is added to 47, the sum is 387. What is the number?

   _____

8. The larger of two numbers is three times the smaller. If the sum of the two numbers is 332, find the numbers.

   _____

9. A mountain cabin and lot are worth $23,400. If the cabin is worth 5.5 times as much as the lot, how much is each worth?

   _____

10. If 14 is added to $\frac{2}{3}$ of a number, the result is 1412. What is the number?

   _____

E. Brain Boosters

1. For what values of a and b is the following equation true?
   $$3x - a(1 - x) = 2x - 4 + b$$

2. Here is a bit of math magic: think of a number; double it; add 6; divide by 2; subtract the number you first thought of. Your answer is 3. Use algebra to explain how this trick works.

3. One of the greatest of Greek mathematicians was Diophantus who lived around 250 A.D. Although his mathematical works still exist, almost all that is known about his personal life is found in his epitaph as quoted by a writer in 500 A.D. Here it is:

   "This tomb holds Diophantus. Ah, how great a marvel! The tomb tells scientifically the measure of his life. God granted him to be a boy for the sixth part of his life, and adding a twelfth part to this, he clothed his cheeks with down; He lit him the light of wedlock after a seventh part, and five years after his marriage He granted him a son. Alas! late-born wretched child; after attaining the measure of half his father's life, chill Fate took him. After consoling his grief by this science of numbers for four years he ended his life."

   Translate this into an algebra equation and solve it to learn the age to which Diophantus lived.

   _____

When you have had the practice you need either return to the preview on page 93 or continue in **19** where you will learn to solve more complex equations.

Date _____

Name _____

Course/Section _____

19   Solving most first-degree equations requires that we use a combination of both the addition–subtraction and the multiplication–division rules. For example, to solve

$4x + 5 - x = 17$

First, simplify by combining like terms:

$3x + 5 = 17$

Second, subtract 5 from both members of the equation:

$3x + 5 \boxed{-5} = 17 \boxed{-5}$

$$3x = 12$$

Third, divide both members of the equation by 3 :

$$\frac{3x}{\boxed{3}} = \frac{12}{\boxed{3}}$$

$x = 4$ The solution is 4.

Finally, **check** the solution by substituting it back into the original equation.

$4(4) + 5 - (4) = 17$
$16 + 5 - 4 = 17$
$21 - 4 = 17$

Solve this equation using both rules:

$6 - 2a = 9 + a$

Check your work in **20**.

---

20 $6 - 2a = 9 + a$

$6 - 2a \boxed{+2a} = 9 + a \boxed{+2a}$ Add $2a$ to each member of the equation.

$6 = 9 + 3a$

$6 \boxed{-9} = 9 + 3a \boxed{-9}$ Subtract 9 from each member of the equation

$-3 = 3a$

$\dfrac{-3}{\boxed{3}} = \dfrac{3a}{\boxed{3}}$ Divide each member of the equation by 3 .

$-1 = a$

or $a = -1$ The solution is -1.

Check: $6 - 2(-1) = 9 + (-1)$
$6 + 2 = 9 - 1$
$8 = 8$

Solving Any First-Degree Equation

The procedure for solving first-degree equations is:

Step 1.   Simplify the equation by removing all parentheses, rearranging terms, and adding like terms. Use the multiplication rule to clear all fractions.

Step 2.   Use the addition-subtraction rule to transform the equation so that all terms containing the variable are on one side of the equation and all terms not containing the variable are on the other side of the equation. If necessary, again combine like terms.

Step 3.   Use the multiplication-division rule to obtain an equation of the form $x = \square$.

121   **2-4 Solving Linear Equations**

Step 4. Check the solution by substituting it into the original equation.

Let's work through another example to be certain you are able to follow these steps.

Solve:

$5p + 2 - p = 12 + p + 5$

Look for our step-by-step solution in **21**.

---

21    $5p + 2 - p = 12 + p + 5$

Step 1. Combine like terms.

$4p + 2 = 17 + p$

Step 2. Use the addition–subtraction rule so as to place all terms containing the variable on the same side of the equation.

$$4p + 2 \boxed{-p} = 17 + p \boxed{-p} \qquad \text{Subtract } \boxed{p}.$$
$$3p + 2 = 17$$
$$3p + 2 \boxed{-2} = 17 \boxed{-2} \qquad \text{Subtract } \boxed{2}.$$
$$3p = 15$$

Step 3. Use the division rule to put the equation into the form $p = \Box$.

$$\frac{3p}{\boxed{3}} = \frac{15}{\boxed{3}} \qquad \text{Divide by } 3 .$$
$$p = 5 \qquad \text{The solution is 5.}$$

Step 4. Check the solution by substituting into the original equation.

Checking

$$5(5) + 2 - (5) = 12 + (5) + 5$$
$$25 + 2 - 5 = 12 + 5 + 5$$
$$22 = 22$$

If the solution does not seem to be correct when it is substituted into the original equation:

1. Go back over each step of the check.
2. If there are no errors in the check, repeat the solution.
3. Check your new solution.

Now exercise your mental muscles by solving these equations.

(a) $6 + 2x = 1 - 3x$ (b) $8 = \dfrac{2y}{3} + 1$

(c) $6K - 5 + 1 = 2K + 8$ (d) $\dfrac{3X}{2} + 4 = 3 + X - 2$

Check your work in **22**.

---

22   (a) $6 + 2x = 1 - 3x$

$$6 + 2x \boxed{+ 3x} = 1 - 3x \boxed{+ 3x} \qquad \text{Add } \boxed{3x}.$$
$$6 + 5x = 1$$
$$6 + 5x \boxed{- 6} = 1 \boxed{- 6} \qquad \text{Subtract } \boxed{6}.$$
$$5x = -5$$

$$\frac{5x}{5} = \frac{-5}{5}$$ Divide by 5.

$$x = -1$$ The solution is -1.

Check: $\quad 6 + 2(-1) = 1 - 3(-1)$
$$6 - 2 = 1 + 3$$

(b) $$8 = \frac{2y}{3} + 1$$

$$8 \;\boxed{-1} = \frac{2y}{3} + 1 \;\boxed{-1}$$ Subtract 1.

$$7 = \frac{2y}{3}$$

$$\boxed{3}\,(7) = \boxed{3}\left(\frac{2y}{3}\right)$$ Multiply by 3 to clear the fractions.

$$21 = 2y$$

$$\frac{21}{\boxed{2}} = \frac{2y}{\boxed{2}}$$ Divide by 2.

$$10\frac{1}{2} = y$$

or $\quad y = 10\frac{1}{2}$ The solution is $10\frac{1}{2}$.

Check: $\quad 8 = \dfrac{2(10\frac{1}{2})}{3} + 1$

$$8 = \frac{21}{3} + 1 = 7 + 1$$

(c) $\quad 6K - 5 + 1 = 2K + 8$

$\quad\quad\;\; 6K - 4 = 2K + 8$ Combine like terms.

$6K - 4 \;\boxed{-2K} = 2K + 8 \;\boxed{-2K}$ Subtract $2K$.

$$4K - 4 = 8$$

$4K - 4 \;\boxed{+4} = 8 \;\boxed{+4}$ Add 4.

$$4K = 12$$

$$\frac{4K}{\boxed{4}} = \frac{12}{\boxed{4}}$$ Divide by 4.

$$K = 3$$ The solution is 3.

Check: $\quad 6(3) - 5 + 1 = 2(3) + 8$
$$18 - 5 + 1 = 6 + 8$$
$$14 = 14$$

(d) $$\frac{3X}{2} + 4 = 3 + X - 2$$

$$\frac{3X}{2} + 4 = X + 1$$ Combine like terms.

$$\frac{3X}{2} + 4 \;\boxed{-4} = X + 1 \;\boxed{-4}$$ Subtract 4.

$$\frac{3X}{2} = X - 3$$

$$\frac{3X}{2} \;\boxed{-X} = X - 3 \;\boxed{-X}$$ Subtract X.

$$\frac{3X}{2} - X = -3$$

$$2\left(\frac{3X}{2}\right) - \boxed{2}\,(X) = \boxed{2}\,(-3) \qquad \text{Multiply by } \boxed{2}.$$
$$3X - 2X = -6$$
$$X = -6 \qquad \text{The solution is } -6.$$

Check:
$$\frac{3(-6)}{2} + 4 = 3 + (-6) - 2$$
$$-\frac{18}{2} + 4 = 3 - 6 - 2$$
$$-9 + 4 = 3 - 8$$
$$-5 = -5$$

Equations with Parentheses

In the language of algebra one of the most useful punctuation marks is a pair of parentheses. As you saw in Unit 1, parentheses often appear when we translate English phrases directly into algebraic equations. But before we can solve an equation that contains parentheses, we usually need to rewrite it as an equivalent equation without parentheses. The distributive law enables us to transform an algebraic equation with parentheses into an equivalent algebraic equation without parentheses.

$$\boxed{\begin{aligned} a(b + c) &= ab + ac \\ \text{or} \qquad a(b - c) &= ab - ac \end{aligned}}$$

Examples:

$3 + (2x - 1) = 10$ is equivalent to $3 + 2x - 1 = 10$

An expression such as $+(\ )$ is equal to $(+1)(\ )$. Simply omit the parentheses.

$11 - (a - 3) = 4$ is equivalent to $11 - a + 3 = 4$

An expression such as $-(\ )$ is equal to $(-1)(\ )$. Omit the parentheses and change the sign of each term inside the parentheses.

$2(1 + y) = 3$ is equivalent to $2 + 2y = 3$

$$\boxed{2}\,(1 + y) = \boxed{2}\,(1) + \boxed{2}\,(y)$$

$-3(2 - x) = 5$ is equivalent to $-6 + 3x = 5$

$$\boxed{-3}\,(2 - x) = \boxed{-3}\,(2) + \boxed{-3}\,(-x)$$

For a review of this, use the distributive law to rewrite the following expressions without parentheses.

(a) $5 + (1 - 3x)$ (b) $4 - (y + 2)$ (c) $9 - (a - 4)$

(d) $5 - (-x - 3)$ (e) $3(1 + a)$ (f) $2(2x - 3)$

(g) $-2(x + 2)$ (h) $-3(2x - 1)$ (i) $-4(3 - 2y)$

(j) $-5(-z - 2)$

Check your answers in **23**.

---

23   (a) $5 + (1 - 3x) = 5 + 1 - 3x = 6 - 3x$

(b) $4 - (y + 2) = 4 - y - 2 = 2 - y$

(c) $9 - (a - 4) = 9 - a + 4 = 13 - a$

(d) $5 - (-x - 3) = 5 + x + 3 = 8 + x$

(e) $\boxed{3}\,(1 + a) = \boxed{3}\,(1) + \boxed{3}\,(a) = 3 + 3a$

(f) $\boxed{2}(2x - 3) = \boxed{2}(2x) + \boxed{2}(-3) = 4x - 6$

(g) $\boxed{-2}(x + 2) = \boxed{-2}(x) + \boxed{-2}(2) = -2x - 4$

(h) $\boxed{-3}(2x - 1) = \boxed{-3}(2x) + \boxed{-3}(-1) = -6x + 3$

(i) $\boxed{-4}(3 - 2y) = \boxed{-4}(3) + \boxed{-4}(-2y) = -12 + 8y$

(j) $\boxed{-5}(-z - 2) = \boxed{-5}(-z) + \boxed{-5}(-2) = 5z + 10$

Solving Equations with Parentheses

To solve an equation containing parentheses, first we usually use the distributive law to remove the parentheses, then solve the resulting equation using the techniques you have already learned. For example, to solve $2(x + 1) = 10$,

remove parentheses,

$2x + 2 = 10.$

Subtract 2 from each member:

$2x + 2 - 2 = 10 - 2$
$\quad\quad\quad 2x = 8$

Divide by 2,

$\quad\quad x = 4 \quad\quad$ The solution is 4.

Check your solution: $\quad 2((4) + 1) = 10$
$\quad\quad\quad\quad\quad\quad\quad\quad 2(5) = 10$

Another example:

$-2(3 - 2x) = 6 \quad\quad$ Use the distributive law to remove parentheses.

$\quad -6 + 4x = 6 \quad\quad$ Add $\boxed{6}$ to each side of the equation.

$\quad\quad\quad 4x = 12 \quad\quad$ Divide each member of the equation by $\boxed{4}$.

$\quad\quad\quad\quad x = 3 \quad\quad$ Check the solution.

Solve:

$3(x - 5) = x + 5$

Look for our step-by-step solution in **24**.

---

24 $\quad 3(x - 5) = x + 5$

Remove parentheses $\quad\quad 3(x) + 3(-5) = x + 5$
$\quad\quad\quad\quad\quad\quad\quad\quad\quad\quad 3x - 15 = x + 5$

Add $\boxed{15}$ $\quad\quad\quad\quad 3x - 15 \boxed{+ 15} = x + 5 \boxed{+ 15}$
$\quad\quad\quad\quad\quad\quad\quad\quad\quad\quad\quad 3x = x + 20$

Subtract $\boxed{x}$ $\quad\quad\quad\quad 3x \boxed{- x} = x + 20 \boxed{- x}$
$\quad\quad\quad\quad\quad\quad\quad\quad\quad\quad 2x = 20$

Divide by $\boxed{2}$ $\quad\quad\quad\quad\quad x = 10 \quad\quad$ The solution is 10.

Check: $\quad 3((10) - 5) = (10) + 5$
$\quad\quad\quad\quad\quad\quad 3(5) = 10 + 5$
$\quad\quad\quad\quad\quad\quad\quad 15 = 15$

Follow these four steps to solve any first-degree equation.

SUMMARY **Step 1.** Simplify the equation by removing all parentheses, clearing fractions, re-arranging terms, and adding like terms.

Step 2. Use the addition–subtraction rule to put all terms containing the variable on one side of the equation and all other terms on the other side of the equation. Combine like terms.

Step 3. Use the multiplication-division rule to obtain an equivalent equation of the form $x = \square$.

Step 4. Check the solution by substituting it into the original equation.

Example Solve: $\qquad\qquad ax + b = c$

Add the opposite of b: $\qquad ax + b - b = c - b$

$$ax = c - b$$

Divide each side by a: $\qquad\qquad \dfrac{ax}{a} = \dfrac{c - b}{a}$

$$x = \dfrac{c - b}{a}$$

Solve the following equations.

(a) $3(a - 4) = 6$

(b) $2(x - 3) = 3x$

(c) $8 = 2(2y - 3)$

(d) $4 = 7 + 3(2x - 5)$

(e) $3z = 21 - 3(5 - 2z)$

(f) $3x + 5(4 - x) = 6$

(g) $4 = 3(2 - x)$

(h) $3 - t = 5(2 - t) - 15$

(i) $5(2 - x) + x = 2$

(j) $x + (x + 1) + (x + 2) = 12$

Check your answers in **25**.

MENTAL MATH QUIZ

Write down the answers to the following problems quickly.
You should not need to do any paper and pencil calculations.

Average time: ___4:42___

Record time: ___1:55___
(These times are for students scoring at least 38 correct in a class in basic algebra in a community college.)

Solve:

1. $2x = 6$	2. $3y = 9$	3. $4a = 8$
4. $5x = 10$	5. $3a = 12$	6. $4q = 16$
7. $2y = 10$	8. $3p = 6$	9. $5t = 25$
10. $6p = 36$	11. $4x = 24$	12. $7x = 14$
13. $8 = 4p$	14. $12 = 3q$	15. $18 = 6y$
16. $21 = 3x$	17. $27 = 9b$	18. $24 = 8a$
19. $0 = 7t$	20. $32 = 2k$	21. $3x = 0$
22. $2y = -4$	23. $3k = -6$	24. $4p = -12$
25. $-6 = -3t$	26. $-5 = -5x$	27. $-3y = 9$
28. $-2x = -24$	29. $\frac{1}{2}x = 5$	30. $\frac{1}{4}P = 2$
31. $\frac{1}{5}t = 6$	32. $3y = -2$	33. $2x - 1 = 7$
34. $3y + 1 = 7$	35. $2p - 2 = 6$	36. $2t + 3 = 5$
37. $2 - 5x = 12$	38. $1 - 2y = 3$	39. $5 - 2p = 3$
40. $2 - 5x = 7$		

The answers are on page 541.

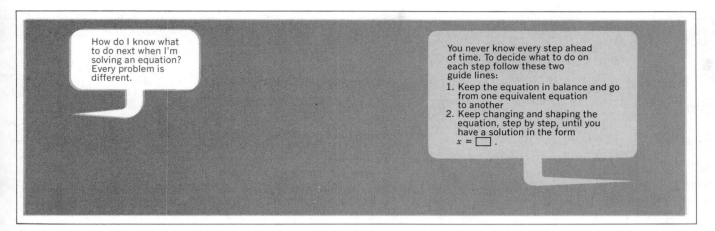

25   (a)   6   (b) -6

         (c) $3\frac{1}{2}$ (d)   2

         (e) -2 (f)   7

         (g) $\frac{2}{3}$ (h) -2

         (i)   2   (j)   3

Now turn to **26** for a set of practice problems.

To solve this equation: $2x - 3 = 10$
Why can't I divide both sides by 2
to get $x - 3 = 5$ so that
$x = 8$. Isn't this ok?

No. When you divide both sides of an equation by a number, you must divide **all** of both sides. Divide every term of the expression on the left, not just part of it. $\frac{2x-3}{2} = \frac{10}{2}$ or $x - \frac{3}{2} = 5$.

$$\frac{3a + 2}{4} = 17$$

$$\frac{3}{4}a = 15 \cdot \frac{4}{3}.$$

$$a = 20$$

Unit 2
Solving
Algebraic Equations

Answers are on page 542.

A. Solve:

1. $4x + 3 - x = 15$ _____

2. $3y + 8 = 5y + 2$ _____

3. $-x = 4 + 2x - 19$ _____

4. $5a + 3a = -24$ _____

5. $4x + 7 = 31$ _____

6. $-15 + 2p = 15$ _____

7. $-9 = 4y - 1$ _____

8. $3x - 1 = 26$ _____

9. $4 + 8m = 52$ _____

10. $5n + 7 + n = 37$ _____

11. $5z - 3z = 22$ _____

12. $8p - 3p + 7 = 87$ _____

13. $27 = 8x - 10x + 10$ _____

14. $2a + 3 - 4a = 7$ _____

15. $4b + 17 = 10b + 5$ _____

16. $0 = 17x - 102$ _____

17. $1 + 2x + 3x + x = 4 - x$ _____

18. $b - 1 = 1 - b$ _____

19. $\dfrac{7a}{2} = 0$ _____

20. $1 + x - 2 - 2x + 3 - 3x = 0$ _____

B. Solve:

1. $\dfrac{2}{5}x + \dfrac{4}{5}x = 36$ _____

2. $y - \dfrac{1}{2}y - 8 = 0$ _____

3. $\dfrac{1}{4}t + \dfrac{1}{2}t = -32$ _____

4. $2 + \dfrac{3C}{3} = -5$ _____

5. $\dfrac{3a}{4} + 2 = 17$ _____

6. $\dfrac{3x}{2} - 7 = 38$ _____

7. $2y - 36 = 8y - 17$ _____

8. $-42 = \dfrac{7p}{10}$ _____

9. $\dfrac{2}{3}C - \dfrac{2}{3} = 0$ _____

10. $0.5x + 0.3x = -8$ _____

11. $29 = \dfrac{2E}{3} + 9$ _____

12. $3a - 5 = 2 - 4a$ _____

13. $x - 2x + 2 = 3x + 4$ _____

14. $1 - x + 6 = 4x + 5$ _____

15. $5d + 7d - 11d = 1 - d$ _____

16. $4H + 2 - H = H - 6$ _____

17. $3 - x = 3 - 2x$ _____

18. $2a + 4 = 1 + 6a$ _____

19. $1 + 2z - 3 = 5z - 7 - z$ _____

20. $7x - 9 = 12x + 5$ _____

Date

Name

Course/Section

21. $5a + 2 = 3a - 7$ _____

22. $0.2x - 0.3 = 0.7x + 0.1$ _____

23. $\dfrac{1}{6} + \dfrac{x}{2} + \dfrac{1}{3} = 2x - 1$ _____

24. $A - 1 = 1 - 2A$ _____

25. $3(y + 1) = 6$ _____

26. $4 = 3(c - 2)$ _____

27. $4(3c - 6) = 0$ _____

28. $48 = 2(A - 1)$ _____

29. $2(x - 1) = 3$ _____

30. $1 - 3(1 + y) = 4$ _____

C. Solve:

1. $2(x - 1) = x - 11$ _____

2. $4(y + 1) = 3$ _____

3. $-(2x - 3) = 7 + x$ _____

4. $2(a - 2) = 5$ _____

5. $1 - (t - 1) = 2t$ _____

6. $6(1 - E) = 11 - E$ _____

7. $x - 3(1 + x) = x$ _____

8. $-2x + 5(1 - x) = 2 - x$ _____

9. $2z + 4(z - 1) = 5$ _____

10. $3(1 - a) = 2(a - 1)$ _____

11. $-a - 2(a - 2) = 2(1 - a)$ _____

12. $1 - 2(2x - 3) = 2 - 3x$ _____

13. $2 + B - (3B - 1) = 6 - 2(1 - B)$ _____

14. $3(2x - 5) = -9$ _____

15. $a - 5a + (1 - a) = 9$ _____

16. $-2(2x - 1) = 3$ _____

17. $-2(2 - a - 1) = 0$ _____

18. $6(2x - 1) + 2 = 2x + 1$ _____

19. $5t - 4(2 - 3t) = 2t - 3$ _____

20. $3(2p - 1) - (p - 2) = 1 - p$ _____

21. $\dfrac{2(x - 1)}{3} = -4$ _____

22. $1 - \dfrac{3(1 - 2E)}{4} = 0$ _____

23. $\dfrac{3(2 - 2Q)}{2} = Q$ _____

24. $\dfrac{3(2 - 2Q)}{2} = Q - 1$ _____

25. $\dfrac{8(x - 1)}{7} = 0$ _____

26. $\dfrac{2}{3}(x - 1) = x - 1$ _____

27. $1 + A - 2(1 + A) = 3(1 - A) + 6A$ _____

28. $1 + \dfrac{2x - 3}{3} = \dfrac{x + 1}{2}$ _____

29. $w - 2(w - 2(w - 1)) = 2$ _____

30. $1 - R - 2(2R + 1) = 0$ _____

D. Translate these problems into algebraic equations in one variable and solve them.

1. The sum of two numbers is 18. One of them is twice as large as the other. What are the numbers?

_____

2. Find two numbers whose sum is 13 and whose difference is 7.

_____

3. For what temperature does the Celsius scale read exactly five times the Fahrenheit scale reading? *Hint:* $5(F + 40) = 9(C + 40)$ and $C = 5F$.

_____

4. Five less than $\frac{1}{2}$ of a number is equal to 6. What is the number?

_____

5. $3x - B = C + 5x$. Solve for x in terms of the other letters.

_____

6. Lynn is 7 years more than twice as old as Mary. The difference in their ages is 20. How old is Lynn?

_____

7. If $\frac{5}{8}$ of a certain number is added to $\frac{5}{16}$ of that number, the sum is 90. What is the number?

_____

8. A 60-in. cord is cut into 6 pieces of equal length and a 9-in. piece remains. How long are each of the 6 equal pieces?

_____

9. The perimeter of a rectangle is 84 ft. What are its dimensions if the length is five times the width?

_____

10. When Bill received his paycheck for working in the college bookstore, he noticed that it was $12.50 less than three times as much as his friend Pam had earned. Their two paychecks totaled $51.50. What did each person earn?

_____

E. Brain Boosters

1. Solve. $10 + x = 10x$ _____

2. Solve for x. $\dfrac{a}{1 - bx} = \dfrac{2b}{1 - ax}$ _____

3. Solve for x. $\dfrac{a + b}{x - b} = \dfrac{2b}{x - a}$ _____

4. John is twice as old as his wife was when he was as old as his wife is now. He is 24. How old is she?

_____

Date _____

Name _____

Course/Section _____

5. To encourage her daughter in learning mathematics, Professor Adams offered to pay her $8 for every equation she solved correctly and to fine her $5 for every incorrect solution. At the end of 26 problems, neither owed any money to the other. How many problems did she solve correctly?

_____

6. Two candles are initially the same length but of different thickness. When they are burned, the thinner one is consumed in 4 hours and the thicker one in 5 hours. If they are lighted at the same time, when will one be three times as long as the other?

_____

When you have had the practice you need either return to the preview on page 93 or continue in **27** where you will learn to solve formulas.

27 A *formula* is a literal equation that contains letters other than the variable and that generally describes some practical situation. Formulas are important in business, science, engineering, and in almost any situation where numbers are used. The area and volume of geometric figures can be calculated from formulas. Physics formulas describe everything from the motion of rockets and atoms to the temperature of a star. A banker calculates your car loan payments from a formula and a nurse uses a formula to determine the amount of medication you should have. When they do these things they look up the formula they need to use from a book, but what they find in the book may not be in the form they need to do the job. Very often algebra is needed to convert the formula to a more useful form.

To *solve a formula* for some letter means to rewrite the formula as an equivalent equation with that letter isolated on the left of the equals sign. If the area of a triangle is given by the formula $A = \dfrac{bh}{2}$ where A is the area, b is the length of the base, and h is the height, then solving for the base gives the equivalent formula $b = \dfrac{2A}{h}$ and solving for the height gives the equivalent formula $h = \dfrac{2A}{b}$.

Solving formulas is an application of algebra that is very useful in other courses and in practical situations. You should be able to solve any formula for any of the letters in it and usually this cannot be done by simply looking at it. Instead you must work through a series of steps.

The procedures used to solve a formula are exactly the same as those used to solve equations.

If $x = 2y$, solve for y.

Try it. Then check your work in **28**.

---

28 $x = 2y$

To solve for y, first divide both members of the equation by 2.

$$\frac{x}{2} = \frac{2y}{2}$$

$$\frac{x}{2} = y$$

or $\quad y = \dfrac{x}{2}$

Notice that the answer to the problem "solve for y" is an equation of the form $y = \square$, where the right member of the equation is an expression containing other letters and perhaps numerals. We solve formulas by using the same rules we used to solve equations: (1) the addition–subtraction rule, (2) the multiplication–division rule, (3) the distributive law.

Solve the physics formula $v_2 = v_1 + gt$ for v_1.

Do it here . . .

. . . then turn to **29** to check your work.

---

29 $v_2 = v_1 + gt$ Use the addition–subtraction rule: if the same quantity is added to or subtracted from both sides of the formula, you will have an equivalent formula.

Subtract the term gt from each side of the equation.

$$v_2 \;\boxed{-\;gt}\; = v_1 + gt \;\boxed{-\;gt}$$

$$v_2 - gt = v_1$$

or $\quad v_1 = v_2 - gt$

Solve the formula $V = iR$ for i.

Check your answer in **30**.

---

30 $\quad V = iR \qquad$ Use the multiplication-division rule: if you multiply or divide both sides of the formula by the same non-zero quantity, you will have an equivalent formula.

Divide by $\boxed{R}$.

$$\frac{V}{\boxed{R}} = \frac{iR}{\boxed{R}}$$

$$\frac{V}{R} = i$$

or $\quad i = \dfrac{V}{R}$

Solve the following formulas as indicated.

(a) Solve $a = bh$ for h.

(b) Solve $s = r + p$ for r.

(c) Solve $v = \dfrac{3k}{t}$ for t.

(d) Solve $q = 1 - r + t$ for r.

(e) Solve $t = \dfrac{a}{b + c}$ for c.

(f) Solve $p = a\left(\dfrac{1}{b} + \dfrac{1}{c}\right)$ for c.

The complete solutions are in **31**. Compare your work with ours when you have solved these.

What is the difference between "solving an equation" and "solving a formula"?

To solve an equation means to find the number or set of numbers that satisfy the equation. To solve a formula means to find an equivalent formula.

---

31 (a) $\quad a = bh \qquad$ To solve for h, divide by $\boxed{b}$:

$$\frac{a}{\boxed{b}} = \frac{bh}{\boxed{b}}$$

$$\frac{a}{b} = h$$

or $\quad h = \dfrac{a}{b}$

(b) $\quad s = r + p \qquad$ To solve for r, subtract $\boxed{p}$:

$$s \;\boxed{-\;p}\; = r + p \;\boxed{-\;p}$$

$$s - p = r$$

or $\quad r = s - p$

(c) $v = \dfrac{3k}{t}$ To solve for t,

multiply by $\boxed{t}$:

$$\boxed{t}\,(v) = \boxed{t}\left(\dfrac{3k}{t}\right)$$

$tv = 3k$

Divide by $\boxed{v}$:

$$\dfrac{tv}{v} = \dfrac{3k}{v}$$

$$t = \dfrac{3k}{v}$$

(d) $q = 1 - r + t$ To solve for r,

add $\boxed{r}$:

$q \; \boxed{+ r} = 1 - r + t \; \boxed{+ r}$

$q + r = 1 + t$

Subtract $\boxed{q}$:

$q + r \; \boxed{- q} = 1 + t \; \boxed{- q}$

$r = 1 + t - q$

(e) $t = \dfrac{a}{b + c}$ To solve for c,

multiply by $(b + c)$:

$t(b + c) = a$

$tb + tc = a$

Subtract tb:

$tc = a - tb$

Divide by t:

$$c = \dfrac{a - tb}{t}$$

(f) $p = a\left(\dfrac{1}{b} + \dfrac{1}{c}\right)$ To solve for c,

divide by a:

$$\dfrac{p}{a} = \dfrac{1}{b} + \dfrac{1}{c}$$

Subtract $\dfrac{1}{b}$:

$$\dfrac{p}{a} - \dfrac{1}{b} = \dfrac{1}{c}$$

or $\dfrac{pb - a}{ab} = \dfrac{1}{c}$

Multiply by abc:

$c(pb - a) = ab$

Divide by $(pb - a)$:

$$c = \dfrac{ab}{pb - a}$$

Usually, solving a formula involves a combination of the steps used in solving an equation. For example, in a chemistry class you may learn that the Fahrenheit and Celsius temperature scales are related by the formula $F = \dfrac{9C}{5} + 32$. Solve this formula for C.

Do it now, and check your work in **32**.

---

32 $F = \dfrac{9C}{5} + 32$ To solve for C,

subtract $\boxed{32}$ from each member of the equation:

$F \; \boxed{- 32} = \dfrac{9C}{5} + 32 \; \boxed{- 32}$

$F - 32 = \dfrac{9C}{5}$

Multiply each member of the equation by $\boxed{5}$:

$$\boxed{5}\,(F - 32) = \boxed{5}\left(\dfrac{9C}{5}\right)$$

$5(F - 32) = 9C$

Divide both members of the equation by 9 :

$$\frac{5(F - 32)}{9} = \frac{9C}{9}$$

$$\frac{5(F - 32)}{9} = C$$

or $\quad C = \dfrac{5(F - 32)}{9}$

Notice that when we multiplied both members of the equation by 5, we placed the left member, $F - 32$, in parentheses: $(F - 32)$. This was done to remind you that *all* of the left member must be multiplied by 5.

Solve the following formulas as indicated.

(a)   Solve $s = \dfrac{m}{2}(a + t)$ for t.

(b)   Solve $A = P(1 + rt)$ for t.

(c)   Solve $v = \pi R^2 h - ab$ for h.

(d)   Solve $y = mx + b$ for x.

(e)   Solve $y = \dfrac{a}{x - b}$ for x.

Check your work in **33**.

---

33   (a) $s = \dfrac{m}{2}(a + t)$ To solve for t,

multiply by 2 to eliminate the fraction:

$$2\,(s) = 2\,\left(\frac{m}{2}\right)(a + t)$$

$$2s = m(a + t)$$

Remove the parentheses using the distributive law:

$$2s = ma + mt$$

Subtract ma from each member of the equation:

$$2s \;-\, ma\; = ma + mt \;-\, ma$$

$$2s - ma = mt$$

Divide by m :

$$\frac{2s - ma}{m} = \frac{mt}{m}$$

$$\frac{2s - ma}{m} = t$$

or $\quad t = \dfrac{2s - ma}{m}$

(b) $A = P(1 + rt)$ To solve for t,

remove the parentheses:

$$A = P + Prt$$

Subtract P :

$$A \;-\, P\; = P + Prt \;-\, P$$

$$A - P = Prt$$

Divide by Pr:

$$\frac{A - P}{Pr} = t$$

or $\qquad t = \frac{A - P}{Pr}$

(c) $\quad v = \pi R^2 h - ab \qquad$ To solve for h,

add ab to both members of the equation:

$v + ab = \pi R^2 h - ab + ab$

$v + ab = \pi R^2 h$

Divide by πR^2:

$$\frac{v + ab}{\pi R^2} = h$$

or $\qquad h = \frac{v + ab}{\pi R^2}$

(d) $\quad y = mx + b \qquad$ To solve for x,

subtract b from each member of the equation:

$y - b = mx + b - b$

$y - b = mx$

Divide by m:

$$\frac{y - b}{m} = \frac{mx}{m}$$

$$\frac{y - b}{m} = x$$

or

$$x = \frac{y - b}{m}$$

(e) $\quad y = \dfrac{a}{x - b} \qquad$ To solve for x,

multiply each member by $x - b$:

$$(x - b)(y) = (x - b)\frac{a}{x - b}$$

$xy - by = a$

Add by to each member:

$xy - by + by = a + by$

$xy = a + by$

Divide by y:

$$\frac{xy}{y} = \frac{a + by}{y}$$

$$x = \frac{a + by}{y}$$

Solving formulas is a very useful algebraic skill. In succeeding units of this book, as you become more expert at working with algebraic expressions, we will work with and solve more complex formulas. At this point, you will profit most by the mental exercise of solving many simple formulas. Turn to **34** for a set of practice problems.

Unit 2
Solving
Algebraic Equations

The answers are on page 542.

A. Solve each of the following formulas for x.

1. $A = 3x$ _____

2. $B = x + 4$ _____

3. $Y = 2bx$ _____

4. $R = \dfrac{ax}{d}$ _____ ✓

5. $Q = 5 - x$ _____

6. $C = 2\pi x + 1$ _____

7. $2ax = 3c$ _____

8. $ax - b^2 = 0$ _____

✓9. $T - x = 3 + A$ _____

10. $A = B - cx$ _____

11. $K = \dfrac{axy}{2b}$ _____

12. $1 - x = E + 4$ _____

13. $2A = 3x - 4$ _____

14. $B - x = A + 2x$ _____

15. $\dfrac{y}{2} = x + 1$ _____

16. $3A = x - A + 2$ _____

17. $2P - 3x = P - 2x + 2$ _____

18. $x - A = 2(x + A)$ _____

19. $3(t - x) + 4 = t + 2x$ _____

20. $x - y = 2(y + 2x) - 2x$ _____

B. Solve each of the following formulas for a or b.

1. $px^2 - a = 2$ _____

2. $3 = 2(x - a)$ _____

3. $a + x = 2(a + x)$ _____

4. $x = y + bt$ _____

5. $c = 2da$ _____

6. $d = 3mbt^2$ _____

7. $d(a - 1) = xc - 1$ _____

8. $q(ca - 2ca) = p$ _____

9. $m(3b - b) = t + p$ _____

10. $s = at - p$ _____

11. $e + 2b = 3x - b$ _____

12. $s + 2p = bp - 2s$ _____

13. $x = \dfrac{2b}{3} - y$ _____

14. $3(x - 2) = 4(a - 2)$ _____

15. $(a - 1) = 2(a - c)$ _____

16. $d(5ca + ca) = 4c^2d^2$ _____

17. $2ax - 5x = 8xy$ _____

18. $t(2ar + ar) = 3r$ _____

19. $x(a - x) = a(2x + x)$ _____

20. $\dfrac{xa + ax}{2x} = d^2 - a$ _____

21. $3(y - a) = y - 2a$ _____

22. $d + 3a = 2(a - t) + t$ _____

Date _____

Name _____

Course/Section _____

23. $5(x - a) = 2(a - 3)$ _____

24. $2(a - p^2) = 3(p - a)$ _____

25. $2b = \dfrac{3b}{2} + x$ _____

26. $1 - a = \dfrac{5a}{3} - y$ _____

27. $\dfrac{a}{x} = \dfrac{c}{d}$ _____

28. $\dfrac{x}{a} = \dfrac{c}{d}$ _____

29. $\dfrac{p}{2t} = \dfrac{1}{a}$ _____

30. $\dfrac{x}{2y - 1} = \dfrac{a}{2x}$ _____

31. $y = \dfrac{x}{1 - a}$ _____

32. $t = \dfrac{2}{1 + b}$ _____

33. $p = \dfrac{1 - 2p}{1 - a}$ _____

34. $3 = \dfrac{1 - b + x}{1 + 2b}$ _____

35. $2 = \dfrac{1}{a} + \dfrac{1}{x}$ _____

36. $x = \dfrac{1}{a} - \dfrac{1}{x}$ _____

37. $y = \dfrac{2}{a} - 1$ _____

38. $t = 3 - \dfrac{2}{b}$ _____

39. $x = 3\left(\dfrac{1}{a} + \dfrac{1}{2}\right)$ _____

40. $y = 2\left(\dfrac{1}{b} - \dfrac{1}{3}\right)$ _____

C. Solve each of the following formulas for the variable indicated.

1. $p = 2a + 2b$ Solve for a. _____

2. $E = mc^2$ Solve for m. _____

3. $Ax + By + c = 0$ Solve for x. _____

4. $Ax + By + c = 0$ Solve for y. _____

5. $C = \dfrac{100B}{L}$ Solve for L. _____

6. $\dfrac{1}{F} = \dfrac{1}{a} + \dfrac{1}{b}$ Solve for F. _____

7. $s = \dfrac{a - rt}{1 - r}$ Solve for r. _____

8. $s = \dfrac{a - rt}{1 - r}$ Solve for t. _____

9. $V = \frac{1}{3}\pi h(3R^2 - h^2)$ Solve for R^2. _____

10. $S = \frac{1}{2}gt^2$ Solve for g. _____

11. $P = i^2 R$ Solve for R. _____

12. $A = 4\pi R^2 h$ Solve for h. _____

13. $E = \frac{1}{2}mv^2$ Solve for m. _____

14. $P = mgh$ Solve for h. _____

15. $a = \dfrac{v^2}{R}$ Solve for R. _____

16. $PV = nRT$ Solve for P. _____

17. $PV = nRT$ Solve for T. _____

18. $V = 2KL\left(\dfrac{R}{r}\right)$ Solve for R. _____

19. $V = KQ\left(\dfrac{1}{a} - \dfrac{1}{b}\right)$ Solve for a. _____

20. $I = \dfrac{E}{R + a}$ Solve for a. _____

D. Word Problems

1. The length of arc of a sector of a circle is given by the formula

$$L = \dfrac{2\pi Ra}{360}$$

where R is the radius of the circle and a is the central angle in degrees. Solve this equation for a.

_____

2. The area of the sector shown in problem 1 is $A = \dfrac{\pi R^2 a}{360}$. Solve this formula for a.

_____

3. When a gas is kept at constant temperature and the pressure on it is changed, its volume changes in accord with the pressure–volume relationship known as Boyle's Law:

$$\dfrac{V_1}{V_2} = \dfrac{P_2}{P_1}$$

where P_1 and V_1 are the beginning volume and pressure, and P_2 and V_2 are the final volume and pressure.

(a) Solve for V_1.   (c) Solve for P_1.

(b) Solve for V_2.   (d) Solve for P_2.

_____

Date _____

Name _____

Course/Section _____

4. Machinists use a formula, known as Pomeroy's formula, to determine roughly the power required by a metal punch machine.

$$P \cong \dfrac{t^2 dN}{3.78}$$

where P is the power needed, in horsepower; t is the thickness of the metal being punched; d is the diameter of the hole being punched; N is the number of holes to be punched at one time. Solve this formula for N. (The symbol $\cong$ means "approximately equal to.")

_____

5. Nurses use a formula known as Young's rule to determine the amount of medicine to give a child under 12 years of age when the adult dosage is known.

$$C = \frac{AD}{A + 12}$$

C is the child's dose; A is the age of the child in years; D is the adult dose. Work backward and find the adult dose in terms of the child's dose. Solve for D.

_____

6. Lens grinders use the following formula to determine the shape of a simple lens:

$$\frac{1}{F} = (n - 1)\left(\frac{1}{R} + \frac{1}{r}\right)$$

where F is the focal length of the lens; n is its index of refraction; R and r are the radii of curvature of the two lens surfaces. Solve for R.

_____

7. The volume of a football is roughly $V = \dfrac{\pi L T^2}{6}$ where L is its length and T is its thickness.
Solve for L.

_____

8. When two resistors R_1 and R_2 are put in series with a battery giving V volts, the current through the resistors is

$$i = \frac{V}{R_1 + R_2}$$

Solve for R_1.

_____

When you have had the practice you need, either return to the preview on page 93 or continue in **35** with the study of inequalities.

_____

35   An equation is an algebraic statement that two expressions are equal. An *inequality* is an algebraic statement that one expression is greater than or less than another algebraic expression. Equations and inequalities may or may not be true, depending on the value of the variables they contain.

$x < 5$ is an inequality. (Read it "x is less than 5.") It is true for many values of x:

$\quad 4 < 5$ is true,

$\quad -3 < 5$ is true,

$\quad \dfrac{1}{2} < 5$ is true, and so on.

The set of all values of the variable x that make this inequality a true statement, is called the solution set, an infinitely large set of numbers rather than one number. Because the solution set contains infinitely many numbers, we cannot list them all but we can show the solution graphically using a number line. The solution set for the inequality $x < 5$ is shown as follows:

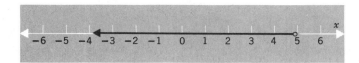

Remember that the inequality sign in $x < 5$ means that any number x is to the left of 5 on the number line. The open circle at $x = 5$ means that we do not include the number 5 in the solution set. (5 is not less than itself.) The shaded arrow covers all numbers that are included in the solution set—all numbers that satisfy the inequality, not only the integers 4, 3, 2, 1, 0, −1, −2, ... that label the line but all fractions and irrational numbers such as $\sqrt{2}$ or $1 - \sqrt{3}$ that are located between integers. The arrowhead on the left means that the shaded part should continue without end along the number line to the left.

Draw the number line graph of the solution set of the inequality $x > -2$.

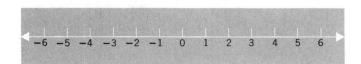

Check your drawing in **36**.

---

36 $x > -2$

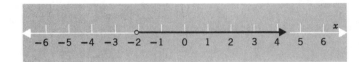

Any number to the right of -2 is greater than -2. Notice the open circle used to show that -2 itself is not a member of the solution set, and the shaded arrow covers all numbers in the solution set.

Checking the Solution of an Inequality

After an equation is solved, we should check the solution by substituting it back into the original equation to be certain it produces a true statement. Obviously we cannot check all of the numbers in the solution set of an inequality in this way, but the graph of the inequality can be checked by testing zero.

First, substitute 0 for x in the inequality: $0 > -2$.
Second, is this a true statement? *Yes, 0 is to the right of -2.*
Third, since 0 satisfies the inequality, it should lie on the shaded part of the number line. *It does.*

Solve the following inequalities by drawing a number line graph for each. Check your work by testing zero.

(a) $x < -1$ (b) $x > 3$ (c) $x < 1$

Check your work in **37**.

37   (a) $x < -1$

(b) $x > 3$

(c) $x < 1$

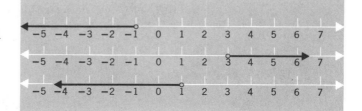

The inequality statement $x \leq 3$ (say "x is less than or equal to 3") is actually a combination of two statements: $x < 3$ or $x = 3$. The solution set of $x \leq 3$ includes the number 3 plus all numbers less than 3.

The filled-in circle at $x = 3$ tells us that 3 is a member of the solution set.

Draw the number line graph of the solution set of the inequality $x \geq -2$.

Check your drawing in **38**.

38 $x \geq -2$

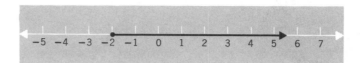

Solving an Inequality

To *solve* an inequality means to find the solution set of numbers that make it true. We solve an inequality such as $3x - 2 < x + 1$ by finding another equivalent inequality of the form $x < \square$ or $x > \square$. To help us in doing this we use addition and multiplication rules similar to (but not exactly the same as) those used in solving equations.

Addition–Subtraction
Rule for Inequalities:

> If any real number is added to or subtracted from both sides of an inequality, the result is an equivalent inequality.

If $x < y$ then $x + a < y + a$ and $x - b < y - b$

The inequality signs are the same.

A few examples from arithmetic will help you understand the rule.

Example 1

$5 < 9$

$5\ \boxed{+2}\ < 9\ \boxed{+2}$ Add $\boxed{2}$.

$7 < 11$, which is also true.

Example 2

$6 \leq 8$

$6\ \boxed{-2}\ \leq 8\ \boxed{-2}$ Subtract $\boxed{2}$.

$4 \leq 6$, which is also true.

This rule can be used to solve inequalities. For example, to solve $x - 3 < 4$, add $\boxed{3}$ to each member of the inequality:

$x - 3\ \boxed{+3}\ < 4\ \boxed{+3}$

$x < 7$

The solution set is the set of all real numbers less than 7.

To check the graph, substitute $x = 0$ into the original inequality: $0 - 3 < 4$ or $-3 < 4$. This inequality is true, -3 is to the left of 4 on the number line; therefore 0 satisfies the original inequality and should be on the shaded part of the number line.

For another check, if the inequality sign in the answer $x < 7$ is replaced with an equals sign to get $x = 7$, then substituting this number into the original inequality should produce an equality. $x - 3 < 4$ becomes $7 - 3 = 4$.

Solve the following inequality using the addition rule:

$3x + 2 < 2x + 1$

Check your work in **39**.

_____

39 $3x + 2 < 2x + 1$

$3x + 2\ \boxed{-2}\ < 2x + 1\ \boxed{-2}$ Subtract $\boxed{2}$.

$3x < 2x - 1$

$3x\ \boxed{-2x}\ < 2x - 1\ \boxed{-2x}$ Subtract $\boxed{2x}$.

$x < -1$ The solution set is the set of all real numbers less than -1.

Check: Substitute $x = -1$ into the equation formed from the original inequality

$3(-1) + 2 = 2(-1) + 1$
$-3 + 2 = -2 + 1$
$-1 = -1$

As a further check, substitute any number $x < -1$ and see if it satisfies the inequality.

For $x = -2$,

$3(-2) + 2 < 2(-2) + 1$

$-6 + 2 < -4 + 1$

$-4 < -3$

. . . which is true.

145   2-6 Inequalities

Try these problems.

(a) $x + 5 > 2$ (b) $-3 + 4x \geq 3x - 1$
(c) $x - 3 + 2x < 1 + 2x - 2$ (d) $2 + x \leq 10$
(e) $2(x - 1) > 3(2 + x)$

Check your solutions in **40**.

---

40   (a) $x + 5 > 2$

$x + 5 \;\boxed{-5}\; > 2 \;\boxed{-5}\;$ Subtract $\boxed{5}$.

$x > -3$ **Check:** $x = -3$ gives $(-3) + 5 = 2$

(b) $-3 + 4x \geq 3x - 1$

$-3 + 4x \;\boxed{+3}\; \geq 3x - 1 \;\boxed{+3}\;$ Add $\boxed{3}$.

$4x \geq 3x + 2$

$4x \;\boxed{-3x}\; \geq 3x + 2 \;\boxed{-3x}\;$ Subtract $\boxed{3x}$.

$x \geq 2$ **Check:** $x = 2$ gives $-3 + 4(2) = 3(2) - 1$
$5 = 5$

(c) $x - 3 + 2x < 1 + 2x - 2$
$3x - 3 < 2x - 1$ Collect like terms.

$3x - 3 \;\boxed{+3}\; < 2x - 1 \;\boxed{+3}\;$ Add $\boxed{3}$.

$3x < 2x + 2$

$3x \;\boxed{-2x}\; < 2x + 2 \;\boxed{-2x}\;$ Subtract $\boxed{2x}$.

$x < 2$ **Check:** $x = 2$ gives $(2) - 3 + 2(2) = 1 + 2(2) - 2$
$3 = 3$

(d) $2 + x \leq 10$

$2 + x \;\boxed{-2}\; \leq 10 \;\boxed{-2}\;$ Subtract $\boxed{2}$.

$x \leq 8$ **Check:** $x = 8$ gives $2 + (8) = 10$

(e) $2(x - 1) > 3(2 + x)$
$2x - 2 > 6 + 3x$ Multiply to remove parentheses.

$2x - 2 \;\boxed{-6}\; > 6 + 3x \;\boxed{-6}\;$ Subtract $\boxed{6}$.

$2x - 8 > 3x$

$2x - 8 \;\boxed{-2x}\; > 3x \;\boxed{-2x}\;$ Subtract $\boxed{2x}$.

$-8 > x$
or $x < -8$ **Check:** $x = -8$ gives $2((-8) - 1) = 3(2 + (-8))$
$2(-9) = 3(-6)$
$-18 = -18$

The second rule for solving inequalities has two separate parts, as shown below and on page 147.

Multiplication–Division
Rule for Inequalities:

> If both sides of an inequality are multiplied or divided by any *positive* real number, an equivalent inequality is produced.

If $x < y$, then $ax < ay$ and $\dfrac{x}{b} < \dfrac{y}{b}$.

Same sign if a and b are positive numbers, not equal to 0.

Example: $5 < 8$

$\boxed{2}\,(5) < \boxed{2}\,(8)$ Multiply by $\boxed{2}$.

$10 < 16$ Which is true.

Example: $6 < 10$

$\dfrac{6}{\boxed{2}} < \dfrac{10}{\boxed{2}}$ Divide by $\boxed{2}$.

$3 < 5$ Which is true.

If both sides of an inequality are multiplied or divided by any *negative* real number, and if the inequality sign is reversed, an equivalent inequality is produced.

If $x < y$, then $ax > ay$ and $\dfrac{x}{b} > \dfrac{y}{b}$.

Reverse sign if a and b are negative numbers, not equal to 0.

Example: $2 < 5$

$\boxed{-3}\,(2) > \boxed{-3}\,(5)$ Multiply by $\boxed{-3}$ and reverse the inequality sign from $<$ to $>$.

$-6 > -15$ True: -6 is to the right of -15 on the number line.

Example: $-6 < 9$

$\dfrac{-6}{\boxed{-3}} > \dfrac{9}{\boxed{-3}}$ Divide by $\boxed{-3}$ and reverse the inequality sign from $<$ to $>$.

$2 > -3$ True: 2 is to the right of -3 on the number line.

Apply this rule to solve the following inequality:

$2x < 6$

Check your work in **41**.

---

41 $2x < 6$

$\dfrac{2x}{2} < \dfrac{6}{2}$ Divide by 2. Since we have divided by a positive number, the inequality sign does not change.

$x < 3$

147   **2-6 Inequalities**

Check: $x = 3$ gives $2(3) = 6$

Now try this problem. Solve: $-3x > 12$.

Remember, you want to get an equivalent inequality of the form $x < \square$ or $x > \square$. Check your work in **42**.

---

42 $-3x > 12$

$\dfrac{-3x}{-3} < \dfrac{12}{-3}$ Divide by -3 and reverse the inequality sign from $>$ to $<$.

$x < -4$ **Check:** $x = -4$ gives $-3(-4) = 12$
$\qquad\qquad\qquad\qquad\qquad\qquad 12 = 12.$

Always check your work.

Multiplying or dividing both members of an inequality by zero is not allowed.

Use these rules to solve the following inequalities.

(a) $2(a + 1) - 3 < 5$

(b) $\dfrac{-x}{2} < 4$

(c) $x - 4 + 3x > 2 + 7x + 3$

(d) $3x \geq 7x - 1$

Check your work in **43**.

---

43 (a) $2(a + 1) - 3 < 5$
$\qquad\quad\ 2a + 2 - 3 < 5$ Multiply to remove parentheses.
$\qquad\quad\ 2a - 1 < 5$ Collect like terms.

$\qquad\quad\ 2a - 1 \;\boxed{+\,1}\; < 5 \;\boxed{+\,1}$ Add $\boxed{1}$.

$\qquad\quad\ 2a < 6$

$\qquad\quad\ \dfrac{2a}{\boxed{2}} < \dfrac{6}{\boxed{2}}$ Divide by $\boxed{2}$.

$\qquad\quad\ a < 3$ **Check:** $a = 3$ gives $2((3) + 1) - 3 = 5$
$\qquad\qquad\qquad\qquad\qquad\qquad\qquad\qquad 2(4) - 3 = 5$
$\qquad\qquad\qquad\qquad\qquad\qquad\qquad\qquad 8 - 3 = 5$

(b) $-\dfrac{x}{2} < 4$

$\qquad \boxed{-2}\left(-\dfrac{x}{2}\right) > \boxed{-2}\,(4)$ Multiply by $\boxed{-2}$ and reverse the inequality sign from $<$ to $>$.

$\qquad x > -8$ **Check:** $x = -8$ gives $-\dfrac{(-8)}{2} = 4$
$\qquad\qquad\qquad\qquad\qquad\qquad\qquad\qquad\qquad\ 4 = 4$

(c) $x - 4 + 3x > 2 + 7x + 3$
$\qquad\ 4x - 4 > 5 + 7x$ Collect like terms.

$\qquad\ 4x - 4 \;\boxed{+\,4}\; > 5 + 7x \;\boxed{+\,4}$ Add $\boxed{4}$.

$\qquad\ 4x > 9 + 7x$

$\qquad\ 4x \;\boxed{-\,7x}\; > 9 + 7x \;\boxed{-\,7x}$ Subtract $\boxed{7x}$.

$\qquad\ -3x > 9$

$\qquad\ \dfrac{-3x}{\boxed{-3}} < \dfrac{9}{\boxed{-3}}$ Divide by $\boxed{-3}$ and reverse the inequality sign from $>$ to $<$.

$$x < -3 \qquad \textbf{Check:} \quad x = -3 \text{ gives } (-3) - 4 + 3(-3) = 2 + 7(-3) + 3$$
$$-3 - 4 - 9 = 2 - 21 + 3$$
$$-16 = -16$$

(d) $\quad 3x \geq 7x - 1$

$$3x \boxed{-7x} \geq 7x - 1 \boxed{-7x} \qquad \text{Subtract } \boxed{7x}.$$

$$-4x \geq -1$$

$$\frac{-4x}{\boxed{-4}} \leq \frac{-1}{\boxed{-4}} \qquad \text{Divide by } \boxed{-4} \text{ and reverse the inequality sign.}$$

$$x \leq \frac{1}{4} \qquad \textbf{Check:} \quad x = \frac{1}{4} \text{ gives } 3\left(\frac{1}{4}\right) = 7\left(\frac{1}{4}\right) - 1$$
$$\frac{3}{4} = \frac{7}{4} - 1$$

Compound Inequality

A *compound inequality* is a statement that several inequalities are true simultaneously. The compound inequality $2 < x < 10$ is equivalent to the pair of inequalities $x < 10$ and $x > 2$.

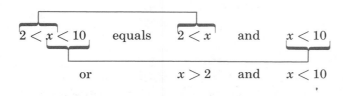

Notice that $2 < x$ is equivalent to $x > 2$.

The solution set of a compound inequality is the single set of numbers for which *both* of its parts are true. The solution set of $2 < x < 10$ is the set of numbers for which both $x < 10$ and $x > 2$ are true.

The graph of the solution set of $x < 10$ is:

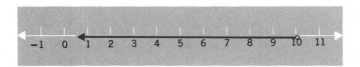

The graph of the solution set of $x > 2$ is:

Combining these gives:

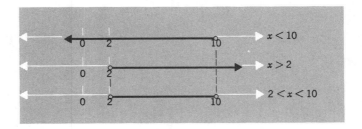

The last graph shows the solution set of the compound inequality $2 < x < 10$. Only numbers in the shaded area satisfy both inequalities. Only these numbers are *both* less

149   2-6 Inequalities

than 10 and greater than 2. Any number outside the shaded area will not satisfy *both* inequalities.

Another example: Draw the graph of the solution set of the compound inequality $-5 < x \leq -1$.

This is equivalent to the two inequalities $x \leq -1$ and $x > -5$.

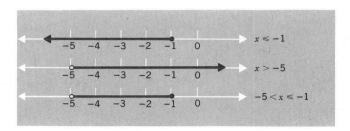

Draw the graph of the solution set of the compound inequality

$-2 < x \leq 4$

Compare your graph with ours in **44**.

---

44 $-2 < x \leq 4$ is equivalent to the two inequalities $x \leq 4$ and $x > -2$.

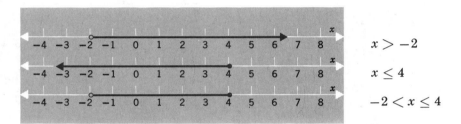

The final graph shows the set of numbers that satisfy both inequalities at the same time.

The third graph is simply a combination of the first two. Both inequalities will be satisfied by the points from the region where the two shaded areas overlap.

For a set of practice problems on solving inequalities, turn to **45**.

Unit 2
Solving
Algebraic Equations

45   The answers are on page 544.

Inequalities

A. Solve and graph each of the following inequalities.

1. $x + 3 < 18$ _____

2. $2 + x > -2$ _____

3. $a - 3 > -4$ _____

4. $2 + d < 4$ _____

5. $y - 1 \leq -1$ _____

6. $8 - x \geq -2$ _____

7. $3 < x - 1$ _____

8. $-5 > 2 - y$ _____

9. $-1 \geq a$ _____

10. $-3 \leq x$ _____

11. $5 - Z > -2$ _____

12. $Q - 3 < -4$ _____

13. $11 < 7 - x$ _____

14. $-6 > P - 3$ _____

15. $-9 \geq t + 5$ _____

16. $-7 \leq -5 - x$ _____

B. Solve each of the following inequalities.

1. $2x - 4 < 7$ _____

2. $3a - 5 > 6$ _____

3. $1 - 2a > -3$ _____

4. $2 - 3x < 4$ _____

5. $-\dfrac{x}{3} > 2$ _____

6. $5 < \dfrac{t}{4}$ _____

7. $6 \leq 2y - 1$ _____

8. $9 \geq 2 - 3a$ _____

9. $4E < -3$ _____

10. $3K > -5$ _____

11. $\dfrac{1}{2}G > \dfrac{3}{2}$ _____

12. $\dfrac{2}{3}m < -1\dfrac{1}{3}$ _____

13. $X + 2 - 3X \geq 1 - 4X + 7$ _____

14. $1 - x - 5 \leq 3x + 2$ _____

15. $-2x > 7$ _____

16. $7y + 2 \leq 2y - 8$ _____

17. $11m + 3 - 2m \leq 7 + m$ _____

18. $-5a < -9$ _____

19. $3(x - 1) < x - 5$ _____

20. $2(3 - Z) + 1 > Z - 2$ _____

21. $2(3X - 5) < 3(2 - X) - 2$ _____

22. $1 - (4Y - 1) > 2(Y - 5)$ _____

23. $5 - 2(3 - X) > 2(2X + 5) + 1$ _____

24. $Q - 3(2Q + 1) < 5(1 + Q) + 2$ _____

25. $-\dfrac{2a}{3} \geq 10$ _____

26. $\dfrac{-3m}{4} \leq 9$ _____

27. $2 - \dfrac{x}{3} < \dfrac{x + 1}{4}$ _____

28. $1 - \dfrac{2a}{3} > \dfrac{1 - a}{2}$ _____

29. $G - \dfrac{1}{4} > 1\dfrac{1}{2}$ _____

30. $\dfrac{1}{2} + p < 2\dfrac{1}{4}$ _____

C. Word Problems

1. One-fourth of a number added to two-thirds of the number is less than or equal to 22. What is the largest number that will satisfy this condition?

   _____

2. The length of a certain rectangle is 6 ft. If the perimeter must be smaller than 17 ft, how large may the width be?

   _____

3. The difference of two numbers is less than 7, and the larger number is three times the smaller. What are the largest possible values for the two numbers?

   _____

4. The sum of the lengths of the sides of a certain triangle is less than 20 inches. The longest side is 2 inches more than double the base and the shortest side is 4 inches long. What are the largest possible dimensions of the triangle?

   _____

5. Professor Hall gives four exams in her algebra course. Her star student, Jane Edge, has scored 88, 85, and 92 on the first three. What must she score on the last exam to average 90 or better?

   _____

6. What numbers satisfy the condition, "Five more than twice the number is greater than or equal to four minus the number."

   _____

7. Three examinations are given in a history course and a student scores 60 and 75 on the first two. Is it possible for this student to average 80 for the three exams if the maximum possible score is 100?

   _____

8. Four times Eric's age, minus three years, is still less than his father's age. If his father is 45 years old, what is the maximum Eric's age could be?

   _____

When you have had the practice you need, either return to the preview on page 93 or turn to **46** where you will learn how to solve algebra word problems.

Date
_____

Name
_____

Course/Section
_____

2-7 WORD PROBLEMS IN ALGEBRA

46 Being able to solve equations does not necessarily mean that you can solve any real problems involving algebra. Many students find word problems or story problems to be the most difficult part of algebra. He reads the problem until he finds a number and then he begins to worry about that number. If he finds a second number, he combines them by adding, multiplying, dividing, or whatever seems to have been happening in the part of the textbook he read last. This random approach to doing problems is not very helpful, but most mathematics textbooks never show the student how to solve word problems. Reading explanations and following examples is not enough. To learn how to do word problems, you must do problems, lots of problems, and you should do them according to some logical plan or strategy. In this section, we will show you a plan that has helped thousands of students.

Our method of solving word problems is called the PQRST method. (Don't bother trying to pronounce it!) The letters stand for

P—Preview

Q—Question

R—Rewrite

S—Solve

T—Test

These letters stand for the five steps you should use in solving any mathematics word problem and they should serve as a reminder of what you are to do. This helpful guide will not guarantee a correct answer to every problem, but it will help you set up the problem, step-by-step, so that you understand it, can solve it correctly, and are more likely to get the correct answer.

Turn to **47** where we will carefully go over each step in PQRST.

Step 1. P—Preview

47 The first letter of PQRST reminds you to *preview* the problem—to read it in an effort to get the big picture.

Reading mathematics is not at all like reading a mystery story or a comedy novel. The ideas in mathematics writing are more concentrated; every word and symbol may be important. You need to read slower. In fact, you should expect to find it necessary to read the material several times—never less than twice.

Your purpose in this first reading of the problem is to learn what the problem is about. What is the general idea of the problem? What kinds of variables appear? What is known and what is unknown? What kind of facts are included that will help you solve the problem? What situation is the problem based on? Later in this section of Unit 2 we will review the main types of word problems and when you first read the problem you'll be trying to identify the problem type.

To read for "the big picture" means that at this first preview step you do not worry about the numbers. For example, you would read "John had $20" as "John had some money" or "The first car travels $\frac{3}{5}$ as fast as the second car" becomes "The first car is traveling slower than the second." Don't worry about the numbers; you're going to read it again—*after* you know what kind of problem you have to work with and after you have a better idea of what must be done to solve it.

A second purpose of this first preview reading is to help you spot any words that you do not recognize or whose meaning you are unsure of.

Circle unfamiliar words or phrases and find out what they mean before you continue. Consult a dictionary or your textbook, ask your instructor or tutor, ask a friend, but do not continue work on the mathematics until you know what all the words and symbols mean.

Be alert for words with special meaning in mathematics or words that can have several meanings. The letter a is a letter of the alphabet and a word, but it also can be a variable, or the name of a set of numbers or a coefficient.

 On this first step you preview the problem; you do no mathematics, no arithmetic, write nothing down, and set up no equations. This is the time you use to see where you are going and what the problem is all about. It is very valuable time.

Preview each of the following problems by (1) reading them, (2) writing down what they are about, and (3) circling any unfamiliar words.

(a) Find three consecutive odd integers whose sum is 33.
(b) An airplane flew for 3 hours with a ground speed of 200 mph and then picked up a tail wind of 100 mph. What was its average speed for a 4-hour trip?
(c) A projectile has an initial velocity of 400 ft/sec and accelerates linearly at 30 ft/sec^2 for 10 sec. What will be its final velocity?
(d) A certain alloy is 3 parts tin, 4 parts lead, and 2 parts antimony. How much tin is needed for a 16-ounce centrifugal casting of this alloy?
(e) Acme Gold Mining's 10-year corporate bonds sell for $98 each and have a current yield at par of 8.4%. What would be your yearly dividend on an investment of $1000?

Check your answers in **48**.

---

48 (a) This problem involves adding odd integers. Unfamiliar words circled: consecutive.
(b) This is a motion problem, involving speed, distances, and times. Unfamiliar words circled: ground speed and tail wind.
(c) A motion problem including a change in speed. Unfamiliar words circled: projectile, initial velocity, accelerates, and linearly.
(d) In this problem different amounts of metals are mixed. Unfamiliar words circled: alloy, antimony, and centrifugal.
(e) This one involves calculating the interest earned. Unfamiliar words circled: yield, par, and dividend.

Step 2. Q—Question

The second letter of PQRST reminds you to *identify the question* you are to answer. What is the question you are to answer? What is it you are to find? What will the correct answer look like?

Usually the key question you are to answer appears at the end of the problem in a sentence that begins "What is _____?" or "Find _____." or "How many _____?"

In many problems the key question you are to answer is not obvious or direct. You may need to write it out yourself. Finding the question is important. If you don't know where you're going, you probably won't get there.

Estimating the Answer

A second part of the Q step is to *estimate the answer*. Once you have identified the question, you should make an educated guess at the size and nature of the answer. In particular, you should be able to recognize the replacement set from which your answer will be taken. Usually the answer will be any number from the set of real numbers, but it may be restricted to the integers or the natural numbers.

When you have completed this second step of the problem solving process, you still have done no mathematics or arithmetic calculation. You will have underlined a question in the problem statement and perhaps have circled a vocabulary word, and you will have written an estimate of the size and nature of the answer.

Apply step 2 to the five problems at the end of **47**, on page 154. Underline the question part of each problem and estimate the answer. Check your work in **49**.

---

49   (a)   Question: "Find three consecutive odd integers."
             Answer: The answer will be a set of three odd integers.
                             The integers will be about 11 (roughly $33 \div 3$).
      (b)   Question: "What was the average speed . . . ?"
             Answer: The speed is between 200 and 300 mph.
      (c)   Question: "What will be the final velocity?"
             Answer: The answer will be a speed greater than 400 ft/sec.
      (d)   Question: "How much tin is needed . . . ?"
             Answer: The answer will be a weight in ounces, roughly 5 oz.
      (e)   Question: "What would be your yearly dividend . . . ?"
             Answer: The answer is an amount of money, roughly 8% of $1000 or about $80.

Step 3.   R—Rewrite

The third letter in PQRST tells you to *rewrite* or *translate* the problem into the language of algebra. This process of translating from English to algebra was discussed in Unit 1, beginning on page 59. Return there if you need a review.

This step involves reading the problem slowly and translating it bit by bit, phrase by phrase. Make the algebra look like the English. Normally the result will be several equations for each problem.

Write out your translation on a separate sheet of paper. Do *not* attempt to do the translation in your head. There is less chance for error if you write it out, as we showed in Unit 1. Work carefully, but do not worry about appearance. This translation will not be turned in to your instructor.

Choose letters for the variables that remind you of the physical meaning of the variable—A for area, d for distance, i for interest, and so on.

Do not solve any equations or do any arithmetic or algebra at this step.

Practice step 3 by rewriting or translating the following problem to algebra equations. "The length of a rectangle is four more than three times the width. Find the width if the length is 13 inches."

Check your answer in **50**.

---

50 $L = 4 + 3W$
$W = ?$
$L = 13$

Step 4.   S—Solve It

The fourth letter in PQRST tells you to use the algebraic skills you have to solve the equations you have written.

First, plan how you will go about solving the problem. Take some time to plan what you will do. Draw a diagram to guide you in your problem solving. We'll show you some diagrams that are useful for solving certain problems when we review the different kinds of problems later in this unit. Use diagrams if they help your algebraic thinking.

Second, do the algebra before you do any arithmetic. It is usually easier to work with letters than to multiply or divide numbers. Many instructors and textbook writers construct problems so that there is much less arithmetic to do if you delay doing any calculations until the very end.

Third, do the arithmetic carefully. It is a tragedy to do every step along the way perfectly and then miss the answer because you made a mistake multiplying 7 times 8.

Turn to **51** for the final step in the PQRST process.

Step 5.   T—Test It

51   The final letter in PQRST reminds you to *test* or *check* your answer.

A very important part of solving every problem is deciding if the answer you find fits the problem. Does the answer make sense? In this book we have showed how to check your arithmetic and algebraic calculations. Use these checks.

Compare your answer with the estimate you made in the Q step. Do the answer and estimate agree reasonably well? If not, you may have made an error.

Return to the original question. Does your result actually answer this question? If the question calls for an answer in dollars, does your result fit? If it calls for an area, do you have an area? If your result is not a reasonable answer to the original question, retrace each step to determine what went wrong.

Here are three answers for each of the questions in **47** on page 154. Decide which answer is best by seeing which fits the question best.

(a)   1.   10, 12, 14   2.   9, 11, 13   3.   9, 10, 11
(b)   1.   225 miles   2.   2.25 hr   3.   225 mph
(c)   1.   700 ft/sec   2.   100 ft/sec   3.   700 ft
(d)   1.   5 parts   2.   5.3 oz   3.   5.3 lb
(e)   1.   840   2.   84   3. $84

Pick an answer for each question and check your answers in **52**.

52   (a)   Answer 2 is best.
    Answer 1 is a set of *even* integers and
    Answer 3 is not a set of *consecutive odd* integers.
(b)   Answer 3 is the only possible correct answer because it is the only answer expressed as a speed.
(c)   Answer 1 must be the best.
    Answer 2 is too small, and
    Answer 3 is not a speed.
(d)   Answer 2 is the only answer expressed in the correct units.
(e)   Answer 3 is the only answer in money units.

The PQRST method is not an automatic problem-solving machine, but it will help you to improve your ability to solve word problems if you use it. Once you get better at doing word problems, you may find them interesting, even fun. Best of all, you will be developing the ability to think analytically and to use the language and operations of algebra as a mental tool, and that is one of the most important reasons for studying algebra in the first place.

Now we are going to examine some of the different types of algebraic word problems. Turn to **53** for the first of these.

Integer Problems

53 The set of integers was defined in Unit 1 as including the positive whole numbers 1, 2, 3, and so on, their negative counterparts -1, -2, -3, and so on, and zero. The follower of any integer n is the integer obtained by adding 1 to n. The follower of 2 is 3. The follower of -4 is $(-4 + 1)$ or -3. The follower of the integer n is the integer *Consecutive Integers* $n + 1$. *Consecutive* integers are any integers in the set that occupy neighboring positions on the number line.

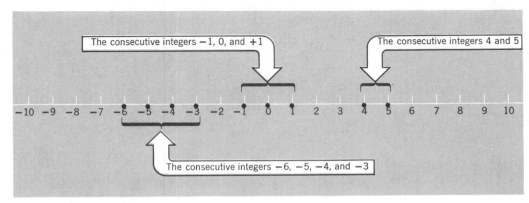

Any integer and its follower are consecutive integers. The integers n, $n + 1$, and $n + 2$ are consecutive integers. The integers n, $n - 1$, and $n - 2$ are also consecutive integers.

Complete each of the following statements.

(a) The follower of -2 is _____.
(b) Four consecutive integers, increasing in size, beginning with -2 are _____.

(c) If x is an integer, then the next smaller consecutive integer is _____.
(d) If the number $3a + 2$ is an integer, then the next two consecutive integers are _____.

Check your answers in **54**.

54 (a) -1 $(-2 + 1 = -1)$ (b) $-2, -1, 0, 1$
(c) $x - 1$ (d) $3a + 3$ and $3a + 4$

Even Integers

An *even* integer is a positive or negative whole number which is evenly divisible by 2. The even integers are

$2, 4, 6, 8, \ldots$ and $-2, -4, -6, -8, \ldots$ and 0

The even integers are multiples of two.

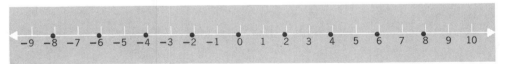

If n is an even integer, then $n + 2$, $n + 4$, $n + 6$, ... and $n - 2$, $n - 4$, $n - 6$, ... are also even integers.

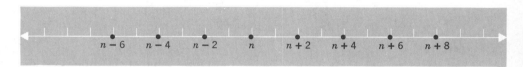

Odd Integers

An *odd* integer is a positive or negative whole number that gives a remainder of 1 when it is divided by two. The odd integers are

$$1, 3, 5, 7, \ldots \qquad \text{and} \qquad -1, -3, -5, -7, \ldots$$

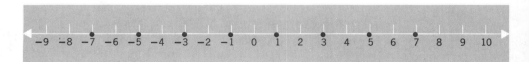

If p is an odd integer, then $p + 2$, $p + 4$, $p + 6$, ... and $p - 2$, $p - 4$, $p - 6$, ... are also odd integers.

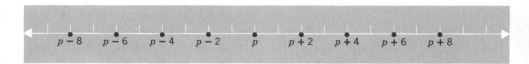

Answer these questions.

(a) If x is an even integer write the next three larger even integers. _____

(b) If y is an odd integer write the next two smaller odd integers. _____

(c) If t is an even integer, write an algebraic expression for the sum of t and the next two larger even integers. _____

Check your answers in **55**.

---

55 (a) $x + 2$, $x + 4$, $x + 6$
(b) $y - 2$, $y - 4$
(c) $t + (t + 2) + (t + 4)$ or $3t + 6$

Problems involving integers are usually easy to solve if you use the PQRST system and if you know how to write algebraic expressions involving integers as we have shown above.

For example, try this problem: Find three consecutive integers whose sum is 78.

Try it and then look in **56** for our step-by-step solution.

---

56 Use the PQRST system.

P—*Preview it*. Read the problem carefully. No unfamiliar words here.
Q—*Question*. What will your answer look like?
          The answer will be a set of three integers somewhere between 20 and 30.
R—*Read* again and translate to an algebraic equation.

If the first integer is n, then the three consecutive integers are

$$n, \qquad (n + 1), \qquad \text{and} \qquad (n + 2)$$

The sum of the three integers is 78.

$$n + (n + 1) + (n + 2) = 78$$

S—*Solve* the equation.
First, collect like terms.

$$n + n + n + 1 + 2 = 78$$
$$3n + 3 = 78$$

Second, solve using the addition–subtraction and multiplication–division rules.

$$3n + 3 - 3 = 78 - 3$$
$$3n = 75$$
$$n = 25$$

Third, write out the solution: $n = 25$, $n + 1 = 26$, $n + 2 = 27$

The three consecutive integers are 25, 26, and 27.

T—*Test* the answer: check it.

$25 + 26 + 27 = 78$,   which is correct.

Now you try a similar problem. Find four consecutive integers whose sum is -82.

Check your work in **57**.

---

57 If x is an integer then $(x + 1)$, $(x + 2)$, and $(x + 3)$ are consecutive integers. The sum is $x + (x + 1) + (x + 2) + (x + 3)$.

The equation to be solved is

$$x + (x + 1) + (x + 2) + (x + 3) = -82$$

Combine terms:

$$4x + 6 = -82$$

Solve:

$$4x + 6 - 6 = -82 - 6$$
$$4x = -88$$
$$x = -22$$

The four integers are $x = -22$
$$x + 1 = -22 + 1 = -21$$
$$x + 2 = -22 + 2 = -20$$
$$x + 3 = -22 + 3 = -19$$

Check: $-22 -21 -20 -19 = -82$ which is correct.

Ready for a few more difficult problems? Try these.

(a) Find three consecutive even integers whose sum is 18.
(b) Find three consecutive integers such that the sum of the smallest and largest is 20.
(c) The larger of three consecutive integers is three more than twice the middle integer. Find the three integers.

Work carefully. Use the PQRST system to help organize your work. The step-by-step solutions are in **58**.

---

58 (a) If the first even integer is n, then the three consecutive even integers are n, $n + 2$, and $n + 4$.

Their sum is

$$n + (n + 2) + (n + 4) = 18$$

Collect like terms:

Solve: $n + n + n + 2 + 4 = 18$
$3n + 6 = 18$
$3n = 12$
$n = 4$

The three integers are $n = 4$, $n + 2 = 6$, $n + 4 = 8$.

(b) If the smallest integer is x, then the three consecutive integers are x, $x + 1$, and $x + 2$.

The sum of the smallest and largest is $x + (x + 2) = 20$.
Collect like terms:

$x + x + 2 = 20$
$2x + 2 = 20$

Solve: $2x = 18$
$x = 9$

The three integers are $x = 9$, $x + 1 = 10$, and $x + 2 = 11$.

(c) If the three consecutive integers are x, $x + 1$, and $x + 2$, the problem can be translated as

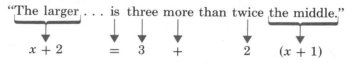

"The larger . . . is three more than twice the middle."

$$x + 2 \qquad = \quad 3 \quad + \qquad 2 \quad (x + 1)$$

or

$$x + 2 = 3 + 2(x + 1)$$

Remove the parentheses: $x + 2 = 3 + 2x + 2$

Collect like terms: $x + 2 = 3 + 2 + 2x$
$x + 2 = 5 + 2x$

Solve: $x + 2 - 5 = 5 + 2x - 5$
$x - 3 = 2x$
$x - 3 - x = 2x - x$
$-3 = x$
or $x = -3$

The three integers are $x = -3$
$x + 1 = -3 + 1 = -2$
$x + 2 = -3 + 2 = -1$

Remember, in each problem check your answer by substituting it back into the original problem.

Now try the following problems involving integers.

1. Find three consecutive integers such that the sum of the largest two is equal to one less than the smallest.
2. Find four consecutive integers where the sum of the second and fourth is 28.
3. Find three consecutive odd integers whose sum is 27.
4. Which three consecutive even integers have a sum equal to five times the middle integer?

5. Show that for any five consecutive integers the sum of the first and fifth equals twice the third integer.
6. Find two consecutive even integers such that four times the larger minus twice the smaller equals twelve.
7. Find the two largest consecutive odd integers whose sum is less than 40.
8. Find the three smallest consecutive even integers whose sum is greater than 15.
9. Find three consecutive integers such that five times the first minus twice the second plus three times the third is equal to 28.
10. Find three consecutive even integers such that the sum of the smallest and largest equals the middle integer.

The correct answers are in **59**. Check your work there.

59
1. $-4, -3, -2$	2. $12, 13, 14, 15$
3. $7, 9, 11$	4. $-2, 0, +2$
5. $n + (n + 4) = 2(n + 2)$	6. $2, 4$
7. $n < 19$; 17 and 19	8. $n > 3$; 4, 6, and 8
9. $4, 5, 6$	10. $-2, 0, 2$

Motion Problems

The motion of any real object is difficult to describe completely. A bicycle being pedaled along the street will be moving forward, slowing for traffic, accelerating, swerving from side to side, bouncing up and down on the rough spots, and all the while its wheels and pedals will be rotating at different rates. Rather than try to describe this kind of complicated set of movements, the scientist or engineer interested in motion usually simplifies the situation by ignoring everything but the simple, straightforward, average motion.

The most obvious aspect of the motion of the bicycle is that as it moves from one point to another, it travels some measurable distance in some measurable time. If the object moves a distance d in time t, then its average speed s during that time is defined as

Average Speed

$$s = \frac{d}{t} \qquad \text{Average speed} = \frac{\text{distance traveled}}{\text{time of travel}}$$

or, rearranging this equation,

$d = st$ distance traveled = average speed × time of travel

If a baseball thrown by an outfielder travels the 330 ft from right field to home plate in 3 seconds, what is its average speed?

Use the formulas above to determine the answer and then check your work in **60**.

---

60 $s = \dfrac{d}{t}$

$s = \dfrac{330 \text{ ft}}{3 \text{ sec}} = 110 \text{ ft/sec}$

Notice that we have included units with our answer—ft/sec. In an algebra course you will naturally be most interested in setting up problems into algebra, solving equations, and doing any necessary arithmetic. In science courses, however, you will be expected to have the correct units attached to your answer.

Distance will have distance units—inches, feet, yards, miles, or centimeters, meters, or kilometers in the metric system. Time will have units of seconds, minutes, or hours. The speed units used will depend on the distance and time units. If the distance is in miles and time in hours, speed will be given in miles per hour, mi/hr, or mph. If the distance is in feet and time in seconds, speed will be given in feet per second or ft/sec.

When you are translating motion problems into algebraic equations, it will usually help if you use a diagram. For example, the baseball problem is diagramed like this:

d–s–t Diagrams

	d	s	t
Ball	330 ft	?	3 sec

$d = st$

$330 \text{ ft} = s \cdot 3 \text{ sec}$

$s = \dfrac{330 \text{ ft}}{3 \text{ sec}} = 110 \text{ ft/sec}$

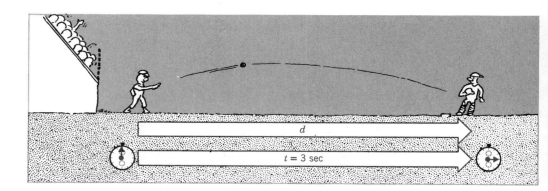

Notice that we set up the box with the spaces labeled d, s, and t from left to right in order to remind you that $d = st$.

Try this motion problem:

An automobile leaves the city at exactly 9:00 A.M. and travels directly south at 50 mph. A bus leaves the same point in the city at 10:00 A.M. and travels north at 30 mph. When will the two vehicles be exactly 130 miles apart?

To help solve this problem, construct two d–s–t boxes, one for each vehicle. Try it, then turn to **61** for a step-by-step solution of the problem.

61

	d	s	t
Car	$d_1 = ?$	50 mph	t
Bus	$d_2 = ?$	30 mph	$t - 1$

$d = st$

$d_1 = 50t$

$d_2 = 30(t - 1)$

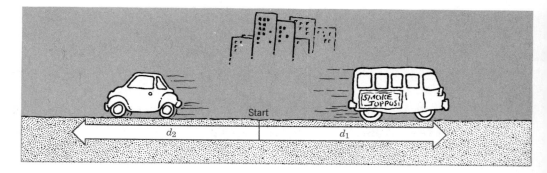

Total distance $= d_1 + d_2$

$50t + 30(t - 1) = 130$ miles

Solve: $50t + 30t - 30 = 130$
$$80t = 160$$
$$t = 2 \text{ hours}$$

Notice the following:

1. Because the speed is in miles per hour, the time must be in hours and the distance in miles.
2. Because the bus starts one hour later than the car, its time of travel is 1 hour less than the time for the car. If t is the time of travel for the car, then $t - 1$ is the time of travel for the bus.
3. Using the PQRST method you should have first estimated the answer: the car would travel 130 miles by itself in less than 3 hours, so the answer will be less than 3 hours.
4. Check your answer. The car travels

$$d_1 = 50 \cdot 2 = 100 \text{ miles}$$

and the bus travels

$$d_2 = 30 \cdot 1 = 30 \text{ miles}$$

for a total of 130 miles.

Putting all of the given information into a nice neat diagram should help you translate the word problem into an algebraic equation. The more difficult the problem, the more valuable the d–s–t boxes will be.

The little 1 in d_1 and the 2 in d_2 are called *subscripts*. Read these as "dee-sub-one" and "dee-sub-two."

To use the d–s–t diagram to set up a problem, follow these steps.

 SUMMARY

Step 1. Write in the numerical value of any quantity given—distance traveled, speed, or time of travel.

Step 2. Write in letters for any space in the d–s–t diagram for which a numerical value is not given.

Step 3. Write down the equation $d = st$ for each row in the d–s–t diagram.

Step 4. Set up an equation by combining the information from each row of the diagram with any additional information in the problem.

Ready for another motion problem? Try this one.

Bob and Eric decide to run a race. Eric's normal running pace is 6 mph and because Bob is faster, Eric gets a 15-minute head start. They finish in a tie, arriving at the end after Eric has run for 45 minutes. How fast did Bob run?

Check your work in **62**.

62

	d	s	t	$d = st$
Eric	$d_1 = ?$	6 mph	$\dfrac{3}{4}$ hr	$d_1 = 6\left(\dfrac{3}{4}\right) = \dfrac{9}{2}$
Bob	$d_2 = ?$	x	$\dfrac{3}{4} - \dfrac{1}{4}$ hr	$d_2 = x\left(\dfrac{3}{4} - \dfrac{1}{4}\right) = \dfrac{1}{2}x$

Eric START FINISH

d_1

Bob

d_2

Since both run the same total distance, $d_1 = d_2$, and the equation to be solved is

$$\frac{9}{2} = \frac{1}{2}x \qquad \text{or} \qquad x = 9 \text{ mph}$$

Notice that we converted the times from minutes to hours to agree with the units for speed: miles per hour.

Be certain to check your answer by substituting it back into the original problem.

$$d_1 = 4\frac{1}{2} \text{ miles}$$

$$d_2 = \frac{1}{2} \cdot 9 = 4\frac{1}{2} \text{ miles}$$

Solve the following motion problems by constructing a table of distances, speeds, and times for each problem. Work carefully.

1. George is driving along the highway at a steady 40 mph. He is passed by a speeding sports car traveling 60 mph in the same direction. In how many hours will the two cars be 50 miles apart?

2. Patty drove from Santa Barbara to Shell Beach at a leisurely 40 mph and later hurried back to Santa Barbara at 60 mph. The total driving time for the round trip was 5 hr. How far apart are the two cities?

3. Elmer drove from his home to Las Vegas in 5 hrs, averaging a steady 50 mph. His friend Bill started later and made the same trip in only 4 hours. How fast did Bill drive?

4. The sheriff and his posse were chasing Big Bad Bruce. Bruce had a 30-minute head start and felt safe because he knew that his horse and the sheriff's horse traveled at the same speed. The sheriff, having watched a lot of cowboy movies, took a 5-mile shortcut through the pass and caught him. How fast did Bruce's horse move?

Our complete solutions are in **63**.

---

63   1.

	d	s	t	$d = st$
George	d_1	40	t	$d_1 = 40t$
Sports car	d_2	60	t	$d_2 = 60t$

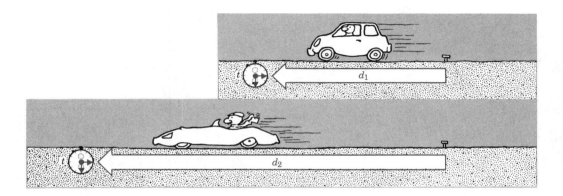

$$d_2 - d_1 = 50 \text{ miles}$$
$$60t - 40t = 50$$
$$20t = 50$$
$$t = 2\frac{1}{2} \text{ hours}$$

Note: The cars travel the same time t from when they pass one another until they are exactly 50 miles apart.

2.

	d	s	t	
1st half of trip	d	40	t	$d = 40t$
2nd half of trip	d	60	$5 - t$	$d = 60(5 - t)$

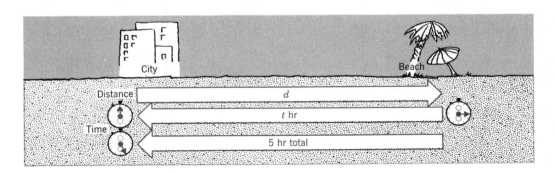

$$40t = 60(5 - t)$$
$$40t = 300 - 60t$$
$$100t = 300$$
$$t = 3 \text{ hr}$$
$$d = 40t = 40(3) = 120 \text{ miles}$$

Note: (a) The distances on each half of the trip are the same.
(b) If the first half of the trip takes t hours, and the entire trip takes 5 hours, then the second half of the trip took $5 - t$ hours.
(c) The problem asks you to find the distance, not the time. After you find t, substitute it back into the d-s-t diagram to find d.

3.

	d	s	t	
				$d = st$
Elmer	d	50	5	$d = 50 \cdot 5 = 250 \text{ miles}$
Bill	d	s	4	$d = 4s$

$$4s = 250 \text{ miles}$$
$$s = 62.5 \text{ mph}$$

4.

	d	s	t
Sheriff	d_1	s	$t - \dfrac{1}{2}$
Bruce	d_2	s	t

$d = st$

$d_1 = s\left(t - \dfrac{1}{2}\right)$

$d_2 = st$

$$d_1 = d_2 - 5$$
$$s\left(t - \frac{1}{2}\right) = st - 5$$
$$st - \frac{1}{2}s = st - 5$$
$$-\frac{1}{2}s = -5$$
$$\frac{1}{2}s = 5$$
$$s = 10 \text{ mph}$$

Note: (a) If Bruce's time is t hours, then the sheriff starting $\frac{1}{2}$ hour later rides for $t - \frac{1}{2}$ hour.

(b) The sheriff takes a 5-mile shortcut. If Bruce rides d_2 miles, then the sheriff rides $d_2 - 5$ miles.

(c) Notice that we convert the time given, 30 minutes, to hours. All units here are miles, hours, and miles per hour.

The more problems you do, the easier they get—we hope! Try this set of motion problems.

1. On a two-day vacation trip, Fred drove 500 miles in a total of 12 hours. On the first day, driving over level roads, he averaged 45 mph. On the second day, driving through the mountains, he averaged 35 mph. How far did he travel each day?

2. Suppose you run at 8 mph for 1 hour and then walk at 4 mph for 3 hours. What is your average speed?

3. Two cars pass one another going in opposite directions. The first car is traveling at 50 mph and the second is traveling at 30 mph. How far apart will they be exactly 90 minutes after passing?

4. In problem 3, suppose the two cars are traveling with the speeds given, but are moving in the same direction. Now how far apart will they be after 90 minutes?

5. Delighted Airlines flight 111 leaves Los Angeles and flys to New York at an average speed of 500 mph. Three hours later an SKG5 Military Jet leaves Los Angeles and follows the same route to New York at 1100 mph. When will the jet overtake the first plane? If New York is 2800 miles from Los Angeles by the route they take, which plane will arrive first?

6. Jack Jogs starts on his morning run along a horse path. He runs at a steady 6 minutes per mile. After three minutes, a horseman started along the same path moving at 20 mph. After what distance will Jack be passed by the horse?

7. After the Friday Afternoon Cheese, Wine, and Hiking Society completed their weekly hike, they calculated that if they had walked the distance twice as fast they would have finished an hour sooner. How much time did the hike take?

8. Lyn takes a "wake-up" run every morning. One day when she was feeling very eager, she increased her speed by $\frac{1}{5}$ and finished the run in exactly 30 minutes. How long does the run usually take?

9. After an airplane had been flying for $1\frac{1}{2}$ hours, a sudden increase in wind speed increased the plane's ground speed by 20 mph. The entire trip of 650 miles took 4 hours. How far did the plane travel in the first $1\frac{1}{2}$ hours?

10. When traffic is light, Jerry needs 20 minutes to drive to work. During the rush hour he must drive at an average of 20 mph less, and the trip home requires 30 minutes. How far does he drive from home to work?

11. Larry Leadfoot is driving 60 mph on a highway where the speed limit is 55 mph. He is being chased by a highway patrol car traveling at 80 mph. If the patrol car was 3 miles behind Larry when the chase started, how far will Larry travel before he is caught?

12. Two bicycle racers training on the same road start at the same time from points 20 miles apart and cycle toward each other. One travels at 15 mph and the other at 20 mph. When will they meet?

When you have finished these problems, check your answers in **64**.

64
1. 360 miles on day one and 140 miles on day two.
2. 5 mph. The total distance is 20 miles and the total time is 4 hours.
3. 120 miles. Remember to change 90 minutes to $1\frac{1}{2}$ hours.
4. 30 miles.
5. 5.5 hours after the first plane leaves. The jet passes at 2750 miles; therefore it arrives in New York first.

6. 1 mile. A speed of 6 minutes per mile is 10 mph. Remember to change 3 minutes to $\frac{1}{20}$ hour.
7. 2 hours.
8. 36 minutes. Increasing the speed s by $\frac{1}{5}$ means she now runs at speed $s + \frac{1}{5}s$ or $\frac{6}{5}s$.
9. 225 miles.
10. 20 miles. Remember to write 20 minutes as $\frac{1}{3}$ hour and 30 minutes as $\frac{1}{2}$ hour.
11. 9 miles.
12. $\frac{4}{7}$ hour or about 34.3 minutes.

Mixture Problems

One common kind of word problem that appears frequently in algebra involves mixing two or more kinds of objects or materials to get some desired combination. A druggist mixes different strength solutions of medicines to obtain exactly the solution he needs. A pet store owner mixes varieties of bird seed to get a special combination. A handful of coins represents a mix of coins of different denominations. Each of these kinds of mixtures can be represented in a diagram similar to the one we used to solve motion problems. For example, if a mixture of various kinds of grass seed contains 10 lb of Kentucky Bluegrass seed selling for 95¢ per lb, we can put that information in a diagram like this:

	Amount	Cost per lb	Total cost
Kentucky Bluegrass	10 lb	95¢/lb	950¢

Amount	$\times$	Cost per lb	$=$	Total cost
10 lb	$\times$	95¢/lb	$=$	950¢ or $9.50

Draw a similar diagram for x lb of brand A tea costing 69¢ per lb. Check your diagram in **65**.

65

	Amount	Cost per lb.	Total cost
Brand A Tea	x lb	69¢/lb	$69x$ ¢

Notice that in these diagrams that the units must be consistent. All amounts are in pounds; all costs are in cents.

Diagrams like these will be very helpful to you in translating word problems to algebraic equations. In fact, most students find that once they have a correct diagram for a problem, solving the final algebraic equation is easy.

Each of the following is a bit of information from a typical mixture problem. Construct a diagram for each one.

(a) A mixture contains N pounds of candy costing $2.10 per pound.
(b) Add in 12 gallons of gasoline at C¢ per gallon.
(c) The cash register contains D dimes.
(d) The nurse prepared K liters of 20% alcohol solution.

Construct a diagram similar to those above for each of these statements. When you are ready to check your work, turn to **66**.

---

66 (a)

	Amount	Cost per lb	Total cost
Candy	N lb	210 ¢/lb	$210N$ ¢

Notice that we wrote the cost per pound in cents so that we need not worry about the decimal point. If you choose to keep the cost in dollars, the total cost would be 2.1N $.

(b)

	Amount	Cost per gal	Total cost
Gasoline	12 gal	C ¢/gal	12C ¢

(c)

	Amount	Value per coin	Total value
Dimes	D coins	10 ¢/coins	10D ¢

In this case the amount is a number of coins rather than a weight in pounds.

(d)

	Amount of solution	Strength of solution	Amount of alcohol
Alcohol	K liters	0.20	0.2 K liters

In this situation we have a solution and we are interested in its strength rather than its cost. The strength is given as 20% and we translate this percent to a decimal 0.20 before we enter it into the diagram. To say that the alcohol solution has a strength of 20% means that 20% of it is alcohol while the remaining 80% is water or some similar solvent.

Now let's use diagrams of this sort to solve the following typical mixture problem. How many pounds of hard candy worth 50¢ per pound must be added to 10 pounds of chocolate candy worth 75¢ per pound to make a mixture worth 60¢ per pound?

First, put this information into the following diagram:

	Amount	Cost per lb	Total cost
Hard candy			
Chocolate			
Mixture			

Check your answer in **67**.

67

	Amount	Cost per lb	Total cost
Hard candy	x	50¢	50x
Chocolate	10	75¢	750
Mixture	$x + 10$	60¢	60(x + 10)

In each case the total cost is the product of the amount and the cost per pound. If the amount of hard candy is x lb and the amount of chocolate candy is 10 lb, then the mixture weighs $(x + 10)$ lb.

The total cost of the mixture is the sum of the total costs of the two kinds of candy being mixed. In other words,

$$50x + 750 = 60(x + 10)$$

Solve this equation in the usual way.

$$50x + 750 = 60x + 600$$
$$10x = 150$$
$$x = 15 \text{ lb}$$

Check: $50(15) + 750 = 60(15 + 10)$
$$750 + 750 = 60(25) = 1500$$

Careful ⟫ Notice that in this diagram we can add weights in the first two *amount* columns to get the weight of the mixture, and we can add costs in the first two *total cost* columns to get the total cost of the mixture. However, we can *never* add the numbers in the middle column.

Try this mixture problem. The school cafeteria manager is making up packages of mixed nuts to sell at the snack counter. He wants to mix peanuts that sell for 85¢ per lb with cashews that sell for $1.15 per lb, to get 12 lb of mixture worth $1.10 per lb. How much of each kind of nut should he use?

Draw your diagram, get an equation by summing the total cost column, and solve the equation.

Our step-by-step solution is in **68**.

68

	Amount	Cost per lb	Total cost
Peanuts	x	85	$85x$
Cashews	12-x	115	$115(12$-$x)$
Mixture	12	110	$110 \cdot 12$

$$85x + 115(12 - x) = 110 \cdot 12$$
$$85x + 1380 - 115x = 1320$$
$$-30x = -60$$
$$x = 2 \text{ lb}$$

Mix 2 lb of peanuts with 10 lb of cashews. Check your answer by substituting it back into the original equation.

Here are two slightly different mixture problems to test your understanding.

(a) Eric's coin collection contains only nickels and dimes and he has three times as many dimes as he has nickels. The total face value of all his coins is $2.45. How many of each coin does he have?

(b) Nurse Hypo must dilute 4 liters of an 80% alcohol solution with pure water to make a 30% alcohol solution. How much water should he add?

Draw the diagrams as before, use the diagrams to write an algebra equation, and solve it. When you have completed these check your work in **69**.

---

69   (a)

	Number of coins 3N	Value of each coin	Total value
Nickels	n	5¢	$5n$
Dimes	$3n$	10¢	$30n$
Mixture	$4n$		$35n$

$$35n = \$2.45 = 245¢$$
$$n = 7$$
$$3n = 21$$

The coin collection contains 7 nickels and 21 dimes.

(b)

	Amount of solution	Strength of solution	Amount of alcohol
Alcohol	4 liters	0.80	0.8(4) = 3.2
Water	x liters	0	0
Mixture	4 + x	0.30	0.3 (4 + x)

$$3.2 + 0 = 0.3(4 + x)$$
$$3.2 = 1.2 + 0.3x$$
$$0.3x = 2.0$$
$$3x = 20$$
$$x = \frac{20}{3} \quad \text{or} \quad 6\tfrac{2}{3} \text{ liters}$$

Add $6\tfrac{2}{3}$ liters of water. Notice that the strength of pure water is 0%. It contains no alcohol at all. If we add 4 liters to x liters in column 1, we have $(x + 4)$ liters for the total mixture.

Here are a few practice problems for you. Each problem requires that you first construct a diagram very similar to the examples given above. Take your time, and be certain that the diagram accurately reflects all of the information given in the problem. Only after you have a correct diagram do you worry about solving an algebra equation.

1. At the Flower Farm Nursery they mix their own grass seed. How many pounds of Bermuda grass seed at 80¢ per lb must be added to 20 lb of Fescue at $1.10 per lb to form a mixture to be sold at 90¢ per lb?
2. In grandma's cookie jar are 20 coins, all dimes and quarters, with a total value of $4.10. How many of each kind of coin are there?
3. A chemist has two different solutions of hydrochloric acid, one 45% and the other 70% in strength. How much of each should be mixed to obtain 5 liters of 60% solution?
4. Laurie bought two kinds of candy for the class Christmas party, Chewie Gooies at 75¢ per lb and Sugar Glops at $1.00 per lb. She bought 2 lbs more of the Sugar Glops than of the Chewie Gooies. The total cost $5.50. How much of each candy did she buy?
5. On the last hand of their regular Saturday night poker game, the math department found that the pot contained $3.10 in nickels, dimes, and quarters. There were 4 more dimes than nickles and 2 less quarters than nickels. How many coins of each kind were in the pot?
6. How many pounds of an alloy containing 20% tin must a metallurgist add to 5 lb of an alloy containing 6% tin to produce an alloy containing 10% tin?
7. A wholesale grocer prepares a special blend of 25 lb of herb tea to be sold for $3.00 per lb. He mixes Rosehip tea selling for $3.75 per pound, Camomile tea selling for $2.10 per lb, and Alfalfa Mint tea for $2.50 per lb. He uses twice as much Camomile tea as Alfalfa Mint tea. How much of each does he use?
8. The Zoo U. auditorium holds 322 people and was filled for the bagpipe rock concert. Student tickets cost $1.75, faculty tickets cost $2.25, and all general admission tickets cost $3.00. The number of general admission tickets sold was 42 less than the number of students. The total ticket income was $736. How many of each kind of tickets were sold?
9. A chemistry student needs to use a 20% salt solution in a certain experiment, and he has 4 liters of a 35% salt solution. How much water should he add to dilute it?

10. At the Happy Critter Pet Store they mix parrot seed to sell at $1.00 per lb. To millet seed at 95¢ per lb they add twice as much organic sunflower seeds at $1.20 per lb. The remainder is parakeet seed mix normally selling for 95¢ per lb. How much of each kind of seed must be mixed to produce 10 lb of the parrot seed?

When you have completed these problems turn to **70** to check your answers.

70
1. 40 lb.
2. 6 dimes, 14 quarters.
3. 2 liters of 45% solution and 3 liters of 70% solution.
4. 2 lb of Chewie Gooies and 4 lb of Sugar Glops.
5. 8 nickels, 12 dimes, and 6 quarters.
6. 2 lb of 20% tin.
7. Roughly 13 lb of Rosehip tea, 8 lb of Camoline, and 4 lbs of Alfalfa Mint.
8. 172 student tickets, 20 faculty, and 130 general admission.
9. 3 liters.
10. 1 lb of millet, 2 lb of sunflower seed, and 7 lb of parakeet mix.

General Rate
Problems

Problems like those involving motion or mixtures are particular examples of a wide variety of very general problems called *rate* problems. Speed of motion is a rate; it is the rate of change or ratio of distance with time. It is the ratio of miles to hours or feet to seconds. The cost of candy in a mixture is a rate or a ratio of cost to amount—cents to pounds. The strength of a salt solution is the ratio of amount of salt to amount of water, the ratio of ounces to pints or grams to liters.

Any rate problem can be solved using the diagrams we have shown you how to construct. For example, suppose a man can paint a room in 6 hours. His rate of painting is

$$\text{Rate of painting} = \frac{1 \text{ room}}{6 \text{ hours}} = \frac{1}{6} \frac{\text{room}}{\text{hour}}$$

His friend, a professional painter, can paint the same room in two hours. His rate of painting is

$$\text{Rate of painting} = \frac{1 \text{ room}}{2 \text{ hours}} = \frac{1}{2} \frac{\text{room}}{\text{hour}}$$

Use this information to set up and solve the following rate problems.

If John can paint a room in 6 hours and if a professional painter can paint it in 2 hours, how long would it take them to do the job if they worked together?

Solve this problem by completing the following diagram:

	Fraction of work done	Rate	Time
John's contribution			
Painter's contribution			
Total job			

Check your work in **71**.

71

	Fraction of work done	Rate	Time
John's contribution	W_J	$\dfrac{1}{6}$	t
Painter's contribution	W_P	$\dfrac{1}{2}$	t
Total job	1		t

Work done = rate·time

$$W_J = \frac{1}{6}t = \frac{t}{6}$$

$$W_P = \frac{1}{2}t = \frac{t}{2}$$

Adding the amounts of work done,

$$W_J + W_P = 1$$

or

$$\frac{t}{6} + \frac{t}{2} = 1$$

Solve for t by multiplying all terms by 6.

$$t + 3t = 6$$
$$4t = 6$$
$$t = 1\frac{1}{2}\text{ hr}$$

Check: $\dfrac{1\frac{1}{2}}{6} + \dfrac{1\frac{1}{2}}{2} = 1$

$$\frac{3}{12} + \frac{3}{4} = \frac{3}{12} + \frac{9}{12} = \frac{12}{12} = 1$$

Notice the following:

(a) W_J is the fraction of the total job done by John.
W_P is the fraction of the total job done by the painter.
Together they do the entire job, so the fraction is 1.
(b) When they work together, each works for the same time t hours.
(c) Each row of the diagram gives an equation as shown.
(d) Check your answer. It should be obvious that if the painter can do the job in two hours by himself, the two men working together should be able to do it in less than two hours. If they worked for one hour together, they would finish $\frac{1}{6} + \frac{1}{2}$ or $\frac{4}{6}$ of the job. The correct answer must be between 1 and 2 hours. Always estimate the answer before you begin and check it afterwards.

Try the following problem to check your understanding.

Tom can mow a certain lawn in 4 hours. His friend, Jerry, can mow the same lawn in 5 hours. How long would it take to mow the lawn if they had two lawn mowers and worked for the same length of time?

Make up a diagram as before and solve it as we showed above. Our solution is in **72**.

72

	Fraction of work done	Rate	Time
Tom	W_T	$\dfrac{1}{4}$	t
Jerry	W_J	$\dfrac{1}{5}$	t
Total	1		t

Work = rate time

$$W_T = \frac{t}{4}$$

$$W_J = \frac{t}{5}$$

$$\frac{t}{4} + \frac{t}{5} = 1$$

Solve. Multiply by 20.

$$5t + 4t = 20$$
$$9t = 20$$
$$t = 2\frac{2}{9} \text{ hr}$$

Check your answer.

You can save time by streamlining your diagram so that it looks like this:

	Work	Rate	Time
Tom	$\frac{t}{4}$	$\frac{1}{4}$	t
Jerry	$\frac{t}{5}$	$\frac{1}{5}$	t

$$\frac{t}{4} + \frac{t}{5} = 1$$

Remember that the entries in the work column are the fraction of the total job that each person does in the time t.

Also remember that the rate for each person is simply one divided by the time he would take to do the job alone. In the problem above, Tom does $\frac{1}{4}$ of a job per hour since the total job takes him 4 hours.

Now for a little practice. Try these.

(a) Two typists are working together on an important report. Ann could type the entire report herself in 10 hours and Al could type it in 12 hours. How long will it take them to complete the job working together?

(b) A large pump working alone can fill a certain water tank in 6 hours. A small pump takes 9 hours to fill the same tank. How long will it take to fill the pool if both pumps are used at the same time?

(c) Bob and Tim work at the Keystone Cone Co. making ice cream cones. Working by himself, Bob needs 8 hours to turn out the daily quota of cones. If Tim helps him they can finish the daily quota in 5 hours. How long would it take Tim to do the job if he worked alone?

Our step-by-step solutions are in **73**. Turn there when you are ready to check your work.

---

73 (a)

	Work	Rate	Time
Ann	$\frac{t}{10}$	$\frac{1}{10}$	t
Al	$\frac{t}{12}$	$\frac{1}{12}$	t

$$\frac{t}{10} + \frac{t}{12} = 1$$

Solve. Multiply by 60.

$$6t + 5t = 60$$
$$11t = 60$$
$$t = 5\frac{5}{11}\text{ hr}$$

(b)

	Work	Rate	Time
Large pump	$\dfrac{t}{6}$	$\dfrac{1}{6}$	t
Small pump	$\dfrac{t}{9}$	$\dfrac{1}{9}$	t

$$\frac{t}{6} + \frac{t}{9} = 1$$

Solve. Multiply by 18.

$$3t + 2t = 18$$
$$5t = 18$$
$$t = 3\frac{3}{5}\text{ hr}$$

(c)

	Work	Rate	Time
Bob	$\dfrac{t}{8}$	$\dfrac{1}{8}$	t
Tim	Rt	R	t
Together	$\dfrac{t}{5}$	$\dfrac{1}{5}$	t

$$\frac{t}{8} + Rt = \frac{t}{5}$$

Solve. Multiply by 40.

$$5t + 40Rt = 8t$$
$$5 + 40R = 8$$
$$40R = 3$$
$$R = \frac{3}{40}$$

Tim's rate is $\dfrac{3}{40}\dfrac{\text{job}}{\text{hour}}$. The time he would need to do the entire job is $\dfrac{40}{3}\dfrac{\text{hour}}{\text{job}}$ or $13\frac{1}{3}$ hours.

If the rate of working is $R\dfrac{\text{job}}{\text{hour}}$, the time needed to do the job is $\dfrac{1}{R}$ hr.

In all of these rate problems we assume that the workers are able to work as fast together as they do alone, no time is lost in coordinating their efforts.

Work these problems for practice in solving general rate problems.

1. Marie, the painter, can paint a certain house in 3 days. Her helper can do the job alone in 5 days. How long would it take them to paint the house working together?
2. One concrete mixer can pour the concrete for a certain job in 30 minutes. A smaller mixer could pour it in 45 minutes. How long would it take to do the job if both mixers were used together?

3. The Ajax Gadget Company has received a special order for a carload of gadgets. Their newest gadget machine can make a carload in 12 hours. Their number two gadget maker can produce a carload in 18 hours. If they use both machines at the same time, how much time will be needed to fill the order?

4. An oil storage tank can be filled in 10 hours. If the outlet pipe is open, the tank will empty in 15 hours. How much time is needed to fill the empty tank if both pipes are operating? (*Hint:* If the tank empties in 15 hours, the pipe has a filling rate of $-\frac{1}{15}$ tank/hour.)

5. Pam and Ed work for the Splash for Cash Paint Co. On a certain job, Ed estimates that he could paint the entire house in 20 hours. Because Ed is busy on another job, Pam starts painting it alone and finishes one-half of the job in 8 hours. They work together to finish the job. How much time was needed for the entire job?

6. A certain water reservoir can be filled in 15 days using its inlet pipe and emptied in 20 days by opening a spillway. With the spillway closed, the reservoir is filled for 10 days. Then the spillway is opened while the inlet pipe continues filling. How much time will be needed to fill the reservoir?

Check your answers in **74**.

---

74   1. $1\frac{7}{8}$ days

2.   18 minutes

3. $7\frac{1}{5}$ hours

4.   30 hours

5. $12\frac{4}{9}$ hours

6.   30 days

Did you make an estimate of each answer *before* you began working the problem? Did you check your answer after you had solved the problem?

Now turn to **75** for a set of practice word problems.

Unit 2
Solving
Algebraic Equations

The answers are on page 544.

Solve:

1. At the end of a week's work as a waitress, Maria found that she had $27.60 in coins from tips. She had 8 times as many quarters as 50¢ pieces and 12 more dimes than quarters. How many of each kind of coin does she have?

   _____

2. At the meat department of the local grocery store, they prepare a meatloaf mixture of protein powder and ground beef. If ground beef sells at 89¢ per lb and the protein powder sells at $1.52 per lb, how much protein powder would be added to 20 lb of ground beef to get a meatloaf mix selling for $1.10 per lb?

   _____

3. Two runners started from opposite ends of a 26-mile marathon course at the same time and ran toward each other. Jan, the faster runner, averaged 8 miles per hour while Paul, the slower runner, averaged 10 minutes per mile. How many minutes after the start did they meet? How far did each person run?

   _____

4. Ann, the best typist in the Company secretarial services office, can type normal copy at 6 pages per hour. Harry is not quite as fast a typist and can type the same material at 4 pages per hour. If they work together, how long will it take them to type a 48-page report?

   _____

5. How many gallons of water must be added to 5 gallons of a 40% solution of detergent concentrate to produce a 25% solution?

   _____

6. In problem 5, suppose you had a dilute the 40% solution by adding 20% solution, now how much must be added to produce a 25% solution?

   _____

7. In problem 5, suppose you had to increase the strength of the 40% solution by adding pure detergent solution. How much detergent must be added to produce a 60% solution?

   _____

8. The total income from a professional Frisbee Tournament was $4393. A ticket

count showed that they sold three times as many general admission tickets at $2.50 each as reserved seat tickets at $4.00 each. What was the total attendance?

_____

9. Two pumps are available for either filling or emptying a swimming pool. Pump A can fill the pool in 12 hours working alone and Pump B can fill it in 15 hours working alone. How much time is needed to fill the pool if both pumps work together?

_____

10. In problem 9, suppose Pump A fills for 4 hours by itself and then Pump B is added. How much time is needed to complete the filling of the pool?

_____

11. In problem 9, if Pump B is reversed it can empty the pool in 20 hours. How much time is needed to fill the pool if both pumps work together, Pump A filling and Pump B emptying?

_____

12. A wild drink called Aligator Aide is prepared by mixing pure wheatgerm oil at $2.00 per pint with carrot juice at 60¢ per pint and apple cider vinegar at 40¢ per pint. The recipe calls for five times as much carrot juice as wheatgerm oil. How much of each part of the mixture would be needed to make up 10 pints of Aligator Aide selling for 66¢ per pint?

_____

13. Ron always drives at the legal speed limit of 55 mph and Susan always drives at 60 mph. If they start together driving in the same direction, how far apart will they be at the end of a 3-hour drive?

_____

14. In problem 13, how far apart would they be at the end of a 3-hour drive if they were driving in opposite directions?

_____

15. Two runners are racing around a one-quarter mile track. The faster one runs at 12 mph and the slower one runs at 10 mph. After how many minutes will the faster runner overtake the slower?

_____

When you have had the practice you need, turn to **76** for a self-test over the work of Unit 2.

_____

Unit 2
Solving
Algebraic Equations

76

1. Solve: $\frac{a}{3} = 13$

   $a = $ _____

2. Solve: $3x - 2 = 8 - x$

   $x = $ _____

3. Solve: $\frac{x}{2} + \frac{1}{4} = 1$

   $x = $ _____

4. Solve: $1 - 2a = a + 10$

   $a = $ _____

5. Solve: $2y - 9 = 3(-1 + y)$

   $y = $ _____

6. Solve: $3 - (1 - 2c) = 4(c - 1)$

   $c = $ _____

7. Solve: $a - 3 + 2a + 4 + a + 1 - a^2 = 3 + a - 2a + 4 - a^2$

   $a = $ _____

8. Solve: $\frac{3x}{4} - \frac{2}{3} = x + 1$

   $x = $ _____

9. Solve for A: $D = 3AM^2$

   _____

10. Solve for R: $I = \frac{E}{R + S}$

    _____

11. Solve for K: $S = 2K(A - 1)$

    _____

12. Solve for M: $B = \frac{2a}{M}$

    _____

13. Solve for Q: $2(1 + Q) = 3(2Q - 1)$

    _____

14. Solve and graph: $2x - 1 > 9 - 3x$

    _____

15. Solve and graph: $2 - \frac{a}{3} \leq 1 + a$

    _____

16. Solve and graph: $-3y > 9$

    _____

17. Solve and graph: $3 < 2(1 + R)$

    _____

18. Find three consecutive integers whose sum is 102.

    _____

19. Two bicycle racers training on the same road start at the same time from points 20 miles apart and cycle toward each other. One travels at 15 mph and the other at 20 mph. When will they meet?

    _____

Date _____

20. Find three consecutive even integers such that the sum of the larger two equals 70.

    _____

Name _____

Course/Section _____

21. In Mrs. Cees Candy Store they want to mix chocolates costing $2.10 per pound with mints costing $1.50 per pound to make a

179   Self-Test 2

10-pound mix costing $1.90 per pound. How many pounds of chocolate should they use?

_____

22. Uncle Willie drives his 1946 clunker to town every Saturday. Last Saturday his nephew, Sam, made the same trip in his sports car, leaving 2 hours later than Uncle Willie, but driving twice as fast and arriving at the same time. How long did the trip take for each of them?

_____

23. An A-1 model pump can fill a certain water tank in 15 hours. The bigger A–2 pump can fill the same tank in 10 hours. How long will it take to fill the tank if both pumps are used at the same time?

_____

24. A nurse dilutes a drug solution from 25% to 10% by adding pure water. How many pints of water should he add to 10 pints of the original solution?

_____

25. Jose can paint the exterior of an average house in 2 days. Mary estimates that she can do the same job in 3 days. How long would the work take if they worked together?

_____

Answers are on page 544.

Unit 2
Solving
Algebraic Equations

Solving Equations by
Adding and
Subtracting Terms

A. **For each of the following equations, does the given number satisfy the equation?**

1. $3x - 4 = 1$ $x = 2$ _____   2. $4a - 11 = 18$ $a = 7$ _____

3. $-2 + 2t = t - 1$ $t = 1$ _____   4. $p - 8 = 2p$ $p = 4$ _____

5. $\frac{1}{2}x - 4x + 7 = 0$ $x = 2$ _____   6. $3q + 2 = -4$ $q = -3$ _____

7. $y + 2 = 3 - 1$ $y = 0$ _____   8. $2x - 2 = 0$ $x = -1$ _____

9. $4x - 7 = 1 - 2x$ $x = -2$ _____   10. $2y - 1 = 3y - 4$ $y = 3$ _____

B. **Solve:**

1. $p + 2 = 12$ _____   2. $x - 2 = 7$ _____

3. $36 - a = 7$ _____   4. $8 = t + 1$ _____

5. $14 = q - 4$ _____   6. $-6 = x - 2$ _____

7. $-y + 3 = 10$ _____   8. $1 + b = -6$ _____

9. $x - 1.3 = -1.7$ _____   10. $2x + 7 = 3x$ _____

11. $13 - y = 0$ _____   12. $0 = 16 + y$ _____

13. $0.4 - d = 0.2$ _____   14. $q - 1.1 = -2.3$ _____

15. $1 - 2a + 3 + 3a = 11$ _____   16. $2x - 8x - x = 0$ _____

17. $2A + 3 = A$ _____   18. $16 - x = 7 - 2x$ _____

19. $5x + 1 = 9 + 4x$ _____   20. $12 + 2y = -2 + y$ _____

21. $-\frac{1}{2} = x - 1\frac{1}{2}$ _____   22. $-t - 2 = -3$ _____

23. $4p + 8 - 2p = p + 1$ _____   24. $8 - w = 3 - 2w$ _____

25. $17 - 2q - 6 = -3q - 1$ _____   26. $-8 - x - 2x - 2 = -4x - 3$ _____

27. $8x + 13 - 3x = 9 + 4x$ _____   28. $1.5y + 4.8 - \frac{1}{2}y = 0$ _____

29. $0 = 1 - t$ _____   30. $0 = 1 + p$ _____

31. $2x = 1 + x$ _____   32. $3a = 2a + 4$ _____

33. $2(y - 1) = y + 3$ _____   34. $3(x + 1) = 1 + 2x$ _____

35. $5(x - 2) - x = 3(x + 2)$ _____   36. $7(p + 1) = 3(2p - 1)$ _____

Date

Name

Course/Section

37. $11x - 7 = 5(3 + 2x)$ _____ **38.** $6 - 9t = 2(1 - 4t)$ _____

39. $a + 2(1 - a) = 1 - 2a$ _____ **40.** $x - 3(x - 2) = 5 - x$ _____

Unit 2
Solving
Algebraic Equations

Solving Equations by
Multiplying and
Dividing

A. Solve:

1. $2x = 12$ _____

2. $3x = 10$ _____

3. $5y = 1$ _____

4. $3x = -1$ _____

5. $2p = -\dfrac{1}{2}$ _____

6. $9a = 3$ _____

7. $5x = 3$ _____

8. $1.2y = 6$ _____

9. $\dfrac{1}{2}q = \dfrac{1}{4}$ _____

10. $3.5 = \dfrac{t}{7}$ _____

11. $6m = 4\dfrac{1}{2}$ _____

12. $-6 = \dfrac{2}{3}c$ _____

13. $-14x = -112$ _____

14. $0.4 = \dfrac{b}{5}$ _____

15. $3d = 24$ _____

16. $1 - x = 0$ _____

17. $2 + q = 0$ _____

18. $-q = 4$ _____

19. $7 = -x$ _____

20. $\dfrac{R}{7} = 11$ _____

21. $\dfrac{y}{-3} = 6$ _____

22. $\dfrac{-z}{-2} = 5$ _____

23. $\dfrac{-p}{3} = -9$ _____

24. $\dfrac{x}{-5} = -10$ _____

25. $\dfrac{x}{3} = -5$ _____

26. $\dfrac{x}{-2} = 4$ _____

27. $-5x = 15$ _____

28. $-6x = -12$ _____

29. $2x = -7$ _____

30. $-6 = -3x$ _____

B. Solve:

1. $3x - 6 = 0$ _____

2. $5a - 1 = 0$ _____

3. $7p - 3 = -4$ _____

4. $q - 5q = 6$ _____

5. $-t + 1 = 2 - t$ _____

6. $1 - 2k = k - 2$ _____

7. $11 - x - 6 = 5 + 6x$ _____

8. $-132 = 5x + 7x$ _____

9. $3c = 24 + c$ _____

10. $n + 1 + 3n = 4 - n$ _____

11. $2q = -4q + 12$ _____

12. $x + 2x + 3x = 5 - x$ _____

13. $-3y - 2 - y - 2y - 4 = 0$ _____

14. $\dfrac{3x}{2} = 6$ _____

15. $\dfrac{-2q}{5} = 10$ _____

16. $\dfrac{-4p}{3} - 12 = 0$ _____

17. $\dfrac{-t}{2} = -8$ _____

18. $x = 1 - x$ _____

19. $-1 - x = -2$ _____

20. $8a + 2 = -1 - a$ _____

21. $1.2x = 3$ _____

22. $-0.91 = 0.7y$ _____

23. $0.6x - 9 = -3x$ _____

24. $-7p + 5 = -4 + 0.5p$ _____

25. $-2x - 14 - 5x = -2 - x$ _____

26. $-23n - 31 - 2n = -n - 1 - 2n$ _____

27. $3q - 1 = q - 1$ _____

28. $4 + x = 1 + x$ _____

29. $2t + 3 = t + 3$ _____

30. $5 + x = 5 - x$ _____

31. $\dfrac{3x}{2} = -5$ _____

32. $\dfrac{-2x}{5} = 4$ _____

33. $\dfrac{5y}{-2} = 4$ _____

34. $\dfrac{-7a}{2} = -5$ _____

35. $\dfrac{x}{2} = 4 - x$ _____

36. $1 - \dfrac{x}{3} = 2x$ _____

37. $\dfrac{t}{3} - 1 = t$ _____

38. $3p + \dfrac{p}{2} = 4$ _____

39. $2(x - 1) + 3x = 1$ _____

40. $2m - 1 = 5(m - 1)$ _____

C. Solve:

1. Four more than a number is three times that number. Find that number.

   _____

2. The sum of two numbers is 16 and their difference is 2. Find the numbers.

   _____

3. A number is multiplied by 6 and when that product is subtracted from 70 the difference is 4. Find the number.

   _____

4. Debbie is four times as old as Angela. The sum of their ages is 45. Find their ages.

   _____

5. If 23 is added to three-fourths of a number, the result is 41. Find the number.

   _____

Date _____

Name _____

Course/Section _____

Unit 2
Solving
Algebraic Equations

Solving Linear Equations

A. Solve:

1. $3x - 10 = 2x$ ___

2. $5p + 3p = 16$ ___

3. $-2y + 4 = -y$ ___

4. $0 = 8x - 1$ ___

5. $-13 + 2a = 1$ ___

6. $b = 4b + 30$ ___

7. $2q = 36 + 5q$ ___

8. $4 - 2t = 6t$ ___

9. $3 - k = 8k$ ___

10. $6x - 7 = 4x + 3$ ___

11. $c + 18 = 6 - 4c$ ___

12. $2x - 7 = 5x - 8$ ___

13. $9z - 3 = 25 + 2z$ ___

14. $w + 4 = 8w + 4$ ___

15. $2y + 1 = 10y - 1$ ___

16. $3t + 4 = 16$ ___

17. $5x + 15 = 0$ ___

18. $-5y + 9 = 14$ ___

19. $0 = -2x - 7$ ___

20. $5 + N - 7 = 8N + 12$ ___

21. $1 - 5p = p - 5$ ___

22. $3 - b = 8b$ ___

23. $x + 12 = 5x$ ___

24. $1.4x = 3 - 0.1x$ ___

25. $1 - x = 5x$ ___

26. $2 - y = -3y$ ___

27. $1.5a = 0.5 - a$ ___

28. $2.2x = x - 0.6$ ___

29. $2 - t = t - 3$ ___

30. $p - 3 = 2p + 3$ ___

B. Solve:

1. $5 - x - 18 = 3x - 1 + 8x$ ___

2. $4q + 16 = 9q - 2q + 8$ ___

3. $x + 2x - 5x + 7 = 1$ ___

4. $-t - 2t = 14 + t$ ___

5. $5 - b = 2b + 29 + 3b$ ___

6. $1 - x + 2x + 4 = 0$ ___

7. $5 = 3(x - 1) - 1$ ___

8. $a + (a - 6) = 10$ ___

9. $6(a + 2) = 12$ ___

10. $5(2 - y) = 26 + y$ ___

11. $-(4 - x) = 2x$ ___

12. $3(2 - 6p) = 2p - 4$ ___

13. $8 = 5x - (1 + 2x)$ ___

14. $18 = y - 2(3 + 5y)$ ___

15. $9 = -3(-1 - 6z) - 10z$ ___

16. $2(1 - x) = 3(1 + x)$ ___

17. $(2 + x) = 1 - (x + 3)$ ___

18. $(2x + 1) = 5(x - 2)$ ___

19. $(a + 5) = (11 - 2a)$ ___

20. $(-6 - 5z) = (1 - z) + 1$ ___

21. $3(y - 5) = 2(2y + 1)$ ___

22. $7(x + 2) = 5(4 + x)$ ___

23. $2(x - 3) - 10 = -3(x + 2)$ ___

24. $5n - (3n + 2) = 18$ ___

25. $\dfrac{3(x - 1)}{5} = 0$ ___

26. $\dfrac{17(2x - 3)}{6} = 0$ ___

27. $1 - \dfrac{3y - 1}{2} = 0$ _____

28. $\dfrac{3(t - 1)}{2} = 4(t + 1) - 1$ _____

29. $\dfrac{q - 2}{3} + \dfrac{q - 3}{2} = 2$ _____

30. $\dfrac{2a + 1}{3} - \dfrac{3a - 1}{5} = 1$ _____

C. **Translate the following problems into algebraic equations and solve them.**

1. Find two numbers whose sum is $\frac{1}{2}$ and whose difference is $\frac{2}{3}$.

_____

2. The perimeter of a rectangle is 68 in. What are its dimensions if its length is 8 in. more than its width?

_____

3. Joe is six years older than Harry. Twelve years ago, Joe was twice as old as Harry was then. How old are they now?

_____

4. Solve for x: $\dfrac{a - x}{b} = \dfrac{b - x}{a}$

_____

5. The sum of two numbers is 85. One of the numbers is four times the other. Find the numbers.

_____

Date _____

Name _____

Course/Section _____

Unit 2
Solving
Algebraic Equations

Solving Formulas

A. Solve each of the following formulas for x.

1. $Q = 8x$ _____

2. $A = 4 - x$ _____

3. $B = 5ax$ _____

4. $R = -3 - x$ _____

5. $T = 2 - 5x$ _____

6. $S = 3\pi x - 1$ _____

7. $A + B = C + x$ _____

8. $2x - q = p$ _____

9. $b = 2ax + 1$ _____

10. $ab^2 x = 3b$ _____

11. $5t = 2x - 4$ _____

12. $2x - q^2 = 0$ _____

13. $2 = 3px$ _____

14. $-Y = 5mx$ _____

15. $12 = \dfrac{kx}{at}$ _____

16. $\dfrac{x}{2} - 1 = t$ _____

17. $A + x = B - x$ _____

18. $x - 3 = 2ax$ _____

19. $a + x - c = 1 - 2x$ _____

20. $7kx - 2 = 5$ _____

21. $4 - 3ax = a$ _____

22. $2x - c = c^2$ _____

23. $1.4x + 6 = B - 2.5x$ _____

24. $1.5 - 3.1x = A - 1.6x$ _____

25. $K = 3.5x - 80$ _____

26. $A = 45 - 1.8x$ _____

27. $M = 4 - x(N - 1)$ _____

28. $T = 2 - x(A + 2)$ _____

29. $B = P + 1 - 3x$ _____

30. $R = 3 - Y + 5x$ _____

B. Solve each of the following formulas for the variable indicated.

1. $p = q - a$ For a _____

2. $2s = 3a - 2p$ For p _____

3. $x = 5kbc$ For b _____

4. $s = \dfrac{1}{2} Gt^2$ For G _____

5. $v^2 = 2as$ For a _____

6. $8(k - 2b) = k$ For b _____

7. $2(t - 1) = 3(2 - x)$ For x _____

8. $a - b = 3(a + b)$ For a _____

9. $P(ab - a) = c$ For a _____

10. $2ny - 3y = k$ For n _____

11. $x = q - \dfrac{2a}{b}$ For a _____

12. $\dfrac{x + y}{x - y} = 2$ For x _____

13. $\dfrac{Px + 1}{2x} = b - a$ For x _____

14. $x(a + b) = a(x + 1)$ For a _____

15. $\dfrac{x + 1}{b} = 2$ For x _____

16. $\dfrac{1}{A} + \dfrac{1}{B} = \dfrac{1}{C}$ For A _____

17. $2AR + 1 = \dfrac{1}{R}$ For A _____

18. $GT + \dfrac{1}{T} = 2$ For G _____

19. $x = \dfrac{2}{R}$ For R _____

20. $X = \dfrac{2}{R + 1}$ For R _____

21. $4(x + A) = 7(x - B)$ For x _____

22. $5(R - a) = 3(a + P)$ For a _____

23. $A = \frac{1}{2}(B - 4)$ For B _____

24. $D = -\frac{2}{3}(1 - E)$ For E _____

25. $\frac{A}{B} - C = 1$ For B _____

26. $\frac{A}{B} + \frac{1}{C} = 2$ For B _____

27. $\frac{p + a}{b} = \frac{b}{a}$ For p _____

28. $\frac{2}{x} = \frac{x + y}{a}$ For y _____

29. $2(A - 3B) = B - A$ For A _____

30. $3(T - Y) = Y + T$ For Y _____

C. Solve for the variable indicated.

1. $s = \frac{a - rt}{1 - r}$ Solve for t. _____

2. $P = mgh$ Solve for m. _____

3. $A = \frac{1}{2}bh$ Solve for b. _____

4. $PV = nRT$ Solve for V. _____

5. $\frac{1}{a} + \frac{1}{b} = \frac{1}{c}$ Solve for c. _____

6. $F = \frac{9}{5}C + 32$ Solve for C. _____

7. $\frac{A}{x} = B + \frac{C}{x}$ Solve for x. _____

8. $A = \frac{h}{2}(b + c)$ Solve for b. _____

9. $P = 2L + 2W$ Solve for W. _____

10. $L = a + (n - 1)d$ Solve for n. _____

Date _____

Name _____

Course/Section _____

Unit 2
Solving
Algebraic Equations

Inequalities

A. Solve and graph each of the following inequalities.

1. $x - 2 > 6$ _____

2. $1 - x \leq 3$ _____

3. $2 \geq x - 4$ _____

4. $2 + x > -1$ _____

5. $7 < 2 + x$ _____

6. $y + 3 \geq 4$ _____

7. $y + 1 < 4$ _____

8. $x + \frac{1}{2} < 2$ _____

9. $z - 4 \leq -8$ _____

10. $-1 > q - 5$ _____

11. $-5 < x + 2$ _____

12. $-1 > y - 1$ _____

B. Solve each of the following inequalities.

1. $3y - 1 \leq 5$ _____

2. $2x > -3$ _____

3. $4 > \frac{t}{2}$ _____

4. $-3 < 4x + 1$ _____

5. $4x - 3 > 17$ _____

6. $-11 \leq 1 + 6y$ _____

7. $6 - 3y \geq -12$ _____

8. $\frac{x}{3} < -1$ _____

9. $6x + 2 < 14 + 8x$ _____

10. $5x + 2 + x > 6x - 8 + x$ _____

11. $x + 5 < x$ _____

12. $-3x \leq -9$ _____

13. $2(x + 2) \geq x - 3$ _____

14. $5(1 - y) \geq y - 5$ _____

15. $2z - \frac{1}{2} > z + \frac{1}{2}$ _____

16. $1 - \frac{y}{2} > 4$ _____

17. $1 - (x - 1) < 2 + (x + 1)$ _____

18. $5p + 2 < p - 10$ _____

19. $x - \frac{1}{4} \geq \frac{1}{2} - x$ _____

20. $-3 < x + (1 + x)$ _____

21. $2 \leq x - (1 - 2x)$ _____

22. $1 - 3x \geq -2(x - 2)$ _____

23. $3(x - 1) > 1 - 2(x + 1)$ _____

24. $2 + 3x < -4(x + 2)$ _____

25. $\frac{1}{2}(x + 3) \geq 2(1 - x)$ _____

26. $\frac{3}{2}(1 - 4x) \leq 2(x - 2)$ _____

27. $\frac{2x - 13}{3} < -2(x + 3)$ _____

28. $-3(1 - 2x) > \frac{4}{5}(7x - 4)$ _____

29. $2 - (x - 4) \geq 3 - 5x$ _____

30. $1 - 2x < 1 - (5 - x)$ _____

C. Solve:

1. Five times a certain positive integer is less than 65. What integers satisfy this condition? _____

2. Twice a number, decreased by four, is less than 24. What numbers satisfy this condition? _____

3. The sum of two numbers is less than 12, and the larger is three times the smaller. What are the largest possible values for the two numbers? _____

_____
Date

_____
Name

_____
Course/Section

Unit 2
Solving
Algebraic Equations

Word Problems **Solve:**

1. If three times a number is increased by 12, the result is 4 less than five times the number. Find the number. _____

2. Separate 45 into two parts such that twice the smaller is 10 less than half the larger. _____

3. The smaller of two numbers is 12 less than the larger. Four times the larger is 10 more than five times the smaller. Find the numbers. _____

4. Find three consecutive even integers whose sum is 54. _____

5. Find four consecutive odd integers whose sum is 80. _____

6. Find three consecutive integers such that the sum of the first and the third is 50. _____

7. Find three consecutive integers such that twice the smallest is 5 more than the largest. _____

8. Find four consecutive odd integers such that the sum of the first three exceeds twice the fourth by 5. _____

9. Two cars start from the same point at the same time and travel in opposite directions. The slower car travels at 30 mph and the faster one at 45 mph. In how many hours will they be 100 miles apart? _____

10. Two cars start at the same time from cities 336 miles apart and they travel toward each other along the same road. The speed of the faster car is twice the speed of the slower. They meet after four hours. Find the speed of each car. _____

11. Fred drives to a neighboring town and returns over the same road. During the first part of the trip he averages 35 mph and on the return he averages 45 mph. If the total driving time is 6 hours, find the total distance of the round trip. _____

12. An airplane made a 1425 mile trip across Alaska in $3\frac{1}{2}$ hours. During the first 2 hours of the trip the weather was excellent, but bad weather during the last $1\frac{1}{2}$ hours caused the pilot to decrease his speed by 100 mph. At what speed did the plane travel on the first part of the trip? _____

13. A bicycle racer on a training spin covered 30 miles in $1\frac{3}{4}$ hr. How fast must she travel on the remaining 20 miles to average 20 mph? _____

14. My change from a shopping trip is $1.95 in nickels, dimes, and quarters. The number of nickels is ten more than the number of dimes. If there are a total of 20 coins, how many coins of each kind are there? _____

15. Terry purchased some 13¢ and 9¢ stamps from the Post Office. For $1.97 she received 17 stamps. How many of each kind did she purchase? _____

16. Scott's coin collection consists entirely of indian head pennies and buffalo nickels. The number of pennies is six more than twice the number of nickels. The face value of the coin collection is $1.67. How many of each coin is in the collection? _____

Date _____

Name _____

Course/Section _____

17. How many pounds of candy worth $1.40 per lb must the owner of the local candy store mix with 30 lb of candy worth $1.80 per lb to produce a mixture to be sold for $1.70 per lb? _____

18. Tickets for the church fund-raising dinner cost $2.50 for adults and 75¢ for children. If a total of $109.75 is collected, how many adults attended the dinner? The number of adults is 27 more than the number of children.

_____

19. A chemist wishes to dilute a 40% solution of a solvent with a 15% solution. How many liters of each should she use to produce 40 liters of a 25% solution?

_____

20. How much pure water must be added to 5 quarts of a 20% salt solution to make a 15% solution? _____

21. In August 1977 the human-powered aircraft *Gossamer Condor* won a prize by flying an out and back course. At its maximum speed it would take 4 minutes to fly the course against a $2\frac{1}{2}$ mph wind and would make the return trip with the wind in 2 minutes and 30 seconds. What was its maximum speed?

_____

22. Seawater has a concentration of 3.5% by weight of dissolved solids, while drinking water has only about 0.05%. To be useful for irrigation, water must generally have no more than 0.15% of dissolved solids. How many gallons of drinking water should be mixed with seawater to obtain 1000 gallons of irrigation water?

_____

23. A water storage tank can be filled in 12 hours. If the outlet valve is opened, it will drain completely in 20 hours. How long will it take to fill the tank if the outlet valve is accidently left open while the tank is being filled?

_____

24. The present labeling machine in the IMD Company takes 4 hours to complete a certain job. A new machine can do the job in $2\frac{1}{2}$ hours. How much time would be needed to do the job if both machines were operated at the same time?

_____

25. In problem 24, a clerk can do the job by hand in 8 hours. How much time would be needed to do the job if the clerk and the present machine were working together? _____

26. For problem 25 and 24, how much time is needed if both machines and the clerk work together? _____

27. Two bicycle racers start at exactly the same time at opposite ends of a 20 mile long bike path. One rider travels 4 mph faster than the other. If they meet 30 minutes after they start, how fast is each cyclist traveling?

_____

Products and Factoring

PREVIEW 3

Objective	Sample Problems	Where To Go for Help

Upon successful completion of this program you will be able to:

			Page	Frame
1. Multiply binomials.	(a) $(a + 1)(2a - 3)$	= _____	207	12
	(b) $(x + 4)^2$	= _____	212	18
	(c) $(y - b)(y + b)$	= _____	211	17
	(d) $3x(x + 1)(3x - 2)$	= _____	207	12
	(e) $(3a - 5b)^2$	= _____	212	18
2. Factor monomials from polynomials.	(a) Factor: $-4a^2 - 8a$	= _____	200	7
	(b) Factor: $2x^2 + ax^2 - 3cx^2$	= _____		
	(c) Factor: $-p^2q + pq^2 - 2pq$	= _____		
3. Factor trinomials.	(a) Factor: $q^2 - 6q + 8$	= _____	221	23
	(b) Factor: $x^2 - 34x - 35$	= _____		
	(c) Factor: $5 - 6x + x^2$	= _____		
	(d) Factor: $2a^2 - 5a - 12$	= _____		
	(e) Factor: $m^2 - n^2$	= _____	211	17
	(f) Factor: $2x^2 + 3xy - 2y^2$	= _____	221	23
	(g) Factor: $1 - 25p^2$	= _____	211	17
4. Solve equations involving products and factors.	(a) Solve: $(a + 1)(a - 2) = 1 + a^2$	_____	228	31
	(b) Solve: $(x - 3)^2 = x^2 + x + 2$	_____		
5. Solve word problems involving products and factors.	My piggybank contains $5.40 in dimes and quarters. There are 12 more dimes than quarters. How many quarters are in my bank? _____		228	31

Date _____

Name _____

Course/Section _____

6. Divide a polynomial by a monomial or a binomial.

    (a) $\dfrac{18x^3 - 12x^2 + 9}{6x}$ = _____   229   **32**

    (b) $(2x^4 + x - 4) \div (x - 1)$ = _____

(Answers to these problems are at the bottom of this page.)

If you are certain you can work all of these problems correctly, turn to page 239 for a self-test. If you want help with any of these objectives or if you cannot work one of the sample problems, turn to the page indicated. Of course, the super-student will want to be certain to learn all of this and will turn to frame **1** and begin work there.

1. (a) $2a^2 - a - 3$ (b) $x^2 + 8x + 16$ (c) $y^2 - b^2$

    (d) $9x^3 + 3x^2 - 6x$ (e) $9a^2 - 30ab + 25b^2$

2. (a) $(-4a)(a + 2)$ (b) $x^2(2 + a - 3c)$ (c) $pq(-p + q - 2)$

3. (a) $(b - 4)(b - 2)$ (b) $(x - 3b)(x + 1)$ (c) $(x - 5)(x - 1)$

    (d) $(2a + 3)(a - 4)$ (e) $(m - n)(m + n)$ (f) $(2x - y)(x + 2y)$

    (g) $(1 - 5p)(1 + 5p)$

4. (a) -3 (b) 1

5. 12 quarters and 24 dimes.

6. (a) $3x^2 - 2x + \dfrac{3}{2x}$ (b) $2x^3 + 2x^2 + 2x + 3 - \dfrac{1}{x - 1}$

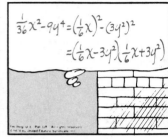

© 1971 United Feature Syndicate, Inc.

3 Products and Factoring

3-1 POLYNOMIALS

1 In your study of mathematics you first learned to multiply simple whole numbers.

$23 \times 47 =$ _____

and decimal numbers

$4.5 \times 3.12 =$ _____

and fractions

$1\frac{2}{3} \times \frac{7}{10} =$ _____

In Unit 1 of this book you learned how to write products of numbers using exponents and to multiply simple algebraic expressions such as:

$(2x)(3x^2) = 6x^3$

or $(4a)(-2a) = -8a^2$

In this unit you will learn how to extend these ideas to include the multiplication of more complicated algebraic expressions.

To begin, work the following review problems.

A. Write the following in exponential form.

1. $2 \cdot 3 \cdot a \cdot a \cdot a \cdot b \cdot b =$ _____

2. $5 \cdot x \cdot x \cdot 3 \cdot y \cdot 2 \cdot x \cdot y \cdot z =$ _____

3. $x^2 \cdot x^3$ _____ $=$ _____

4. $q \cdot q^5$ _____ $=$ _____

5. $a(a^2)$ _____ $=$ _____

6. $(a^2)^5$ _____ $=$ _____

7. $t^2 \cdot t^2 \cdot t^2$ _____ $=$ _____

8. $(a^2b^3)^2$ _____ $=$ _____

B. Multiply.

1. $3(2d^2)$ _____ $=$ _____

2. $(ab)(ab)$ _____ $=$ _____

3. $(a^2)(4a^3)$ _____ $=$ _____

4. $(2p^2q)(5p^3q)$ _____ $=$ _____

5. $(-x^2)(2x)$ = _____   6. $-6(3t^2)$ = _____

7. $(2a)(-3a^3)$ = _____   8. $(3m)(2m^2n)$ = _____

Check your answers in **2**

---

2   A.   1. $6a^3b^2$ 2. $30x^3y^2z$
             3. x^5 4. q^6
             5. a^3 6. a^{10}
             7. t^6 8. a^4b^6

If you answered any of these first eight problems incorrectly, you should return to frame **25** on page 29 and review the multiplication of exponent numbers before you continue in this unit.

B.   1. $6d^2$ 2. a^2b^2
      3. $4a^5$ 4. $10p^5q^2$
      5. $-2x^3$ 6. $-18t^2$
      7. $-6a^4$ 8. $6m^3n$

If you answered any of these second eight problems incorrectly, you should return to frame **40** on page 46 and review the multiplication of simple expressions before you continue in this unit.

If you answered all of the review quiz questions correctly, you are a quiz whiz and you may take a giant step forward.

Polynomial

A *polynomial* is defined as an expression containing only non-negative integer powers of literal numbers. For example,

$x + 1$ and $2x^2 + 2x + 1$ are polynomials in x.

$t^3 - \frac{1}{2}t^2 + \sqrt{3}at - 1.24a$ is a polynomial in t and a.

5 is a polynomial since $5 = 5x^0$ and 0 is a non-negative integer.

A polynomial may *never* contain negative powers of a variable. Expressions such as

$\frac{1}{x}$ or $\frac{1}{a + b}$ or t^{-3} are *not* polynomials.

A polynomial cannot contain square roots or other irrationals in x, such as $\sqrt{x}$.

It is often very convenient to describe a polynomial by telling how many terms it contains. Any polynomial consisting of only one term is called a *monomial*.

Monomials

For example, x, $2a$, 4, and $5x^3$ are all monomials.

Binomials

Any polynomial consisting of only two terms is called a *binomial*. For example, $a + b$, $x - 1$, and $3ax^3 - 12a^2tx^5$ are all binomials.

Trinomials

Any polynomial consisting of only three terms is called a *trinomial*. For example, $1 + x + t$, $2x^2 - 3x + 4$, and $2at^3 + \frac{1}{2}a^2t^3 - t$ are all trinomials.

Identify each of the following as either a monomial, binomial, or trinomial.

(a) 2   (b) t
(c) $3x - 1$ (d) $x + y$
(e) $17a^2b^3c^4x$ (f) $3 - 2x + y$
(g) $ax^2 + bx + c$ (h) $\frac{1}{2}x$

Check your answers in **3**.

---

3   (a)   monomial   (b)   monomial
    (c)   binomial   (d)   binomial
    (e)   monomial   (f)   trinomial
    (g)   trinomial   (h)   monomial

In earlier sections of this book you learned how to add and subtract polynomials by combining like terms. In this unit you will learn how to multiply and factor them.

Multiplying Monomials

As you should recall from Unit 1, the multiplication of monomials is a very simple two-step process.

Example

$$(2x^2)(3x^3) = 6x^5$$

Step 1. Multiply coefficients

$2 \cdot 3 = 6$

Step 2. Multiply the variable parts by adding the exponents of like bases

$x^2 \cdot x^3 = x^5$

Of course, this is review for you, but it is important because all polynominal multiplications can be reduced to this simple two-step operation.

Copyright © 1982 by John Wiley & Sons, Inc.

The algebra involved in multiplying a binomial by a monomial will be easier to understand if we start with an arithmetic problem.

Calculate: $3(2 + 4) = $ _____

There are two ways to do this multiplication. First,

$$3(2 + 4) = 3(6) = 18$$

and, second,

$$3(2 + 4) = 3(2) + 3(4) = 6 + 12 = 18$$

The second multiplication uses the distributive law you worked with in Unit 1.

Distributive Law

$$\boxed{a(b + c) = ab + ac}$$ The *distributive law*

where a, b, and c are any numbers.

To multiply a monomial by a binomial, use the distributive law. For example, to find the product

$2x \ (x + 2)$

First, apply the distributive law,

$2x \ (x + 2) = \ 2x \ (x) + \ 2x \ (2)$

Second, multiply the monomials.

$$2x(x + 2) = 2x^2 + 4x$$

Try this one: Multiply $2a(3a - 2c)$

Check your work in **4**.

---

4 **Step 1.** Apply the distributive law.

$2a \ (3a - 2c) = \ 2a \ (3a) + \ 2a \ (-2c)$

Notice that we treat the binomial as a *sum* of terms: $3a - 2c = (3a) + (-2c)$. This will avoid confusion with negative signs.

Step 2. Multiply the monomials.

$$2a(3a - 2c) = 6a^2 - 4ac$$

where

$$2a(3a) = 6a^2$$

and

$$2a(-2c) = -4ac$$

Easy? Try these.

(a) $4(t - 2)$ (b) $2p(p + 1)$
(c) $x(2x - 3)$ (d) $-3a(2 - 4a^2)$

Check your answers in **5**.

---

5   (a) $4\,(t-2) = \boxed{4}\,(t) + \boxed{4}\,(-2)$

$$= 4t - 8$$

(b) $\boxed{2p}\,(p+1) = \boxed{2p}\,(p) + \boxed{2p}\,(1)$

$$= 2p^2 + 2p$$

(c) $\boxed{x}\,(2x-3) = \boxed{x}\,(2x) + \boxed{x}\,(-3)$

$$= 2x^2 - 3x$$

(d) $\boxed{-3a}\,(2-4a^2) = \boxed{-3a}\,(2) + \boxed{-3a}\,(-4a^2)$

$$= -6a + 12a^3$$

A bit of practice on this kind of multiplication will get it grooved into your mathematical muscles. Work these problems.

Multiply:

1. $3(8w + 2)$
2. $-4(1 + 2t)$
3. $5(2x - 7)$
4. $-2(3y - x)$
5. $x^2(x - 2)$
6. $2\pi(R^2h - a^3)$
7. $-3a^2(a - b)$
8. $2pq(p^2 - q^2)$
9. $2h(h^2 - 3h + 3)$
10. $-3t(a^2t + at - t^2)$

Multiply and simplify:

11. $3x - 2(x - 1)$
12. $ax + a(1 - x)$
13. $2(x + 1) + 2(x - 1)$
14. $3(y + x) - 2x(1 + y)$
15. $3a(a + 2) - 2a(1 + a)$
16. $x(x^2 + 2x + 1) - x^2(x - 1)$
17. $5a(2 - a) + a(5a + 1)$
18. $2a^2b(a + b + a^2) + ab(ab + a^2)$
19. $3x^2y(x - y - xy) + x^2y^2$
20. $(x + 2)2x - x$

Solve:

21. $-3(1 - x) = 9$
22. $3(x - 1) = 0$
23. $2(3x - 4) = 1$
24. $2x(x - 1) = x(1 + 2x) + 3$
25. $3a(1 - a) = a(2 - 3a) - 1$
26. $2a(3a + 2) - 6a^2 = 0$
27. $3x - 2x(x - 4) + 2x^2 = x + 1$
28. $-x^2(1 - x) + x(x - x^2) + x = 4$

When you have completed these turn to **6** to check your answers.

6
1. $24w + 6$
2. $-4 - 8t$
3. $10x - 35$
4. $-6y + 2x$
5. $x^3 - 2x^2$
6. $2\pi R^2h - 2\pi a^3$
7. $-3a^3 + 3a^2b$
8. $2p^3q - 2pq^3$
9. $2h^3 - 6h^2 + 6h$
10. $-3a^2t^2 - 3at^2 + 3t^3$
11. $x + 2$
12. a
13. $4x$
14. $3y + x - 2xy$
15. $a^2 + 4a$
16. $3x^2 + x$
17. $11a$
18. $3a^3b + 3a^2b^2 + 2a^4b$
19. $3x^3y - 2x^2y^2 - 3x^3y^2$
20. $2x^2 + 3x$
21. $x = 4$
22. $x = 1$
23. $x = 1\frac{1}{2}$
24. $x = -1$
25. $a = -1$
26. $a = 0$
27. $x = \frac{1}{10}$
28. $x = 4$

Factors

In the product $2 \cdot 3 = 6$, the numbers 2 and 3 are called the *factors* of 6.

If two algebraic expressions are multiplied together to form another expression, the original expressions are called *factors* of the final product. For example,

$$x(2x + 1) = 2x^2 + x$$

The factors of the expression $2x^2 + x$ are x and $2x + 1$.

Factoring

The process of rewriting a given polynomial as a product of other polynomials is called *factoring*. Factoring a polynomial is the exact reverse of multiplying polynomials.

Factor the expression $3x^2 + 2x$.

Try it. Write this expression as a product.

Check your answer in **7**.

---

7 $3x^2 + 2x = x(3x + 2)$

Factoring a Monomial from a Polynomial

To factor a monomial from a polynomial we must first examine each term of the original polynomial to determine what they have in common.

$$3x^2 + 2x = 3 \cdot x \cdot \boxed{x} + \quad 2 \cdot \boxed{x}$$

common factor

In this case, each term of the expression contains the variable x as a factor. Using the distributive law in reverse,

$$3x^2 + 2x = 3 \cdot x \cdot \boxed{x} + 2 \cdot \boxed{x} = \boxed{x} \, (3x + 2)$$

common factors

If each term in the polynomial to be factored is written out as a product in this way, it is easy to find the common factors. For example, to factor the expression:

$$4x^2y - 6x$$

Rewrite it as

$$2 \cdot 2 \cdot x \cdot x \cdot y - 2 \cdot 3 \cdot x$$

Now search each term for the common factors. Notice that each term contains both 2 and x.

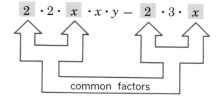

common factors

The first term is equal to $2x \cdot 2xy$.
The second term is equal to $2x \cdot 3$.
Therefore

$$4x^2y - 6x = 2x(2xy - 3)$$

To check your work multiply $2x$ by $2xy - 3$. Factoring is simply the reverse of multiplying.

Factor the expression

$$2x^3 + 4x^2 - 6x$$

Check your answer in **8**

DEGREE OF A POLYNOMIAL

The *degree* of a term in a polynomial in one variable is the exponent of the variable in the term. For example,

$2x$ is of degree 1 in the variable x.
$3a^2$ is of degree 2 in the variable a.
$12t^4$ is of the fourth degree in the variable t.
4 is of degree 0 since $4 = 4x^0$.

If more than one variable is present in the term, then the degree of the term is the sum of the exponents of the variables in the term. For example,

xy^2 has degree 3 in the variables x and y.
$2a^3b^5$ has degree 8 in a and b.

The *degree of a polynomial* is the degree of the highest degree term in the polynomial. For example, the polynomial

$x^2 + 1$ has degree 2.
$3x^3 + 2x - 1$ is of degree 3
$2x^3 + 5x^2y^3 - y^2$ is of degree 5.

---

8    $2x^3 + 4x^2 - 6x = \underbrace{\boxed{2} \cdot \boxed{x} \cdot x \cdot x} + \underbrace{\boxed{2} \cdot 2 \cdot \boxed{x} \cdot x} - \underbrace{\boxed{2} \cdot 3 \cdot \boxed{x}}$

The common factor is $2x$

Check your answer by multiplying the factors to obtain the original polynomial.

With a little practice it will be possible for you to find the common factor without actually rewriting each term. You will find it quicker to mentally test each term in the expression. For example, to factor the expression

$$2x^3 + 4x^2 - 6x$$

first, test each term for divisibility by 2, the numerical part of the first term. Notice that each of the three terms is evenly divisible by 2, so that 2 is a factor of the entire polynomial.

Second, test each term for divisibility by x. Notice that each of the three terms is evenly divisible by x, so that x is also a factor of the entire polynomial.

Third, test each term for divisibility by x^2. The last term is not divisible by x^2.

Because both 2 and x evenly divide all terms of the expression, $2x$ is a factor of the original expression, and we can write it as

$$2x^3 + 4x^2 - 6x = 2x(x^2 + 2x - 3)$$

Remember to check your answer by multiplying the factors to be certain their product is the original polynomial.

Factor the following polynomials:

(a) $4x - 4y$ (b) $3a - 12$

(c) $x - xy + x^2$ (d) $2x^3 - 4xy$

(e) $-ab - abc - bc$ (f) $xy^2 - x^2y - 2x^2y^2$

Look in **9** for our step-by-step solutions.

WRITING ALGEBRAIC EQUATIONS—600 A.D.

A mathematician from India about 1300 years ago would find the mathematics taught in this book very interesting and up-to-date, but he would use a very different algebraic language than we use. For example, he would write the equation

$$x^2 - 10x = -9$$

like this

$$ya \quad v \, | \, ya \ 10$$
$$ru \quad \dot{9}$$

ya means "unknown"
v means "squared"
The dot indicates a negative number.
The left-hand member of the equation is on the first line.
The right-hand member of the equation is on the second line.
ru means that the number following is an ordinary number.

    . . . and you thought today's algebra was difficult!

---

9   (a) $4x - 4y = 4 \cdot x - 4 \cdot y = 4(x - y)$

(b) $3a - 12 = 3 \cdot a - 3 \cdot 4 = 3(a - 4)$

(c) $x - xy + x^2 = x \cdot 1 - x \cdot y + x \cdot x = x(1 - y + x)$

(d) $2x^3 - 4xy = 2 \cdot x \cdot x \cdot x - 2 \cdot 2 \cdot x \cdot y = 2x(x^2 - 2y)$

(e) $-ab - abc - bc = (-1) \cdot a \cdot b + (-1) \cdot a \cdot b \cdot c + (-1) \cdot b \cdot c$
$$= (-1) \cdot b \cdot (a + ac + c)$$
$$= -b(a + ac + c)$$

Notice that (-1) is a factor in all three terms of the expression in (e).

(f) $xy^2 - x^2y - 2x^2y^2 = x \cdot y \cdot y - x \cdot x \cdot y - 2 \cdot x \cdot x \cdot y \cdot y$
$$= xy(y - x - 2xy)$$

A common mistake on this last problem is to omit part of the common monomial factor and write $xy^2 - x^2y - 2x^2y^2 = x(y^2 - xy - 2xy^2)$. This is *not* complete because the expression in the parentheses may be factored again. Each term contains the variable y. A polynomial has been factored completely only when *all* of the common factors have been identified and separated out.

The following diagram may help you to understand this process of factoring polynomials.

Polynomial to be factored	Factors		Polynomial factored
$4x - 4y$	4	$x - y$	$4(x - y)$
$3a - 12$	3	$a - 4$	$3(a - 4)$
$x - xy + x^2$	x	$(1 - y + x)$	$x(1 - y + x)$
$2x^3 - 4xy$	$2x$	$x^2 - 2y$	$2x(x^2 - 2y)$
$-ab - abc - bc$	$-b$	$(a + ac + c)$	$-b(a + ac + c)$
$xy^2 - x^2y - 2x^2y^2$	xy	$(y - x - 2xy)$	$xy(y - x - 2xy)$

Try these practice problems.

Factor the following polynomials:

1. $2a + 2b$
2. $3t - 6$
3. $x^3 - 3x^2 + 5x$
4. $5x^2 - 10x$
5. $2y^2 - 2y + 2$
6. $6ax^2 + 9x$
7. $18ab - 12a + 6b$
8. $-2a^2 - 2a$
9. $-ab - b^2c - abc$
10. $-2x^2 + 4xy - 2x$
11. $-t + t^2$
12. $x^2y + xy^2 - 2xy$
13. $-xy^2 - x^2y - 3xy$
14. $Q + 2QT$
15. $3ax - x^2 - xy$
16. $ab^2 - c^2b^2 + 2b^3 - ab^2c$
17. $4a^2xy^2 - 6ax^2y + 8axy^2$
18. $4p^3q^5 - 2p^2q^4 + 6p^4q^3$
19. $a(x + 1) + b(x + 1)$
20. $p(p - 4) - 2(p - 4)$
21. $4x(2x - 1) + 2(2x - 1)$
22. $3y(y^2 + 2) - (y^2 + 2)$
23. $2x(x - 1) + x^2(x - 1)$
24. $4a^2(b + 2) - 2a(b + 2)$

Look in **10** to check your answers.

10

1. $2(a + b)$
2. $3(t - 2)$
3. $x(x^2 - 3x + 5)$
4. $5x(x - 2)$
5. $2(y^2 - y + 1)$
6. $3x(2ax + 3)$
7. $6(3ab - 2a + b)$
8. $-2a(a + 1)$
9. $-b(a + bc + ac)$
10. $2x(-x + 2y - 1)$
11. $-t(1 - t)$
12. $xy(x + y - 2)$
13. $-xy(y + x + 3)$
14. $Q(1 + 2T)$
15. $x(3a - x - y)$
16. $b^2(a - c^2 + 2b - ac)$
17. $2axy(2ay - 3x + 4y)$
18. $2p^2q^3(2pq^2 - q + 3p^2)$
19. $(x + 1)(a + b)$
20. $(p - 4)(p - 2)$
21. $(2x - 1)(4x + 2)$
    or $2(2x + 1)(2x - 1)$
22. $(y^2 + 2)(3y - 1)$
23. $x(x + 2)(x - 1)$
24. $2a(2a - 1)(b + 2)$

Answer 10 could also be written $(-1)2x(x - 2y + 1)$ or $-2x(x - 2y + 1)$.
Answer 11 could also be written $(-1)t(1 - t)$ or $t(t - 1)$.

In problems 19 through 24 the common factor is the quantity in parentheses. Factor out this common factor, then simplify the factors if possible.

Problem 23 gives

$$2x(x - 1) + x^2(x - 1) = (x - 1)(2x + x^2)$$
$$= (x - 1)x(2 + x)$$
$$= x(x - 1)(x + 2)$$

Problem 24 gives

$$4a^2(b + 2) - 2a(b + 2) = (b + 2)(4a^2 - 2a)$$
$$= (b + 2)2a(2a - 1)$$
$$= 2a(2a - 1)(b + 2)$$

Remember:

1. Always double check the factor in parentheses to be certain that its terms have no common factors.
2. Check your answer by multiplying the factors in the answer to obtain the original polynomial.

Now, turn to **11** for a set of problems on multiplying polynomials by monomials and factoring monomials from polynomials.

MENTAL MATH QUIZ

Write down the answers to the following problems quickly. You should not need to do any paper and pencil calculations.

Average time: _5:41_

Record time: _3:21_

(These times are for students scoring at least 38 correct in a class in basic algebra in a community college.)

Factor:

1. $2x + 4$	2. $3y - 9$	3. $5p + 15$	4. $8 - 4q$
5. $2x + ax$	6. $3y - py$	7. $y - ay$	8. $x - ax$
9. $a - a^2$	10. $k^2 - 2k$	11. $p^3 - 3p$	12. $t^3 - t^2$
13. $2d^2 - 3d$	14. $2a^2 - 3a$	15. $2t^4 - 4t$	16. $x - xy$
17. $q + q^2$	18. $ab + b^2$	19. $m^2n - mn$	20. $a^2b - ab$
21. $3a^3 - 15$	22. $-2y - y^3$	23. $-4p - 8p^3$	24. $-3ab - 9b^2$
25. $3x - 3y$	26. $2ab - 3b$	27. $5xy - xy^5$	28. $abc + a$
29. $xyz - xy$	30. $2axt - a$	31. $3x^2y - y$	32. $ab^2x + aby$
33. $-x - y$	34. $-ab - t$	35. $-2x - 3$	36. $-3t^2 - 2$
37. $-2 - x^2$	38. $-x + xy$	39. $-2x - 3xyz$	40. $3t - 6k$

The answers are on page 545.

Unit 3
Products
and Factoring

The answers are on page 545.

A. Multiply:

1. $2(t + 1)$ _____

2. $x(2x + 3)$ _____

3. $5(3 - y)$ _____

4. $-2(p + 3)$ _____

5. $-a(x + 3)$ _____

6. $-b(2 - b)$ _____

7. $p(x - p^2)$ _____

8. $4(c - c^2)$ _____

9. $2x(3x - 1)$ _____

10. $3n(2n + 2)$ _____

11. $-y(1 - y)$ _____

12. $-2t(t - t^2)$ _____

13. $7p(3q + 2p)$ _____

14. $-3(x^2 - 2x)$ _____

15. $-2a(-4ax + 3x^2)$ _____

16. $x^2(1 - x)$ _____

17. $2pt(t - p)$ _____

18. $m^2n(m + n)$ _____

19. $(3x - 5)x$ _____

20. $(y^2 - 6y)2y$ _____

B. Multiply:

1. $3(x + y - 1)$ _____

2. $2(x^2 - 2x + 5)$ _____

3. $5(3a^2 - 2a + 1)$ _____

4. $4(2b^2 - b - 3)$ _____

5. $x(x^2 - x - 1)$ _____

6. $y(x^3 + 2x - 2)$ _____

7. $2ab(a + b - 1)$ _____

8. $3pq(p^2 - q^2 - q)$ _____

9. $-4ax(a^2 + a - x)$ _____

10. $-2x^2(x + a - 2)$ _____

11. $-b^2(a^2 + 2ab - b^2)$ _____

12. $-q^2(-x - q - 1)$ _____

13. $x^3(1 + x - 2x^2)$ _____

14. $2y^3(3 - 2y + y^2)$ _____

15. $(3x - 4y + 2xy)x^2$ _____

16. $(1 - 2x + 4y)y$ _____

17. $5t^2(r - t + t^2 + rt - 3)$ _____

18. $ab(a + b - ab + a^2b - ab^2)$ _____

19. $-a^2(1 - a - a^2 - a^3 + a^5)$ _____

20. $-x(x^3 - x^2 + x - 1)$ _____

C. Factor out the common monomials:

1. $x^2 - x$ _____

2. $a - 2a^2$ _____

3. $6y - 3$ _____

4. $8 + 4t$ _____

5. $2a + 4b$ _____

6. $6a^2bc + 9b^3$ _____

Date _____

Name _____

Course/Section _____

7. $8xy^2 - 12x^3$ _____

8. $2ab - 4bc$ _____

9. $27t^2x + 18tx^2$ _____

10. $21ab^2c + 14a^2b$ _____

11. $-3a^3 - 6a$ _____

12. $-10b^3 - 2bc$ _____

13. $4x^2 + 2x$ _____

14. $8a^3 + 4a^2$ _____

15. $30a^4 - 10a^2$ _____

16. $24t^5 - 36t^2$ _____

17. $x^2 + 2x + x^3$ _____

18. $2m^3 - 3m^2 + 5m$ _____

19. $2a^2 - 2a + 6$ _____

20. $3b^3 - 3b + 9$ _____

21. $-12x + 4x^2 - 20x^4$ _____

22. $-6n^2 - 4n$ _____

23. $-20pq - 15p^2$ _____

24. $-6y - 4y^2 - 12y^3$ _____

25. $6a^3b^2c + 10a^2b^5c^2$ _____

26. $16xy^3t^2 - 12x^4yz^3$ _____

27. $21t^2r^3 + 28tr^2$ _____

28. $9p^5q^2r - 12p^2q^4r^2$ _____

29. $a^4 - a^3 + a^2c^2 - ac^3$ _____

30. $x^7 - 4x^3 - 2x^2 - 3x$ _____

31. $x - 2xt$ _____

32. $ax^2 + bx^2 - cx^2$ _____

33. $p(2x - 3) + 3p(2x - 3)$ _____

34. $x^2(x + 3) - 3x(x + 3)$ _____

35. $xy(y - 5) - y^2(y - 5)$ _____

36. $2a^2(2a - 7) + 4a(2a - 7)$ _____

D. Brain Boosters

1. Show that the sum of any three consecutive integers is always divisible by 3.

_____

2. Show that the sum of any three consecutive even integers is always divisible by 6. (*Hint:* Let $2n$ be the first even integer in the set of three.)

_____

3. The Rhind Papyrus, written in Egypt more than 4000 years ago, is the oldest existing mathematics text. Problem 28 in this book reads: "A quantity and its $\frac{2}{3}$ are added together and from the sum $\frac{1}{3}$ of the sum is subtracted and 10 remains. What is the quantity?" Solve this problem.

_____

When you have had the practice you need, either return to the preview on page 193 or turn to **12** where you will learn to multiply binomials.

12 The distributive law shows us how to multiply a binomial by a monomial, and it can be extended to allow us to multiply by polynomials with more than one term.

Multiplying by a monomial,

$$x\,(a + b) = x\,(a) + x\,(b)$$

In the same way, multiplying by a binomial,

$$(x + y)\,(a + b) = (x + y)\,(a) + (x + y)\,(b)$$

The multiplication of a binomial by another binomial is an operation that is used very often in algebra and one that you must be able to perform quickly and correctly. In this section you will learn how to use the distributive law to multiply binomials.

First, perhaps it will help you to work through a binomial multiplication from arithmetic. Multiply $(20 + 4)(30 + 2)$.

Try it this way:

1. Arrange the factors vertically:

$$\begin{array}{r} (20 + 4) \\ \times (30 + 2) \end{array}$$

2. Multiply by one of the terms in the multiplying factor:

$$\begin{array}{r} 20 + 4 \\ 30 + 2 \\ \hline 40 + 8 \end{array}$$

3. Multiply by the second term in the multiplying factor:

$$\begin{array}{r} 20 + 4 \\ 30 + 2 \\ \hline 600 + 120 \end{array}$$

4. Add: $40 + 8 + 600 + 120 = 768$

Check:

$$\begin{array}{r} 24 \\ \times 32 \\ \hline 48 \\ 72 \\ \hline 768 \end{array}$$

768 which agrees with our answer.

The procedure for multiplying binomial algebraic expressions is exactly the same. Use the same procedure to multiply $(x + 2)(2x + 3)$.

Check your work in **13**.

---

13 1. Arrange the factors vertically:

$$\begin{array}{r} (2x + 3) \\ \times (x + 2) \end{array}$$

2. Multiply by the rightmost term:
$2 \cdot 3 = 6$
$2 \cdot 2x = 4x$

$$\begin{array}{r} 2x + 3 \\ x + 2 \\ \hline 4x + 6 \end{array}$$

3. Multiply by the second term:
$x \cdot 3 = 3x$
$x \cdot 2x = 2x^2$

$$\begin{array}{r} 2x + 3 \\ x + 2 \\ \hline 2x^2 + 3x \end{array}$$

4. Add by combining like terms: $4x + 6 + 2x^2 + 3x = 2x^2 + 7x + 6$

Notice that the terms in the answer, $2x^2 + 7x + 6$, are arranged in order of decreasing powers of the variable x, with the x^2 term on the left and numerical terms on the right. This is a convention that most mathematicians follow—not necessary, but neat and orderly.

Try another. Multiply $(2a + b)(a - 2b)$.

Our work is in **14**.

---

14 1. $2a + b$ or better $2a + b$
      $\underline{a - 2b}$ $\underline{a + (-2b)}$

It will help you avoid confusion to write each binomial factor as a *sum* of terms.

2. $2a + b$
$\underline{a + (-2b)}$
$(-2b) \cdot (2a) + (-2b) \cdot (b) = -4ab - 2b^2$

3. $2a + b$
$\underline{a + (-2b)}$
$a \cdot 2a + a \cdot b = 2a^2 + ab$

4. Add: $-4ab - 2b^2 + 2a^2 + ab = 2a^2 - 3ab - 2b^2$

Writing the binomial multiplication vertically, as we have done above, makes it easy to see the sequence of operations to be performed, but it is actually a slow and clumsy way to do the job. It is possible to do the multiplication with no rewriting—"in your head" or "by inspection" as mathematicians say.

The FOIL Method of Multiplying Binomials

Multiplying a binomial by another binomial in this shortcut way involves five simple steps. Here they are: Multiply $(2x + 3)(x + 2)$.

Step 1. $(2x + 3)(x + 2)$

First, multiply the first terms $2x \cdot x = 2x^2$

Step 2. $(2x + 3)(x + 2)$

Second, multiply the outer terms $2x \cdot 2 = 4x$

Step 3. $(2x + 3)(x + 2)$

Third, multiply the inner terms $3 \cdot x = 3x$

Step 4. $(2x + 3)(x + 2)$

Fourth, multiply the last terms $3 \cdot 2 = 6$

Step 5. Finally, add the products by combining like terms.
$2x^2 + 4x + 3x + 6 = 2x^2 + 7x + 6$

This five-step shortcut is exactly the same sequence of multiplications as in the vertical method, and both methods are actually the distributive law at work. You should use the five-step shortcut shown above when you multiply binomials.

Need a memory helper? Some students use the following memory-aide to help them remember this sequence of steps.

FOIL stands for F . . . multiply the first terms
O . . . multiply the outer terms
I . . . multiply the inner terms
L . . . multiply the last terms

Ready for some practice? Multiply these using the five-step process.

(a) $(x + 1)(x + 2) = $ _____

(b) $(2a + b)(a + 3b) = $ _____

(c) $(y - 1)(2y + 3) = $ _____

Work carefully. Our step-by-step multiplications are shown in **15**.

---

15   (a) $(x + 1)(x + 2)$

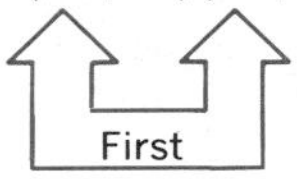

First $x \cdot x = x^2$

$(x + 1)(x + 2)$

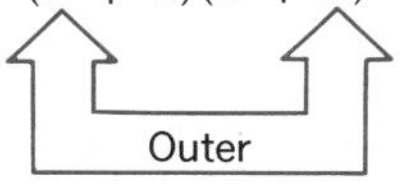

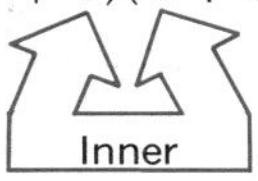

Inner $1 \cdot x = x$

$(x + 1)(x + 2)$

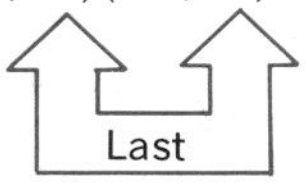

Outer $x \cdot 2 = 2x$

$(x + 1)(x + 2)$

Last $1 \cdot 2 = 2$

Add: $x^2 + 2x + x + 2 = x^2 + 3x + 2$

(b) $(2a + b)(a + 3b)$

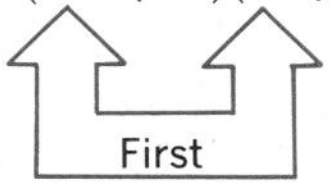

First $2a \cdot a = 2a^2$

$(2a + b)(a + 3b)$

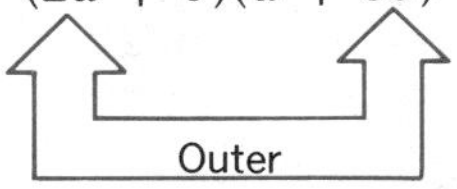

Outer $2a \cdot 3b = 6ab$

$(2a + b)(a + 3b)$

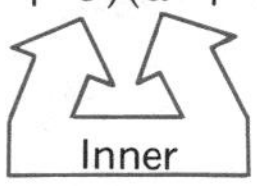

Inner $b \cdot a = ab$

$(2a + b)(a + 3b)$

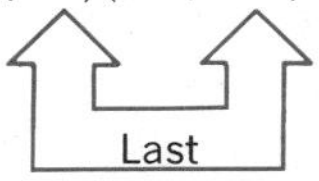

Last $b \cdot 3b = 3b^2$

Add: $2a^2 + 6ab + ab + 3b^2 = 2a^2 + 7ab + 3b^2$

(c) $(y - 1)(2y + 3)$

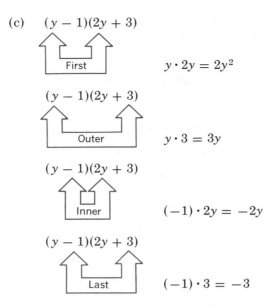

$y \cdot 2y = 2y^2$

$(y - 1)(2y + 3)$

$y \cdot 3 = 3y$

$(y - 1)(2y + 3)$

$(-1) \cdot 2y = -2y$

$(y - 1)(2y + 3)$

$(-1) \cdot 3 = -3$

Add: $2y^2 + 3y - 2y - 3 = 2y^2 + y - 3$

Notice in the last problem that we treat the multiplier $(y - 1)$ as a sum $(y + (-1))$.

When you develop confidence and quickness at this process, you will be able to write the final answer without writing any of the intermediate steps. While you are learning the process you'll want to be careful and probably need to write out each step. Your actual work will probably look something like this:

$$(y - 1)(2y + 3) = 2y^2 + 3y - 2y - 3$$
$$= 2y^2 + y - 3$$

Here are a few practice problems.

1. $(p + 3)(p + 4)$ 2. $(x + 1)(x + 1)$
3. $(a + 2)(a - 5)$ 4. $(y + 3)(2y - 1)$
5. $(x - 2)(x - 2)$ 6. $(a - 4)(a + 4)$
7. $(2q + 1)(q + 4)$ 8. $(3c - 1)(2c + 3)$
9. $(1 - x)(2 - x)$ 10. $(3 - y)(3y + 2)$
11. $(2a^2 - 3b)(a + 2b^2)$ 12. $(x^2 + 2x)(x + 2)$

Our answers are in **16**.

---

16   1. $p^2 + 7p + 12$ 2. $x^2 + 2x + 1$
3. $a^2 - 3a - 10$ 4. $2y^2 + 5y - 3$
5. $x^2 - 4x + 4$ 6. $a^2 - 16$
7. $2q^2 + 9q + 4$ 8. $6c^2 + 7c - 3$
9. $x^2 - 3x + 2$ 10. $-3y^2 + 7y + 6$
11. $2a^3 + 4a^2b^2 - 3ab - 6b^3$ 12. $x^3 + 4x^2 + 4x$

Several kinds of binomial products appear so often in algebraic problems that they should receive special attention.

Can you find the pattern that links these problems?

(a) $(a - 1)(a + 1) = $ _____

(b) $(x + 2)(x - 2) = $ _____

(c) $(y - 3)(y + 3) = $ _____

(d) $(5 - a)(a + 5) = $ _____

Multiply each pair of binomials and find the pattern. Turn to **17** to continue.

17

(a) $(a - 1)(a + 1) = a^2 - 1$
(b) $(x + 2)(x - 2) = x^2 - 4$
(c) $(y - 3)(y + 3) = y^2 - 9$
(d) $(5 - a)(a + 5) = 25 - a^2$

In each of these problems one of the binomial multipliers is the *sum* of two terms and the other binomial is the *difference* of the same two terms. Their product is always the difference of two perfect squares.

$$(a + b)(a - b) = a^2 - b^2$$

The product of the sum and difference of terms is the difference of their squares.

$$\underbrace{(a + b)}\underbrace{(a - b)} = \underbrace{a^2 - b^2}$$

The sum   The difference   The difference
of terms   between terms between a^2 and b^2
a and b a and b

Notice that the difference of squares on the right follows the order of the difference on the left. If the difference of terms is $(a - b)$, the differences of squares is $a^2 - b^2$. If the difference of terms is $(b - a)$, the differences of squares is $b^2 - a^2$.

$$(a + b)\underbrace{(a - b)} = \underbrace{a^2 - b^2}$$

or

$$(a + b)\underbrace{(b - a)} = \underbrace{b^2 - a^2}$$

If you recognize a given binomial multiplication as the product of the sum and difference of two terms, you can immediately write down the answer. For example,

$$\underbrace{(3x + 2y^2)}\ \underbrace{(3x - 2y^2)} = \underbrace{9x^2} - \underbrace{4y^4}$$

The sum   The dif-
of $3x$ ference be-
and $2y^2$ tween $3x$
         and $2y^2$ $(3x)^2 - (2y^2)^2$

Here are a few practice problems to help you remember this special binomial product.

(a) $(a + 8)(a - 8) = $ _____

(b) $(2x - 3t)(3t + 2x) = $ _____

(c) $(q^2 + p^2)(q^2 - p^2) = $ _____

211   **3-4 Multiplying Binomials**

(d) $(1 - 2x^3)(1 + 2x^3) = $ _____

(e) $(-3x - 2c)(-3x + 2c) = $ _____

Check your work in **18**.

---

18   (a) $(a + 8)(a - 8) = a^2 - 8^2 = a^2 - 64$
        (b) $(2x - 3t)(3t + 2x) = (2x)^2 - (3t)^2 = 4x^2 - 9t^2$
        (c) $(q^2 + p^2)(q^2 - p^2) = (q^2)^2 - (p^2)^2 = q^4 - p^4$
        (d) $(1 - 2x^3)(1 + 2x^3) = 1^2 - (2x^3)^2 = 1 - 4x^6$
        (e) $(-3x - 2c)(-3x + 2c) = [(-3x) - (2c)][(-3x) + (2c)]$
$$= (-3x)^2 - (2c)^2$$
$$= 9x^2 - 4c^2$$

Square of a Binomial

Another special binomial product occurs when we multiply a binomial by itself. Again, find the pattern in each of the following products:

(a) $(x + 1)^2 = $ _____

(b) $(a + 2)^2 = $ _____

(c) $(p + q)^2 = $ _____

(d) $(2a + 3)^2 = $ _____

Multiply them using the normal five-step process you have already learned, then try to discover the pattern that exists in the answers. Check your guess in **19**.

---

19   (a) $(x + 1)^2 = x^2 + 2x + 1$
        (b) $(a + 2)^2 = a^2 + 4a + 4$
        (c) $(p + q)^2 = p^2 + 2pq + q^2$
        (d) $(2a + 3)^2 = 4a^2 + 12a + 9$

In each problem the product is a trinomial—it has three terms. The first term in the product is the square of the first term in the binomial. The third term in the product is the square of the second term in the binomial. The middle term of the product is exactly twice the product of the two terms in the binomial.

$$\boxed{(a + b)^2 = a^2 + 2ab + b^2}$$

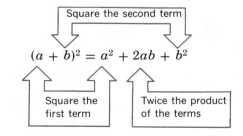

Knowing this pattern allows you to write the square of any binomial quickly and with only mental calculations.

If the binomial to be squared contains negative terms, write it as a sum of terms and then continue as before. For example,

$$(a - b)^2 = (a + (-b))^2 = a^2 + 2a(-b) + (-b)^2 = a^2 - 2ab + b^2$$

Twice the product of a and $(-b)$

Remember that $(-b)^2 = b^2$.

More examples:
$$(2x - 5t)^2 = (2x)^2 + 2(2x)(-5t) + (-5t)^2 = 4x^2 - 20xt + 25t^2$$
$$(-3a - 4c)^2 = (-3a)^2 + 2(-3a)(-4c) + (-4c)^2 = 9a^2 + 24ac + 16c^2$$

Practice squaring binomials by completing this set of problems. Write each product as a polynomial.

1. $(2 + x)^2$ 2. $(\triangle + \square)^2$ 3. $(3x + y)^2$

4. $(a - 3b)^2$ 5. $(5s + 2t)^2$ 6. $(2m - 3n)^2$

7. $(-2x + 6y)^2$ 8. $(-4p - 5q)^2$ 9. $(2a^2 + 3ab)^2$

10. $\left(\dfrac{1}{2}x - 2y\right)^2$ 11. $(3k^2 - k)^2$ 12. $(-2d - d^2)^2$

Check your answers in **20**.

BINOMIAL PRODUCTS AS AREAS

Some students find it helpful to have a geometric picture of a binomial product.

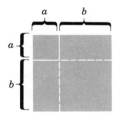

The area of the large square is $(a + b)^2$.

The large square may be divided into four parts:

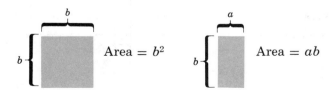

Because these four parts make up the whole square, we may write.

$$(a + b)^2 = a^2 + b^2 + ab + ab = a^2 + 2ab + b^2$$

20

1. $(2 + x)^2 = 4 + 4x + x^2$
2. $(\triangle + \square)^2 = \triangle^2 + 2\triangle\square + \square^2$
3. $(3x + y)^2 = 9x^2 + 6xy + y^2$
4. $(a - 3b)^2 = a^2 - 6ab + 9b^2$
5. $(5s + 2t)^2 = 25s^2 + 20st + 4t^2$
6. $(2m - 3n)^2 = 4m^2 - 12mn + 9n^2$
7. $(-2x + 6y)^2 = 4x^2 - 24xy + 36y^2$
8. $(-4p - 5q)^2 = 16p^2 + 40pq + 25q^2$
9. $(2a^2 + 3ab)^2 = 4a^4 + 12a^3b + 9a^2b^2$
10. $\left(\dfrac{1}{2}x - 2y\right)^2 = \dfrac{1}{4}x^2 - 2xy + 4y^2$
11. $(3k^2 - k)^2 = 9k^4 - 6k^3 + k^2$
12. $(-2d - d^2)^2 = 4d^2 + 4d^3 + d^4$

Multiplying a General Polynomial by a Binomial

The distributive law can be used to help us multiply trinomials or larger polynomials by binomials. For example, as you learned earlier in this unit,

$$\boxed{3}\,(x - 2y + 2) = \boxed{3}\,(x) + \boxed{3}\,(-2y) + \boxed{3}\,(2) = 3x - 6y + 6$$

Multiply each term in the parentheses by 3.

or

$$\boxed{x}\,(x^2 - 2x - 3) = \boxed{x}\,(x^2) + \boxed{x}\,(-2x) + \boxed{x}\,(-3) = x^3 - 2x^2 - 3x$$

Multiply each term in the parentheses by x.

or even

$$\boxed{-2a}\,(2a^2 - ab + 3b^2) = \boxed{(-2a)}\,(2a^2) + \boxed{(-2a)}\,(-ab) + \boxed{(-2a)}\,(3b^2)$$
$$= -4a^3 + 2a^2b - 6ab^2$$

Multiply each term in the parentheses by $(-2a)$.

If the multiplier is a binomial, the procedure is only slightly more difficult. For example,

$$\boxed{(x + 2)}\,(x^2 + 3x + 4)$$

$$= \boxed{x}\,(x^2 + 3x + 4) + \boxed{2}\,(x^2 + 3x + 4) \qquad \text{using the distributive law}$$

$$= \boxed{x}\,(x^2) + \boxed{x}\,(3x) + \boxed{x}\,(4) + \boxed{2}\,(x^2) + \boxed{2}\,(3x) + \boxed{2}\,(4)$$

multiplying each term in each parentheses

$$= x^3 + 3x^2 + 4x + 2x^2 + 6x + 8$$

Finally, simplify by combining like terms.

$$= x^3 + 5x^2 + 10x + 8$$

Repeat the process above on each of the following problems.

(a) $(x - y)(x^2 + 2xy - y^2)$
(b) $(a - 1)(a - b + 1)$
(c) $(2x + 3t)(x + xt - t)$
(d) $(x + 1)^3$
(e) $t(t + 1)^2$
(f) $x(x - 3)(x^2 - x - 1)$

Work carefully. Our step-by-step explanations are in **21**.

---

21 (a) $\boxed{(x - y)}\,(x^2 + 2xy - y^2)$

$$= \boxed{x}\,(x^2 + 2xy - y^2) - \boxed{y}\,(x^2 + 2xy - y^2)$$

$$= \boxed{x}\,(x^2) + \boxed{x}\,(2xy) + \boxed{x}\,(-y^2) + \boxed{(-y)}\,(x^2) + \boxed{(-y)}\,(2xy) + \boxed{(-y)}\,(-y^2)$$

$$= x^3 + 2x^2y - xy^2 - x^2y - 2xy^2 + y^3$$

$$= x^3 + x^2y - 3xy^2 + y^3 \qquad \text{combining like terms}$$

(b) $(a - 1)\ (a - b + 1)$

   $=\boxed{a}\ (a - b + 1) +\ \boxed{(-1)}\ (a - b + 1)$

   $=\boxed{a}\ (a) +\ \boxed{a}\ (-b) +\ \boxed{a}\ (1) +\ \boxed{(-1)}\ (a) +\ \boxed{(-1)}\ (-b) +\ \boxed{(-1)}\ (1)$

   $= a^2 - ab + a - a + b - 1$

   $= a^2 - ab + b - 1$ combining like terms

(c) $(2x + 3t)\ (x + xt - t)$

   $=\boxed{2x}\ (x + xt - t) +\ \boxed{3t}\ (x + xt - t)$

   $=\boxed{2x}\ (x) +\ \boxed{2x}\ (xt) +\ \boxed{2x}\ (-t) +\ \boxed{3t}\ (x) +\ \boxed{3t}\ (xt) +\ \boxed{3t}\ (-t)$

   $= 2x^2 + 2x^2t - 2xt + 3xt + 3xt^2 - 3t^2$

   $= 2x^2 + 2x^2t + xt + 3xt^2 - 3t^2$

(d) $(x + 1)^3 =\ \boxed{(x + 1)}\ (x + 1)(x + 1)$

   $=\ \boxed{(x + 1)}\ (x^2 + 2x + 1)$

   $=\ \boxed{x}\ (x^2 + 2x + 1) +\ \boxed{1}\ (x^2 + 2x + 1)$

   $= x^3 + 2x^2 + x + x^2 + 2x + 1$

   $= x^3 + 3x^2 + 3x + 1$

(e) $\boxed{t}\ (t + 1)^2 =\ \boxed{t}\ (t^2 + 2t + 1)$

   $=\ \boxed{t}\ (t^2) +\ \boxed{t}\ (2t) +\ \boxed{t}\ (1)$

   $= t^3 + 2t^2 + t$

(f) $x(x - 3)(x^2 - x - 1) = x[\ \boxed{x}\ (x^2 - x - 1)\ \boxed{-3}\ (x^2 - x - 1)]$

   $= x[x^3 - x^2 - x - 3x^2 + 3x + 3]$

   $= x[x^3 - 4x^2 + 2x + 3]$

   $= x^4 - 4x^3 + 2x^2 + 3x$

Now turn to **22** for a set of practice problems on binomial products.

I multiplied $(a + b)^2$ and got $a^2 + b^2$. What did I do wrong?

$(a + b)^2$ is equal to $(a + b)(a + b)$ and when you multiply these binomials you get $a^2 + 2ab + b^2$. Don't forget to multiply the inner and outer terms.

MORE BINOMIAL PRODUCTS AS AREAS

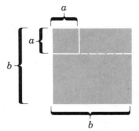

The area of the large square is b^2. This square may be divided into three parts:

 Area $= a^2$

 Area $= a(b - a)$

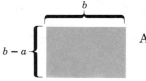 Area $= b(b - a)$

Adding the two rectangles together,

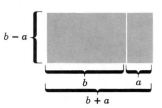

Area $= (b - a)(b + a)$

Therefore the large area is

$$b^2 = a^2 + (b - a)(b + a) \qquad \text{or} \qquad b^2 - a^2 = (b - a)(b + a)$$

Unit 3
Products
and Factoring

Answers are on page 546.

A. Write each of the following products as a polynomial.

1. $(a + 2)(a + 3)$ $a^2 + 6a + 6$

2. $(x - 2)(x - 3)$ _____

3. $(t - 3)(t + 4)$ _____

4. $(p + 5)(p - 4)$ $p^2 + p - 20$

5. $(1 - x)(4 - x)$ _____

6. $(2 - y)(5 + y)$ _____

7. $(3R + 1)(2R - 1)$ $6R^2 - R - 1$

8. $(5Q + 1)(2 - 3Q)$ _____

9. $(a + 2)(2a - 2)$ _____

10. $(c - 3)(3c - 1)$ $3c^2 - 10c + 3$

11. $(4x - 2a)(x + 2a)$ _____

12. $(5t - 2p)(t - 3p)$ _____

13. $(1 - 2x)(2a + 4x)$ $-8x^2 - 4ax + 2a + 4x$

14. $(3 - 2t)(3t + 5)$ _____

15. $(a + b)(x + y)$ _____

16. $(a - b)(x + y)$ $ax + ay - bx - by$

17. $(a - b)(x - y)$ _____

18. $(a + b)(x - y)$ _____

19. $(2ax - x)(a + 2x)$ $2a^2x + 4ax^2 - ax - 2x^2$

20. $(3p + q)(q - 2p)$ _____

21. $(x^2 - x)(x - 1)$ _____

22. $(a^2 + 2a)(3a + 2)$ $3a^3 + 8a^2 + 4a$

23. $(2t^2 - 3t)(t^2 + 2t)$ _____

24. $(2m^2 + n^2)(3n^2 - m^2)$ _____

25. $(a - 1)(2ax - 1)$ $2a^2x + a - 2ax + 1$

26. $(2R - 3)(4 + Rt)$ _____

27. $(1 - 2x)(2 - 3x)$ _____

28. $(G - 3F)(3G - F)$ $3G^2 - 10FG + 3F^2$

29. $(c^2 - 1)(c - 1)$ _____

30. $(n^2 + 1)(n^3 - 1)$ _____

B. Write each of the following products as a polynomial.

1. $(x - 1)^2$ $(x-1)(x-1)$

2. $(Q + 2)^2$ _____

3. $(T - 3)^2$ _____

4. $(1 + q)^2$ $q^2 + 2q + 1$

5. $(p + q)^2$ _____

6. $(m - n)^2$ _____

7. $(2a - b)^2$ $4a^2 - 4ab + b^2$

8. $(3x + 2y)^2$ _____

9. $(-2 - x)^2$ _____

10. $(-c - 3)^2$ $c^2 + 6c + 9$

11. $(-d + 4)^2$ _____

12. $(-5 + x)^2$ _____

13. $(2m + 3n)^2$ $4m^2 + 12mn + 9n^2$

14. $(m^2 + n^2)^2$ _____

15. $(2ax - 3a^2)^2$ _____

16. $(-5bc + 2c^2)^2$ $25b^2c^2 - 20bc^3 + 4c^4$

17. $2(x - 4)^2$ _____

18. $3(3 - y)^2$ _____

Date _____

Name _____

Course/Section _____

19. $a(a + 1)^2$ $\underline{a^3 + 2a^2 + a}$

20. $p(2p + 3)^2$ _____

21. $2(x - 2)(y + 3)$ _____

22. $5(1 - a)(1 - 2a)$ $\underline{5 - 15a + 10a^2}$

23. $t(2t - 1)(1 - t)$ _____

24. $x(2x - a)(3x + a)$ _____

25. $(a + 1)(a - 1)$ $\underline{a^2 - 1}$

26. $(x + 2)(2 - x)$ _____

27. $(2t - 3x)(3x + 2t)$ _____

28. $(2t - 3x)(2t + 3x)$ $\underline{4t^2 - 9x^2}$

29. $(2p + 2q)(2p - 2q)$ _____

30. $(ax^2 - y^2)(y^2 + ax^2)$ _____

31. $2(p + 1)(p - 1)$ _____

32. $p(p + 2)(p - 2)$ _____

C. Multiply.

1. $(x + 2)(x^2 - 3x - 5)$ _____

2. $(y - 3)(y^2 - y - 2)$ _____

3. $(a - 3)(2a + 1 - a^2)$ _____

4. $(c + 2)(2 - 3c + 4c^2)$ _____

5. $(2x - y)(2x + y + 2)$ _____

6. $(1 + x)(1 + x + x^2)$ _____

7. $(1 + x)(1 - x + x^2)$ _____

8. $n(n + 1)(2n + 1)$ _____

9. $n^2(n + 1)^2$ _____

10. $(a - x)(a^2 + ax + x^2)$ _____

11. $(a + x)^3$ _____

12. $(2 + p)^3$ _____

13. $(a + x)(a^2 + x^2)$ _____

14. $(a + b)(a^2 + ax + x^2)$ _____

15. $(2x + 3y)(x^2 - 4xy + y^2)$ _____

16. $(a - 2b)(3b^2 - 3ab - a^2)$ _____

17. $(1 + x)(x + 2)^2$ _____

18. $(t - 2)(t - 1)^2$ _____

19. $(1 - x)(1 + x + x^2 + x^3)$ _____

20. $(a^2 + x^2)^2 - 2a^2x$ _____

21. $(x + 2)(x^2 - 2x - 1)$ _____

22. $(a - 2b)(2a^2 - ab - b^2)$ _____

23. $y(y - 1)(y + 4)$ _____

24. $x(2x - 3)(2x - 5)$ _____

25. $(x - 1)(x + 2)(x - 3)$ _____

26. $(a - 2)(a - 3)(a + 4)$ _____

27. $(x - 1)(x^4 - x^2 - 1)$ _____

28. $(2t - 1)(3t^3 - t + 2)$ _____

29. $(2a - 1)^2(3a - 2)$ _____

30. $(5x - 3)(2x - 3)^2$ _____

D. Brain Boosters.

1. Denny has the unfortunate habit of forgetting to include parentheses when he writes algebra expressions. When he was asked to multiply $(a - b)(c + d)$ he wrote the problem as $a - b(c + d)$. Find the difference between the two expressions. (*Hint:* Write them as polynomials and subtract.)

_____

2. Show that $(1 + x)(1 + x + x^2)(1 - x + x^2)$ equals $1 + x + x^2 + x^3 + x^4 + x^5$.

_____

3. $1^2 + 2^2 = 1 + 4 = 5$

   $1^2 + 2^2 + 3^2 = 1 + 4 + 9 = 14$

   and, in general,

   $$1^2 + 2^2 + 3^2 + \cdots + n^2 = \frac{n}{6}(n+1)(2n+1)$$

   (a) Write this last product as a polynomial.

   (b) Find the sum of the squares of the first 50 positive integers (up to $n = 50$) using this formula.

   _____

4. Find $(a + b + c)^2 = $ _____

   $(a + b - c)^2 = $ _____

   $(a - b - c)^2 = $ _____

   $(1 + x)^4 = $ _____

5. About 300 B.C., Euclid, a very famous ancient Greek mathematician, wrote a mathematics textbook that was used for more than 2000 years. Part two of Euclid's book includes the following ten theorems (written here in modern algebra notation). Show that these are true by writing both sides of each equation as a polynomial.

   (1) $a(b + c + d + e + \ldots) = ab + ac + ad + ae + \ldots$ _____

   (2) $(a + b)a + (a + b)b = (a + b)^2$ _____

   (3) $(a + b)a = a^2 + ab$ _____

   (4) $(a + b)^2 = a^2 + b^2 + 2ab$ _____

   (5) $(a + b)(a - b) + b^2 = a^2$ _____

   (6) $(2a + b)b + a^2 = (a + b)^2$ _____

   (7) $a^2 + b^2 = 2ab + (a - b)^2$ _____

   (8) $4(a + b)a + b^2 = (2a + b)^2$ _____

   (9) $(a + b)^2 + (a - b)^2 = 2a^2 + 2b^2$ _____

   (10) $(2a + b)^2 + b^2 = 2(a + b)^2 + 2a^2$ _____

6. Sally Smith has some strange ideas about how to do arithmetic. For example, when she was asked to evaluate

   $$\frac{37^3 + 13^3}{37^3 + 24^3}$$

   she "canceled" the exponents to get $\dfrac{37 + 13}{37 + 24}$. Now cancelling like this is illegal, but the answer just happens to be correct. Show that for any integers a and b,

$$\frac{a^3 + b^3}{a^3 + (a - b)^3} = \frac{a + b}{a + (a - b)}$$

(*Hint:* Cross multiply the fractions to get

$(a^3 + b^3)(a + (a - b)) = (a + b)(a^3 + (a - b)^3)$.

Then multiply the expressions on both sides of this equation to see if the equation is true.)

7. You can show, by doing the arithmetic, that $1\frac{1}{2} \cdot 3 = 1\frac{1}{2} + 3$ and that $1\frac{1}{3} \cdot 4 = 1\frac{1}{3} + 4$.

   (a) Use your imagination to find the pattern in these equations.

   (b) Use your knowledge of algebra to show that
   $$\left(1 + \frac{1}{n}\right)(n + 1) = \left(1 + \frac{1}{n}\right) + (n + 1).$$
   (*Hint:* the first arithmetic equation comes from putting $n = 2$ in the algebra formula. The second arithmetic equation comes from putting $n = 3$ in the algebra formula.)

8. Select any two numbers whose difference is one. (They need not be integers, but may be any real or imaginary, rational or irrational, numbers.) Show that the difference of their squares always equals the sum of the two numbers.

9. If a and b are two non-zero integers, for what values of a and b is the following equation true?
   $(x - a)(3x - 2) = 3x^2 + bx + 10$

When you have had the practice you need either return to the preview on page 193 or turn to **23** where you will learn more about factoring polynomials.

3-5 WORKING WITH
POLYNOMIALS

23

In the first section of this unit you learned to factor monomials from polynomials. For example, you learned that

$$2a^2 - 4a = 2a(a - 2)$$

and

$$3x^3 - 2x^2 + x = x(3x^2 - 2x + 1)$$

This kind of factoring is easy to do. You simply inspect the polynomial term by term to see if the terms have any factor in common.

Factoring Trinomials

It is possible to factor trinomials in a more general way. Some trinomials, but not *all*, can be rewritten as a product of binomials, as the following diagram shows.

Trinomial to be factored	Factors		Trinomial factored
$x^2 + 3x + 2$	$x + 1$	$x + 2$	$(x + 1)(x + 2)$
$2a^2 + ab - b^2$	$2a - b$	$a + b$	$(2a - b)(a + b)$
$8x^2 + 10xy - 3y^2$	$4x - y$	$2x + 3y$	$(4x - y)(2x + 3y)$

In this chapter we will consider only factors whose numerical coefficients are rational numbers, in other words, factors such as $(x - 2)$, $(3a + b)$, or $(2x^2 - 5)$, but never factors such as $(x - \sqrt{2})$ or $(2a + \sqrt{3})$.

Being able to write a polynomial as a product of its factors is a very useful skill in algebra, a skill that is used very often, but one that some students find difficult to learn. The method you will learn here is a simple, step-by-step process designed for beginning algebra students.

To begin, factor each of the following expressions:

$x^2 + 2x = $ _____

and

$5x + 10 = $ _____

When you have completed these check your answers in **24**.

---

24 $x^2 + 2x = x(x + 2))$

$5x + 10 = 5(x + 2)$

Notice that the binomial $(x + 2)$ is a factor in both expressions.

Because these expressions are factorable in this way, the polynomial $x^2 + 2x + 5x + 10$ can be written as

$x(x + 2) + 5(x + 2)$

Using the distributive law in reverse, this can be rewritten as

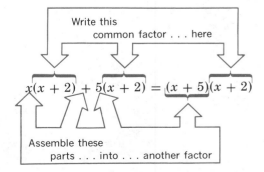

221   3-5 Factoring Trinomials

Copyright © 1982 by John Wiley & Sons, Inc.

According to the distributive law,

$(x + 5)\ (x + 2) =\ x\ (x + 2) +\ 5\ (x + 2)$

Therefore, the reverse is true also:

$x\ (x + 2) +\ 5\ (x + 2) =\ (x + 5)\ (x + 2)$

The factors of the polynomial $x^2 + 2x + 5x + 10$ are $(x + 5)$ and $(x + 2)$.

Use the method shown above to factor the expression $a^2 + 7a + 2a + 14$.

Check your work in **25**

---

25 $a^2 + 7a + 2a + 14 = (a^2 + 7a) + (2a + 14)$

and $a^2 + 7a = a(a + 7)$
$\qquad\quad 2a + 14 = 2(a + 7)$

so that

$a^2 + 7a + 2a + 14 = a(a + 7) + 2(a + 7) = (a + 2)(a + 7)$

The factors of $a^2 + 7a + 2a + 14$ are $(a + 2)$ and $(a + 7)$.

As you will see in a few frames, this procedure of breaking the polynomial to be factored into parts, factoring the parts, and regrouping to get the final factors, is the key to factoring trinomials. To be certain you can do this correctly, try the following problems.

Factor

(a) $p^2 + 3p + 4p + 12$ (b) $x^2 + 2x - 5x - 10$

(c) $y^2 - y + 4y - 4$ (d) $t^2 - 4t - 5t + 20$

(e) $6x^2 + 3x + 4x + 2$ (f) $9a^2 + 15a - 12a - 20$

Our step-by-step answers are in **26**

---

26   (a) $p^2 + 3p + 4p + 12 = (p^2 + 3p) + (4p + 12)$
$\qquad\qquad\qquad\qquad\quad = p(p + 3) + 4(p + 3)$
$\qquad\qquad\qquad\qquad\quad = (p + 4)(p + 3)$

(b) $x^2 + 2x - 5x - 10 = (x^2 + 2x) + (-5x - 10)$
$\qquad\qquad\qquad\qquad\quad = x(x + 2) + (-5)(x + 2)$
$\qquad\qquad\qquad\qquad\quad = (x - 5)(x + 2)$

Writing the polynomial as a sum of two binomials as shown in the first line helps you to keep careful track of negative signs. Check to be certain that

$(-5)(x + 2) = (-5x - 10)$

(c) $y^2 - y + 4y - 4 = (y^2 - y) + (4y - 4)$
$\qquad\qquad\qquad\qquad = y(y - 1) + 4(y - 1)$
$\qquad\qquad\qquad\qquad = (y + 4)(y - 1)$

(d) $t^2 - 4t - 5t + 20 = (t^2 - 4t) + (-5t + 20)$
$\qquad\qquad\qquad\qquad\quad = t(t - 4) + (-5)(t - 4)$
$\qquad\qquad\qquad\qquad\quad = (t - 5)(t - 4)$

Of course, $t + (-5) = t - 5$.

(e) $6x^2 + 3x + 4x + 2 = (6x^2 + 3x) + (4x + 2)$
$\qquad\qquad\qquad\qquad\quad = 3x(2x + 1) + 2(2x + 1)$
$\qquad\qquad\qquad\qquad\quad = (3x + 2)(2x + 1)$

When factoring $6x^2 + 3x$ be certain you factor out all of the common factor $3x$ and not only the 3 or the x.

(f) $9a^2 + 15a - 12a - 20 = (9a^2 + 15a) + (-12a - 20)$
$= 3a(3a + 5) + (-4)(3a + 5)$
$= (3a - 4)(3a + 5)$

Using this method, the trinomial $x^2 + 6x + 8$ is easy to factor if you write it as

$x^2 + 4x + 2x + 8.$

But how do you know to rewrite it in this way? What flash of intuition tells you to rewrite $6x$ as $4x + 2x$?

The answer is that to factor any trinomial $ax^2 + bx + c$, where $a = 1$, you must find two integers whose *sum* is b, the coefficient of the x term, and whose *product* is c, the constant term. Then use these two new integers in rewriting the trinomial. For example,

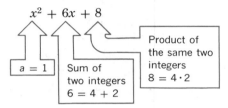

Write $x^2 + 6x + 8$ as $x^2 + 4x + 2x + 8$ and then factor as before.

$x^2 + 6x + 8 = x^2 + 4x + 2x + 8$
$= x(x + 4) + 2(x + 4)$
$= (x + 2)(x + 4)$

Use this method to factor the trinomial $7p + 12 + p^2$.

Our explanation is in **27**.

---

27   Follow these three steps.

Step 1. Arrange the terms of the polynomial so that the squared term appears first, the first power term is second, and the constant term last, if they are not already in this order. Identify the coefficient of the first power term and the constant term.

Example: If the trinomial to be factored is given as $7p + 12 + p^2$ write it as $p^2 + 7p + 12$. The coefficient of the first power term is 7 and the constant term is 12.

Step 2. Find two integers whose sum is the coefficient of the first power term and whose product is the constant term.

Example: Find two integers whose sum is 7 and whose product is 12. The integers are 3 and 4.

Factors of 12	Sum of Factors
$12 = 1 \cdot 12$	$1 + 12 = 13$
$12 = 2 \cdot 6$	$2 + 6 = 8$
$12 = 3 \cdot 4$	$3 + 4 = 7$

$\Leftarrow$ 3 and 4 are the integers we need.

Step 3. Rewrite the trinomial using the two integers found in step 2 and factor it. The first power term is rewritten as a sum of two terms and the two integers found in step 2 are the coefficients of these two terms.

Example: $p^2 + 7p + 12 = p^2 + 3p + 4p + 12$
$$= p(p + 3) + 4(p + 3)$$
$$= (p + 4)(p + 3)$$

In step 2 the two "sum and product" integers may be either positive or negative.

As a final step, always check your answer by multiplying the factors you have found to be certain their product equals the original trinomial.

Check: $(p + 4)(p + 3) = p^2 + 7p + 12$

$\boxed{\text{Important}}\!\!>$ This procedure is useful only when the coefficient of the second power term is one, that is when the trinomial begins with x^2 or a^2 or t^2 or a similar term but never with $2x^2$ or $5t^2$ or $6a^2$.

Now, factor these expressions using this method.

(a) $10x + x^2 + 24$ (b) $a^2 - 16 + 6a$

(c) $q^2 - 10q + 21$ (d) $y^2 - 20 - 8y$

(e) $x^2 - 30 - x$

Check your work in **28.**

---

28   (a)   **Step 1.** $10x + x^2 + 24 = x^2 + 10x + 24$

Step 2.   Find two integers whose sum is 10 and whose product is 24. The integers are 6 and 4.

$$6 + 4 = 10$$
$$6 \cdot 4 = 24$$

Step 3. $x^2 + 10x + 24 = x^2 + 6x + 4x + 24$
$$= x(x + 6) + 4(x + 6)$$
$$= (x + 4)(x + 6)$$

To find the two integers 4 and 6 it may be necessary for you to examine all of the possible divisors of 24.

$$1 \cdot 24 = 24 \quad \text{but} \quad 1 + 24 \neq 10$$
$$2 \cdot 12 = 24 \quad \text{but} \quad 2 + 12 \neq 10$$
$$3 \cdot 8 \; = 24 \quad \text{but} \quad 3 + 8 \; \neq 10$$
$$4 \cdot 6 \; = 24 \quad \text{and} \quad 4 + 6 \; = 10 \quad \text{Jackpot!}$$

Be sure to check your answer.

(b)   **Step 1.** $a^2 - 16 + 6a = a^2 + 6a - 16$

Step 2.   Find two integers whose sum is 6 and whose product is -16. The integers are $+8$ and -2.

$$+8 - 2 = 6$$
$$(8)(-2) = -16$$

Step 3. $a^2 + 6a - 16 = a^2 + 8a - 2a - 16$
$$= a(a + 8) - 2(a + 8)$$
$$= (a - 2)(a + 8)$$

(c) $q^2 - 10q + 21$

Find two integers whose sum is -10 and whose product is 21. The integers are -3 and -7.

$$(-3) + (-7) = -10$$
$$(-3)(-7) = 21$$

$$q^2 - 10q + 21 = q^2 - 3q - 7q + 21$$
$$= q(q - 3) - 7(q - 3)$$
$$= (q - 7)(q - 3)$$

(d) $y^2 - 20 - 8y = y^2 - 8y - 20$

Find two integers whose sum is -8 and whose product is -20. The integers are -10 and $+2$.

$$(-10) + 2 = -8$$
$$(-10)(2) = -20$$

$$y^2 - 8y - 20 = y^2 - 10y + 2y - 20$$
$$= y(y - 10) + 2(y - 10)$$
$$= (y + 2)(y - 10)$$

(e) $x^2 - 30 - x = x^2 - x - 30$

Find two integers whose sum is -1 and whose product is -30. The integers are -6 and $+5$.

$$(-6) + 5 = -1$$
$$(-6)(5) = -30$$

$$x^2 - x - 30 = x^2 - 6x + 5x - 30$$
$$= x(x - 6) + 5(x - 6)$$
$$= (x + 5)(x - 6)$$

If you need to refresh your memory on the addition and multiplication of signed numbers, you should return to frame **14** on page 15 for a review.

If you cannot immediately identify the two "sum and product" integers, try all of the divisors of the constant term, including negative integers, testing the sum of each pair of divisors.

Trinomials that Cannot Be Factored

If no "sum and product" integers can be found, the original trinomial is simply not factorable. For example, a bit of searching will show you that the following trinomials cannot be factored into binomial products with integer coefficients:

$x^2 - x - 1$
$x^2 + 2x + 3$
$x^2 + 2x - 1$ Of course there are many more unfactorable trinominals.

So far we have factored only polynomials where the coefficient of the squared term is equal to 1. Can you factor the trinomial

$3x^2 + 11x + 6$?   Try it.

Check your answer in **29**.

---

29   If you are imaginative and very persistent, you might have factored the trinomial

$3x^2 + 11x + 6$ into $(3x + 2)(x + 3)$

by trial and error. But, happily, our three-step "sum and product" method can be changed slightly so that it enables us to solve *any* factorable trinomial.

Step 1.   Write the trinomial to be factored as $ax^2 + bx + c$, where a, b, and c are integers. Form the product ac.

Example: $3x^2 + 11x + 6$

$ax^2 + bx + c$ $b = 11$ $a = 3$ $c = 6$ $ac = 3 \cdot 6 = 18$

Step 2. Find two integers whose sum is the coefficient b of the first power term and whose product is ac, the product of the constant and the coefficient of the squared term.

Example: Find two integers whose sum is 11 and whose product is 18. The integers are 9 and 2.

$$9 + 2 = 11$$
$$9 \cdot 2 = 18.$$

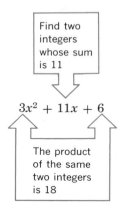

Step 3. Rewrite the trinomial using the two integers found in step 2 and factor it.

Example:
$$
\begin{aligned}
3x^2 + 11x + 6 &= 3x^2 + 9x + 2x + 6 \\
&= 3x(x + 3) + 2(x + 3) \\
&= (3x + 2)(x + 3)
\end{aligned}
$$

Finally, check your answer by multiplying the factors to get the original trinomial.

Check: $(3x + 2)(x + 3) = 3x^2 + 11x + 6$

Ready to try it yourself? Factor $8x^2 + 10x + 3$.

Our step-by-step solution is in **30**.

I don't see why $a(x + y) + b(x + y) = (a + b)(x + y)$.

Think of it this way: if $x + y = N$ then the sum is $aN + bN$ which equals $N(a + b)$ or $(a + b)N$ or $(a + b)(x + y)$. OK?

30 **Step 1.** $8x^2 + 10x + 3$, $\quad b = 10$, $\quad ac = 8 \cdot 3 = 24$.

Step 2. Find two integers whose sum is 10 and whose product is 24. The integers are 4 and 6.

$$4 + 6 = 10$$
$$4 \cdot 6 = 24.$$

Step 3. Rewrite the trinomial and factor it.

$$8x^2 + 10x + 3 = 8x^2 + 4x + 6x + 3$$
$$= 4x(2x + 1) + 3(2x + 1)$$
$$= (4x + 3)(2x + 1)$$

In step 2, if the "sum and product" integers are not easy to find, try testing all of the divisors of 24.

$1 \cdot 24 = 24$ but $1 + 24 \neq 10$
$2 \cdot 12 = 24$ but $2 + 12 \neq 10$
$3 \cdot 8 = 24$ but $3 + 8 \neq 10$
$4 \cdot 6 = 24$ and $4 + 6 = 10$ . . . That's it!

Don't forget to check your answer by multiplying the factors to be certain they produce the original trinomial.

A bit of practice will help you to understand this process.

Factor these trinomials.

(a) $6x^2 + 13x + 6$

(b) $6a^2 - a - 12$

(c) $2y^2 + 7y - 15$

(d) $8p^2 - 10p + 3$

Our step-by-step solutions are in **31**.

WHY DOES THE "SUM AND PRODUCT" METHOD WORK?

For trinomials whose x^2 coefficient is 1, if the trinomial is factorable at all we can write

$$(x + a)(x + b) = x^2 + ax + bx + ab$$
$$= x^2 + \underline{(a + b)}x + ab$$

Sum of a and b Product of a and b

For trinomials whose x^2 coefficient is not equal to 1, if the trinomial can be factored at all we can write

$$(ax + b)(cx + d) = (ax)(cx) + (ax)(d) + (b)(cx) + (b)(d)$$
$$= acx^2 + adx + bcx + bd$$
$$= (ac)x^2 + \underline{(ad + bc)}x + (bd)$$

Sum of ad and bc

This product is $(ac)(bd)$, which is the product of ad and bc

---

31   (a) $6x^2 + 13x + 6$ $b = 13$ $ac = 6 \cdot 6 = 36$

Find two integers whose sum is 13 and whose product is 36. The integers are 4 and 9.

$4 + 9 = 13$
$4 \cdot 9 = 36$

Rewrite the trinomial and factor it.

$$6x^2 + 13x + 6 = 6x^2 + 4x + 9x + 6$$
$$= 2x(3x + 2) + 3(3x + 2)$$
$$= (2x + 3)(3x + 2)$$

Check it.

(b) $6a^2 - a - 12$ $b = -1$ $ac = (6)(-12) = -72$

Find two integers whose sum is -1 and whose product is -72. The two integers are -9 and $+8$.

$(-9) + (8) = -1$
$(-9)(8) = -72$

Rewrite the trinomial and factor it.

$$6a^2 - a - 12 = 6a^2 - 9a + 8a - 12$$
$$= 3a(2a - 3) + 4(2a - 3)$$
$$= (3a + 4)(2a - 3)$$

To find the "sum and product" integers, test pairs of divisors until you find that $8 \cdot 9 = 72$, then arrange negative signs so that $(-9) + 8 = -1$ and $(-9)(8) = -72$.

(c) $2y^2 + 7y - 15$ $b = 7$ $ac = (2)(-15) = -30$

Find two integers whose sum is 7 and whose product is -30. The integers are 10 and -3.

$(10) + (-3) = 7$
$(10)(-3) = -30$

Rewrite the trinomial and factor it.

$$2y^2 + 7y - 15 = 2y^2 + 10y - 3y - 15$$
$$= 2y(y + 5) - 3(y + 5)$$
$$= (2y - 3)(y + 5)$$

(d) $8p^2 - 10p + 3$ $b = -10$ $ac = (8)(3) = 24$

Find two integers whose sum is -10 and whose product is 24. The two integers are -6 and -4.

$(-6) + (-4) = -10$
$(-6)(-4) = 24$

Rewrite the trinomial and factor it.

$$8p^2 - 4p - 6p + 3 = 4p(2p - 1) - 3(2p - 1)$$
$$= (4p - 3)(2p - 1)$$

Don't forget to check your answer by multiplying the factors.

Solving Equations Involving Products

If an equation includes binomial or other products, it may be possible to multiply to remove parentheses and then solve the resulting equation as before. For example, to solve

$(x - 1)(x + 3) = x^2 + 1$

first, multiply the binomials on the left:

$x^2 + 2x - 3 = x^2 + 1$

Second, rearrange to place all terms containing the variable on the left:

$x^2 + 2x - 3 \quad - x^2 = x^2 + 1 \quad - x^2$

or

$2x - 3 = 1$ If the x^2 terms drop out we can solve as before.

Collect all constant terms on the right:

$2x - 3 \quad + 3 = 1 \quad + 3$

or

$2x = 4$
$x = 2$ The solution is 2.

Of course you should check your answer by substituting it back into the original equation.

Check: $((2) - 1)((2) + 3) = (2)^2 + 1$
$(1)(5) = 4 + 1$
$5 = 5$

Ready to try it on your own?

Solve: $(2x + 1)(x + 3) = x(2x + 1) - 9$

Check your answer in 32.

---

32 $(2x + 1)(x + 3) = x(2x + 1) - 9$

Multiply $2x^2 + 7x + 3 = 2x^2 + x - 9.$

Collect terms containing the variable on the left and terms containing the constant on the right:

$2x^2 + 7x + 3 \quad - 2x^2 - x - 3 = 2x^2 + x - 9 \quad - 2x^2 - x - 3$

Combine like terms $6x = -12$
$x = -2$

The solution is -2.

Check: $(2(-2) + 1)((-2) + 3) = (-2)(2(-2) + 1) - 9$
$(-3)(1) = (-2)(-3) - 9$
$-3 = 6 - 9$
$-3 = -3$

To solve word problems, translate the English phrases and sentences directly into algebraic expressions as explained in Units 1 and 2 and solve the equations as shown above.

To divide a polynomial by a monomial, divide each term of the polynomial separately.

*Dividing by
a Monomial*

Example: $(6x^3 + 9x^2 - x + 2) \div 3x = \dfrac{6x^3 + 9x^2 - x + 2}{3x}$

$$= \frac{6x^3}{3x} + \frac{9x^2}{3x} - \frac{x}{3x} + \frac{2}{3x}$$

$$= 2x^2 + 3x - \frac{1}{3} + \frac{2}{3x}$$

Example: $(4x^5 - x^3 + 6x^2 + 2x - 1) \div 2x^2 = \dfrac{4x^5 - x^3 + 6x^2 + 2x - 1}{2x^2}$

$$= \frac{4x^5}{2x^2} - \frac{x^3}{2x^2} + \frac{6x^2}{2x^2} + \frac{2x}{2x^2} - \frac{1}{2x^2}$$

$$= 2x^3 - \frac{x}{2} + 3 + \frac{1}{x} - \frac{1}{2x^2}$$

Try it. Divide each of the following.

(a) $8x^4 - 3x^2 + 4x - 1 \div 2x$

(b) $\dfrac{6x^5 + x^4 - 3x^2 - x + 12}{3x^2}$

Check your work in **33**.

---

33 (a) $\dfrac{8x^4 - 3x^2 + 4x - 1}{2x} = \dfrac{8x^4}{2x} - \dfrac{3x^2}{2x} + \dfrac{4x}{2x} - \dfrac{1}{2x}$

$$= 4x^3 - \frac{3x}{2} + 2 - \frac{1}{2x}$$

(b) $\dfrac{6x^5 + x^4 - 3x^2 - x + 12}{3x^2} = \dfrac{6x^5}{3x^2} + \dfrac{x^4}{3x^2} - \dfrac{3x^2}{3x^2} - \dfrac{x}{3x^2} + \dfrac{12}{3x^2}$

$$= 2x^3 + \frac{x^2}{3} - 1 - \frac{1}{3x} + \frac{4}{x^2}$$

*Dividing by
a Binomial*

To divide a polynomial by a binomial, we may use a process very similar to long division in arithmetic. For example, to divide

$$\frac{2x^2 - x - 10}{x + 2} \qquad \text{or} \qquad \underbrace{(2x^2 - x - 10)}_{\text{Dividend}} \div \underbrace{(x + 2)}_{\text{Divisor}}$$

First, write the division horizontally, with the terms of both polynomials written in descending order.

$$x + 2 \overline{)2x^2 - x - 10}$$

Second, divide the first term $2x^2$ of the dividend by the first term x of the divisor. Write the answer in the answer space above the first term of the dividend.

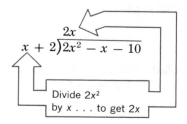

Divide $2x^2$
by x . . . to get $2x$

Third, find the product of the divisor $(x + 2)$ and the first term of the answer $2x$.

$$2x \cdot (x + 2) = 2x^2 + 4x$$

Subtract this product from the dividend polynomial.

$$
\begin{array}{r}
2x \phantom{{}-x-10} \\
x + 2 \overline{)\ 2x^2 - x - 10} \\
-(2x^2 + 4x) \\
\hline
0 - 5x
\end{array}
$$

$2x \cdot (x + 2) = 2x^2 + 4x$

$-x - (4x) = -5x$

Fourth, bring down the next term -10 in the dividend, and repeat the process.

Quotient

$$
\begin{array}{r}
2x - 5 \phantom{{}- 10} \\
x + 2 \overline{)\ 2x^2 - x - 10} \\
-2x^2 - 4x \\
\hline
-5x - 10 \\
-(-5x - 10) \\
\hline
0
\end{array}
$$

$(-5)(x + 2) = -5x - 10$

Remainder

The quotient, the answer to the division, is $2x - 5$, and the remainder is 0.

Fifth, check the division by multiplying.

$$(x + 2)(2x - 5) = 2x^2 - x - 10$$

A remainder of zero means that the divisor is a factor of the polynomial being divided.

If the divisor does not divide the polynomial exactly, a non-zero remainder will be obtained. For example,

$$(x^2 + x - 15) \div (x - 4)$$

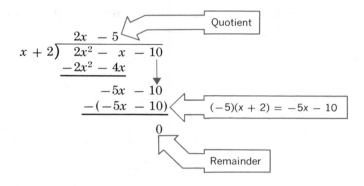

x^2 divided by x equals x

$5x$ divided by x equals 5

$x(x - 4) = x^2 - 4x$

$5(x - 4) = 5x - 20$

Remainder

Save time by writing the division this way:

$$
\begin{array}{r}
x + 5 \phantom{{}- 15} \\
x - 4 \overline{)\ x^2 + x - 15} \\
-x^2 + 4x \\
\hline
5x - 15 \\
-5x + 20 \\
\hline
5
\end{array}
$$

3-6 Division of Polynomials

Write the answer to
the division this way:

$$\frac{x^2 + x - 15}{x - 4} = (x + 5) + \frac{5}{x - 4}$$

Always check the division by multiplying by the divisor.

$$(x - 4)\left[x + 5 + \frac{5}{x - 4}\right] = (x - 4)(x + 5) + 5$$

$$= x^2 + x - 20 + 5$$

$$= x^2 + x - 15 \dots \text{the original polynomial}$$

If there are any missing terms in the polynomial being divided, insert zero coefficients. For example, the division

$$\frac{x^4 - 1}{x + 2} = (x^4 - 1) \div (x + 2) \qquad \text{would be set up like this:}$$

$$
\begin{array}{r}
x^3 - 2x^2 + 4x - 8 \\
x + 2 \overline{\smash{\big)}\ x^4 + 0x^3 + 0x^2 + 0x - 1} \\
\underline{-x^4 - 2x^3} \\
-2x^3 + 0x^2 \\
\underline{+2x^3 + 4x^2} \\
4x^2 + 0x \\
\underline{-4x^2 - 8x} \\
-8x - 1 \\
\underline{8x + 16} \\
15
\end{array}
$$

$$\begin{cases} x^4 \div x = x^3 \\ x^3 \cdot (x + 2) = x^4 + 2x^3 \end{cases}$$

$$\begin{cases} -2x^3 \div x = -2x^2 \\ (-2x^2)(x + 2) = -2x^3 - 4x^2 \end{cases}$$

$$\begin{cases} 4x^2 \div x = 4x \\ 4x \cdot (x + 2) = 4x^2 + 8x \end{cases}$$

$$\begin{cases} -8x \div x = -8 \\ (-8)(x + 2) = -8x - 16 \end{cases}$$

These are the terms of the quotient

The quotient is $x^3 - 2x^2 + 4x - 8$ and the remainder is 15.
We can write the division as

$$\frac{x^4 - 1}{x + 2} = x^3 - 2x^2 + 4x - 8 + \frac{15}{x + 2} \qquad \textit{Check it} \text{ by multiplying by } (x + 2).$$

Practice polynomial division with the following problems.

(a) $\dfrac{y^2 - 8y + 12}{y - 2} = \underline{\hspace{2cm}}$ (b) $\dfrac{3x^2 + 5x - 10}{x + 3} = \underline{\hspace{2cm}}$

(c) $(x^3 - 2) \div (x + 1) = \underline{\hspace{2cm}}$

The complete solutions are in **34**.

---

34   (a)

$$
\begin{array}{r}
y - 6 \\
y - 2 \overline{\smash{\big)}\ y^2 - 8y + 12} \\
\underline{-y^2 + 2y} \\
-6y + 12 \\
\underline{6y - 12} \\
0
\end{array}
$$

$$\begin{cases} y^2 \div y = y \\ y \cdot (y - 2) = y^2 - 2y \end{cases}$$

$$\begin{cases} -6y \div y = -6 \\ (-6)(y - 2) = -6y + 12 \end{cases}$$

Since the remainder is zero, we know that the divisor $(y - 2)$ is a factor of the polynomial $(y^2 - 8y + 12)$.

$(y - 2)(y - 6) = y^2 - 8y + 12$

(b)

$$\begin{array}{r} 3x \quad - 4 \\ x + 3 \overline{)\ 3x^2 + 5x - 10} \\ \underline{-3x^2 - 9x} \\ -4x - 10 \\ \underline{4x + 12} \\ 2 \end{array}$$

$\begin{cases} 3x^2 \div x = 3x \\ 3x \cdot (x + 3) = 3x^2 + 9x \end{cases}$

$\begin{cases} -4x \div x = -4 \\ (-4)(x + 3) = -4x - 12 \end{cases}$

The quotient is $(3x - 4)$ and the remainder is 2.
Write the division as

$$\frac{3x^2 + 5x - 10}{x + 3} = 3x - 4 + \frac{2}{x + 3}$$

Check it by multiplying by $(x + 3)$.

(c) $(x^3 - 2) \div (x + 1)$ should be written as

$$\begin{array}{r} x^2 - x \ + 1 \\ x + 1 \overline{)\ x^3 + 0x^2 + 0x - 2} \\ \underline{-x^3 - \ x^2} \\ -x^2 + 0x \\ \underline{x^2 + \ x} \\ x - 2 \\ \underline{-x - 1} \\ -3 \end{array}$$

$\begin{cases} x^3 \div x = x^2 \\ x^2 \cdot (x + 1) = x^3 + x^2 \end{cases}$

$\begin{cases} -x^2 \div x = -x \\ (-x)(x + 1) = -x^2 - x \end{cases}$

$\begin{cases} x \div x = 1 \\ 1 \cdot (x + 1) = x + 1 \end{cases}$

$$\frac{x^3 - 2}{x + 1} = x^2 - x + 1 - \frac{3}{x + 1}$$

Now turn to **35** for a set of practice problems on the work of this section of Unit 3.

Unit 3
Products
and Factoring

PROBLEM SET 3-3

35

Working with
Polynomials

The answers are on page 547.

A. Factor:

1. $x^2 + 10x + 16$ _____
2. $a^2 + 2a + 1$ _____
3. $p^2 + 6p + 8$ _____
4. $y^2 - 8y + 12$ _____
5. $c^2 - 5c + 6$ _____
6. $x^2 - x - 6$ _____
7. $q^2 + 9q + 20$ _____
8. $a^2 + 15a + 56$ _____
9. $k^2 + 3k - 40$ _____
10. $x^2 - 9x - 22$ _____
11. $y^2 + 15y + 36$ _____
12. $p^2 + 7p + 6$ _____
13. $m^2 - 5m - 50$ _____
14. $y^2 + 11y - 60$ _____
15. $x^2 - 5xy + 6y^2$ _____
16. $e^2 - 12e + 35$ _____
17. $a^2 - 2ab + b^2$ _____
18. $x^2 - xy - 2y^2$ _____
19. $a^2 - b^2$ _____
20. $x^2 - 4y^2$ _____
21. $t^2 + at - 6a^2$ _____
22. $x^2 + 8x + 16$ _____
23. $p^2 - 12p + 20$ _____
24. $t^2 + 17t + 72$ _____

B. Factor:

1. $2a^2 + 7a + 3$ _____
2. $2x^2 + 6x - 8$ _____
3. $2t^2 + 10t + 8$ _____
4. $3c^2 + 16c + 5$ _____
5. $3x^2 + 4x + 1$ _____
6. $2p^2 + 3p + 1$ _____
7. $2q^2 - 3q + 1$ _____
8. $3m^2 + 7m + 2$ _____
9. $3a^2 + 2a - 5$ _____
10. $2y^2 + 4y - 6$ _____
11. $6t^2 - t - 1$ _____
12. $1 - 5q + 4q^2$ _____
13. $16c^2 - 5 - 16c$ _____
14. $4z^2 + 1 - 4z$ _____
15. $6a^2 - 29a + 35$ _____
16. $3x^2 + x - 4$ _____
17. $9x^2 + 9xy - 4y^2$ _____
18. $2x^2 + 5xy - 3y^2$ _____
19. $4p^2 + 16pq + 15q^2$ _____
20. $15t^2 + 17at - 4a^2$ _____
21. $4x^2 - 9y^2$ _____
22. $16a^2 - 1$ _____
23. $9y^2 - 4$ _____
24. $25t^2a^2 - 4b^2$ _____

Date

Name

Course/Section

235 **Problem Set 3-3**

C. Factor completely. If a monomial factor is part of each term, factor it out first. If the polynomial cannot be factored, say so.

1. $3x^2 + 27x + 60$ _____
2. $2a^2 - 22a + 60$ _____
3. $x^3 + 2x^2 + 4x$ _____
4. $t^3 - 13t^2 + 42t$ _____
5. $y^3 - 16y$ _____
6. $4p^3 - 9p$ _____
7. $a^2 + 4a + 7$ _____
8. $6x^3 + 4x^2 - 10x$ _____
9. $12q^3 - 24q^2 - 15q$ _____
10. $2a^2b + 14ab + 12b$ _____
11. $3a^2t^2 + 24a^2t + 21a^2$ _____
12. $x^2 + 5x + 11$ _____
13. $2x^3y + x^2y - 15xy$ _____
14. $12x^2y - 2xy - 80y$ _____
15. $4y^2 - 9a^2b^2$ _____
16. $8p^2q - 4q^3$ _____
17. $x^2 + 2$ _____
18. $a^3b - b^3a$ _____
19. $50p^2t - 5pt - 105t$ _____
20. $2 - 6x - 4x^2$ _____

D. Solve:

1. $2(x - 1) = x + 1$ _____
2. $4(1 - 2x) = 2(x - 4)$ _____
3. $x(x + 1) = x^2 - 4$ _____
4. $2x(x - 2) = 2x^2 - 8$ _____
5. $(x + 1)(x - 1) = x(x - 2)$ _____
6. $(x + 2)(x + 3) = x(x - 1)$ _____
7. $(2x - 1)(x - 2) = (1 - x)(5 - 2x)$ _____
8. $3x(x - 1) = (x - 2)(3x + 4)$ _____
9. $(1 - x)(x + 4) = x(3 - x)$ _____
10. $(5 - x)(3 - x) = (x + 4)(x + 2)$ _____

E. Divide:

1. $(2x^2 + 5x) \div 2x$ _____
2. $(3x^2 - 4x) \div 4x$ _____
3. $\dfrac{a^2b - a^2b^2}{a}$ _____
4. $(2x^2y + xy^3) \div xy$ _____
5. $\dfrac{y^2 - 3xy}{-y^2}$ _____
6. $(x^2y + 2x^4) \div x^3$ _____
7. $\dfrac{ab^2 + 3a^2b^2 + a^2b}{a^2b}$ _____
8. $\dfrac{x^3y^4 - x^2}{x^2y^2}$ _____
9. $\dfrac{3xy - 4xy^2}{6x^2}$ _____
10. $\dfrac{4xy^2z - 2xyz^2 + 6x^2y}{-8x^2yz}$ _____
11. $(x^3 - 7x^2 + 6x + 18) \div (x - 3)$ _____
12. $(2x^2 + 15x + 28) \div (x + 4)$ _____
13. $\dfrac{2x^3 + 11x^2 - 2x - 15}{x + 5}$ _____
14. $\dfrac{3y^3 - 4y + 5}{y - 1}$ _____

236 **Products and Factoring**

15. $\dfrac{x^4 - 2x^3 - x^2 + 3x - 2}{x - 2}$ _____

16. $\dfrac{x^3 - x + 1}{x + 2}$ _____

17. $(x^4 - 2) \div (x - 1)$ _____

18. $(x^5 + 3) \div (x + 1)$ _____

19. $(2x^4 - 3x^2 + 1) \div (2x + 1)$ _____

20. $(2t^3 + 5t^2 + 4t + 3) \div (2t + 1)$ _____

21. $\dfrac{x^3 - a^3}{x - a}$ _____

22. $\dfrac{x^5 - a^5}{x + a}$ _____

23. $(x^3 + 3x^2 - x - 6) \div (x^2 + x - 3)$ _____

24. $(x^3 - 4x^2 - 5x + 20) \div (x^2 - x + 1)$ _____

25. $(x^3 + x^2 - 5x + 1) \div (x + 3)$ _____

26. $(x^4 - 5x^3 - x^2 + 6x + 3) \div (x - 5)$ _____

27. $(x^4 + 2) \div (x - 1)$ _____

28. $(1 + x^3) \div (x - 1)$ _____

F. Brain Boosters

1. At the end of a busy day, the cash register at the Big Mac Pizza Place contained $18.40 in quarters, dimes, and nickels. The number of dimes was three more than five times the number of quarters and the number of nickels was four times the number of dimes. How many of each kind of coin was in the cash register? Who cares?

   _____

2. Show that for any four even consecutive integers the product of the middle two minus the product of the first and fourth is always equal to 8.

   _____

3. John is three times as old as Bill was four years ago and Tom is twice as old as Bill was three years ago. If John and Tom are the same age, how old is Bill?

   _____

4. If you multiply one sum of two squares by another sum of two squares, the product is always the sum of two squares. By multiplying, show that each of the following is true.

   (a) $(a^2 + b^2)(c^2 + d^2) = (ac + bd)^2 + (ad - bc)^2$
   (b) $(a^2 + b^2)(c^2 + d^2) = (ac - bd)^2 + (ad + bc)^2$

   _____

5. The product of any four consecutive integers increased by one is always equal to a perfect square. For example,

   $(3 \cdot 4 \cdot 5 \cdot 6) + 1 = 361 = 19^2$

   Show by multiplying that $(a - 1)(a)(a + 1)(a + 2) + 1 = (a^2 + a - 1)^2$

   _____

Date _____

Name _____

Course/Section _____

6. Factor $x^4 + 4$ (*Hint:* Write this expression as $x^4 + 4 + 4x^2 - 4x^2$ and then factor it.)

_____

7. Show by multiplying that

    (a) $(a^2 + b^2)^3 = (a^3 + ab^2)^2 + (-a^2b - b^3)^2$
    (b) $(a^2 + b^2)^3 = (a^3 - 3ab^2)^2 + (3a^2b - b^3)^2$

8. Simplify. $\left(\dfrac{45a^3b^2c^4}{27a^2b^2c}\right)^2 \times \left(\dfrac{243a^4b^4c^4d}{180a^2bc}\right)^2$

_____

9. Simplify. $[(m + 1)a + (n + 1)b][(m - 1)a + (n - 1)b]$
                      $+ [(m + 1)a - (n + 1)b][(m - 1)a - (n - 1)b] = ?$

_____

When you have had the practice you need, turn to **36** for a self-test on Products and Factoring.

_____

Unit 3
Products
and Factoring

Self-Test 3

36

1. Multiply: $(1 - x)(x + 1) =$ _____

2. Multiply: $(a + 3)^2 =$ _____

3. Multiply: $(2x - y)(x + 3y) =$ _____

4. Multiply: $3q(2 + q)(2q - 1) =$ _____

5. Multiply: $(a + b)(a^2 + b^2) =$ _____

6. Multiply: $(2c - 3k)^2 =$ _____

7. Multiply: $(q - 2)(q + 2) =$ _____

8. Multiply: $(x - 1)(x - 2)(x - 3) =$ _____

9. Factor: $3x^2 - 6x =$ _____

10. Factor: $ax^2 - ax + x^2 =$ _____

11. Factor: $-3ps + 6s - 3ps^2 =$ _____

12. Factor: $2x + x^2 - 5x - 2x^2 + x =$ _____

13. Factor: $y^2 + 7y + 10 =$ _____

14. Factor: $1 - x^2 =$ _____

15. Factor: $-2 + x + 6x^2 =$ _____

16. Factor: $4a^2c - cb^2 =$ _____

17. Factor: $2b^2 - 2a^2 + 3ab =$ _____

18. Factor: $a + b - ax - bx =$ _____

19. Solve: $(1 - x)(2x + 1) = 2(1 - x^2)$ _____

20. Solve: $(p - 2)^2 - p^2 = 1$ _____

21. Solve: $t^2 - (t + 2)(t + 1) = 0$ _____

22. Find three consecutive even integers such that four times the middle integer minus the smallest integer equals 32.

_____

23. Find three consecutive odd integers such that the product of the largest two integers minus the square of the smallest equals 50.

_____

Date _____

Name _____

Course/Section _____

Self-Test 3

Copyright © 1982 by John Wiley & Sons, Inc.

24. At closing time the manager of the Happy Pickle vegetarian restaurant found that the cash register contained $4.10 in pennies, nickels, and dimes, with eight more nickels than pennies and five more dimes than pennies. How many pennies are in the cash register?

_____

25. The Girl Scout bake sale earned $7.30. They sold cookies at 3¢ each, cupcakes at 7¢ each, and doughnuts at 10¢ each. They sold 10 more cookies than cupcakes and 10 less doughnuts than cupcakes. How many of each did they sell?

_____

26. Divide: $\dfrac{8x^3 - 12x + 2}{4x} =$ _____

27. Divide: $(2x^3 - 7x^2 - 6x + 10) \div (x - 4) =$ _____

28. Divide: $\dfrac{x^5 + 2}{x - 1} =$ _____

Answers are on page 548.

Unit 3
Products
and Factoring

A. Multiply:

1. $2(x - 1)$ _____

2. $y(x + 2)$ _____

3. $-3(p - 5)$ _____

4. $-b(3 - b)$ _____

5. $5(x - x^2)$ _____

6. $2t(t + 2)$ _____

7. $3x(x^2 - 2x)$ _____

8. $q(a - q)$ _____

9. $3(b - 1)$ _____

10. $a(a + b)$ _____

11. $a(a - b)$ _____

12. $-a(a + b)$ _____

13. $-a(a - b)$ _____

14. $-a(-a - b)$ _____

15. $a(-a - b)$ _____

16. $4(x^2 - 5x)$ _____

17. $3(x + a - 2)$ _____

18. $2(x + x^2 + x^3 + 1)$ _____

19. $4(2p + p^2 + c)$ _____

20. $2ab(a^2 - b^2)$ _____

21. $-x^2(-x - 2 + x^2)$ _____

22. $pq(p - q + 2pq - pq^2)$ _____

23. $-x^3(1 - x - x^2)$ _____

24. $a^2b(a - b)$ _____

25. $2mn(m - n^2)$ _____

26. $-5(3x - 5)x$ _____

27. $(3x^2y - xy^2)2x$ _____

28. $17pa^2(-2p + 4pa)$ _____

29. $-z(-z - z^2 + 1 - z^3)$ _____

30. $-t(3 - 2t^2)$ _____

31. $(3ax)^2$ _____

32. $(-2x^2)^2$ _____

33. $\left(\frac{1}{2}ab^2\right)^3$ _____

34. $(-8x)(-2y)$ _____

35. $(-3z)(8xy^2)$ _____

36. $(-2m)^2(3n^2)$ _____

37. $(-x^2)^2(-x)^2$ _____

38. $(2ab^3)^0(4a)$ _____

39. $(3xy^3)(-4x^4y)$ _____

40. $(-18at^3)(2a^2)$ _____

B. Factor out the common monomials:

1. $a^3 - a$ _____

2. $t^2 - 3t^4$ _____

3. $8b - 2$ _____

4. $3x + 6y$ _____

5. $8a^2 - 6b^2$ _____

6. $4pq - 5ps$ _____

7. $t - t^2$ _____

8. $x^3 + x$ _____

9. $2a^2b - b^2a$ _____

10. $mn - nm^2$ _____

11. $xyz - x$ _____

12. $12x^3 - 6x^2 + 3x - 3$ _____

13. $y - 3ay$ _____

14. $7x^2 - ax^2 + pq^2x^2$ _____

15. $-5z - 10z^2 - 5z^3$ _____

16. $abc^2 - a^2b^2 - 4ab^2x$ _____

17. $16d^2 - 40d$ _____

18. $-a^2b + b^2a$ _____

19. $x^3 + 2x^2 - 2x$ _____

20. $y^4 - 6y^2$ _____

21. $-a^8 - a^6b^2 + a^4b^4$ _____

22. $\pi R^2 L - 2\pi R$ _____

23. $ax + ba - at - qa$ _____

24. $3x^2y - 7xy^2 - 4x^2y^2$ _____

25. $\frac{1}{2}ak + ak^2$ _____

26. $5x^2 + 10$ _____

27. $56xy^2 - 63xt$ _____

28. $4xy - 5y^2x^2$ _____

29. $20p^3q^2 - 15p^2q$ _____

30. $abcd - a^2$ _____

31. $8x^4 - 56x^2$ _____

32. $7y^3 + 42y$ _____

33. $9p^3 + 54p^2$ _____

34. $8x^5 + 72x$ _____

35. $6x^3 - 18x^2 + 3x$ _____

36. $7x^4 + 28x^2 - 14x$ _____

37. $y^4 + 5y^2 + y$ _____

38. $15x^3 - 5x^2 + 20x$ _____

39. $3x^2 - 90$ _____

40. $5a^3 + 20a$ _____

41. $8t^3 + 4t$ _____

42. $2xy^3 - 4x^2y^2 + 12x^2y^3$ _____

43. $4y^5 - 8x^4 + 6x^2y + 10$ _____

44. $ab - 2a^2b + 6ab^3 - 4a$ _____

45. $7x^5 + 28x^2 - x$ _____

46. $ax^3 - 2x^4 - x^2 + cx^5$ _____

47. $2a^2x - 6ax^4 - 8x^5$ _____

48. $20a^2x^5 - 10abx^4 + 30ax^3$ _____

49. $24x^7 - 18x^{11} + 9x^9$ _____

50. $32y^{12} + 36y^{10} - 28y^9$ _____

Unit 3
Products
and Factoring

A. Write each of the following products as a polynomial.

1. $(x + 1)(x + 5)$ _____

2. $(x - 1)(x + 4)$ _____

3. $(p + 2)(p + 7)$ _____

4. $(a + 5)(a + 8)$ _____

5. $(q - 11)(q + 1)$ _____

6. $(t - 2)(t - 6)$ _____

7. $(-w - 1)(w + 5)$ _____

8. $(2 - x)(3 - x)$ _____

9. $(a + b)(x + y)$ _____

10. $(2x - 4)(x - 6)$ _____

11. $(2x - a)(x + 2a)$ _____

12. $(7y - p)(p + 5y)$ _____

13. $(8x - 9y)(2x - y)$ _____

14. $(4a - 3b)(2b + 2a)$ _____

15. $(x^2 + x)(a^2 + a)$ _____

16. $(q^2 - 2)(q - 2)$ _____

17. $(1 - 2x)(1 - 2y)$ _____

18. $(x + y)^2$ _____

19. $(a - b)^2$ _____

20. $(1 + t)^3$ _____

21. $(3 - a)^3$ _____

22. $(p + 1)(p - 1)$ _____

23. $(x + a)(a - x)$ _____

24. $2(b + 2)(b - 2)$ _____

25. $4(2x - 3y)^2$ _____

26. $(-7abc - p^2)^2$ _____

27. $(x + y)(xy - 2y)$ _____

28. $(p - q^2)(2p - 3q)$ _____

29. $(-m - n)(n^2 - m^2)$ _____

30. $(1 + x)(1 + x^2)$ _____

B. Multiply:

1. $(x + 1)(x^2 - 2x - 3)$ _____

2. $(y + 5)(1 - y - y^2)$ _____

3. $(p + 3)(p^2 + 2p + 5)$ _____

4. $(x - 4)(1 + x - 2x^2)$ _____

5. $(a + b)(a - b - 1)$ _____

6. $(x - y)(1 + y - x^2)$ _____

7. $(2 + q)(2 + q + q^2)$ _____

8. $n(n + 1)(n + 2)$ _____

9. $(a + x)^3(a - x)$ _____

10. $(2a - 7b)(b^3 + b^2 - b)$ _____

11. $\left(x + \dfrac{1}{2}\right)(x + 1)$ _____

12. $\left(2y - \dfrac{1}{2}\right)(y - 1)$ _____

13. $(3x - 2y)^0(x - y)^2$ _____

14. $(x - y)(x + y - 3xy + 1)$ _____

Date _____

15. $(2 + z)(-1 + z + z^2 + z^3)$ _____

16. $(x^2 + y^2)(x^2 - y^2)$ _____

Name _____

17. $(a^2 + b^2)^2$ _____

18. $(p^3 + 2t^2)(3p^2 - 2t^3)$ _____

Course/Section _____

19. $(5 - ax)(5 + ax - x)$ _____

20. $(-y - 4)(-3y - 5)$ _____

21. $(x - 3)(x - 12)$ _____

22. $(x + 5)(x - 11)$ _____

23. $(y + 5)(y + 8)$ _____

24. $(y - 6)(y - 9)$ _____

25. $(2x - 1)(x + 9)$ _____

26. $(3x + 1)(x - 10)$ _____

27. $(3y - 2)(y - 2)$ _____

28. $(3x + 4)(x - 6)$ _____

29. $(x - 4)(2x - 7)$ _____

30. $(y - 8)(2y - 9)$ _____

31. $(x - 11)(2x - 1)$ _____

32. $(x - 3)(3x + 5)$ _____

33. $(3x - 2)(2x - 7)$ _____

34. $(5y - 3)(2y + 5)$ _____

35. $(4x - 1)(2x + 5)$ _____

36. $(2x + 7)(3x - 4)$ _____

37. $(x - 1)(x^2 - x - 2)$ _____

38. $(x - 3)(x^2 + 2x - 3)$ _____

39. $(y - 5)(y^3 - y - 1)$ _____

40. $(y - 4)(2y^2 - y + 3)$ _____

41. $(2x + 3)(x^2 + 2x - 1)$ _____

42. $(4x + 3)(x^2 - 3x - 4)$ _____

43. $(5x - 2)(x^2 + 4x + 2)$ _____

44. $(3x - 4)(x^2 - 5x + 1)$ _____

45. $(3x - 5)(x^2 - 3x - 3)$ _____

46. $(2x - 3)(2x^2 - 2x - 3)$ _____

47. $(7x - 1)(x^2 - x + 1)$ _____

48. $(8x + 3)(x^2 - 2x - 2)$ _____

49. $(x - 2)(x^3 - x^2 - x - 4)$ _____

50. $(x - 4)(x^3 - 2x^2 + x - 1)$ _____

Unit 3
Products
and Factoring

A. Factor completely:

1. $24 - 11t + t^2$ _____

2. $x^2 + 6x + 5$ _____

3. $p^2 + 5p + 4$ _____

4. $10 + 7z + z^2$ _____

5. $y^2 + 11 + 12y$ _____

6. $x^2 + 3x + 2$ _____

7. $x^2 - 6x + 5$ _____

8. $4 - a^2$ _____

9. $x^2 - 2x + 1$ _____

10. $p^2 - 8p + 7$ _____

11. $y^2 + 6 - 5y$ _____

12. $m^2 - 10m + 21$ _____

13. $x^2 - 9y^2$ _____

14. $p^2 + 16 - 8p$ _____

15. $x^2 + 6x + 9$ _____

16. $a^2b^2c^2 - 1$ _____

17. $t^2 - 6 + t$ _____

18. $x^2 - 4x - 21$ _____

19. $\dfrac{x^2}{4} - x^4$ _____

20. $\dfrac{1}{4}x - \dfrac{1}{8} + x^2$ _____

21. $2a^2 - 3a + 1$ _____

22. $3ab + b^2 + 2a^2$ _____

23. $2r^2 + 5r + 2$ _____

24. $x^2 + 4xy - 5y^2$ _____

25. $y^2 + 3xy + 2x^2$ _____

26. $t^2 - 3tu - 10u^2$ _____

27. $6x^2 - 7xy + 2y^2$ _____

28. $2a^2 - 7ab + 5b^2$ _____

29. $2p^2q - 2pq - pq^2 + q^2$ _____

30. $ab^2 + b^2 + a^2b + ab$ _____

B. Factor each of the following expressions completely. If a monomial factor is part of each term, factor it out first. If the polynomial cannot be factored, say so.

1. $x^3 - x$ _____

2. $a^3 - 3a^2 + 2a$ _____

3. $a^3b + 2a^2b + ab$ _____

4. $2y^2 - 6y - 20$ _____

5. $2t^3 - 18t^2 + 36t$ _____

6. $6p^2 - 14pq + 4q^2$ _____

7. $-2b + 3a^2b - 5ab$ _____

8. $-8y + 10xy + 12x^2y$ _____

9. $36x^2y - 4y^3$ _____

10. $y^2 - 2$ _____

11. $x^3y - y^3x$ _____

12. $3 - 12p^2$ _____

13. $4 - 6x + 2x^2$ _____

14. $-p^2 + 2 - 3p$ _____

15. $-q^2 - 4q - 3$ _____

16. $-x^3 - 4x^2 - 4x$ _____

17. $x^2 + 1$ _____

18. $1 - x^2$ _____

19. $y^2 + 3y - 88$ _____

20. $p^2 - 24p + 119$ _____

21. $x^2 - 13x + 40$ _____

22. $x^2 - 16x + 63$ _____

23. $x^2 - 12x + 11$ _____

24. $x^2 - 15x + 26$ _____

25. $x^3 - 12x^2 + 20x$ _____

26. $x^3 + 6x^2 - 27x$ _____

27. $2x^2 - 2x - 84$ _____

28. $3x^2 + 9x - 84$ _____

29. $2x^3 + 2x^2 - 24x$ _____

30. $3a^3 + 21a^2 + 30a$ _____

C. Divide:

1. $(x^4 - 12x^3 + 3x^2) \div x^2$ _____

2. $(2x^3 - 6x^2 + 4x) \div 2x$ _____

3. $\dfrac{20x^3 - 15x^2 + 5x}{5x}$ _____

4. $\dfrac{12y^4 - 18y^2 + 6y}{3y}$ _____

5. $\dfrac{8x^3y - 2x^2y^2 - 4xy^3}{2xy}$ _____

6. $\dfrac{-20xy^3 + 30xy^2 - 15xy}{-5xy}$ _____

7. $\dfrac{-6x^5 + 4x^3 - x^2}{-x^2}$ _____

8. $\dfrac{-9y^6 - 6y^4 + 4y^3}{-2y^2}$ _____

9. $\dfrac{-5x^4 + 3x^2 - 2x}{x^2}$ _____

10. $\dfrac{-8x^5 + 6x^3 - 4x}{2x^2}$ _____

11. $\dfrac{x^2 + 4x - 21}{x - 3}$ _____

12. $\dfrac{x^2 + 4x - 45}{x - 5}$ _____

13. $\dfrac{x^2 - 9x - 30}{x + 3}$ _____

14. $\dfrac{x^2 - 16x + 50}{x - 6}$ _____

15. $\dfrac{2x^2 - 13x + 20}{x - 2}$ _____

16. $\dfrac{3x^2 - 22x + 40}{x - 5}$ _____

17. $\dfrac{3x^2 + 16x + 6}{x + 5}$ _____

18. $\dfrac{4x^2 - 27x + 15}{x - 6}$ _____

19. $\dfrac{6x^2 - 29x + 20}{2x - 3}$ _____

20. $\dfrac{15x^2 + 13x - 10}{3x - 1}$ _____

Equations in Two Variables

Objective	Sample Problems	Where To Go for Help	

Upon successful completion of this program you will be able to:

		Page	Frame

1. Draw the graph for a set of ordered pairs of numbers.

   (a) Graph: $\{(2, 1), (1, 0), (-1, 4), (3, -2), (-2, -1)\}$ — 253 — 1

   (b) Graph:

x	1	2	3	0	-1	-2	-3	-4	-6
y	$2\frac{1}{2}$	3	$3\frac{1}{2}$	2	$1\frac{1}{2}$	1	$\frac{1}{2}$	0	-1

   (c) Find the coordinates of each point on the following graph. — 257 — 5

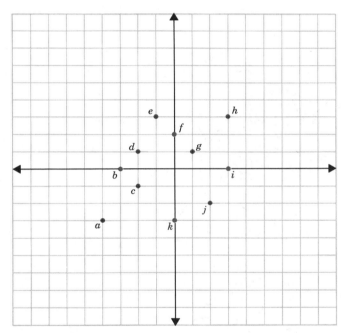

2. Graph linear equations.

   (a) $x = 2$ _____ — 263 — 8

   (b) $y = -2$ _____

   (c) $2x - 1 = 0$ _____

   (d) $y = 2x - 3$ _____

   (e) $2x + 3y = 6$ _____

3. Identify and work with linear equations.

   Indicate which of the following equations are linear: — 263 — 8

   (a) $x = 2$ _____

(b) $y = 10$ _____

(c) $y = 1 - 2x$ _____

(d) $3x - 4y = 11$ _____

(e) $y = 3 + x^2$ _____

(f) $\dfrac{y}{x} = 1 + x$ _____

Write each of the following equations in the form $y = mx + b$, find its slope and y intercept.

270   11

(g) $y + 7 = 3x - 4$ _____

(h) $2x = 3y - 1$ _____

(i) $y = -2$ _____

(j) $-x = 3y - 1$ _____

4. Solve systems of two simultaneous linear equations.

(a)   Solve: $\begin{cases} 2x + y = 1 \\ x = 2y + 8 \end{cases}$ _____

291   24

(b)   Solve: $\begin{cases} 3x - 3y = 5 \\ 2x + 3y = 0 \end{cases}$ _____

(c)   Solve: $\begin{cases} 2a + 3b = 5 \\ 4a + 7b = 11 \end{cases}$ _____

5. Solve and graph first degree inequalities in two variables.

(a)   Graph: $x + 2y > 6$

281   17

(b)   Graph: $2x + 3y \leq 12$

(c)   Graph the set of simultaneous inequalities:

$$\begin{cases} x + y \leq 2 \\ x - y \geq 0 \end{cases}$$

_____

6. Solve word problems involving equations in two variables.

(a)   The sum of two numbers is 22. The difference of the same two numbers is 6. What are the numbers?   _____

313   39

(b)   The relationship between the Fahrenheit and Celsius temperature scale is $9C = 5F - 32$. When does the Fahrenheit temperature equal twice the Celsius temperature?   _____

(c)   On an engineering design job, the artist charged $8 per hour and the printer charged $10 per hour. The

_____

Date

_____

Name

_____

Course/Section

total bill was $68. The artist worked three times as many hours as the printer. How many hours did each work? _____

(d) Bill and Jim live 50 miles apart. If they start at the same time and cycle toward each other, they meet in 2 hours. If they both ride in the same direction, Bill will overtake Jim in 5 hours. How fast are they cycling? _____

(Answers to these problems are on page 250.)

If you are certain you can work all of these problems correctly, turn to page 325 for a self-test. If you want help with any of these objectives or if you cannot work one of the sample problems, turn to the page indicated. Naturally the super-student will want to learn all of this and will turn to frame 1 and begin work there.

Date _____

Name _____

Course/Section _____

1. (a) 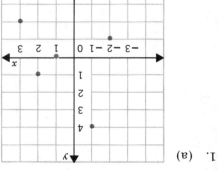 (b)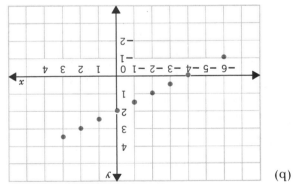

(c) $a(-4, -3)$ $b(-3, 0)$
    $c(-2, -1)$ $d(-2, 1)$
    $e(-1, 3)$ $f(0, 2)$
    $g(1, 1)$ $h(3, 3)$
    $i(3, 0)$ $j(2, -2)$ $k(0, -3)$

2. (a) 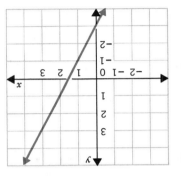 (b) 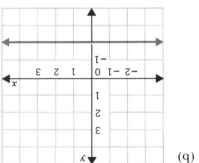 (c)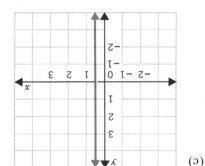

(d) 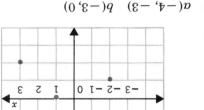 (e)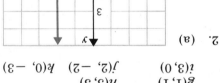

3. (a), (b), (c), and (d) are linear.

(g) $y = 3x - 11$ slope $= 3$; y intercept $= -11$

(h) $y = \dfrac{1}{3} x + \dfrac{2}{3}$;   slope $= \dfrac{2}{3}$; y intercept $= \dfrac{1}{3}$

(i) $y = 0 \cdot x - 2$ slope $= 0$; y intercept $= -2$

(j) $y = -\dfrac{1}{3} x + \dfrac{1}{3}$ slope $= -\dfrac{1}{3}$; y intercept $= \dfrac{1}{3}$

4. (a) $(x, y) = (2, -3)$ (b) $(x, y) = \left(1, -\dfrac{2}{3}\right)$ (c) $(a, b) = (1, 1)$

5. (a)

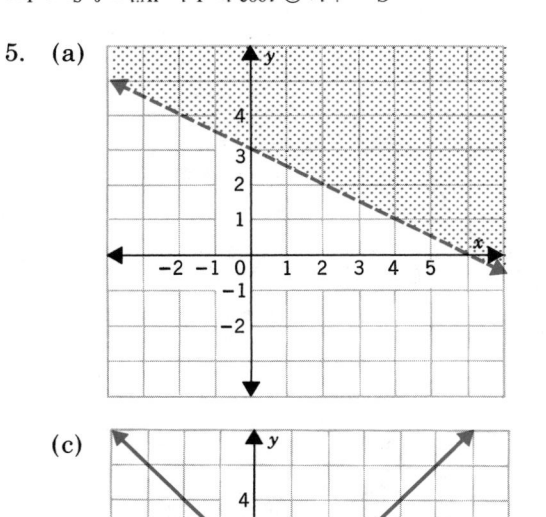

(b)

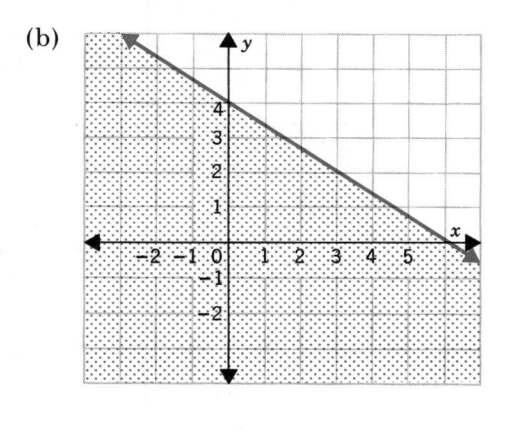

(c)

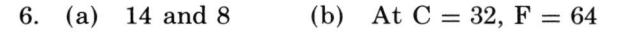

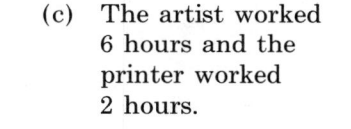

6. (a)   14 and 8   (b)   At C = 32, F = 64   (c)   The artist worked
                                                                       6 hours and the
                                                                       printer worked
                                                                       2 hours.

    (d)   Bill goes 17.5 mph; Jim goes 7.5 mph.

4 Equations in Two Variables

© 1965 United Feature Syndicate, Inc.

4-1 SOLVING EQUATIONS IN TWO VARIABLES

1   Peppermint Patty has found a problem that cannot be solved very easily with the algebra we have learned so far. The problems we have considered up to this point lead to equations in only one variable and these can be solved using the addition and multiplication rules. However, many algebraic problems translate into equations containing two variables, and many practical situations are described best in terms of a relationship between two variables. In this unit we will work with equations in two variables and before the unit is completed you will be able to solve Peppermint Patty's problem with ease.

One of the most powerful features of algebra is that it can be used to describe and analyze the relationship between two or more variables. For example, since the time of the great Italian scientist Galileo in the seventeenth century, it has been known that if a ball is dropped from rest and allowed to fall freely with no air resistance, its speed increases during each second of fall. We can describe how the speed of the falling ball is related to the time it has been falling by

(a) A sentence: The ball "*acquires during equal time intervals equal increments of speed.*"—Galileo *Dialogues Concerning Two New Sciences,* 1638

(b) A picture-graph:

| Start 0 sec | After 1 sec | After 2 sec | After 3 sec |

(c) An equation: $s = 10t$ where s is the approximate speed in meters per second and t is the time in seconds.

or

(d) A set of measurement numbers:

s, speed (m/sec)	0	10	20	30	40	50
t, time (sec)	0	1	2	3	4	5

The sentence description is interesting and helpful, but it does not allow us to analyze the relationship. The best description for the scientist to use is the algebraic equation.

In Unit 2 we solved an equation in one variable by finding one or more values of the variable for which the equation was a true statement. The equation $s = 10t$ is an equation in two variables, s and t. A solution of this equation would be any pair of numbers which, when substituted in the equation for s and t, make the equation a true statement.

Find a solution of the equation $s = 10t$.

$s = $ _____ $t = $ _____ .

Try it, then check your answer in **2**.

---

2 The pair of numbers given by $s = 10$ and $t = 1$ are a solution of $s = 10t$, because $10 = (10)(1)$ is a true statement.

The pairs of numbers given by

$t = 2 \qquad s = 20$
$t = 3 \qquad s = 30$
$t = 4.5 \qquad s = 45 \qquad$ and so on

are also solutions of this equation.

You should verify that each of these number pairs is a solution of the equation by substituting them into the equation.

It should be obvious that there are infinitely many pairs of numbers that satisfy this equation, so that the complete solution of the equation is an infinite set of *pairs* of numbers.

For neatness and simplicity, qualities much prized in mathematics, we can abbreviate the solutions given above as $(1, 10)$, $(2, 20)$, $(3, 30)$, and $(4.5, 45)$.

For each of the pairs of numbers in parentheses it is understood that the t value is to be written first and the s value second.

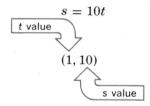

Ordered Pair
Component

Any pair of numbers written in this way, with an order given by some rule or equation, is called an *ordered pair*. The two numbers that make up the ordered pair are called the *components* of the ordered pair.

  1 and 10 are the components of the ordered pair $(1, 10)$.
  2 and 20 are the components of the ordered pair $(2, 20)$.
  3 is the first component of the ordered pair $(3, 30)$.
45 is the second component of the ordered pair $(4.5, 45)$.

Is $(60, 6)$ a solution of $s = 10t$?

Check your answer in **3**.

3 The ordered pair $(60, 6)$ is *not* a solution of the equation $s = 10t$. Remember, the t value comes *first* in an ordered pair—the order is very important.

According to tradition, algebraic equations in two variables are most often written using the letters x and y to represent the variables. In this case, mathematicians have agreed that the ordered pair representing a solution of an equation in x and y will be written with the x value first in the parentheses and the y value second. For example, the equation $y = 2x + 5$ has the ordered pair $(3, 11)$ as a solution.

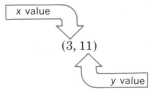

x value

$(3, 11)$

y value

$$y = 2x + 5$$
$$(11) = 2(3) + 5$$

The ordered pair is (x, y), where the x and y values are always written in order.

Complete the following ordered pairs for the equation $y = 3x - 5$.

(a) $(1, \quad)$ (b) $(2, \quad)$ (c) $(3, \quad)$ (d) $(10, \quad)$
(e) $(0, \quad)$ (f) $(-1, \quad)$ (g) $(-2, \quad)$ (h) $(-10, \quad)$

Look in **4** for the correct ordered pairs.

4 $y = 3x - 5$

(a) $(1, -2)$ $y = 3(1) - 5 = 3 - 5 = -2$ replace x by 1 to get $y = -2$
(b) $(2, 1)$ $y = 3(2) - 5 = 6 - 5 = 1$ replace x by 2 to get $y = 1$
(c) $(3, 4)$ $y = 3(3) - 5 = 9 - 5 = 4$ replace x by 3 to get $y = 4$

(d) $(10, 25)$ $y = 3(10) - 5 = 30 - 5 = 25$
(e) $(0, -5)$ $y = 3(0) - 5 = 0 - 5 = -5$
(f) $(-1, -8)$ $y = 3(-1) - 5 = -3 - 5 = -8$
(g) $(-2, -11)$ $y = 3(-2) - 5 = -6 - 5 = -11$
(h) $(-10, -35)$ $y = 3(-10) - 5 = -30 - 5 = -35$

In each case, knowing that the ordered pair must have the form (x, y), we can substitute the given component into the equation to find the second component of each pair.

Being able to write the ordered pairs associated with a given algebraic equation is a very valuable skill. Try these problems for practice. In each problem complete the ordered pairs for the equation given.

1. $y = x - 1$ (a) $(0,\ )$ (b) $(1,\ )$ (c) $(2,\ )$
                      (d) $(-1,\ )$ (e) $(-2,\ )$

2. $y = 2x + 3$ (a) $(1,\ )$ (b) $(2,\ )$ (c) $(0,\ )$
                      (d) $(-2,\ )$ (e) $(-1,\ )$

3. $y = 6 - 2x$ (a) $(0,\ )$ (b) $(1,\ )$ (c) $(3,\ )$
                      (d) $(-1,\ )$ (e) $(-3,\ )$

4. $y = x$ (a) $(1,\ )$ (b) $(0,\ )$ (c) $(2,\ )$
                      (d) $(-1,\ )$ (e) $(-2,\ )$

5. $y = -x$ (a) $(1,\ )$ (b) $(0,\ )$ (c) $(2,\ )$
                      (d) $(-1,\ )$ (e) $(-2,\ )$

6. $y = x^2 - 1$ (a) $(0,\ )$ (b) $(1,\ )$ (c) $(-1,\ )$
                      (d) $(2,\ )$ (e) $(-2,\ )$

7. $2y = 3x + 6$ (a) $(0,\ )$ (b) $(1,\ )$ (c) $(2,\ )$
                      (d) $(-2,\ )$ (e) $(-1,\ )$

8. $y - 1 = 2x + 3$ (a) $(0,\ )$ (b) $(1,\ )$ (c) $(2,\ )$
                      (d) $(-2,\ )$ (e) $(-3,\ )$

9. $y = 2x - 1$ (a) $(0,\ )$ (b) $(\ , 0)$ (c) $(1,\ )$
                      (d) $(\ , 2)$ (e) $(\ , -3)$

10. $3y = x - 6$ (a) $(\ , 0)$ (b) $(0,\ )$ (c) $(\ , 2)$
                      (d) $(\ , -1)$ (e) $(9,\ )$

Our answers are in **5**.

REWRITING EQUATIONS IN TWO VARIABLES

When an equation in two variables is written in the form

$y =$ (a polynomial in x)

we say that it has been solved *explicitly* for y.

For example, the equations

$y = 2x - 1$
$y = x^2 - x + 4$ and
$y = 2 + x$

have all been solved explicitly for y.

If an equation is not given with y appearing only on the left in this way, we can usually rewrite it in this form. For example,

$2y = 3x + 6$

becomes

continued . . .

$y = \dfrac{3}{2}x + 3$ if we divide each term by 2 and

$9x - 5 = 3y + 1$

becomes

$9x - 5 - 1 = 3y$ if we add -1 to each side.

or

$3y = 9x - 6$
$y = 3x - 2$ if we divide each term by 3.

As you will see later in this unit, it is often helpful to have an equation written explicitly in terms of y. Practice by solving each of the following equations explicitly for y. Check your answers on page 548.

1. $2y - 1 = 1 - 2x$ 2. $x - 3 = 2y$ 3. $y = 2 + x - y$

4. $2y = 3 - xy$ 5. $x = 3 + \dfrac{y}{2}$

5

1. $y = x - 1$ (a) $(0, -1)$ (b) $(1, 0)$ (c) $(2, 1)$
                      (d) $(-1, -2)$ (e) $(-2, -3)$

2. $y = 2x + 3$ (a) $(1, 5)$ (b) $(2, 7)$ (c) $(0, 3)$
                      (d) $(-2, -1)$ (e) $(-1, 1)$

3. $y = 6 - 2x$ (a) $(0, 6)$ (b) $(1, 4)$ (c) $(3, 0)$
                      (d) $(-1, 8)$ (e) $(-3, 12)$

4. $y = x$ (a) $(1, 1)$ (b) $(0, 0)$ (c) $(2, 2)$
                      (d) $(-1, -1)$ (e) $(-2, -2)$

5. $y = -x$ (a) $(1, -1)$ (b) $(0, 0)$ (c) $(2, -2)$
                      (d) $(-1, 1)$ (e) $(-2, 2)$

6. $y = x^2 - 1$ (a) $(0, -1)$ (b) $(1, 0)$ (c) $(-1, 0)$
                      (d) $(2, 3)$ (e) $(-2, 3)$

7. $2y = 3x + 6$ (a) $(0, 3)$ (b) $(1, 4.5)$ (c) $(2, 6)$
                      (d) $(-2, 0)$ (e) $(-1, 1.5)$

8. $y - 1 = 2x + 3$ (a) $(0, 4)$ (b) $(1, 6)$ (c) $(2, 8)$
                      (d) $(-2, 0)$ (e) $(-3, -2)$

9. $y = 2x - 1$ (a) $(0, -1)$ (b) $\left(\dfrac{1}{2}, 0\right)$ (c) $(1, 1)$
                      (d) $(1.5, 2)$ (e) $(-1, -3)$

10. $3y = x - 6$ (a) $(6, 0)$ (b) $(0, -2)$ (c) $(12, 2)$
                      (d) $(3, -1)$ (e) $(9, 1)$

In problem 7 a few extra steps are needed.

$2y = 3x + 6$

(a) $2y = 3(0) + 6$ substituting 0 for x
      $2y = 6$
       $y = 3$

(b) $2y = 3(1) + 6$ substituting 1 for x
      $2y = 3 + 6 = 9$

       $y = \dfrac{9}{2} = 4.5$ and so on

In parts (b), (d) and (e) of problem 9 you must solve for x rather than y to complete the ordered pair.

$y = 2x - 1$

(b) $(0) = 2x - 1$ substituting 0 for y

$1 = 2x$

$\dfrac{1}{2} = x$ or $x = \dfrac{1}{2}$

(d) $(2) = 2x - 1$ substituting 2 for y

$3 = 2x$

$\dfrac{3}{2} = x$ or $x = 1\dfrac{1}{2}$

(e) $(-3) = 2x - 1$ substituting -3 for y

$-2 = 2x$

$-1 = x$ or $x = -1$

4-2 GRAPHS

We can describe a relationship with words or with an algebraic equation, but we can also *see* it. A *graph* of an equation is a visual picture or visual description of the relationship represented by the equation. In this unit you will learn how to draw a graph of an algebraic equation.

First, you should recall from Unit 1 that a number line is a visual way of uniquely associating numbers with points on a straight line. Every point on the number line shown has a number associated with it.

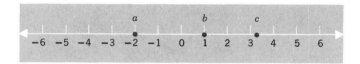

For example,
Point a is located at $x = -2$,
Point b is located at $x = +1$,
Point c is located at $x = 3.4$,
and so on.

If we draw a second number line perpendicular to the first and crossing it at some point, then every point in the plane of these lines can be uniquely associated with a *pair* of numbers. For example, the point at p is associated with the pair of numbers $(3, 4)$ as shown.

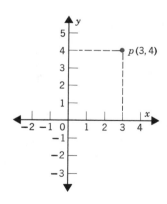

Mathematicians have developed some very special vocabulary words to describe graphs, and it is important that you learn to use the correct words. The two number lines are called *axes*. The horizontal number line is called the x axis and is labeled with the letter x. The vertical number line is called the y *axis* and is labeled with the letter y. Because the two axes are set at right angles, the entire arrangement is called a *rectangular coordinate system*.

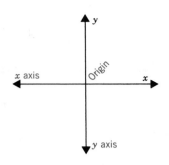

Origin

The point $x = 0, y = 0$ where the two axes cross is called the *origin* of the coordinate system.

Quadrants

Notice that the two axes divide the surrounding area into four parts called *quadrants,* which are numbered counter-clockwise with Roman numerals I, II, III, and IV.

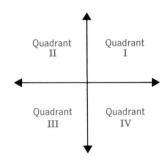

Now, to test yourself on these new vocabulary words, draw a pair of coordinate axes and label the *x* axis, *y* axis, origin, and four quadrants. Do it here . . .

. . . then continue in **6**.

---

6   The numbers in any ordered pair give the location of a particular point. For example, the ordered pair $(3, 4)$ corresponds to the point at $x = 3, y = 4$.

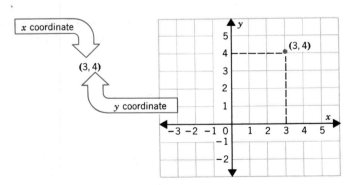

x and y Coordinates

We say that the *x coordinate* of the point is 3 and the *y coordinate* of the point is 4.

The number pair enables you to locate the corresponding point by giving travel instructions starting from the origin.

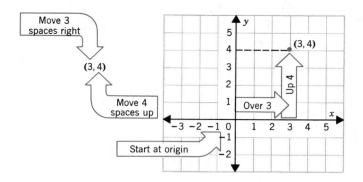

Plotting a Point

To *plot* a point means to use an ordered pair of numbers to locate the point on a rectangular coordinate system.

Plot the points   (a)   (1, 4)   (b)   (4, 1)

                     (c)   (0, 5)   (d)   (2, 0)

                     (e)   (3, 6)

Use this coordinate system.

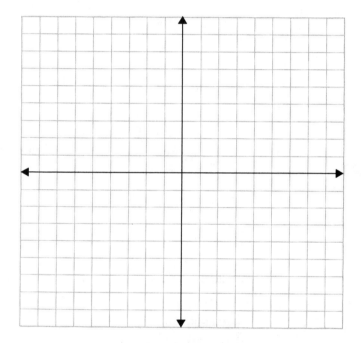

When you have located these points check your answer in **7**.

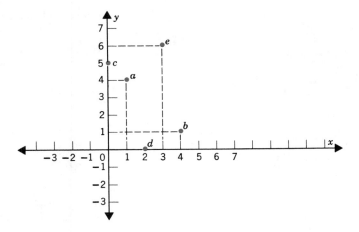

Abscissa
Ordinate

Mathematicians often refer to the x and y numbers in the ordered number pair as the *abscissa* and *ordinate*.

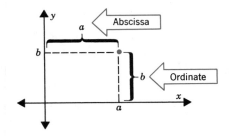

Negative number coordinates indicate distances to the left of the origin on the x axis and below the origin on the y axis. For example, the point $(-4, 3)$ is located as shown.

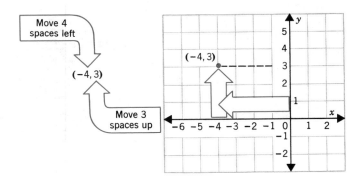

The point $(-4, -3)$ is located in a similar way.

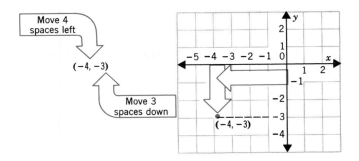

Ready for some practice in using rectangular coordinate systems?

1. Plot the following points.

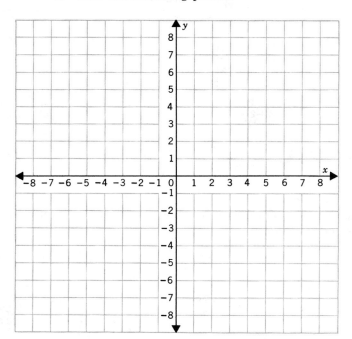

(a) (1, 1)	(b) (2, −1)
(c) (0, −4)	(d) (−2, 0)
(e) (−1, −3)	(f) (2, −5)
(g) (−4, 1)	(h) (0, 0)
(i) (0, 5)	(j) (8, 0)
(k) (8, −1)	(l) (7, 7)
(m) (−6, 7)	(n) (8, −6)
(o) (−6, −8)	

2. Determine the coordinates of each of the points shown.

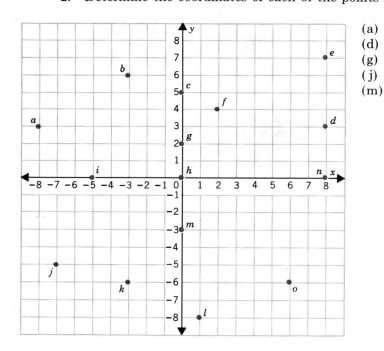

(a)	(b)	(c)
(d)	(e)	(f)
(g)	(h)	(i)
(j)	(k)	(l)
(m)	(n)	(o)

Check your work in **8**.

8   1.

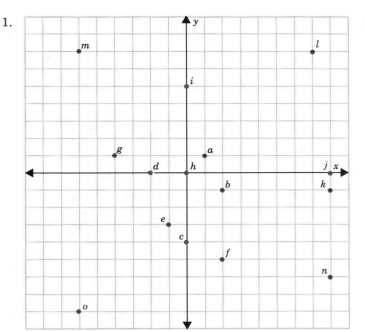

2.   (a) $(-8, 3)$ (b) $(-3, 6)$
       (c) $(0, 5)$ (d) $(8, 3)$
       (e) $(8, 7)$ (f) $(2, 4)$
       (g) $(0, 2)$ (h) $(0, 0)$
       (i) $(-5, 0)$ (j) $(-7, -5)$
       (k) $(-3, -6)$ (l) $(1, -8)$
       (m) $(0, -3)$ (n) $(8, 0)$
       (o) $(6, -6)$

Plotting an Equation

The equation $2y = x + 2$ has the ordered pair $(4, 3)$ as a solution. The ordered pairs $(-2, 0)$, $(0, 1)$, $(1, 1.5)$, and $(2, 2)$ are also solutions of this equation. We may plot these points on a rectangular coordinate axis as shown.

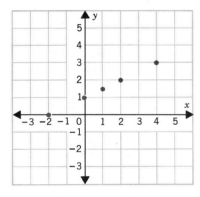

The points plotted are only a few of the infinite number of points whose coordinates satisfy the equation. Obviously we cannot plot every individual point that satisfies the equation, since that would require finding the y value corresponding to every possible value of x—all possible integers, fractions, and decimal numbers. However, we can plot these few points and search for a pattern. In the graph here the pattern is easy to see; the points plotted lie in a straight line.

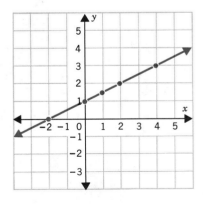

Draw a straight line through the plotted points. The complete graph of the solution set of the equation is a straight line of infinite length.

It is possible to prove that the graph of any first-degree equation is a straight line. (You should remember from Unit 2 that a first-degree equation is one that contains only first powers of the variables. $y = 2x + 1$ is a first-degree equation and so are the equations $2y = 5x - 3$, $y = 0$, and $x = 2$.)

Graphing a Linear Equation

To graph a first-degree equation follow these steps:

Step 1. Construct a table with two columns headed x and y.

Example: $y = 2x - 1$

x	y

Step 2. Select at least three values of x or y and enter them in the proper column.

x	y
0	
1	
	3

Step 3. Use the equation to find the x or y value that completes each row.

x	y
0	−1
1	1
2	3

$$y = 2(0) - 1 = -1$$
$$y = 2(1) - 1 = 2 - 1 = 1$$
$$(3) = 2x - 1$$
$$2x = 4, \ x = 2$$

Step 4. Each row of the table is an ordered pair. Plot these points on a rectangular coordinate system.

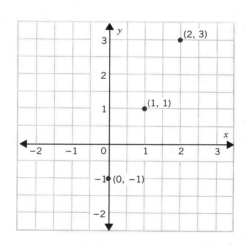

Step 5. Draw a straight line through the points using a straight edge.

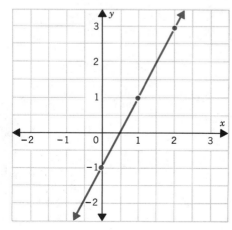

264   Equations in Two Variables

Only two points are needed to determine exactly the position and direction of a straight line, but you should plot at least three points (not too close together, please) to be certain you have made no mistakes.

Your turn. Use the step-by-step method shown above to plot the graph of the equation $y = 4 - 2x$

Take your time and work carefully. Our work is shown in **9**.

9

x	y
0	4
1	2
2	0

$y = 4 - 2(0) = 4$

$y = 4 - 2(1) = 4 - 2 = 2$

$y = 4 - 2(2) = 4 - 4 = 0$

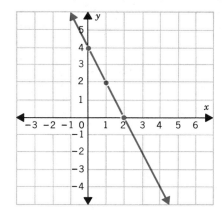

For this equation it is easiest to start by letting x take the values 0, 1, and 2.

Using the equation, find the value of y corresponding to each value of x and place each number pair in the table. Finally, plot the three ordered pairs $(0, 4)$, $(1, 2)$, and $(2, 0)$ on the coordinate axes and draw a straight line through these points.

Any values of x or y may be selected at the start of the process—you get to choose them. However, some values of x or y are easier to work with than others. For example, in the equation $2y + 3x = 12$ the number pair with $x = 0$ is easy to calculate:

$2y + 3(0) = 12$
$\quad 2y = 12$
$\quad\; y = 6$ The point is at $(0, 6)$.

The number pair with $y = 0$ is also easy to find:

$2(0) + 3x = 12$
$\quad\;\; 3x = 12$
$\quad\;\; x = 4$ The point is at $(4, 0)$.

The number pair with $x = -2$ is found in the same way:

$2y + 3(-2) = 12$
$\quad 2y - 6 = 12$
$\quad\;\;\; 2y = 18$
$\quad\;\;\; y = 9$ The point is at $(-2, 9)$.

Now you need some practice to make certain that you can graph first-degree equations quickly and accurately. Try these problems. For each equation construct a table of at least three number pairs, plot the number pairs, and draw a straight line through them.

1. $y = 2x$ 2. $y = x - 2$ 3. $y = -2x + 1$

4. $2x - 5y = 10$ 5. $y = 4$ 6. $x = 4$

7. $3y = 6 - x$ 8. $x = -2$ 9. $y = -5$

10. $y = -x$

Our graphs are in **10**. Check your work when you have completed these.

10

1. $y = 2x$

x	y
0	0
1	2
2	4

$y = 2(0) = 0$

$y = 2(1) = 2$

$y = 2(2) = 4$

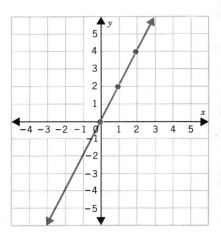

2. $y = x - 2$

x	y
0	-2
1	-1
2	0

$y = (0) - 2 = -2$

$y = (1) - 2 = -1$

$y = (2) - 2 = 0$

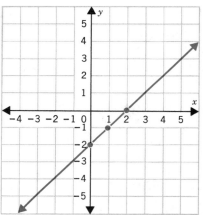

3. $y = -2x + 1$

x	y
0	1
1	-1
2	-3

$y = -2(0) + 1 = 1$

$y = -2(1) + 1 = -2 + 1 = -1$

$y = -2(2) + 1 = -4 + 1 = -3$

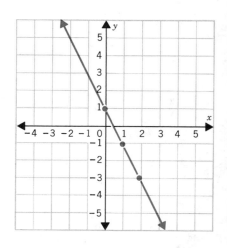

4. $2x - 5y = 10$

x	y
0	-2
5	0
-5	-4

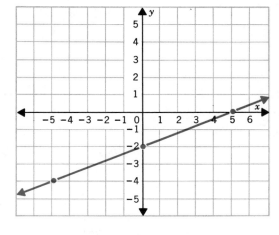

$2(0) - 5y = 10$
$-5y = 10$
$y = -2$

$2x - 5(0) = 10$ $2(-5) - 5y = 10$
$2x = 10$ $-10 - 5y = 10$
$x = 5$ $-5y = 20$
                          $y = -4$

The points $(0, \ )$ and $(\ , 0)$ were chosen because they are easiest to calculate.

5. $y = 4$

x	y
0	4
1	4
2	4

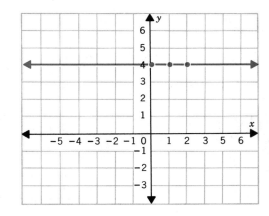

$y = 4$ for all values of x. The graph of the equation $y = 4$ is a straight line parallel to the x axis.

6. $x = 4$

x	y
4	0
4	1
4	2

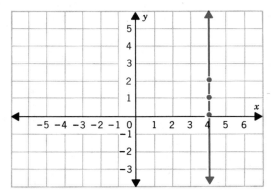

$x = 4$ for all values of y. The graph of the equation $x = 4$ is a straight line parallel to the y axis.

7. $3y = 6 - x$

x	y
0	2
3	1
6	0

$3y = 6 - (0)$
$3y = 6$
$y = 2$

$3y = 6 - (3)$ $3(0) = 6 - x$
$3y = 3$ $0 = 6 - x$
$y = 1$ $x = 6$

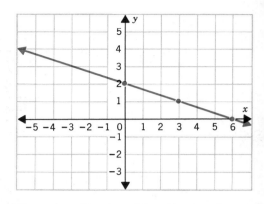

The points (0,   ) and (, 0) were chosen because they are easiest to calculate.

8. $x = -2$

x	y
−2	0
−2	1
−2	2

$x = -2$ for all values of y.

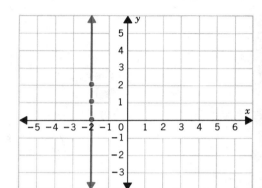

9. $y = -5$

x	y
0	−5
1	−5
2	−5

$y = -5$ for all values of x.

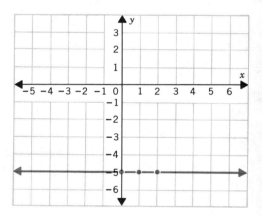

10. $y = -x$

x	y
1	−1
2	−2
3	−3

$y = -(1) = -1$
$y = -(2) = -2$
$y = -(3) = -3$

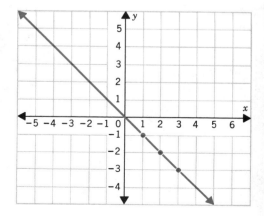

Notice that the graph of any equation of the form $y = a$, where a is any constant number, is a straight line parallel to the x axis.

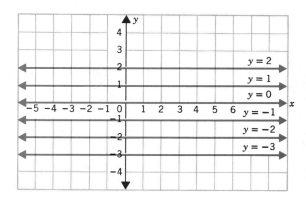

The graph of any equation of the form $x = b$, where b is any constant number, is a straight line parallel to the y axis.

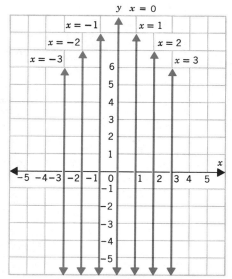

Quickly now, with no calculations, sketch the graphs of the following equations.

(a)　$x = -3$　　　(b)　$y = 2$　　　(c)　$y = 0$　　　(d)　$x = 0$

Our work is in **11**.

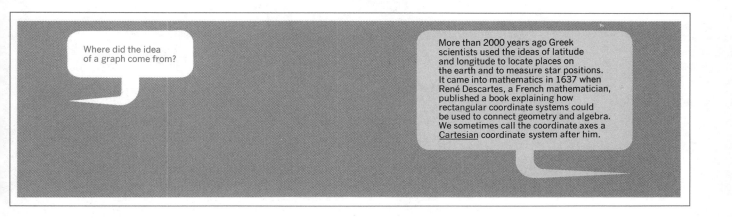

11   (a)   (b)   (c)   (d)

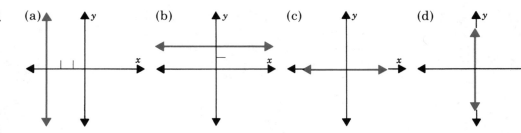

Any first-degree or linear equation in two variables can be written

$$Ax + By + C = 0$$

where A, B, and C are constants.

Slope-Intercept Form

An equation in the form $y = mx + b$ is said to be in standard or *slope-intercept* form. For example, the equation

$2y = x + 4$ can be rewritten in this form:

$y = \dfrac{x}{2} + \dfrac{4}{2}$ Divide by 2.

$y = \left(\dfrac{1}{2}\right)x + (2)$

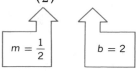

$m = \dfrac{1}{2}$ $b = 2$

The equation $3y + 4x = -2$ can be rewritten in this form:

$3y = -4x - 2$

$y = -\dfrac{4x}{3} - \dfrac{2}{3}$ Divide by 3.

$y = \left(-\dfrac{4}{3}\right)x + \left(-\dfrac{2}{3}\right)$

$m = -\dfrac{4}{3}$ $b = -\dfrac{2}{3}$

The equation $y = -4$ can be rewritten in this form:

$y = (0)x + (-4)$

$m = 0$ $b = -4$

Because the graph of a first-degree or linear equation is always a straight line, this way of writing the equation gives us valuable information about the graph of a line.

Rewrite each of the following equations in the form $y = mx + b$ and find m and b for each equation.

(a) $y + 2 = 3x$ (b) $2x + y = 6$
(c) $3x + 2y = 6$ (d) $2y - x + 6 = 0$
(e) $y = 5$

Check your work in **12**.

12

(a) $y + 2 = 3x$

$$y = 3x - 2$$

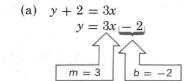

$m = 3$ $b = -2$

(b) $2x + y = 6$

$$y = -2x + 6$$
$$y = (-2)x + 6$$

$m = -2$ $b = 6$

(c) $3x + 2y = 6$

$$2y = -3x + 6$$
$$y = -\frac{3x}{2} + \frac{6}{2}$$
$$y = \left(-\frac{3}{2}\right)x + (3)$$

$m = -\dfrac{3}{2}$ $b = 3$

(d) $2y - x + 6 = 0$

$$2y = x - 6$$
$$y = \frac{x}{2} - \frac{6}{2}$$
$$y = \left(\frac{1}{2}\right)x + (-3)$$

$m = \dfrac{1}{2}$ $b = -3$

(e) $y = 5$

$$y = (0)x + 5$$

$m = 0$ $b = 5$

Intercept and slope

If we let $x = 0$ in the equation $y = mx + b$, then

$$y = m(0) + b$$

or $y = b$

y-intercept

and the point $(0, b)$ is a point on the graph line. The point $(0, b)$ is where the graph line crosses the y axis. We call this point the *y-intercept*

The line crosses the y axis at $y = b$.

Slope

The number m is called the *slope* of the straight line graph of the equation. The slope of the line is the ratio of the vertical change to the horizontal change between any two points on the line.

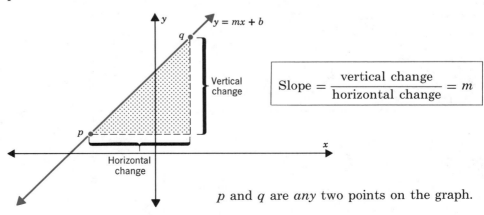

$$\text{Slope} = \frac{\text{vertical change}}{\text{horizontal change}} = m$$

p and q are *any* two points on the graph.

For example, plotting the equation $y = 2x + 1$, we get this graph.

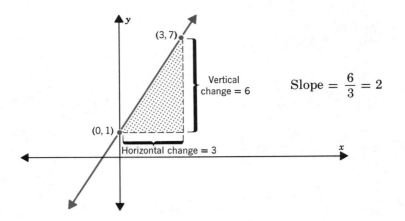

Slope $= \dfrac{6}{3} = 2$

Choose any two points on the graph line. Here we have chosen points $(0, 1)$ and $(3, 7)$.

From the equation itself we see that

$$y = 2x + 1$$

$m = 2$ which agrees with the slope as calculated from the graph.

When the absolute value of the slope is large, the line is tilted steeply up from the x axis. When the absolute value of the slope is small, the line is tilted up only slightly from the x axis. A slope of zero means that the line is parallel to the x axis.

Zero Slope

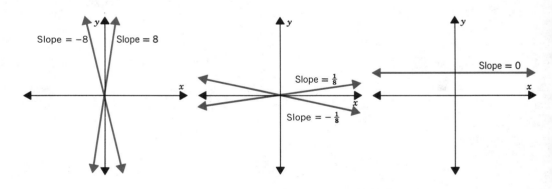

Negative Slope

A positive slope number means that the line is tilted upward to the right. A negative slope number means that the line is tilted upward to the left.

Slope of a Vertical Line

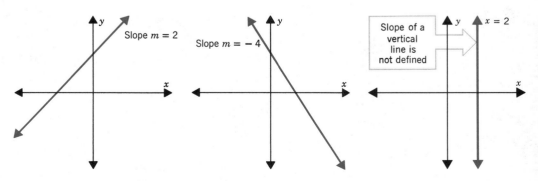

The concept of slope of a graph line is very important in many scientific and technical applications of algebra.

Use these problems to test your understanding of the concepts of slope and y intercept.

1. If slope $m = 2$ and y intercept $b = 4$ for a certain graph, write the equation of the graph and plot it.
2. For the following set of measurements on a group of children in a psychology experiment,
    (a) Plot the graph.
    (b) Find m, the slope of the graph.
    (c) Find b, the y-intercept of the graph.
    (d) Write the equation of the graph using these values of m and b.

y, test score	25	35	75	85	105	15	55
x, age, yrs	2	3	7	8	10	1	5

Check your answers to these problems in **13**.

---

13   1. $m = 2$, $b = 4$

$y = mx + b$
$y = 2x + 4$

x	y
0	4
1	6
2	8

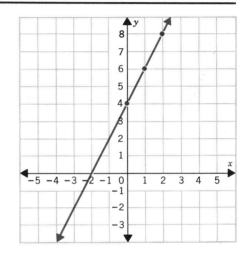

2. (a)

x	y
2	25
3	35
7	75
8	85
10	105
1	15
5	55

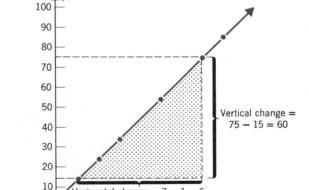

(b) Slope $m = \dfrac{60}{6} = 10$

(c) y intercept $b = 5$
(d) $y = mx + b$
     $y = 10x + 5$

Finding the Equation of a Line Given Two Points on the Line

Any two points define one and only one straight line. If we place the two points on a rectangular coordinate plane, a straight line graph may be drawn through the points and the equation of the line may be found from the coordinates of the two points. For example, if any two points (x_1, y_1) and (x_2, y_2) are plotted on the coordinate axes, a straight line may be drawn as shown below.

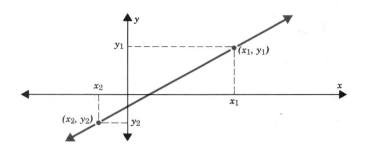

The slope m of the line drawn through these two points is given by the equation

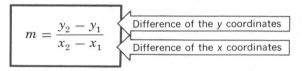

$$m = \frac{y_2 - y_1}{x_2 - x_1}$$

Difference of the y coordinates

Difference of the x coordinates

and the y intercept is

$$b = y_1 - mx_1$$

The equation of the straight line is $y = mx + b$

Example:

If we wish to find the equation of the straight line graph through the points $(3, 4)$ and $(-3, -1)$ we can use these formulas. Here $x_1 = 3$, $y_1 = 4$, $x_2 = -3$, and $y_2 = -1$.

$$m = \frac{(-1) - (4)}{(-3) - (3)} = \frac{-5}{-6} = \frac{5}{6}$$

$$b = 4 - \left(\frac{5}{6}\right)3 = 4 - 2\frac{1}{2} = \frac{3}{2}$$

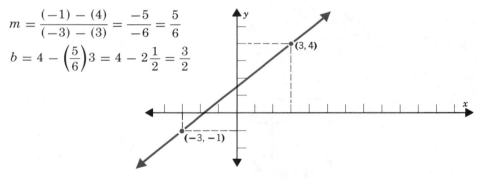

The equation of the line is $y = \frac{5}{6}x + \frac{3}{2}$ or, if we multiply by 6 to clear fractions,

$6y = 5x + 9$

Try it yourself: for each of the following pairs of points find m, b, and the equation of the straight line through the points.

(a) $(4, 1)$ and $(1, 4)$ (b) $(2, 3)$ and $(0, -2)$
(c) $(3, 5)$ and $(2, 0)$ (d) $(1, -1)$ and $(-4, -3)$

Check your work in **14**.

14

(a) $m = \dfrac{(4) - (1)}{(1) - (4)} = \dfrac{3}{-3} = -1$ $x_1 = 4, y_1 = 1$

                                                    $x_2 = 1, y_2 = 4$

     $b = 1 - (-1)4 = 1 + 4 = 5$

     The equation is $y = -x + 5$

(b) $m = \dfrac{(-2) - (3)}{(0) - (2)} = \dfrac{-5}{-2} = \dfrac{5}{2}$ $x_1 = 2, y_1 = 3$

                                                    $x_2 = 0, y_2 = -2$

     $b = 3 - \left(\dfrac{5}{2}\right)2 = 3 - 5 = -2$

     The equation is $y = \dfrac{5}{2}x + (-2) = \dfrac{5x}{2} - 2$

     or, if we multiply by 2 to clear fractions,
     $2y = 5x - 4$

(c) $m = \dfrac{(0) - (5)}{(2) - (3)} = \dfrac{-5}{-1} = 5$ $x_1 = 3, y_1 = 5$

                                                    $x_2 = 2, y_2 = 0$

     $b = 5 - (5)3 = 5 - 15 = -10$

     The equation is $y = 5x + (-10) = 5x - 10$

(d) $m = \dfrac{(-3) - (-1)}{(-4) - (1)} = \dfrac{-3 + 1}{-4 - 1} = \dfrac{-2}{-5} = \dfrac{2}{5}$ $x_1 = 1, y_1 = -1$

                                                          $x_2 = -4, y_2 = -3$

     $b = -1 - \left(\dfrac{2}{5}\right)1 = -1 - \dfrac{2}{5} = -\dfrac{7}{5}$

     The equation is $y = \dfrac{2}{5}x + \left(-\dfrac{7}{5}\right) = \dfrac{2x}{5} - \dfrac{7}{5}$

     or, if we multiply by 5 to clear the fractions,
     $5y = 2x - 7$

*Finding the Equation
Given the Slope and
a Point on the Line*

If we know the slope of a line and the coordinates of any point on that line, we can find b, the y-intercept, and the equation of that line.

Example: Find the equation of the line with slope $m = -4$ that intersects the point $(3, -2)$.

First, we have $x_1 = 3, y_1 = -2, m = -4$

Then, $b = y_1 - mx_1 = -2 - (-4)(3) = -2 + 12 = 10$

The equation of the line can be found by substituting these values of b and m into the slope-intercept equation of a line.

$y = mx + b = (-4)x + 10$ or $y = -4x + 10$

Another Example: Find the equation of the line passing through the point $(-2, 5)$ with slope -3.

$x_1 = -2, y_1 = 5, m = -3$

$b = y_1 - mx_1 = 5 - (-3)(-2) = 5 - 6 = -1$

The equation of the line is

$y = mx + b = (-3)x + (-1)$ or $y = -3x - 1$

For practice, solve the following problems. Find the equation of the line passing through the given point and having the slope indicated.

(a) (3, 4); $m = 3$

(b) (−4, 1); $m = -2$

(c) (−2, −3); $m = 2$

(d) (−1, 3); $m = \dfrac{1}{2}$

Check your work in 15.

---

15 (a) $x_1 = 3, y_1 = 4, m = 3$

$b = y_1 - mx_1 = 4 - (3)(3) = 4 - 9 = -5$

The equation is $y = mx + b = (3)x + (-5)$
or $y = 3x - 5$

(b) $x_1 = -4, y_1 = 1, m = -2$

$b = y_1 - mx_1 = 1 - (-2)(-4) = 1 - 8 = -7$

The equation is $y = mx + b = (-2)x + (-7)$

or $y = -2x - 7$

(c) $x_1 = -2, y_1 = -3, m = 2$

$b = y_1 - mx_1 = -3 - (2)(-2) = -3 + 4 = 1$

The equation is $y = mx + b = (2)x + (1)$
or $y = 2x + 1$

(d) $x_1 = -1, y_1 = 3, m = \dfrac{1}{2}$

$b = y_1 - mx_1 = 3 - \left(\dfrac{1}{2}\right)(-1) = 3 + \dfrac{1}{2} = \dfrac{7}{2}$

The equation is $y = mx + b = \left(\dfrac{1}{2}\right)x + \dfrac{7}{2}$

or $y = \dfrac{x}{2} + \dfrac{7}{2}$

Multiplying each term by 2 to clear the fractions, we have $2y = x + 7$

Now turn to 16 for a set of problems on graphing equations.

Unit 4
Equations
in Two Variables

16   The answers are on page 549.

Graphing Equations

A. Plot the following points using the coordinate axes shown.

1. (a) $(1, 2)$ (b) $(2, -3)$
    (c) $(0, 4)$ (d) $(3, 0)$
    (e) $(-1, 2)$ (f) $(5, -6)$
    (g) $(-4, -3)$ (h) $(-5, 0)$
    (i) $(7, 7)$ (j) $(-2, -2)$

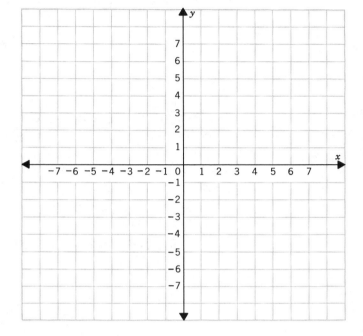

2. (a) $(40, 40)$ (b) $(40, 0)$
    (c) $(0, 40)$ (d) $(-40, 40)$
    (e) $(40, -40)$ (f) $(-40, -40)$
    (g) $(0, 0)$ (h) $(40, 20)$
    (i) $(40, -10)$ (j) $(-30, 50)$

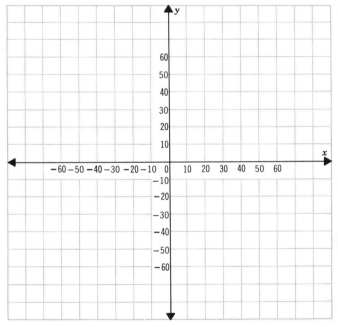

Date _____

Name _____

Course/Section _____

B. Determine the coordinates of each of the points shown.

1.

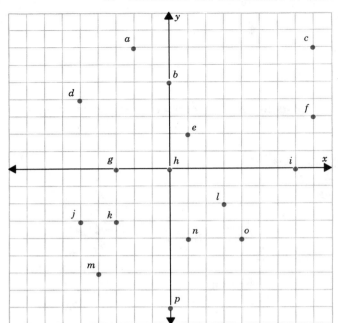

(a) _____   (b) _____

(c) _____   (d) _____

(e) _____   (f) _____

(g) _____   (h) _____

(i) _____   (j) _____

(k) _____   (l) _____

(m) _____   (n) _____

(o) _____   (p) _____

C. Plot graphs for each of the following equations.

1. $y = x + 3$ 2. $y = 2x - 4$

3. $y = -3x + 3$ 4. $y = -3x$

5. $y = 3$ 6. $x = -5$

7. $2y - 4x = 12$ 8. $5x - 2y = 10$

9. $y = \dfrac{1}{2}x$ 10. $3x + 2y = 4$

11. $x = 0$ 12. $3x - 6 = 4y$

13. $2y = -x - 3$ 14. $3x + y = 5$

15. $y = -\dfrac{1}{4}x + 1$ 16. $x - y = 2$

17. $y = 20x$ 18. $10y - 25x = 50$

19. $F = \dfrac{9}{5}C + 32$ 20. $s = 10t$

21. $s = 20 + 10t$ 22. $D = 20t + 10$

23. $y = 100 + 10x$ 24. $x = 100 + 10y$

D. Write each of the following equations in slope-intercept form and identify the slope and y intercept for each equation.

1. $y = 3 + x$ _____   2. $y = 2x - 1$ _____

3. $y = \dfrac{1}{2}x - 5$ _____   4. $2y = x + 1$ _____

5. $3y = x + 2$ _____   6. $x - y = 4$ _____

7. $2y = -x - 4$ _____   8. $y = 2x$ _____

9. $y = -5x$ _____

10. $y = 6$ _____

11. $2y + 3x = 60$ _____

12. $5y - 4x = -20$ _____

13. $F = \dfrac{9}{5}C + 32$ _____

14. $s = 10t$ _____

15. $y = -4$ _____

16. $10y = 4x - 2$ _____

17. $x - 3y = 5$ _____

18. $8 - 2x - 4y = 0$ _____

19. $3x - y = y + x + 6$ _____

20. $x - y = y - x$ _____

21. $3(x - y) - x = 2y + 1$ _____

22. $2 - 2(x - 1) = 3y - 5$ _____

23. $1 - 3(y + 1) = 2(1 - 3x)$ _____

24. $x - 2(x - 1) = y - (3y + 2)$ _____

25. $x + 2(1 - y) - 3(y - x) = 0$ _____

26. $4(y + x) = 1$ _____

27. $\dfrac{x}{2} + 3y = \dfrac{1}{4}$ _____

28. $\dfrac{2y}{3} - 1 = 4x$ _____

29. $\dfrac{1}{2}(2x + 1) = y - 1$ _____

30. $\dfrac{1}{4}(y + 1) = 1 - x$ _____

E. **Determine the equation of the line passing through each of the following pairs of points.**

1. $(2, 5)$ and $(1, 7)$

2. $(-1, 4)$ and $(4, -1)$

3. $(3, -4)$ and $(-2, 5)$

4. $(-4, -5)$ and $(3, 1)$

5. $(-2, 2)$ and $(4, 0)$

6. $(-2, -4)$ and $(0, 0)$

7. $(-1, -3)$ and $(4, 2)$

8. $(5, -3)$ and $(-3, 1)$

9. $\left(1, \dfrac{1}{2}\right)$ and $\left(2, -\dfrac{1}{2}\right)$

10. $\left(\dfrac{1}{8}, 1\right)$ and $\left(\dfrac{1}{4}, \dfrac{3}{4}\right)$

11. $\left(1\dfrac{2}{3}, 2\right)$ and $\left(\dfrac{2}{3}, -1\right)$

12. $\left(-\dfrac{4}{5}, \dfrac{4}{5}\right)$ and $\left(1, -1\right)$

13. $(-6, -7)$ and $(-4, -9)$

14. $(-3, 2)$ and $(-4, -5)$

15. $(0, 6)$ and $(3, 0)$

16. $(3, 0)$ and $(0, -9)$

Determine the equation of the line passing through the given point and having the given slope.

17. $(1, 4)$; $m = 3$

18. $(4, -2)$; $m = 2$

19. $(3, -2)$; $m = 4$

20. $(-2, -3)$; $m = 1$

21. $(2, 6)$; $m = -3$

22. $(5, -1)$; $m = -4$

23. $(4, -1)$; $m = -2$

24. $(-5, -2)$; $m = -1$

Date _____

Name _____

Course/Section _____

279   **Problem Set 4-1**

25. $(-6, -4); m = \dfrac{1}{2}$

26. $(10, -8); m = \dfrac{2}{5}$

27. $(-6, -7); m = -\dfrac{3}{4}$

28. $(-9, -5); m = -\dfrac{2}{3}$

29. $\left(\dfrac{1}{2}, -\dfrac{1}{4}\right); m = -1$

30. $\left(-\dfrac{4}{5}, \dfrac{2}{5}\right); m = 2$

F. Brain Boosters

1. Graph the following equations on the same coordinate axis.
   (a) $y = 2x + 1$ (b) $y = 2x + 2$ (c) $y = 2x + 3$

2. Graph the following equations on the same coordinate axis.
   (a) $y = 2x + 1$ (b) $y = 3x + 1$ (c) $y = 4x + 1$

3. Graph the following equations on the same coordinate axis.

   (a) $y = x$ (b) $y = 2x$ (c) $y = 3x$ (d) $y = \dfrac{1}{2}x$

4. By measuring the heights and weights of many people, physiologists have found that the following equation gives the best fitting relationship between the weight of a person in lb and their height in in.

   $W = 5.5H - 220$ for $H \geq 60$ in.

   (a) Graph this equation.
   (b) Plot your height and weight on the same graph. Do you fit the "average" graph?

5. The atmospheric temperature on a typical day as you travel outward from the earth is as follows:

Altitude(ft)	Sea level	1000	5000	10,000	15,000	20,000	30,000	35,000
Temperature($°F$)	59	56	41	23	5	-15	-4	-70

   Plot a graph of this information.

When you have had the practice you need, either return to the preview on page 247 or move ahead to **17** where you will study the graphing of inequalities.

17 In Unit 2 we used the number line to graph inequalities containing one variable. (If you need to review the basic concepts of inequalities return to frame **35** on page 143.) To graph an inequality means to display on a number line the set of numbers that satisfy the inequality. For example, the graph of the inequality $y < 3$ looks like this:

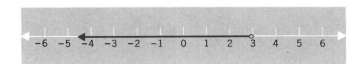

An inequality of this kind is a statement or rule that divides the number line into two portions. All values of y that make the inequality a true statement are in the shaded portion of the line to the left of 3. All numbers that do not satisfy the inequality are in the unshaded portion of the line.

Linear Inequalities   In a very similar way, a first-degree or *linear* inequality in two variables divides the plane of a rectangular coordinate system into two portions. For example, the inequality $2x + y \geq 4$ divides the plane of the coordinate axes as shown here:

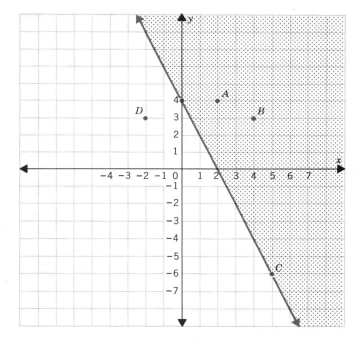

The shaded area of the plane contains all points whose coordinates (x, y) make the inequality a true statement. Point A, for example, has coordinates $(2, 4)$. Substituting this pair of numbers into the inequality, we have

$$2x + y \geq 4$$
$$2(2) + (4) \geq 4$$
$$4 + 4 \geq 4$$

or $8 \geq 4$ which is certainly true.

Show that the coordinates of points B and C satisfy the inequality and that the coordinates of point D do not.

Continue in **18** after you have shown this.

---

18   For point $B(4, 3)$ $2x + y \geq 4$
$$2(4) + (3) \geq 4$$
$$8 + 3 \geq 4$$
$$11 \geq 4 \qquad \text{which is true}$$

For point $C(5, -6)$ $2x + y \geq 4$
$$2(5) + (-6) \geq 4$$
$$10 - 6 \geq 4$$
$$4 \geq 4 \qquad \text{which is true; they are equal.}$$

For point $D(-2, 3)$ $2x + y \geq 4$
$$2(-2) + (3) \geq 4$$
$$-4 + 3 \geq 4$$
$$-1 \geq 4 \qquad \text{which is not true.} \ -1 \text{ is to the left}$$
of 4 on the number line and therefore -1 is *less than* 4.

Notice that the solid line $2x + y = 4$ is part of the graph of the inequality $2x + y \geq 4$. Points on the solid line satisfy this inequality.

The graph line separates the plane into three sets of points:

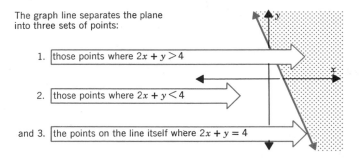

1. those points where $2x + y > 4$

2. those points where $2x + y < 4$

and 3. the points on the line itself where $2x + y = 4$

The graph of an inequality containing the symbols $\leq$ or $\geq$ includes the graph line. The graph of an inequality containing the symbols $<$ or $>$ does not include the graph line. For example,

the graph of $3y - 2x > 6$ is   the graph of $3y - 2x \geq 6$ is

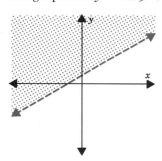   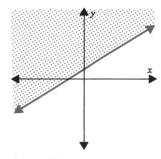

the graph of $3y - 2x < 6$ is

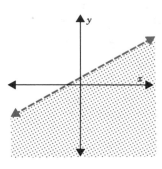

the graph of $3y - 2x \leq 6$ is

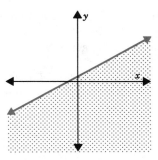

Which of the following is the correct graph for the inequality $y \leq 3x - 6$?

(a) 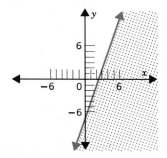   (b)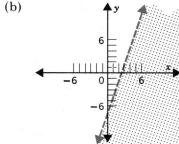

If your answer is (a) go to **20**.
If your answer is (b) go to **19**.

---

19   Sorry, your answer is incorrect.

The dotted graph line means that the points on this line have *not* been included in the graph of the inequality. But the symbol $\leq$ in $y \leq 3x - 6$ tells us to include all points (x, y) where $y < 3x - 6$ *and* all points (x, y) where $y = 3x - 6$.

A dotted graph line appears only with $<$ or $>$.

Now, return to **18** and follow the other answer.

---

20   Right you are. Very good.

In order to find the graph of a linear inequality in two variables, follow these three easy steps:

Step 1.   Replace the inequality symbol with an equals sign and plot the graph of the resulting equation. This graph will be a straight line.

Example: To graph $2x + 3y < 4$, replace $<$ by $=$ and plot $2x + 3y = 4$.

x	y
0	$\dfrac{4}{3}$
2	0
-1	2

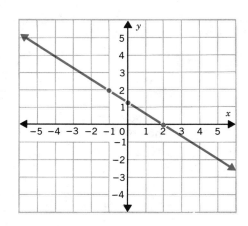

Step 2. If the original inequality contains a $<$ or $>$ symbol, the straight line graph should be a dotted line. If the original inequality contains a $\leq$ or $\geq$ symbol, the straight line graph should be a solid line.

Example: The inequality $2x + 3y < 4$ contains the symbol $<$; therefore the straight line graph should be a dotted line.

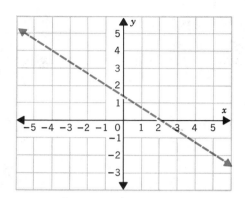

Step 3. Test to decide which half of the plane should be shaded. Pick any point not on the graph line and substitute its coordinates into the original inequality. If the substitution produces a true statement, shade in the half plane that contains the test point. If the substitution produces a false statement, shade in the other half plane.

Example: Use the origin $(0, 0)$ as the test point. Substitute $(0, 0)$ into the inequality.

$$2x + 3y < 4$$
$$2(0) + 3(0) < 4$$
$$0 < 4 \qquad \text{which is true.}$$

$(0, 0)$ is one of the points in the solution. Shade in the half plane containing the test point $(0, 0)$.

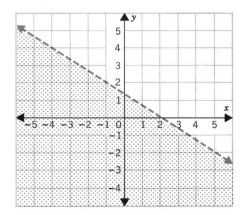

The origin $(0, 0)$ is a very convenient point to use as a test point in step 3, but any point not on the graph line will do.

Your turn. Use this three-step process to find the graph of the inequality $3y \geq 12 - 2x$.

Check your work in **21**.

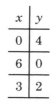

What does the dotted border line on an inequality graph mean?

If that graph line is dotted it means that the points on the line are not included in the graph and only points in the shaded area satisfy the inequality. The points on a solid line are included in the graph.

21   **Step 1.**   Replace $\geq$ in the inequality $3y \geq 12 - 2x$ by $=$ and plot $3y = 12 - 2x$.

x	y
0	4
6	0
3	2

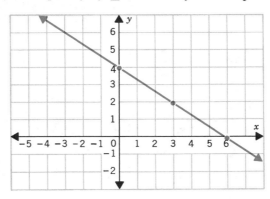

Step 2.   The original inequality $3y \geq 12 - 2x$ contains the symbol $\geq$; therefore the points on $3y = 12 - 2x$ are included in the graph. The graph of $3y = 12 - 2x$ should be a *solid* line to indicate this.

Step 3.   Substitute the coordinates of the origin $(0, 0)$ into the inequality.

$$3y \geq 12 - 2x$$
$$3(0) \geq 12 - 2(0)$$
$$0 \geq 12$$

This is a *false* statement; therefore the half plane containing the test point $(0, 0)$ should not be shaded. Shade in the other half plane. The final graph is

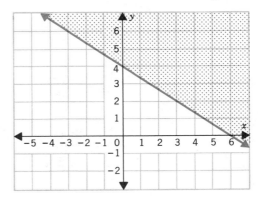

More practice will help make the process clear. Graph each of the following inequalities.

(a) $x - y > 5$ (b) $x \geq 3y - 8$ (c) $y > 2$

(d) $x \leq -2$ (e) $x < y$ (f) $5y + 2x \leq 10$

Work carefully. Check your graphs in **22**.

---

22   (a)

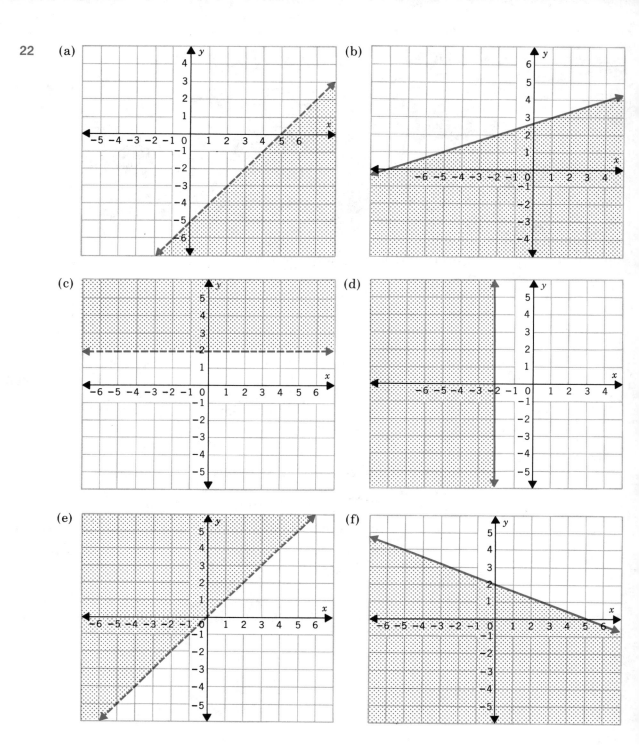

In problem (e), because the origin $(0, 0)$ is a point on the graph $x = y$, we must use some other point to test for shading the graph. If we use the point $A = (0, 1)$ the inequality becomes $0 < 1$, which is true. Shade the graph to include point A.

If several inequalities are true simultaneously, we may combine their graphs to find the set of points that satisfies *all* of the inequalities.

For example, what single set of points satisfies *all* of the following inequalities?

$$x > -1 \quad \text{and} \quad y > 2 \quad \text{and} \quad x + y \leq 6$$

If we graph each inequality separately we obtain the following three graphs:

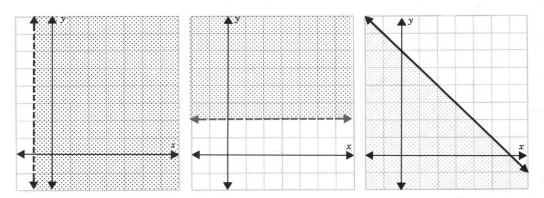

We can combine these graphs to find the set of points that satisfy all three inequalities at the same time. To do this imagine the three graphs superimposed over one another.

The only points that satisfy all three inequalities are those points that the three graphs have in common. The solution of the set of three inequalities is this graph:

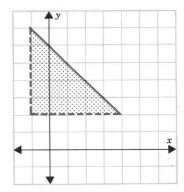

Notice that, as before, the boundary is part of the solution only when the boundary is a solid line.

Another example: Graph the following system of inequalities:

$$y \geq -2, \qquad 2y < x - 2, \qquad y \leq 5 - x$$

First, plot each inequality separately.

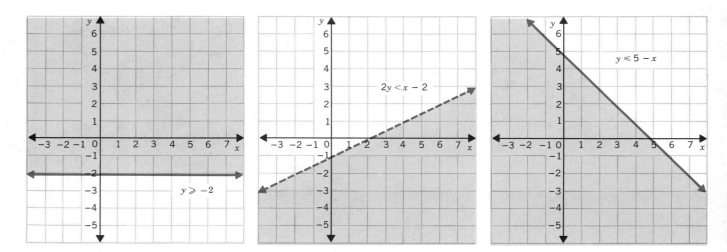

In the first graph, the graph line is solid to indicate the $\geq$ symbol. If we substitute the coordinates of the origin in the first inequality, we get $0 \geq -2$ which is true. Therefore the origin is a point in the shaded area.

In the second graph, the line is dotted to indicate the $<$ symbol. Points on this line are not included in the graph. Substituting the coordinates of the origin in the inequality, we get $0 < 0 - 2$ which is false. Therefore the shaded area does not include the origin.

In the third graph, the graph line is solid to indicate the $\leq$ symbol. Points on this line are included in the graph. Substituting the coordinates of the origin, we get $0 \leq 5 - 0$ which is true. Therefore the shaded area must include the origin.

Combining these three graphs, we have

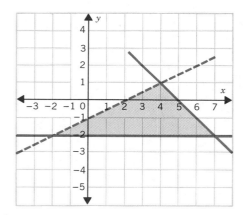

Now turn to **23** for a set of practice problems on graphing inequalities.

Unit 4
Equations
in Two Variables

The answers are on page 552.

A. Graph each of the following inequalities:

1. $x + y > 4$
2. $y - x \leq 5$
3. $y \leq -1$
4. $x > 2$
5. $2x < 5$
6. $x \geq y + 8$
7. $y \geq x + 2$
8. $2x < 1 - y$
9. $x < 3 - y$
10. $y \geq 2x + 1$
11. $y > 4$
12. $x \leq 8$
13. $x < -3$
14. $y \geq 0$
15. $x \geq 0$
16. $y > -2$
17. $10 \leq x - 2y$
18. $16 \geq y - 2x$
19. $y + x < 7$
20. $x - y < 8$

B. Graph each of the following inequalities:

1. $2x + 5y < -2$
2. $2y - 4x > 7$
3. $3x - 5y \geq 30$
4. $7y + 2x \leq 21$
5. $2x \leq y + 6$
6. $4y \geq 2x - 10$
7. $9 > 2y - 3x$
8. $18 < 2x - 3y$
9. $12 - 3x \leq 2y$
10. $20 - 5y \geq 2x$
11. $x \geq -y$
12. $y < -x$
13. $2x < 3y$
14. $4y \geq 3x$
15. $x - y \leq x + y - 2$
16. $3 + x \leq y - x + 7$
17. $y > -2 - x$
18. $x < -y - 5$
19. $x - \frac{1}{2}y > 4$
20. $y + \frac{1}{2}x \geq 6$

C. Word Problems

1. If L is the length of a rectangle and W is its width, for what values of L and W will the perimeter of the rectangle be less than 30 meters?

2. Find all possible pairs of numbers whose sum is greater than or equal to 7. Show them graphically.

3. Find all possible pairs of positive numbers such that the sum of the first and twice the second is less than 20.

Date

Name

Course/Section

4. Graph the pair of inequalities $x + y \leq 2$ and $2y + 6 > 3x$.

5. Graphically show all rectangles whose perimeter is less than 20 meters and whose sides differ by more than 2 meters.

6. Every Martian frammis is more than twice as tall as he is wide. The sum of the height and width of a frammis is 6 Mars bars.
   (a) Show in a graph all the possible frammis sizes.
   (b) From this graph find the height of the shortest possible frammis.

When you have had the practice you need either return to the preview on page 247 or turn to **24** and continue with the study of systems of equations.

24 Many very practical applications of algebra to science, business, or technology produce problems involving two equations in two variables. In Unit 2 you learned to solve a linear equation in one variable and in this section of Unit 4 you will learn to solve a pair of linear equations each containing the same two variables.

The equation $2x + y = 5$ has an infinite number of solutions, each an ordered pair of numbers. For example, each of these number pairs satisfy the equation: $(0, 5)$, $(1, 3)$, $(2, 1)$, $(3, -1)$, $(4, -3)$, $(-1, 7)$, $(-2, 9)$, and so on. The equation $4y - x = 11$ also has an infinite number of solutions, here are a few: $(-3, 2)$, $(5, 4)$, $(1, 3)$, $(-7, 1)$, $(9, 5)$, and so on. In these two infinitely large sets of number pairs is there one pair of numbers that satisfies *both* equations? By comparing the number pairs in the two lists you will find that the solution $(1, 3)$ is in both lists.

Show that the ordered pair $(1, 3)$ is a solution of the pair of equations:

$$2x + y = 5$$
$$4y - x = 11$$

by substituting 1 for x and 3 for y in both equations. Do it now, then continue in **25**.

25
$$
\begin{array}{ll}
2x + y = 5 & \qquad 4y - x = 11 \\
2(1) + (3) = 5 & \qquad 4(3) - (1) = 11 \\
2 + 3 = 5 & \qquad 12 - 1 = 11 \\
5 = 5 & \qquad 11 = 11
\end{array}
$$

The number pair $(1, 3)$ is a solution of both equations.

A set of two or more equations to be considered together is called a *system of simultaneous equations*. To *solve* a system of simultaneous equations means to find the solutions that satisfy *all* the equations in the system. In a system of two linear equations in two variables, the solution will be a number pair. In a system of three linear equations in three variables, the solution will be a triplet of numbers, and so on. In this book we will consider only systems of two linear equations in two variables, so that a solution will usually be a single number pair.

Solution by Graphing

One way to find the solution to a system of simultaneous equations is to graph the two equations on the same set of coordinate axes. For example, the system of equations

$$4x + 3y = 24$$
$$3y - 2x = 6$$

can be graphed as shown.

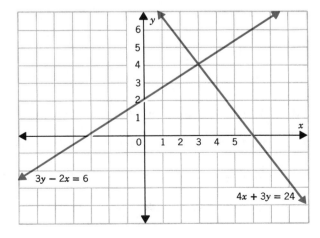

Each equation is represented on the graph by a straight line and the point of intersection of the two straight lines gives the solution of the system of equations. In this case, the two straight lines intersect at the point $(3, 4)$ so that the solution of the system of equations is $x = 3$, $y = 4$, or $(3, 4)$.

Check this solution by substituting these values back into the original equations.

Check your work in **26**.

26

$$4x + 3y = 24 \qquad\qquad 3y - 2x = 6$$
$$4(3) + 3(4) = 24 \qquad\quad 3(4) - 2(3) = 6$$
$$12 + 12 = 24 \qquad\qquad\quad 12 - 6 = 6$$
$$24 = 24 \qquad\qquad\qquad\quad 6 = 6$$

Because the number pair $(3, 4)$ satisfies both equations, it is a solution of this system of simultaneous equations.

Solve the following system of simultaneous equations by graphing the equations on the same set of coordinate axes.

$$-4x + y = 5$$
$$x - y = 4$$

Check your graph in **27**.

27

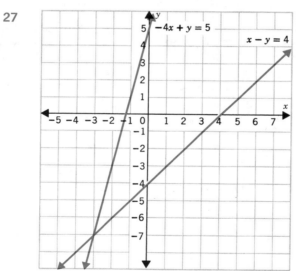

The two graph lines intersect at $(-3, -7)$, or $x = -3$ and $y = -7$.

The solution of this system of simultaneous equations is $(-3, -7)$.

Check:

$$-4x + y = 5 \qquad\qquad x - y = 4$$
$$-4(-3) + (-7) = 5 \qquad (-3) - (-7) = 4$$
$$12 - 7 = 5 \qquad\qquad -3 + 7 = 4$$
$$5 = 5 \qquad\qquad\qquad 4 = 4$$

There are three kinds of systems of simultaneous equations:

Three Kinds of Systems of Equations

1. *Consistent and Independent* equations: the graph is a pair of intersecting straight lines.
2. *Inconsistent* equations: the graph is a pair of straight lines that do not intersect but are parallel.
3. *Dependent* equations: the graph is a single straight line.

For example, the equations $-4x + y = 5$ and $x - y = 4$ are consistent. As shown in the graph on page 292, they produce a pair of intersecting straight lines.

The equations $y - 2x = 3$ and $2y - 4x = 10$ are inconsistent equations.

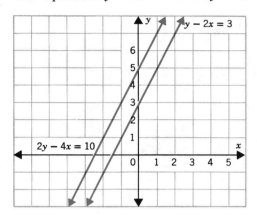

Graphing these equations, we see that they form two parallel straight lines. There is no ordered pair that will satisfy both equations. This system of equations has no solution.

The equations $2y - x = 5$ and $6y = 3x + 15$ are dependent. Both equations have the same graph.

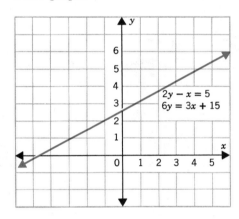

Any ordered pair that satisfies one of these equations will also satisfy the other. There are infinitely many solutions for this system of equations. Dependent equations are equivalent forms of the same equation.

Graph each of the following systems of equations; label each as consistent, inconsistent, or dependent, and find the solution if the equations are consistent.

(a) $x - y = 2$
     $x + y = 6$

(b) $x + y = 2$
     $2x + 2y = 8$

(c) $x - 2y = 6$
     $y = \frac{1}{2}x - 3$

(d) $2x - y = 6$
     $3y - x = 12$

(e) $y = 2x$
     $x = 4$

(f) $x = -2y$
     $2x = 3y$

Work carefully. The correct graphs are in **28**.

28 (a)

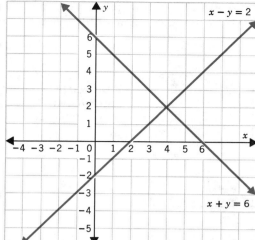

$x - y = 2$

$x + y = 6$

Consistent equations,
solution is $x = 4$, $y = 2$

(b)

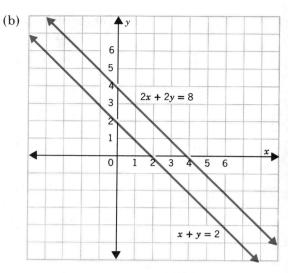

$2x + 2y = 8$

$x + y = 2$

Inconsistent equations,
no solution

(c)

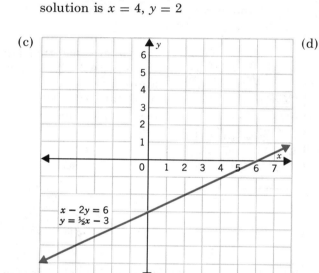

$x - 2y = 6$
$y = \frac{1}{2}x - 3$

Dependent equations,
no unique solution

(d)

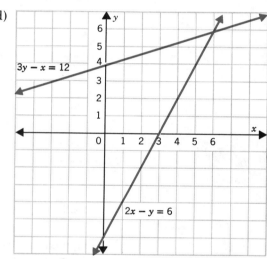

$3y - x = 12$

$2x - y = 6$

Consistent equations,
solution is $x = 6$, $y = 6$

(e)

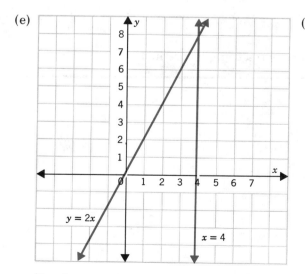

$y = 2x$

$x = 4$

Consistent equations,
solution is $x = 4$, $y = 8$

(f)

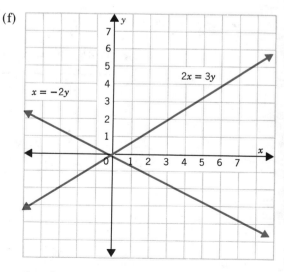

$x = -2y$

$2x = 3y$

Consistent equations,
solution is $x = 0$, $y = 0$

Graphing as a method of solving a system of simultaneous equations is time-consuming and gives only approximate answers. If the coordinates of the intersection point are not integers, we must estimate their values from the graph. If we want an exact, precise answer, we must use algebraic methods rather than graphic methods to solve the system of equations.

Notice in the system of equations

$y = x + 1$
$2x + y = 4$

that the first equation gives y in terms of x; it has been solved explicitly for y. We may solve this system of equations by substituting this value of y into the second equation.

$y = \boxed{x + 1}$

$2x + y = 4$ becomes $2x + \boxed{x + 1} = 4$

or $3x + 1 = 4$
$3x = 3$
$x = 1$

Now substitute this value of x back into either of the original equations. We choose the first equation.

$y = x + 1$
$y = (1) + 1$
$y = 2$

The solution is $(1, 2)$.

Check: $y = x + 1$ $2x + y = 4$
$(2) = (1) + 1$ $2(1) + (2) = 4$
$2 = 2$ $2 + 2 = 4$
$4 = 4$

 Always check your solution by substituting into *both* equations.

Solve the following system of equations by the method of substitution:

$y = 3 - x$
$3x + y = 11$

Check your answer in **29**.

---

29 $y = 3 - x$ Substitute $\boxed{3 - x}$ for y in the second equation.

$3x + y = 11$ becomes $3x + \boxed{3 - x} = 11$

or $2x + 3 = 11$
$2x = 8$
$x = 4$

Substituting this value of x into the first equation gives

$y = 3 - x$
$y = 3 - (4)$
$y = 3 - 4 = -1$

The solution is $(4, -1)$.

Check: $y = 3 - x$ $3x + y = 11$
$(-1) = 3 - (4)$ $3(4) + (-1) = 11$
$-1 = -1$ $12 - 1 = 11$
$11 = 11$

To solve any system of equations using the method of substitution, follow these steps:

Step 1. Solve one of the equations explicitly for either x or y.

Example: $y - x = 2 \longrightarrow y = \boxed{2 + x}$ We solve for a variable whose coeffi-
$2x + y = 5$ cient is equal to 1.

Step 2. Substitute this expression into the other equation to get an equation in one variable.

Example: $2x + \boxed{2 + x} = 5$

Step 3. Solve the new equation to get a numerical value for this variable.

Example: $3x + 2 = 5$
$3x = 3$
$x = 1$

Step 4. Substitute the value of this variable into one of the original equations and solve to get a numerical solution for the second variable.

Example: $y - (1) = 2$
$y = 3$ The solution is $(1, 3)$.

Step 5. Check the solution by substituting it into both equations.

Example: $y - x = 2$ $2x + y = 5$
$(3) - (1) = 2$ $2(1) + (3) = 5$
$2 = 2$ $5 = 5$

Use these five steps to solve the following systems of equations.

(a) $x = 1 + y$ (b) $3x + y = 1$ (c) $x - 3y = 4$
    $2y + x = 7$ $y + 5x = 9$ $3y + 2x = -1$

(d) $y + 2x = 1$ (e) $x - y = 2$ (f) $2x - y = 5$
    $3y + 5x = 1$ $y + x = 1$ $2y - 4x = 3$

When you have solved these systems of equations turn to **30** to check your work.

---

30   (a) $x = 1 + y$.   Substitute $\boxed{1 + y}$ into the second equation for x.

$2y + x = 7$ becomes $2y + \boxed{1 + y} = 7$

$3y + 1 = 7$
$3y = 6$
$y = 2$

$x = 1 + (2) = 3$ The solution is given by the equations $x = 3, y = 2$, or $(3, 2)$.

Check: $x = 1 + y$ $2y + x = 7$
           $(3) = 1 + (2)$ $2(2) + (3) = 7$
           $3 = 3$ $7 = 7$

(b) $3x + y = 1$ or $y = 1 - 3x$.   Substitute $\boxed{1 - 3x}$ into the second equation for y.

$y + 5x = 9$ becomes $\boxed{1 - 3x} + 5x = 9$

$$1 + 2x = 9$$
$$2x = 8$$
$$x = 4$$

Substitute this value of x into the first equation

$y = 1 - 3x$
$y = 1 - 3(4)$
$y = 1 - 12$
$y = -11$ The solution is given by the equations $x = 4$, $y = -11$, or $(4, -11)$.

Check:

$3x + y = 1$	$y + 5x = 9$
$3(4) + (-11) = 1$	$(-11) + 5(4) = 9$
$12 - 11 = 1$	$-11 + 20 = 9$
$1 = 1$	$9 = 9$

(c) $x - 3y = 4$ or $x = 4 + 3y$.   Substitute $\boxed{4 + 3y}$ into the second equation for x.

$3y + 2x = -1$ becomes $3y + 2(\boxed{4 + 3y}) = -1$

$$3y + 8 + 6y = -1$$
$$9y + 8 = -1$$
$$9y = -9$$
$$y = -1$$

Substitute this value of y into the first equation

$x = 4 + 3y$
$x = 4 + 3(-1)$
$x = 4 - 3$
$x = 1$ The solution is given by the equations $x = 1$, $y = -1$, or $(1, -1)$.

Check:

$x - 3y = 4$	$3y + 2x = -1$
$(1) - 3(-1) = 4$	$3(-1) + 2(1) = -1$
$1 + 3 = 4$	$-3 + 2 = -1$
$4 = 4$	$-1 = -1$

(d) $y + 2x = 1$ or $y = 1 - 2x$ Substitute $\boxed{1 - 2x}$ into the second equation for y.

$3y + 5x = 1$ becomes $3(\boxed{1 - 2x}) + 5x = 1$

$$3 - 6x + 5x = 1$$
$$-x = -2$$
$$x = 2$$

Substitute this value of x into the other equation.

$y = 1 - 2x$ becomes $y = 1 - 2(2) = 1 - 4 = -3$.   The solution is given by $x = 2$, $y = -3$, or $(2, -3)$.

Check:

$y + 2x = 1$	$3y + 5x = 1$
$(-3) + 2(2) = 1$	$3(-3) + 5(2) = 1$
$-3 + 4 = 1$	$-9 + 10 = 1$
$1 = 1$	$1 = 1$

(e) $x - y = 2$ or $x = 2 + y$.   Substitute $\boxed{2 + y}$ into the second equation for x.

$$y + x = 1 \qquad \text{becomes} \qquad y + \boxed{2 + y} = 1$$
$$2y + 2 = 1$$
$$2y = -1$$
$$y = -\frac{1}{2}$$

Substitute this value of y into the other equation.

$x = 2 + (-\frac{1}{2}) = 2 - \frac{1}{2} = 1\frac{1}{2}$ The solution is given by $x = 1\frac{1}{2}$, $y = -\frac{1}{2}$, or $(1\frac{1}{2}, -\frac{1}{2})$.

Check: $x - y = 2$ $y + x = 1$

$$\left(1\frac{1}{2}\right) - \left(-\frac{1}{2}\right) = 2 \qquad \left(-\frac{1}{2}\right) + \left(1\frac{1}{2}\right) = 1$$
$$2 = 2 \qquad\qquad\qquad 1 = 1$$

(f) $2x - y = 5$ or $y = 2x - 5$.   Substitute $\boxed{2x - 5}$ into the second equation for y.

$$2y - 4x = 3 \qquad \text{becomes} \qquad 2(\ \boxed{2x - 5}\ ) - 4x = 3$$
$$4x - 10 - 4x = 3$$

or $-10 \neq 3$.   There is no solution for this system of equations. The equations are inconsistent.

Solution by Elimination

Solving a system of equations by substitution is quick and simple if one of the two equations can be solved explicitly for either x or y without creating fractional coefficients. However, the most general, foolproof method of solving a system of equations is to eliminate one of the two variables by combining the equations. For example, to solve the system of equations:

$$x + y = 8$$
$$\underline{x - y = 2} \qquad \text{Add them to get}$$
$$2x + 0 = 10 \qquad +y - y = 0$$
$$2x = 10$$
$$x = 5 \qquad \text{Substitute 5 for } x \text{ in the first equation to get } y.$$

The first equation gives $(5) + y = 8$
$$y = 8 - 5$$
$$y = 3 \qquad \text{The solution is given by } x = 5, y = 3, \text{ or } (5, 3).$$

Check: $x + y = 8$ $x - y = 2$
$$(5) + (3) = 8 \qquad (5) - (3) = 2$$
$$8 = 8 \qquad\qquad 2 = 2$$

By adding the two equations we are able to eliminate one of the variables and obtain an equation in one variable that is easily solved. This value of one variable is then substituted back into one of the original equations and it is solved for the second variable.

Try it. Solve the following system of equations by adding as shown above.

$$2x - y = 3$$
$$y + x = 9$$

Check your work in **31**.

31 $2x - y = 3$
       $y + x = 9$

The first step is to arrange the terms in the equations so that the variables appear in the same order in both equations.

$$2x - y = 3$$
$$\underline{x + y = 9}$$ The x variable term now appears first in both equations.
$$3x + 0 = 12 \longleftarrow \text{Adding } x \text{ terms, adding } y \text{ terms, and adding constants.}$$
$$3x = 12$$
$$x = 4$$

Substitute 4 for x in the second equation.

$$(4) + y = 9$$
$$y = 5$$ The solution is given by the equations $x = 4$, $y = 5$, or $(4, 5)$.

Check: $2x - y = 3$ $y + x = 9$
         $2(4) - (5) = 3$ $(5) + (4) = 9$
           $8 - 5 = 3$ $9 = 9$
             $3 = 3$

It is important to rearrange the terms in the equations so that the x and y terms appear in the same order in both equations.

$$2\ x\ -\ y\ =\ 3$$
$$x\ +\ y\ =\ 9$$

constant terms

y column

x column

Solve the following systems of equations by this process of elimination.

(a) $x + 5y = 17$
    $-x + 3y = 7$

(b) $x + y = 16$
    $x - y = 4$

(c) $3x - y = -5$
    $y - 5x = 9$

(d) $3x - 4y = 30$
    $4y + 3x = -6$

(e) $6x - y = 5$
    $y - x = -5$

(f) $\frac{1}{2}x + 2y = 10$
    $y + 1 = \frac{1}{2}x$

Our step-by-step solutions are in **32**.

32 (a)
$$x + 5y = 17$$
$$\underline{-x + 3y = 7}$$
$$0 + 8y = 24 \qquad \text{adding like terms.}$$
$$8y = 24$$
$$y = 3$$

Substitute 3 for y in the first equation.

$$x + 5(3) = 17$$
$$x + 15 = 17$$
$$x = 2 \qquad \text{The solution is } (2, 3).$$

Check:

$x + 5y = 17$	$-x + 3y = 7$
$(2) + 5(3) = 17$	$-(2) + 3(3) = 7$
$2 + 15 = 17$	$-2 + 9 = 7$
$17 = 17$	$7 = 7$

(b)
$$x + y = 16$$
$$\underline{x - y = 4}$$
$$2x + 0 = 20 \qquad \text{adding like terms.}$$
$$2x = 20$$
$$x = 10$$

Substitute 10 for x in the first equation.

$$(10) + y = 16$$
$$y = 6 \qquad \text{The solution is } (10, 6).$$

Check:

$x + y = 16$	$x - y = 4$
$(10) + (6) = 16$	$(10) - (6) = 4$
$16 = 16$	$4 = 4$

(c) $3x - y = -5$
$y - 5x = 9$

Rearrange the order of the terms in the second equation.

$$3x - y = -5$$
$$\underline{-5x + y = 9}$$
$$-2x + 0 = 4 \qquad \text{adding like terms.}$$
$$-2x = 4$$
$$x = -2$$

Substitute -2 for x in the first equation.

$$3(-2) - y = -5$$
$$-6 - y = -5$$
$$-y = -5 + 6 = 1$$
$$y = -1 \qquad \text{The solution is } (-2, -1).$$

Check:

$3x - y = -5$	$y - 5x = 9$
$3(-2) - (-1) = -5$	$(-1) - 5(-2) = 9$
$-6 + 1 = -5$	$-1 + 10 = 9$
$-5 = -5$	$9 = 9$

(d) $3x - 4y = 30$
$4y + 3x = -6$

Rearrange the order of terms in the second equation.

$$3x - 4y = 30$$
$$\underline{3x + 4y = -6}$$
$$6x + 0 = 24 \quad \text{adding like terms.}$$
$$6x = 24$$
$$x = 4$$

Substitute 4 for x in the first equation.

$$3(4) - 4y = 30$$
$$12 - 4y = 30$$
$$-4y = 18$$
$$y = -4\frac{1}{2} \quad \text{The solution is } \left(4, -4\frac{1}{2}\right).$$

Check:
$$3x - 4y = 30 \qquad\qquad 4y + 3x = -6$$
$$3(4) - 4\left(-4\frac{1}{2}\right) = 30 \qquad 4\left(-4\frac{1}{2}\right) + 3(4) = -6$$
$$12 + 18 = 30 \qquad\qquad -18 + 12 = -6$$
$$30 = 30 \qquad\qquad\qquad -6 = -6$$

(e) $\quad 6x - y = 5$
$\qquad\quad y - x = -5$

Rearrange terms in the second equation.

$$6x - y = 5$$
$$\underline{-x + y = -5}$$
$$5x + 0 = 0 \quad \text{adding like terms.}$$
$$5x = 0$$
$$x = 0$$

Substitute 0 for x in the first equation.

$$6(0) - y = 5$$
$$0 - y = 5$$
$$-y = 5$$
$$y = -5 \quad \text{The solution is } (0, -5).$$

Check:
$$6x - y = 5 \qquad\qquad y - x = -5$$
$$6(0) - (-5) = 5 \qquad (-5) - (0) = -5$$
$$0 + 5 = 5 \qquad\qquad -5 - 0 = -5$$
$$5 = 5 \qquad\qquad\qquad -5 = -5$$

(f) $\quad \dfrac{1}{2}x + 2y = 10$

$\qquad\quad y + 1 = \dfrac{1}{2}x$

Rearrange terms in the second equation so that they are in the same order as in the first equation.

$$\frac{1}{2}x + 2y = 10$$
$$\underline{-\frac{1}{2}x + y = -1}$$
$$0 + 3y = 9 \quad \text{adding like terms.}$$
$$3y = 9$$
$$y = 3$$

Substitute 3 for y in the first equation.

$$\frac{1}{2}x + 2(3) = 10$$

$$\frac{1}{2}x + 6 = 10$$

$$\frac{1}{2}x = 4$$

$$x = 8 \qquad \text{The solution is } (8, 3).$$

Check: $\qquad \frac{1}{2}x + 2y = 10 \qquad y + 1 = \frac{1}{2}x$

$$\frac{1}{2}(8) + 2(3) = 10 \qquad (3) + 1 = \frac{1}{2}(8)$$

$$4 + 6 = 10 \qquad 3 + 1 = 4$$

$$10 = 10 \qquad 4 = 4$$

The Multiply and Add Method

With some systems of equations, neither x nor y can be eliminated by simply adding like terms. For example, in the system

$$3x + y = 17$$
$$x + y = 7$$

adding like terms will not eliminate either variable. To solve this system of equations, multiply all terms of the second equation by -1 so that

$$\boxed{x + y = 7} \quad \fbox{becomes} \Rightarrow \boxed{-x - y = -7}$$

and the system of equations becomes

$$3x + y = 17$$
$$-x - y = -7$$

The system of equations may now be solved by adding like terms as before.

$$3x + y = 17$$
$$\underline{-x - y = -7}$$
$$2x + 0 = 10 \qquad \text{adding like terms}$$
$$2x = 10$$
$$x = 5$$

Substitute 5 for x in the first equation.

$$3(5) + y = 17$$
$$15 + y = 17$$
$$y = 17 - 15 = 2 \qquad \text{The solution is given by the equations } x = 5, y = 2, \text{ or } (5, 2).$$

Check: $\qquad 3x + y = 17 \qquad x + y = 7$

$$3(5) + (2) = 17 \qquad (5) + (2) = 7$$

$$15 + 2 = 17 \qquad 5 + 2 = 7$$

$$17 = 17 \qquad 7 = 7$$

Use this "multiply and add" procedure to solve the following system of equations:

$$2x + 7y = 29$$
$$2x + y = 11$$

Check your solution in **33**.

---

33 $\qquad 2x + 7y = 29$
$\qquad 2x + y = 11$

Multiply all terms in the second equation by -1.

$$2x + 7y = 29$$
$$\underline{-2x - y = -11}$$
$$0 + 6y = 18 \qquad \text{adding like terms.}$$
$$6y = 18$$
$$y = 3$$

Substitute 3 for y in the first equation.

$$2x + 7(3) = 29$$
$$2x + 21 = 29$$
$$2x = 8$$
$$x = 4 \qquad \text{The solution is } (4, 3).$$

Check:

$2x + 7y = 29$	$2x + y = 11$
$2(4) + 7(3) = 29$	$2(4) + (3) = 11$
$8 + 21 = 29$	$8 + 3 = 11$
$29 = 29$	$11 = 11$

Notice that we use the original pair of equations to check the solution.

Solving by the multiply-and-add procedure may involve multiplying by constants other than -1, of course. For example, use this method to solve the pair of simultaneous equations:

$$2x + 4y = 26$$
$$3x - 2y = 7$$

Look for our solution in **34**.

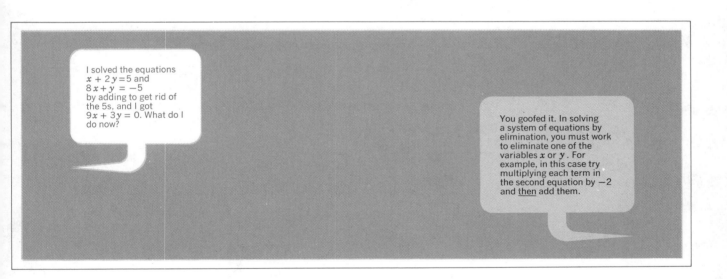

34 $2x + 4y = 26$
$3x - 2y = 7$

First, look at these equations carefully. Notice that the y terms can be eliminated easily if we multiply all terms in the second equation by 2.

The y column $\begin{array}{c} +4y \\ -2y \end{array}$ becomes $\begin{array}{c} +4y \\ -4y \end{array}$ when we multiply the second equation by **2** .

Sum $= 0$

The second equation becomes

$$\boxed{2}\,(3x) - \boxed{2}\,(2y) = \boxed{2}\,(7) \qquad \text{or} \qquad 6x - 4y = 14$$

and the system of equations is converted to the equivalent system

$$
\begin{aligned}
2x + 4y &= 26 \\
\underline{6x - 4y} &= \underline{14} \\
8x + 0 &= 40 \qquad \text{adding like terms.} \\
8x &= 40 \\
x &= 5
\end{aligned}
$$

Substitute 5 for x in the first equation.

$$
\begin{aligned}
2(5) + 4y &= 26 \\
10 + 4y &= 26 \\
4y &= 16 \\
y &= 4 \qquad \text{The solution is given by the equations } x = 5,\ y = 4,\ \text{or } (5, 4).
\end{aligned}
$$

Check the solution by substituting it back into the original equations. Try these problems to make certain you understand this procedure.

(a) $\quad 3x - 2y = 14$
$\qquad\ \ 5x - 2y = 22$

(b) $\quad 5x + 6y = 14$
$\qquad\ \ 3x - 2y = -14$

(c) $\quad\ \ 5x - y = 1$
$\qquad\ \ 2y + 3x = 11$

(d) $\quad -x - 2y = 1$
$\qquad\ \ 19 - 2x = -3y$

Our solutions are in **35**.

---

35 (a) $\quad 3x - 2y = 14$
$\qquad\quad 5x - 2y = 22$

Multiply each term in the first equation by -1.

$$
\begin{aligned}
(-1)(3x) - (-1)(2y) &= (-1)(14) \\
-3x + 2y &= -14
\end{aligned}
$$

The system of equations is therefore

$$
\begin{aligned}
-3x + 2y &= -14 \\
\underline{5x - 2y} &= \underline{22} \\
2x + 0 &= 8 \qquad \text{adding like terms.} \\
2x &= 8 \\
x &= 4
\end{aligned}
$$

Substitute 4 for x in the first equation.

$$
\begin{aligned}
3(4) - 2y &= 14 \\
12 - 2y &= 14 \\
-2y &= 2 \\
-y &= 1 \\
y &= -1 \qquad \text{The solution is } (4, -1).
\end{aligned}
$$

Check:

$$
\begin{array}{ll}
3x - 2y = 14 & 5x - 2y = 22 \\
3(4) - 2(-1) = 14 & 5(4) - 2(-1) = 22 \\
12 + 2 = 14 & 20 + 2 = 22 \\
14 = 14 & 22 = 22
\end{array}
$$

(b) $\quad 5x + 6y = 14$
$\qquad\ \ 3x - 2y = -14$

Multiply each term in the second equation by 3.

$$(3)(3x) - (3)(2y) = (3)(-14)$$
$$9x - 6y = -42$$

The system of equations is now

$$5x + 6y = 14$$
$$\underline{9x - 6y = -42}$$
$$14x + 0 = -28 \qquad \text{adding like terms.}$$
$$14x = -28$$
$$x = -2$$

Substitute -2 for x in the first equation.

$$5(-2) + 6y = 14$$
$$-10 + 6y = 14$$
$$6y = 24$$
$$y = 4 \qquad \text{The solution is } (-2, 4).$$

Be certain to check your solution.

(c) $\quad 5x - y = 1$
$\quad 2y + 3x = 11$

Rearrange to put the terms in the second equation in the same order as they are in the first equation.

$$5x - y = 1$$
$$3x + 2y = 11$$

Multiply each term in the first equation by 2.

$$10x - 2y = 2$$
$$\underline{3x + 2y = 11}$$
$$13x + 0 = 13 \qquad \text{adding like terms.}$$
$$13x = 13$$
$$x = 1$$

Substitute 1 for x in the first equation.

$$5(1) - y = 1$$
$$5 - y = 1$$
$$-y = 1 - 5 = -4$$
$$y = 4 \qquad \text{The solution is } (1, 4).$$

Check it.

(d) $\quad -x - 2y = 1$
$\quad 19 - 2x = -3y$

Rearrange the order of the terms in the second equation so that they are in the same order as they are in the first equation.

$$-x - 2y = 1$$
$$-2x + 3y = -19$$

Multiply each term in the first equation by -2.

$$(-2)(-x) - (-2)(2y) = (-2)(1)$$
$$2x + 4y = -2$$

The system of equations is now

$$2x + 4y = -2$$
$$\underline{-2x + 3y = -19}$$
$$0 + 7y = -21 \quad \text{adding like terms.}$$
$$7y = -21$$
$$y = -3$$

Substitute -3 for y in the first equation in the original problem.

$$-x - 2(-3) = 1$$
$$-x + 6 = 1$$
$$-x = 1 - 6 = -5$$
$$x = 5 \quad \text{The solution is } (5, -3).$$

Check the solution.

If you examine the system of equations

$$3x + 2y = 7$$
$$4x - 3y = -2$$

you will find that there is no single integer that we can use as a multiplier that will allow us to eliminate one of the variables when the equations are added. Instead we must convert each equation separately to an equivalent equation, so that when the new equations are added one of the variables is eliminated. For example, with the system of equations above, if we wish to eliminate the y variable we must multiply the first equation by 3 and the second equation by 2.

First equation: $\boxed{3x + 2y = 7}$ $\boxed{\text{Multiply by 3}}\!>$ $\boxed{9x + 6y = 21}$

Second equation: $\boxed{4x - 3y = -2}$ $\boxed{\text{Multiply by 2}}\!>$ $\boxed{8x - 6y = -4}$

The new system of equations is

$$9x + 6y = 21$$
$$\underline{8x - 6y = -4}$$
$$17x + 0 = 17 \quad \text{adding like terms.}$$
$$17x = 17$$
$$x = 1$$

Substitute 1 for x in the original first equation.

$$3x + 2y = 7$$
$$3(1) + 2y = 7$$
$$3 + 2y = 7$$
$$2y = 4$$
$$y = 2$$

The solution is $(1, 2)$.

Check this solution by substituting it back into the original pair of equations.

Check:
$$3x + 2y = 7 \qquad 4x - 3y = -2$$
$$3(1) + 2(2) = 7 \qquad 4(1) - 3(2) = -2$$
$$3 + 4 = 7 \qquad 4 - 6 = -2$$
$$7 = 7 \qquad -2 = -2$$

Use this same procedure to solve the following system of equations:

$$2x - 5y = 9$$
$$3x + 4y = 2$$

Check your work in **36.**

36 We can eliminate x from the two equations as follows:

First equation: $\boxed{2x - 5y = 9}$ $\boxed{\text{Multiply by } -3} \Rightarrow$ $-6x + 15y = -27$

Second equation: $\boxed{3x + 4y = 2}$ $\boxed{\text{Multiply by } 2} \Rightarrow$ $6x + 8y = 4$

The new system of equations is:

$$\begin{array}{l} -6x + 15y = -27 \\ \underline{6x + 8y = 4} \\ 0 + 23y = -23 \qquad \text{adding like terms.} \\ 23y = -23 \\ y = -1 \end{array}$$

Substitute -1 for y in the first original equation.

$$\begin{array}{l} 2x - 5y = 9 \\ 2x - 5(-1) = 9 \\ 2x + 5 = 9 \\ 2x = 4 \\ x = 2 \qquad \text{The solution is given by } x = 2, \ y = -1, \text{ or } (2, -1). \end{array}$$

Check this solution.

Of course we could have chosen to eliminate the y variable and we would have arrived at the same solution. Try it.

When you are ready to continue, practice your new skills by solving the following systems of equations.

(a) $2x + 2y = 4$
$5x + 7y = 18$

(b) $5x + 2y = 11$
$6x - 3y = 24$

(c) $3x + 2y = 10$
$2x = 5y - 25$

(d) $-7x - 13 = 2y$
$3y + 4x = 0$

When you have solved these problems check your work in **37**.

When I solved the equations
$x + 3y = 6$
$x + 2y = 5$
I <u>subtracted</u> the second equation from the first to get
$y = 1$. Isn't that OK?

It's fine. Most students are less likely to make mistakes if they multiply by -1 and add like terms, but subtracting has the same effect. When you solve more difficult problems the multiply-and-add procedure will be easiest.

---

37 (a) $2x + 2y = 4$
$5x + 7y = 18$

To eliminate the x variable, multiply the first equation by -5:

$$(-5)(2x) + (-5)(2y) = (-5)(4)$$
$$-10x - 10y = -20$$

and multiply the second equation by 2.

$$(2)(5x) + (2)(7y) = (2)(18)$$
$$10x + 14y = 36$$

The new equations are

$$-10x - 10y = -20$$
$$\underline{10x + 14y = 36}$$
$$0 + 4y = 16 \quad \text{adding like terms.}$$
$$4y = 16$$
$$y = 4$$

Substitute 4 for y in the first original equation.

$$2x + 2(4) = 4$$
$$2x + 8 = 4$$
$$2x = -4$$
$$x = -2 \quad \text{The solution is } (-2, 4).$$

Check the solution.

(b) $5x + 2y = 11$
     $6x - 3y = 24$

To eliminate the y variable, multiply the first equation by 3.

$$(3)(5x) + (3)(2y) = (3)(11)$$
$$15x + 6y = 33$$

and the second equation by 2

$$(2)(6x) - (2)(3y) = (2)(24)$$
$$12x - 6y = 48$$

The new set of equations is

$$15x + 6y = 33$$
$$\underline{12x - 6y = 48}$$
$$27x + 0 = 81 \quad \text{adding like terms.}$$
$$27x = 81$$
$$x = 3$$

Substitute 3 for x in the first equation.

$$5(3) + 2y = 11$$
$$15 + 2y = 11$$
$$2y = -4$$
$$y = -2 \quad \text{The solution is } (3, -2).$$

Always check your answer.

(c) $3x + 2y = 10$
     $2x = 5y - 25$

Rearrange the order of terms in the second equation to agree with the first equation.

$$3x + 2y = 10$$
$$2x - 5y = -25$$

To eliminate the x variable, multiply the first equation by -2 and the second equation by 3.

$$-6x - 4y = -20$$
$$\underline{6x - 15y = -75}$$
$$0 - 19y = -95 \quad \text{adding like terms.}$$
$$-19y = -95$$
$$y = 5$$

Substitute 5 for y in the first equation.

$$3x + 2(5) = 10$$
$$3x + 10 = 10$$
$$3x = 0$$
$$x = 0 \qquad \text{The solution is } (0, 5).$$

Check it.

(d) $\quad -7x - 13 = 2y$
$\qquad 3y + 4x = 0$

Rearrange the order of terms in the first equation to agree with the second equation.

$$-2y - 7x = 13$$
$$3y + 4x = 0$$

To eliminate the y variable, multiply the first equation by 3 and the second equation by 2.

$$-6y - 21x = 39$$
$$\underline{6y + 8x = 0}$$
$$0 - 13x = 39 \qquad \text{adding like terms.}$$
$$-13x = 39$$
$$-x = 3$$
$$x = -3$$

Substitute -3 for x in the first equation.

$$-7(-3) - 13 = 2y$$
$$21 - 13 = 2y$$
$$8 = 2y$$
$$4 = y$$
$$y = 4 \qquad \text{The solution is } (-3, 4).$$

Check the solution.

For a set of practice problems in solving systems of simultaneous linear equations turn to **38**.

Unit 4
Equations
in Two Variables

38

Systems of Equations

The answers are on page 556.

A. Solve each of the following systems of equations by graphing. Label each system as consistent, inconsistent, or dependent.

1. $x + y = 10$
   $2x - y = -4$ ___

2. $x = y$
   $x + y = 10$ ___

3. $3x - y = 5$
   $2x + y = 15$ ___

4. $y = -3$
   $2x - y = 7$ ___

5. $x = 3y - 1$
   $y = 2x - 8$ ___

6. $y = x - 3$
   $y = x - 13$ ___

7. $x = 2$
   $y = 3x + 1$ ___

8. $y = 3x + 5$
   $5x - y = -1$ ___

9. $4y - 2x = 8$
   $2y - x = 4$ ___

10. $2y + x = 9$
    $x + 4 = 2y$ ___

B. Solve each of the following systems of equations using the method of substitution. If the system has no solution, say so.

1. $y = 10 - x$
   $2x - y = -4$ ___

2. $y = -3$
   $2x - y = 7$ ___

3. $3x - y = 5$
   $2x + y = 15$ ___

4. $y = 3x + 5$
   $5x - y = -1$ ___

5. $x + 5 = 2y$
   $y + 2x = 20$ ___

6. $2x - y = 3$
   $x - 2y = -6$ ___

7. $2y - 4x = -3$
   $y = 2x + 4$ ___

8. $3x + 5y = 26$
   $x + 2y = 10$ ___

9. $x = 4 - 3y$
   $2y + x = 7$ ___

10. $x = 10y + 1$
    $y = 10x + 1$ ___

11. $3x + 4y = -3$
    $x + y = 1$ ___

12. $2x + 3y + 4 = 0$
    $x + 3y = 1$ ___

13. $3x + 2y = 9$
    $3x - y = -9$ ___

14. $3x + 2y = -1$
    $2y + 4x = 2$ ___

15. $5y - x = -6$
    $x + 3y = -10$ ___

16. $6x - 2y = 16$
    $3x + 5y = -4$ ___

17. $5x + 2y = 3$
    $2x - 3y = -14$ ___

18. $5x - 2y = -1$
    $4x - 3y = 9$ ___

C. Solve each of the following systems of equations. If the system has no solution, say so.

1. $x + y = 5$
   $x - y = 13$ ___

2. $2x + y = 9$
   $-2x + y = 1$ ___

3. $2x + 2y = 10$
   $3x - 2y = 10$ ___

4. $5x - 3y = 13$
   $-2x + 9y = -13$ ___

5. $2y = 3x + 5$
   $2y = 3x - 7$ ___

6. $0 = x - 3y$
   $y + 3x = 20$ ___

7. $2y = 2x + 2$
   $4y = 5 + 4x$ ___

8. $10y - x = 11$
   $15y + 2x = 13$ ___

9. $x = 3y + 7$
   $x + y = -5$ ___

10. $4x - y = 12$
    $2y - 8x = -24$ ___

11. $3x - 2y = -11$
    $x + y = -2$ ___

12. $2x + 3y = 2$
    $5x - 2y = 24$ ___

13. $5x - 4y = 1$
    $3x - 6y = 6$ ___

14. $5y - 3x = 7$
    $1 + 7y = -12x$ ___

15. $y = 3x - 5$
    $6x - 3y = 3$ ___

Date ___

Name ___

Course/Section ___

311   Problem Set 4-3

Copyright © 1982 by John Wiley & Sons, Inc.

16. $4x + 3y = 1$
$8x + 9y = 11$ _____

17. $y - 2x = -8$
$x - \frac{1}{2}y = 4$ _____

18. $3y + 2x = 2 + 2y$
$3x + 2y = 2x - 2$ _____

19. $x + y = a$
$x - y = b$ _____

20. $3x - 3y = 2 - 2x$
$4y = 7x - 3$ _____

21. $4x + 4y = 5$
$8y - 6x = 3$ _____

22. $6x + 9y = 1$
$3x + 3y = 1$ _____

23. $4y + x = -1$
$5x + 12y = -9$ _____

24. $4x + 2y = 1$
$8x + 3y = 0$ _____

D. Brain Boosters

1. Solve by substitution: $\begin{cases} x = 3 \\ x + y = 5 \\ z + y = 10 \end{cases}$ _____

2. Solve: $\begin{cases} \dfrac{x}{2} + \dfrac{y}{3} = 5 \\ x - y = 10 \end{cases}$ _____

3. Solve: (a) $\begin{cases} x + y = a \\ y - 2x = b \end{cases}$ _____   (b) $\begin{cases} ax + by = e \\ cx + dy = f \end{cases}$ _____

4. Solve: $\begin{cases} 6751x + 3249y = 26751 \\ 3249x + 6751y = 23249 \end{cases}$ _____   (*Hint:* You may need a calculator for this one.)

5. Two dozen eggs and a loaf of bread cost $1.75. Half a dozen eggs and two loaves of bread cost $1.33. What is the cost of a loaf of bread?

_____

6. A man weighs twice as much as his wife. His wife weighs twice as much as their daughter. Their total weight is 371 lb. How much does each person weigh?

_____

When you have had the practice you need, either return to the preview on page 247 or continue in **39** with the study of word problems involving equations in two variables.

39 In Unit 1 you learned how to translate simple English phrases and sentences into mathematical expressions or equations. By now you should be fairly skillful at this process of translating English to algebra, but if you need to refresh your memory return to frame **48** on page 59 and take time out for a review.

With some word problems, the translation may produce two equations in two variables, and these equations must be solved as a system of simultaneous equations. In fact, it is often *easier* to solve many problems if we translate them into a system of equations rather than try to put all the problem information into a single equation. For example, in the following problem

"The sum of two numbers is 25 and their difference is 15. What are the two numbers?"

the first phrase should be translated as

"The sum of two numbers is 25 . . ."

$$x + y = 25 \qquad \text{(The two numbers are } x \text{ and } y.\text{)}$$

and the second phrase should be translated as

". . . their difference is 15 ."

$$x - y = 15$$

This word problem is equivalent to the system of equations

$x + y = 25$
$x - y = 15$

As you have already learned in this unit, this system of equations can be solved easily. Add the equations to obtain

$2x = 40 \qquad \text{or} \qquad x = 20$

Substitute this value of x back into the first equation to obtain a value for y.

$(20) + y = 25$
$\qquad y = 5 \qquad$ The solution is $(20, 5)$.

Check this solution by substituting it back into both of the original equations and into the original word problem.

Translate the following problem into a system of equations and solve it.

"The difference of two numbers is 14 and the larger number is three more than twice the smaller number."

Our step-by-step solution is in **40**.

---

40 The first phrase in the problem should be translated as

"The difference of two numbers is 14 . . ."

$$L - S = 14$$

and the second phrase should be translated as

"... the larger number is three more than twice the smaller..."

$$L = 3 + 2S$$

The system of simultaneous equations is

$L - S = 14$
$L = 3 + 2S$

To solve this system of equations, substitute the value of L from the second equation into the first equation so that the first equation becomes

$(3 + 2S) - S = 14$ or $3 + S = 14$
$S = 11$

Now substitute this value of S into the first equation to find L.

$L - (11) = 14$
$L = 25$ The solution is $L = 25$ and $S = 11$.

Check the solution by substituting it back into both of the original equations and into the word problem. Never neglect to check your answer.

Translating Word Problems into Systems of Equations

Translating word problems into systems of equations is a very valuable and very practical algebraic skill. Translate each of the following problems into a system of equations.

(a) Bob is four times as old as his son Eric. In five years Bob will be three times as old as Eric. Find the present age of each person.

(b) In a collection of old coins the number of nickels is four more than twice the number of dimes. The total face value of the coins is $1.40. How many of each kind of coin are in the collection?

(c) The perimeter of a rectangular lot is 350 ft. The length of the lot is 10 ft more than twice the width. Find the dimensions of the lot.

(d) Shopping for the mathematician's annual picnic, Professor Khayyam noticed that a loaf of bread, a jug of wine, and a hunk of cheese together cost $1.80. The wine cost 80¢ more than the bread alone and the bread cost exactly twice as much as the cheese. How much did each cost?

Write each problem as a system of equations and solve. Check your answers in **41**.

---

41   (a)   "Bob is four times as old as Eric."

$$B = 4 \times E$$ (B is Bob's age; E is Eric's age.)

"In five years Bob will be three times as old as Eric,"

$$B + 5 = 3 \times (E + 5)$$ ($E + 5$ is Eric's age in 5 years.)

The system of equations is

$B = 4E$
$B + 5 = 3(E + 5)$

Solve by substituting the value of B from the first equation into the second equation.

$(4E) + 5 = 3(E + 5)$

The solution is given by the equations $E = 10$ and $B = 40$. Check it.

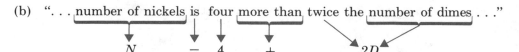

"... number of nickels is four more than twice the number of dimes ..."

$$N = 4 + 2D$$

(N is the number of nickels; D is the number of dimes.)

"... total value of the coins is $1.40."

$$5N + 10D = 140$$

Value of   Value of   Total value
N nickels D dimes   in cents
in cents   in cents

The system of equations is

$N = 4 + 2D$
$5N + 10D = 140$

Solve this system of equations by substituting the value of N from the first equation into the second equation.

$5(4 + 2D) + 10D = 140$ or $20 + 20D = 140$

The solution is given by the equations $D = 6$, $N = 16$. Check it.

(c)   "... perimeter of a rectangular lot is 350 ft."

$$2L + 2W = 350$$

(L is the length and W is the width.)

The perimeter of a rectangle is $L + W + L + W$ or $2L + 2W$

"... length is 10 ft more than twice the width."

$$L = 10 + 2W$$

The system of equations is

$2L + 2W = 350$ or $2L + 2W = 350$
$L = 10 + 2W$ $L - 2W = 10$
Add to get $3L = 360$

The solution is $L = 120$ ft, $W = 55$ ft. Be sure to check your solution by substituting it into *both* of the equations in the system of equations, and into the original word problem.

(d)   "... bread ... wine ... cheese together cost $1.80."

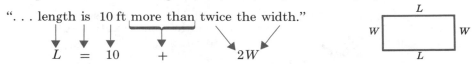

$$B + W + C = 180$$

where B is the cost of a loaf of bread, W is the cost of a jug of wine, and C is the cost of a hunk of cheese.

"... the wine cost 80¢ more than the bread ..."

$$W = 80 + B$$

"... bread cost exactly twice as much as the cheese."

$$B = 2C$$

The system of equations is

$B + W + C = 180$
$W = 80 + B$
$B = 2C$

This system of three equations in three variables can be reduced to a system of two equations in two variables by substituting the value of B in the third equation into the first two equations. The first two equations become

$(2C) + W + C = 180$ or $W + 3C = 180$
$W = 80 + (2C)$ $W = 80 + 2C$

This system of equations can be solved by substituting W from the second equation into the first equation.

$(80 + 2C) + 3C = 180$
or $80 + 5C = 180$
            $C = 20¢$

Using this value of C, we can find the value of B and W from the original equations.

$B = 2C$ gives $B = 2(20)$ or $B = 40¢$
$W = 80 + B$ gives $W = 80 + (40)$ or $1.20.

The solution is $C = 20¢$, $B = 40¢$, $W = \$1.20$. Check it.

We have separated the problem into parts, written an equation describing the information in each part, and solved the system of equations. For most students, solving this last problem by translating it into a system of equations is by far the easiest way to answer the question. Divide and conquer.

Using Diagrams to Solve Word Problems

In Unit 2 you learned to solve difficult word problems by arranging the information in a table or diagram. In frame **60** on page 161 this procedure was introduced to solve motion problems. For example, this problem can be solved with a table of information.

"George is driving at a steady speed of 40 mph on a straight road and he is passed by Jose driving in the same direction at 60 mph. How many hours after they pass will they be 50 miles apart?"

This information can be put into the following table:

	Distance, miles	Speed, mph	Time, hr
George	G	40	t
Jose	J	60	t

where G is the distance George drives in time t and J is the distance Jose drives in time t. The variable t is the time in hours that they travel from when they first pass one another until they are 50 miles apart.

Each row of this table produces an equation since distance = speed × time.

$G = 40t$
$J = 60t$

If they are 50 miles apart after t hours, $J = 50 + G$. The problem is therefore equivalent to the system of equations:

$G = 40t$
$J = 60t$
$J = 50 + G$

This system of equations can be solved quickly and easily by substituting the values of G and J from the first two equations into the third equation.

$(60t) = 50 + (40t)$ or $60t = 50 + 40t$

Solving this equation we find that $t = 2\frac{1}{2}$ hr. This value of t can be substituted back into the first two equations to find the distances G and J.

Again the problem becomes much easier to solve if we first translate it into a set of simple equations and then solve these equations as a system.

Solve the following problems using this procedure.

(a) Jack Jogs starts out from his home and runs at 6 mph toward the home of his friend Terry Trotter. Two hours later Terry starts from her home and runs at 10 mph to meet Jack. If they live 28 miles apart, how far does each run before they meet?

(b) A health food store operator wishes to blend peanuts selling for 90¢ per pound with cashews selling for $1.30 per pound to get a mix that he plans to sell for $1.00 per pound. How many pounds of each kind of nut must he use to get 40 lb of the mix?

(c) A chemist wishes to mix a 5% salt solution and a 15% salt solution to obtain 4 liters of a 12% salt solution. How many liters of each solution must be added?

(d) On a trip in her new car, Patty travels a total of 80 miles, driving at 50 mph on the open highway and 20 mph in the city. She drives for a total of two hours. How many miles does she drive in the city?

(e) Laurie's airplane cruises at a steady 120 mph in still air. On a trip across California she flies for two hours with the wind and then reverses direction and flies for two more hours against the wind. Traveling with the wind she goes 100 miles farther than against the wind. How far does she travel and how fast is the wind blowing?

Our complete solutions are in **42**.

(If you need to review word problems of this kind hop back to page 153.)

42

(a)

	d	s	t	$d = st$
Jack	d_1	6	t	$\longrightarrow d_1 = 6t$
Terry	d_2	10	$t - 2$	$\longrightarrow d_2 = 10(t-2)$

$d_1 = 6t$
$d_2 = 10(t-2)$ } System of equations
$d_1 + d_2 = 28$

d_1 is the distance Jack jogs in miles.
d_2 is the distance Terry trots in miles.
t is the time Jack jogs in hours.

Estimate the answer: Jack could jog all the way in less than 5 hours, so the answer will be somewhat less than 5 hours.

Combine this set of equations by substituting the first two equations into the third.

$(6t) + 10(t - 2) = 28$

Combining terms

$16t - 20 = 28$
$16t = 48$
$t = 3$ hours   Then $d_1 = 6t$ or 18 miles and $d_2 = 10(t - 2)$ or 10 miles.

Don't forget to check your answer.

317   4-5 Word Problems Involving Two Variables   Copyright © 1982 by John Wiley & Sons, Inc.

(b)

	Amount, lb	Cost per lb	Total cost
Peanuts	P	90	$90P$
Cashews	C	130	$130C$

P is the amount of peanuts and C is the amount of cashews.

Total amount: $P + C = 40$ lb
Total cost: $90P + 130C = 100 \cdot 40$

The system of equations to be solved is

$$P + C = 40$$
$$9P + 13C = 400 \qquad \text{(Each term in the second equation has been divided by 10.)}$$

To solve this system of equations multiply each term in the first equation by -9 and add the equations to get

$$4C = 40$$
$$C = 10 \text{ lb}$$
$$P = 30 \text{ lb}$$

(c)

	Amount	Salt fraction	Total salt
5%	A	0.05	$0.05A$
15%	B	0.15	$0.15B$

A is the amount of 5% solution, in liters.
B is the amount of 15% solution, in liters.

Total amount of solution: $A + B = 4$ liters
Total amount of salt: $0.05A + 0.15B = (0.12)4$

The system of equations to be solved is

$$A + B = 4$$
$$5A + 15B = 48 \qquad \text{(Each term in the second equation was multiplied by 100.)}$$

To solve this system of equations, multiply each term in the first equation by -5 and add to get

$$10B = 28$$
$$B = 2.8 \text{ liters}$$
$$A = 1.2 \text{ liters}$$

(d)

	Distance	Speed	Time
Highway	H	50	t_H
City	C	20	t_C

$\left. \begin{array}{l} d = st \\ H = 50t_H \\ C = 20t_C \\ H + C = 80 \\ t_H + t_C = 2 \end{array} \right\}$ System of Equations

H is the distance in miles driven on the highway.
C is the distance in miles driven in the city.
t_H and t_C are the times, in hours, driven on the highway and in the city.

Substitute H and C from the first equation into the third equation to get

$$50t_H + 20t_C = 80 \qquad \text{or} \qquad 5t_H + 2t_C = 8$$

The system of equations to be solved is

$$t_H + t_C = 2$$
$$5t_H + 2t_C = 8$$

To solve this system of equations, multiply the first equation by -2 and add to get

$$t_H = 1\frac{1}{3}\,\text{hr}$$

$$t_C = \frac{2}{3}\,\text{hr}$$

Substitute these times back into the original equations to get

$$H = 50\left(1\frac{1}{3}\right) = 66\frac{2}{3}\ \text{miles on the highway}$$
$$C = 20\left(\frac{2}{3}\right) = 13\frac{1}{3}\ \text{miles in the city}$$

(e)

	Distance	Speed	Time
With the wind	W	$120 + v$	2
Against the wind	A	$120 - v$	2

$d = st$

$W = (120 + v)2$

$A = (120 - v)2$

$W = 100 + A$

W is the distance traveled with the wind, in miles.
A is the distance traveled against the wind, in miles.
v is the speed of the wind.

Solve this system of equations by substituting the values of W and A in the first two equations into the third equation.

$$(120 + v)2 = 100 + (120 - v)2 \qquad \text{or} \qquad 4v = 100$$
$$v = 25\ \text{mph}$$

Substituting this value of v back into the first two equations,

$W = (120 + 25)2 = 290$ mph with the wind
$A = (120 - 25)2 = 190$ mph against the wind

Notice that the speed of the plane traveling with the wind is equal to the speed of the plane in still air (120 mph) *plus* the speed of the wind. The speed of the plane traveling against the wind is equal to the speed of the plane in still air (120 mph) *minus* the speed of the wind.

For some practice in solving word problems involving two variables, turn to **43**.

MIXTURE PROBLEMS

Mixture problems have tested mathematics students since the time of the ancient Egyptians. Here is a way of solving them that is more than 200 years old. And it still works!

Sample problem: Mix candy A at 49¢ per pound with candy B at 89¢ per pound to make 20 pounds of a mix that will sell for 79¢ per pound.

Solution: Write the given prices at the left corners of a square and the mixture price at the center.

Candy A 49¢

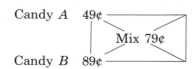

Candy B 89¢

At the right corners put the cost differences.

Candy A 49¢

Candy B 89¢

The numbers in the right corners tell us the ratio of the amounts of candy A and candy B to be mixed.

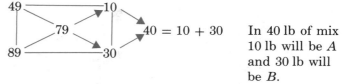

$40 = 10 + 30$ In 40 lb of mix 10 lb will be A and 30 lb will be B.

$$\frac{\text{Amount of mix}}{\text{Amount of candy } A} = \frac{40}{10} = \frac{20 \text{ lb}}{A}$$ Actual amount of mix Amount of candy A

$$A = 5 \text{ lb}$$

$$\frac{\text{Amount of mix}}{\text{Amount of candy } B} = \frac{40}{30} = \frac{20 \text{ lb}}{B}$$

$$B = 15 \text{ lb}$$

Translating the problem to an algebraic equation and solving the equation is a much better way to solve any mixture problem, but the "gimmick" shown above is the way George Washington might have solved a mixture problem.

Unit 4
Equations
in Two Variables

The answers are on page 557.

A.   Translate each problem statement into a system of equations and solve.

1. The sum of two numbers is 39 and their difference is 7. What are the numbers?

    _____

2. The sum of two numbers is 14. The larger is two more than three times the smaller. What are the numbers?

    _____

3. The sum of two numbers is $8\frac{1}{2}$. The larger is 3 more than the smaller. Find the numbers.

    _____

4. The cash register in the cafeteria has $4.50 in nickels and dimes. If it has twice as many nickels as dimes, how many of each does it contain?

    _____

5. The difference between two numbers is 21. The smaller is equal to three times the larger. Find the numbers.

    _____

6. Separate a collection of 20 objects into two parts so that twice the larger amount equals three times the smaller amount.

    _____

7. The average of two numbers is 25 and their difference is 8. What are the numbers?

    _____

8. Four bleebs and three freems cost $11. Three bleebs and four freems cost $10. What does a bleeb cost?

    _____

9. Maria's bank contains 17 coins, all pennies and nickels. If the total value of the coins is 61¢, how many of each coin does she have?

    _____

Date

_____

Name

_____

Course/Section

10. The postmistress at Upper Fig Newton has 200 postage stamps in 13¢ and 20¢ denominations worth a total of $32.16. How many of each kind of stamp does she have?

_____

11. The perimeter of a rectangular window is 14 ft and its length is 2 ft less than twice the width. What are the dimensions of the window?

_____

12. Harold exchanged a $1 bill for change and received his change in nickels and dimes, with seven more dimes than nickels. How many of each coin did he receive?

_____

13. If four times the larger of two numbers is added to three times the smaller, the result is 26. If three times the larger number is decreased by twice the smaller, the result is 11. Find the numbers.

_____

14. A rectangle has a perimeter of 100 ft and the difference between its length and width is 6 ft. Find the dimensions of the rectangle.

_____

15. Two numbers are such that the sum of the first plus twice the second is equal to 93. Furthermore, the sum of twice the first plus the second is 78. Find the two numbers.

_____

16. When Jill cashed her $210 paycheck, she received it in ten and twenty dollar bills. The number of twenty dollar bills was three more than twice the number of ten dollar bills. How many of each kind of bill did she receive?

_____

17. The Goodie Shop bakery sells five pound bags of mixed cookies for $2.55 per bag. These mixtures are made up of ginger snaps that sell for 45¢ per pound and sugar cookies that sell for 60¢ per pound. How much of each kind of cookie should be in the mixture?

_____

18. Mr. Brown bought five cans of peas and four cans of corn, but he forgot what each cost. He knows that the total cost was $2.90 and he recalls that a can of peas cost 5¢ less than a can of corn. How much did each can cost?

_____

19. A boat traveled for four hours against a 5 mph current and then turned and traveled for three hours with the same current. If the total distance traveled was 44 miles, what was the speed of the boat in still water?

_____

20. A superjet left Los Angeles and flew with the wind four hours, landing at Washington, D.C. 2400 miles away. Later it returned to Los Angeles in five hours over the same route and against the same wind. What was the cruising speed of the plane and what was the wind speed?

_____

21. An airplane flew 800 miles in 3 hours with an aiding tailwind. The return trip against the same wind took 5 hours. What was the wind speed?

_____

22. A jet aircraft flew on an out and back trip with a 120 mph wind. The journey into the wind took $3\frac{1}{2}$ hours and the return with the wind took 2 hours. How far was the trip?

_____

23. Three years ago Steve was four times as old as Tom was then. Five years from now Steve will be twice as old as Tom will be then. How old is each boy now?

_____

24. How much of a 40% acid solution must be added to a 25% acid solution to make up 20 liters of a 30% solution?

_____

25. The sum of the digits of a two-digit number is 13. If the digits are reversed, the new number is greater than the original number by 9. What was the original number? (*Hint:* let the original number have digits a and b so that the number is $10a + b$.)

_____

26. Conrad leaves work and bicycles toward home at 15 mph. One-half hour later Mike leaves from the same place and drives his car at 40 mph in the same direction. How far will Conrad travel before Mike overtakes him?

_____

Date

_____

Name

_____

Course/Section

27. A total of 548 people attended the Santa Barbara Junior High track championships. Admission was $1.50 for adults and $1.00 for children. A total of $719 was collected from ticket sales. How many adults and how many children attended?

_____

28. A sportscar speeding along at 65 mph is being chased by a highway patrol car at 80 mph. The patrol car is two miles behind the motorist. How long will it take the patrol car to overtake the motorist?

_____

29. Bob leaves home every day at the same time on his bicycle ride to work. If he cycles to work at 15 mph he arrives 10 minutes early, and if he goes at a leisurely 10 mph, he arrives 10 minutes late. How far is his trip to work?

_____

30. How much pure water must be added to 20 gallons of a 10% salt solution to dilute it to a 8% solution?

_____

When you have had the practice you need, turn to **44** for a self-test covering the work of Unit 4, Equations in Two Variables.

Unit 4
Equations
in Two Variables

44   Graph:

   1. $y = -x$

   2. $y = 2x - 2$

   3. $3x + 4y = 12$

   4. $x = -4$

Find the y intercept and slope for each equation.

   5. $7x + 5y = 35$

   6. $4y = -15x - 30$

Plot each pair of points, draw a straight line through them, and find its slope algebraically.

   7. $(3, -1)$ and $(1, 4)$ _____

   8. $(2, 4)$ and $(-1, -1)$ _____

Write an equation with integer coefficients having the following slope and y intercept.

   9. $m = 5, \ b = -4$ _____

   10. $m = \dfrac{1}{3}, \ b = 2$ _____

Write the equation of the straight line passing through the following points.

   11. $(4, 0)$ and $(-2, 3)$ _____

   12. $(-3, -4)$ and $(6, 2)$ _____

Graph the following inequalities.

   13. $2x - y > 5$

   14. $x \leq y - 7$

   15.   Find the set of points that satisfies all three of these inequalities simultaneously: $y \geq 0, \ y + 2x < 10, \ 2y - x < 10$

   _____

Solve these systems of equations.

   16. $2x + \ y = 3$
         $\ x + 2y = 0$

   17. $2a + 5b = 8$
         $\ a - 3b = 4$

   18. $\ y = 3 - x$
         $3x + 2y = 2$ _____

Date _____

Name _____

Course/Section _____

   19. $2x = -8y + 6$
         $4y = -x + 3$ _____

20. $2x + 3y = 11$
    $3x - 2y = 10$ _____

21. $7x - 3y = 17$
    $5y + 3x = 23$ _____

22. Find two numbers whose sum is 21 and whose difference is 13.

    _____

23. A motor boat can travel 56 miles downstream in 4 hours and 40 miles upstream in 5 hours. What is the speed of the current and what is the cruising speed of the boat in still water?

    _____

24. How much each of a 50% salt solution and a 20% salt solution must be mixed to produce 5 liters of a 29% solution?

    _____

25. Two packages together weigh a total of 35 pounds. The heavier package weighs 8 pounds more than twice the weight of the lighter package. How much does each weigh?

    _____

Answers are on page 557.

Unit 4
Equations
in Two Variables

Graphing Equations

A. Plot the following points.

1. (a) $(0, 4)$ (b) $(-4, 0)$ (c) $(2, -2)$ (d) $(-4, -4)$
   (e) $(-10, 2)$ (f) $(7, 5)$ (g) $(5, -7)$ (h) $(-5, -7)$
   (i) $(-3, 10)$ (j) $(6, 6)$ (k) $(0, -5)$ (l) $(6, 0)$

2. (a) $(10, 10)$ (b) $(10, 0)$ (c) $(0, 10)$ (d) $(0, -10)$
   (e) $(-10, 0)$ (f) $(-10, -10)$ (g) $(1, 1)$ (h) $(-2, -3)$
   (i) $(-3, 5)$ (j) $(5, -3)$ (k) $(-5, 7)$ (l) $(4, 2)$

B. Determine the coordinates of each of the points shown.

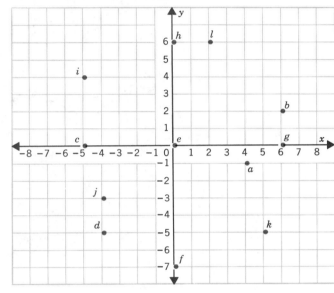

(a) _____   (b) _____

(c) _____   (d) _____

(e) _____   (f) _____

(g) _____   (h) _____

(i) _____   (j) _____

(k) _____   (l) _____

C. Plot graphs for each of the following equations

1. $y = 2x$ 2. $y = x - 4$ 3. $y = 3x + 2$
4. $y = 5$ 5. $y = -4$ 6. $x = 2$
7. $y = 2 - 3x$ 8. $y = 6 - \dfrac{1}{2}x$ 9. $y = 8x$
10. $2y - x = 4$ 11. $y = 5 + 3x$ 12. $x - 3y + 6 = 0$
13. $5x - 30 = 3y$ 14. $20 - 2x = 5y$ 15. $12 - 3y = -4x$
16. $2y - 5 = 0$

D. Write each of the following equations in slope-intercept form and identify the slope and y-intercept for each equation. Plot each equation.

1. $y = x - 2$ _____   2. $y = 3x + 1$ _____

3. $y = \dfrac{1}{2}x + 3$ _____ 4. $4y = 3x - 2$ _____

5. $3x + y = x - y + 2$ _____   6. $y = 5x$ _____

7. $x = 2y + 3$ _____   8. $x - y = 6$ _____

9. $y - x = 2x$ _____   10. $x + y = 4 - x$ _____

Date _____

Name _____

Course/Section _____

Unit 4
Equations
in Two Variables

Graphing Inequalities

A. Graph each of the following inequalities.

1. $x < 5$ 2. $y \leq 2$ 3. $x \geq -3$

4. $y > -4$ 5. $x + y > 3$ 6. $y \geq x - 5$

7. $y - x \leq -3$ 8. $y < x + 6$ 9. $y < 4 - x$

10. $y + x > -6$ 11. $8 < x + 2y$ 12. $-4 \geq y + 2x$

13. $2x + 3y \geq -12$ 14. $18 < 2y - 3x$ 15. $9 + x \leq y$

16. $y \leq -x$ 17. $2x \geq 5y$ 18. $x - 14 < 7y$

19. $y - x \geq -y - 6$ 20. $-x + y > x - y + 4$ 21. $y - 2 \geq 2x$

22. $2x - 6y < 0$ 23. $x + y < -6$ 24. $3x \geq 8y$

B. Solve:

1. If L is the length of a rectangle and W is its width, for what values of L and W will the perimeter of the rectangle be less than 10 inches?

_____

2. Graph the following set of inequalities:
   $y \leq -2, 2x - y \geq 6, x > -1$

3. Graph the following set of inequalities:
   $5y - 4x \leq 20, y > -3, 2y > 10 - 5x$

4. Find all possible pairs of numbers whose sum is less than or equal to 10. Show them graphically.

Date

Name

Course/Section

Unit 4
Equations
in Two Variables

Systems of Equations

A. Solve each of the following systems of equations by graphing.

1. $x = y$
   $y = 2x + 3$

2. $y = -x$
   $y = x - 1$

3. $y = 3x - 1$
   $y = 2 - 3x$

4. $x + y = 3$
   $2x - y = -9$

5. $x + y = 5$
   $x - y = 1$

6. $x + 2y = -4$
   $x - y = 2$

7. $y = 4$
   $y = 2x$

8. $x - 2y = 0$
   $x - 4y = -4$

9. $y = 3x - 5$
   $y = x + 4$

10. $2x - y = 3$
    $y + 3x = 7$

11. $y = x + 2$
    $y = 3x - 4$

12. $x = y + 4$
    $y = 2x - 4$

B. Solve each of the following systems of equations using the substitution method. If the equation has no solution, say so.

1. $x + y = 12$
   $x - y = 4$ ___

2. $x - 2y = 6$
   $x + 3y = 1$ ___

3. $4x - y = 10$
   $2x + 3y = 12$ ___

4. $3x - y = 3$
   $x + 3y = 11$ ___

5. $x + y = 4$
   $x + y = 3$ ___

6. $x + 3y = 4$
   $2x - y = 1$ ___

7. $x - y = -1$
   $3x - 2y = 3$ ___

8. $5x - 3y = 3$
   $2x - y = 0$ ___

9. $4x - 6y = 15$
   $6x - 4y = 10$ ___

10. $y = 2x + 3$

    $y = 2x - 3$ ___

11. $\frac{1}{2}x + \frac{1}{6}y = -2$

    $\frac{3}{4}x + \frac{2}{3}y = 2$ ___

12. $\frac{1}{3}x + \frac{1}{4}y = 10$

    $\frac{1}{3}x - \frac{1}{2}y = 4$ ___

C. Solve each of the following systems of equations. If the system has no solution, say so.

1. $x - 2y = -2$
   $2x - y = 5$ ___

2. $2x + 3y = 3$
   $3x + 4y = 3$ ___

3. $x + y = 10$
   $3x - 2y = 15$ ___

4. $x = y + 4$
   $2x - 5y = 8$ ___

5. $y = 1 - 3x$
   $y = 4 - 3x$ ___

6. $x - 2y = -2$
   $5 + y = 2x$ ___

7. $y = 3x + 1$
   $5x - 2y = 1$ ___

8. $y = x + 2$
   $y = 3x - 4$ ___

9. $2x = 3y$
   $4x = 12 + 3y$ ___

10. $5x + 8y = 1$
    $3x + 1 = -4y$ ___

11. $3x + y = 6$
    $x - 10 = -3y$ ___

12. $y = x + 1$
    $y = x - 6$ ___

Date _____

Name _____

Course/Section _____

Unit 4
Equations
in Two Variables

Word Problems
Involving Two
Variables

Translate each problem statement into a system of equations and solve.

1. The sum of two numbers is 22 and their difference is 4. Find the numbers.

   _____

2. The sum of two numbers is 49. Twice the smaller is 2 more than the larger. Find the numbers.

   _____

3. The difference of two numbers is 9. The larger is 3 less than twice the smaller. Find the numbers.

   _____

4. The average of two numbers is 11 and their difference is 6. Find the numbers.

   _____

5. When Don cashed his $127 pay check he received the money in $1, $5, and $10 bills. The number of $5 bills was one less than the number of $1 bills and the number of $10 bills was two more than the number of $1 bills. How many bills of each kind did Don receive?

   _____

6. If five times the larger of two numbers is added to twice the smaller, the sum is 43. If twice the larger is decreased by the smaller, the result is 10. Find the numbers.

   _____

7. How many pounds of birdseed at 90¢ per lb should be mixed with 30 lb of seed worth 70¢ per lb to produce a mix to sell for 75¢ per lb?

   _____

8. If four avocados and three grapefruit cost $3.17 and if two avocados and five grapefruit cost $2.11, then what does one avocado cost?

   _____

Date

_____

9. A powerboat can travel 20 miles downstream in 1 hour, but needs $2\frac{1}{2}$ hours to make the return trip. What would be the speed of the boat in still water?

Name

_____

Course/Section

10. The perimeter of a rectangle is 28 ft. The length of the rectangle is three times its width. Find the dimensions of the rectangle.

_____

11. The sum of Al's age and Sally's age is 41. Ten years ago Al was twice as old as Sally was then. How old is each person now?

_____

12. Laurie bicycled to the university at 12 mph and returned home in a car traveling the same route at 48 mph. The total travel time for the round trip was $37\frac{1}{2}$ minutes. How far did she bicycle?

_____

5 Fractions and Radicals

Objective	Sample Problems	Where To Go for Help

Upon successful completion of this program you will be able to:

Sample Problems

Where To Go for Help

	Page	Frame

Reduce to lowest terms: 340   **4**

1. Work with algebraic fractions.

(a) $\dfrac{-5a^2b^3}{10a^3b} =$ _____

(b) $\dfrac{9x + 3y}{12x + 4y} =$ _____

(c) $\dfrac{x^2 - y^2}{x + y} =$ _____

(d) $\dfrac{4x^2 - 3x - 1}{x - 1} =$ _____

Write as an equivalent fraction: 345   **8**

(e) $\dfrac{5}{3x} = \dfrac{?}{6x}$ _____

(f) $\dfrac{-x^2}{y^2} = \dfrac{?}{3y^3}$ _____

2. Add and subtract algebraic fractions.

(a) $\dfrac{2}{x} + \dfrac{1}{x - 1} =$ _____   351   **12**

(b) $\dfrac{5}{2x} - \dfrac{5}{6x} =$ _____

(c) $\dfrac{x + 1}{x + 2} - \dfrac{x - 1}{x + 2} =$ _____

3. Multiply and divide algebraic fractions.

(a) $\left(\dfrac{6x^3}{8}\right)\left(\dfrac{2}{3x}\right) =$ _____   363   **22**

(b) $\left(\dfrac{3a^2b}{4x^2y}\right)\left(\dfrac{8x^2y^3}{6ab^2}\right) =$ _____

(c) $\left(\dfrac{4x - 8}{6x + 3}\right)\left(\dfrac{12x + 6}{2x - 4}\right) =$ _____

(d) $\dfrac{4x^2}{3y} \div \dfrac{2x^3}{2y} =$ _____

(e) $\dfrac{a + \dfrac{1}{b}}{1 + \dfrac{1}{b}} =$ _____

Date _____

Name _____

Course/Section _____

Solve:

4. Solve algebraic problems involving fractions.

(a) $\dfrac{x}{3} - 2 = \dfrac{4}{5}$ $\qquad x = $ _____

(b) $\dfrac{x + 8}{6} - \dfrac{x + 5}{4} = 1$ $\quad x = $ _____

(c) If $\dfrac{2}{3}$ of a certain number is added to twice that number, the result is 16. Find the number.

_____

(d) If it takes 3 hr to make 48 Thingies, how many Thingies will be made in 5 hr at the same rate?

_____

Simplify:

405   **47**

5. Work with radical expressions.

(a) $\sqrt{45} = $ _____

(b) $\sqrt{a^5} = $ _____

(c) $\sqrt{8x^3y^2} = $ _____

(d) Add:
$2\sqrt{3x} + 3\sqrt{12x} = $ _____

(e) Multiply:
$\sqrt{2x} \cdot \sqrt{6x} = $ _____

(f) Rationalize:
$\dfrac{\sqrt{3}}{\sqrt{2}} = $ _____

(g) $\dfrac{\sqrt{12}}{\sqrt{3x}} = $ _____

(h) $\sqrt{\dfrac{a^3}{ab}} = $ _____

(Answers to these problems are on page 337)

If you are certain that you can work all of these problems correctly, turn to page 433 for a self-test. If you want help with any of these objectives or if you cannot work one of the sample problems, turn to the page indicated. Of course, the super-student will want to learn all of this and will turn to frame **1** and begin work there.

PREVIEW 5

Answers to Sample Problems

1. (a) $-\dfrac{b^2}{2a}$ (b) $\dfrac{3}{4}$ (c) $x - y$ (d) $4x + 1$ (e) $\dfrac{10}{6x}$ (f) $\dfrac{-3x^2 y}{3y^3}$

2. (a) $\dfrac{3x - 2}{x(x - 1)}$ (b) $\dfrac{5}{3x}$ (c) $\dfrac{2}{x + 2}$

3. (a) $\dfrac{x^2}{2}$ (b) $\dfrac{ay^2}{b}$ (c) 4 (d) $\dfrac{4}{3x}$ (e) $\dfrac{1 + ab}{1 + b}$

4. (a) $8\dfrac{2}{5}$ (b) -11 (c) 6 (d) 80

5. (a) $3\sqrt{5}$ (b) $a^2\sqrt{a}$ (c) $2xy\sqrt{2x}$ (d) $8\sqrt{3x}$ (e) $2x\sqrt{3}$
   (f) $\dfrac{\sqrt{6}}{2}$ (g) $\dfrac{2\sqrt{x}}{x}$ (h) $\dfrac{a\sqrt{b}}{b}$

5 Fractions and Radicals

5-1 ALGEBRAIC FRACTIONS

1 A fraction, as you should recall from your study of arithmetic, is a number formed as a result of a division. The division $a \div b$ can be written as the fraction $\frac{a}{b}$ where a and b are any real numbers and b cannot be equal to zero. If a and b are algebraic expressions, the fraction $\frac{a}{b}$ is called an *algebraic fraction* or *rational expression*.

The operations used with other kinds of algebraic expressions can be applied to algebraic fractions: addition, subtraction, multiplication, division, powers and roots, and equation solving. Many students find the topic of algebraic fractions to be difficult because they must remember and use so many skills from beginning algebra and from arithmetic. If you need to review any of these arithmetic skills, pause here and return to page 15 for a review of signed numbers, page 29 for a quick look at working with exponents, or Unit 3, starting on page 195, for a review of the basic algebra. To check your need for a review of arithmetic fractions, try the following quiz.

1. Reduce to lowest terms:

  (a) $\dfrac{8}{12} = $ _____

  (b) $-\dfrac{21}{35} = $ _____

2. Change to an equivalent fraction:

  (a) $\dfrac{5}{9} = \dfrac{?}{27}$ _____

  (b) $\dfrac{-7}{12} = \dfrac{?}{48}$ _____

3. Find the least common denominator of 6 and 8. _____

4. Add or subtract:

  (a) $\dfrac{1}{3} + \dfrac{2}{5} = $ _____

  (b) $\dfrac{5}{7} - \dfrac{2}{3} = $ _____

5. Multiply:

    (a) $\dfrac{2}{3} \cdot \dfrac{5}{8} =$ _____
           (b) $\left(-\dfrac{3}{5}\right) \cdot \left(\dfrac{1}{2}\right) =$ _____

6. Divide:

    (a) $\dfrac{2}{3} \div \dfrac{2}{5} =$ _____
           (b) $\left(\dfrac{5}{12}\right) \div \left(-\dfrac{1}{4}\right) =$ _____

Turn to **2** to check your answers.

---

2   Answers to arithmetic fractions quiz:

1. (a) $\dfrac{2}{3}$ (b) $\dfrac{-3}{5}$ 2. (a) $\dfrac{15}{27}$ (b) $\dfrac{-28}{48}$

3. 24   4. (a) $\dfrac{11}{15}$ (b) $\dfrac{1}{21}$

5. (a) $\dfrac{5}{12}$ (b) $\dfrac{-3}{10}$ 6. (a) $\dfrac{5}{3}$ (b) $\dfrac{-5}{3}$

If you answered any of these questions incorrectly, turn to the Fractions Review on page 521 at the back of this book.

If you were able to answer all of these correctly, turn to frame **3** and continue.

Algebraic
Fraction

3   An *algebraic fraction* or rational expression is the quotient of two algebraic expressions. It is a way of showing the division of algebraic quantities. For example, the division $a \div b$ can be written as the fraction $\dfrac{a}{b}$ where a and b may be any literal numbers, variables, or algebraic expressions, except that b, just as in arithmetic, cannot equal zero. The algebraic fraction $\dfrac{x}{y}$ is a real number for any values of x and y except $y = 0$.

The algebraic fraction $\dfrac{2}{x-1}$ is a real number for any values of x except $x = 1$.

When $x = 1$ the denominator of this fraction is equal to $1 - 1$ or 0, and we have no real number equal to $\dfrac{2}{0}$.

The algebraic fraction $\dfrac{2x^2 - x + 5}{x^2 + x - 6}$ represents a real number for any values of x except $x = 2$ and $x = -3$. If you evaluate the polynomial $x^2 + x - 6$ for $x = 2$ or $x = -3$, you will find that for these values of x the polynomial is equal to zero and, therefore, the fraction does not represent a real number at these points.

Write each of the following divisions as an algebraic fraction and indicate the values of the variables for which the fraction represents a real number.

(a) $x \div 2$ (b) $y \div 2x$ (c) $-3x \div y$
(d) $2x \div (x - 1)$ (e) $(x + y) \div (x - y)$ (f) $(x^2 + 2x - 5) \div (x + 2)$

Check your fractions in **4**.

---

4   (a) $\dfrac{x}{2}$ defined for all values of x.

       (b) $\dfrac{y}{2x}$ defined for all values of x and y except $x = 0$.

(c) $\dfrac{-3x}{y}$ defined for all values of x and y except $y = 0$.

(d) $\dfrac{2x}{x - 1}$ defined for all values of x except $x = 1$.

(e) $\dfrac{x + y}{x - y}$ defined for all values of x and y except $x = y$.

(f) $\dfrac{x^2 + 2x - 5}{x + 2}$ defined for all values of x except $x = -2$.

Equivalent Algebraic Fractions

Two algebraic fractions are said to be *equivalent* if they are alternative ways of writing the same algebraic expression. Equivalent algebraic fractions name the same real number. For example, the algebraic fractions

$$\frac{x}{y}, \frac{2x}{2y}, \frac{3x}{3y}, \frac{4x}{4y}, \ldots \frac{ax}{ay}, \ldots \text{ and so on}$$

are all equivalent algebraic fractions. Any algebraic fraction can be written in an infinite number of equivalent forms.

When we work with algebraic fractions it is often very helpful to be able to rewrite any fraction in its simplest equivalent form. It is certainly not easy to see that

$$\frac{(x^2 - 9)(x + 2)}{2(x + 3)(x^2 - x - 6)} = \frac{1}{2}$$

but these two fractions are equivalent and $\frac{1}{2}$ is the simplest equivalent form of the first fraction.

Reducing to Lowest Terms

The process of finding the simplest fraction equivalent to a given fraction is known as *reducing to lowest terms*. To reduce an algebraic fraction to lowest terms, follow this procedure.

 First: Write the numerator and denominator of the given fraction as products of their factors.

Example: Reduce $\dfrac{6x^2}{4x}$ to lowest terms.

$$\frac{6x^2}{4x} = \frac{2 \cdot 3 \cdot x \cdot x}{2 \cdot 2 \cdot x}$$

Second: Regroup the factors to form fractions of the form $\dfrac{a}{a}$. Find the common factors, those that appear in *both* top and bottom of the fraction.

$$\frac{6x^2}{4x} = \frac{\boxed{2 \cdot x \cdot} \; 3 \cdot x}{\boxed{2 \cdot x \cdot} \; 2}$$

Third: Set each factor of the form $\dfrac{a}{a}$ equal to 1. Simplify the fraction.

$$\frac{6x^2}{4x} = 1 \cdot 1 \cdot \frac{3x}{2} = \frac{3x}{2}$$

Reducing a fraction to lowest terms is a matter of finding factors common to both the numerator and denominator of that fraction and "removing" them by dividing.

The process in the example above is usually written as

$$\frac{6x^2}{4x} = \frac{2 \cdot 3 \cdot x \cdot \cancel{x}}{\cancel{2} \cdot 2 \cdot \cancel{x}} = \frac{3x}{2}$$

The "cancel" marks are a reminder that we have divided both numerator and denominator of the fraction by $2x$.

Another example: $\dfrac{-3ab}{12b^2} = \dfrac{(-1) \cdot 3 \cdot a \cdot b}{2 \cdot 2 \cdot 3 \cdot b \cdot b} = \dfrac{\boxed{3b} \cdot (-a)}{\boxed{3b} \cdot (4b)} = \dfrac{-a}{4b}$

or in shorthand notation,

$$\frac{-3ab}{12b^2} = \frac{(-1) \cdot \cancel{3} \cdot a \cdot \cancel{b}}{2 \cdot 2 \cdot \cancel{3} \cdot \cancel{b} \cdot b} = \frac{-a}{4b}$$

▷ If the fraction to be reduced includes a negative sign, you will need to write (-1) as a factor.

Reduce this fraction to lowest terms:

$$\frac{15x^2y^2}{3x}$$

Check your work in **5**.

---

5 $\quad \dfrac{15x^2y^2}{3x} = \dfrac{3 \cdot 5 \cdot x \cdot x \cdot y \cdot y}{3 \cdot x} = \boxed{\dfrac{3 \cdot x}{3 \cdot x}} \cdot \dfrac{5 \cdot x \cdot y \cdot y}{1} = 5xy^2$

or, in the shorthand notation,

$$\frac{15x^2y^2}{3x} = \frac{\cancel{3} \cdot 5 \cdot \cancel{x} \cdot x \cdot y \cdot y}{\cancel{3} \cdot \cancel{x}} = 5xy^2$$

A fraction is in "lowest terms" when the numerator and denominator contain *no* common factors.

Reduce each of the following algebraic fractions to lowest terms.

(a) $\dfrac{3x^2}{12x^3}$ (b) $\dfrac{-2a}{16a^4}$ (c) $\dfrac{18}{3y^2}$

(d) $-\dfrac{x^2y^3}{xy^2}$ (e) $\dfrac{xy^3}{x^3y}$ (f) $\dfrac{-2a^3b^2c}{-4abc^2}$

Work carefully. Our answers are in **6**.

---

6   (a) $\dfrac{3x^2}{12x^3} = \dfrac{\cancel{3} \cdot \cancel{x} \cdot \cancel{x}}{2 \cdot 2 \cdot \cancel{3} \cdot \cancel{x} \cdot \cancel{x} \cdot x} = \dfrac{1}{4x}$

   (b) $\dfrac{-2a}{16a^4} = \dfrac{(-1) \cdot \cancel{2} \cdot \cancel{a}}{\cancel{2} \cdot 2 \cdot 2 \cdot 2 \cdot \cancel{a} \cdot a \cdot a \cdot a} = \dfrac{-1}{8a^3}$

   (c) $\dfrac{18}{3y^2} = \dfrac{2 \cdot \cancel{3} \cdot 3}{\cancel{3} \cdot y \cdot y} = \dfrac{6}{y^2}$

   (d) $-\dfrac{x^2y^3}{xy^2} = \dfrac{-x^2y^3}{xy^2} = \dfrac{(-1) \cdot \cancel{x} \cdot x \cdot \cancel{y} \cdot \cancel{y} \cdot y}{\cancel{x} \cdot \cancel{y} \cdot \cancel{y}} = -xy$

   (e) $\dfrac{xy^3}{x^3y} = \dfrac{\cancel{x} \cdot \cancel{y} \cdot y \cdot y}{\cancel{x} \cdot x \cdot x \cdot \cancel{y}} = \dfrac{y^2}{x^2}$

   (f) $\dfrac{-2a^3b^2c}{-4abc^2} = \dfrac{(-\cancel{1}) \cdot \cancel{2} \cdot \cancel{a} \cdot a \cdot a \cdot \cancel{b} \cdot b \cdot \cancel{c}}{(-\cancel{1}) \cdot \cancel{2} \cdot 2 \cdot \cancel{a} \cdot \cancel{b} \cdot \cancel{c} \cdot c} = \dfrac{a^2b}{2c}$

Negative Signs on Fractions   Notice in problems (d) and (f) that when the fraction has a negative sign or when either the numerator or denominator of the fraction is negative, it is least confusing to rewrite the fraction so that only the numerator has a negative sign.

$-\dfrac{a}{b}$ is equal to $\dfrac{-a}{b}$

$\dfrac{a}{-b}$ is equal to $\dfrac{-a}{b}$

$-\dfrac{-a}{b}$ is equal to $\dfrac{a}{b}$

$-\dfrac{a}{-b}$ is equal to $\dfrac{a}{b}$

$\dfrac{-a}{-b}$ is equal to $\dfrac{a}{b}$

342   **Fractions and Radicals**

and

$$-\frac{-a}{-b} \quad \text{is equal to} \quad \frac{-a}{b}$$

Notice than an *even* number of negative signs (two of them) makes the fraction positive, while an *odd* number of negative signs (one or three of them) makes the fraction negative.

Of course, an algebraic fraction may have any algebraic expression as numerator or denominator, including polynomials and products of polynomials. For example, the algebraic fraction

$$\frac{(x-1)(x+2)}{(x-1)(x+3)} \quad \text{when reduced to lowest terms becomes} \quad \frac{x+2}{x+3}.$$

Both numerator and denominator of the original fraction contain the factor $(x-1)$. To reduce to lowest terms, divide top and bottom of the fraction by this common factor.

Another example: $\dfrac{x^2 + 2x}{2x^3 + x} = \dfrac{\cancel{x}(x+2)}{\cancel{x}(2x^2+1)} = \dfrac{x+2}{2x^2+1}$

First, we factor both numerator and denominator, then we divide out the common factors.

Reduce the following algebraic fractions to lowest terms.

(a) $\dfrac{3(x+y)}{5(x+y)}$ (b) $\dfrac{4x - 2y}{2x - y}$ (c) $-\dfrac{x}{x + xy}$

(d) $\dfrac{x^2 - 4}{-(x+2)}$ (e) $\dfrac{x^2 + x - 2}{x^2 + 5x + 6}$ (f) $\dfrac{x^2 - 1}{x^2 + 4x - 5}$

Check your work in **7**.

---

7 (a) $\dfrac{3\cancel{(x+y)}}{5\cancel{(x+y)}} = \dfrac{3}{5}$

(b) $\dfrac{4x - 2y}{2x - y} = \dfrac{2\cancel{(2x-y)}}{\cancel{(2x-y)}} = 2$ First factor 2 from each term in the numerator.

(c) $-\dfrac{x}{x+xy} = \dfrac{-x}{x+xy} = \dfrac{(-1)\cancel{x}}{\cancel{x}(1+y)} = \dfrac{-1}{1+y}$ Factor $x + xy$ into $x(1+y)$.

(d) $\dfrac{x^2 - 4}{-(x+2)} = \dfrac{-(x^2-4)}{(x+2)} = \dfrac{(-1)\cancel{(x+2)}(x-2)}{\cancel{(x+2)}} = -(x-2)$

The answer may also be written as $-x + 2$ or $2 - x$.

You should remember from Unit 3 that the difference of two squares can always be factored into the product of a sum and a difference.

(e) $\dfrac{x^2 + x - 2}{x^2 + 5x + 6} = \dfrac{(x-1)\cancel{(x+2)}}{\cancel{(x+2)}(x+3)} = \dfrac{x-1}{x+3}$ Again the first step is to factor the polynomials in the numerator and denominator.

(f) $\dfrac{x^2 - 1}{x^2 + 4x - 5} = \dfrac{\cancel{(x-1)}(x+1)}{\cancel{(x-1)}(x+5)} = \dfrac{x+1}{x+5}$

Reducing to lowest terms may involve subtle rearrangement of the terms of the numerator or denominator expressions. For example, it is easy to see that

$$\frac{x+y}{y+x} = \frac{\cancel{x+y}}{\cancel{x+y}} = 1$$

but not so easy to see that $\quad \dfrac{x-y}{y-x} = \dfrac{(-1)\cancel{(y-x)}}{\cancel{(y-x)}} = -1$

In this last example, factoring (-1) from the numerator produces a common factor in both numerator and denominator.

$$x - y = -y + x = (-1)(y - x)$$

Practice this tricky business on these similar problems.

(a) $\dfrac{a - b}{b - a}$ (b) $-\dfrac{2(x - y)}{(y - x)}$

(c) $\dfrac{1 - x^2}{x + 1}$ (d) $-\dfrac{b - a}{xa - xb}$

Our answers are in **8**.

CANCELLING MISTAKES
or
Three Quick Ways to Fail Algebra

We can reduce an algebraic fraction to lowest terms by dividing out common factors in numerator and denominator. Many students do this using the shortcut known as cancelling. For example,

$$\frac{2\cancel{x}y}{3\cancel{x}} = \frac{2y}{3}$$

Here we divided numerator and denominator by x. The cancel lines remind us that it is a division.

The short cut is legal, but you must be very careful to divide *each term* of the numerator and *each term* of the denominator by the common factor.

NEVER, NEVER do this:

$$\frac{x + \cancel{6}}{\cancel{6}} \qquad \text{to get } x + 1$$

or this: $\dfrac{\cancel{6}x + 2}{\cancel{6}} \qquad \text{to get } x + 2$

or this: $\dfrac{5x + 1\overset{2}{\cancel{2}}}{\cancel{6}} \qquad \text{to get } 5x + 2$

In each case, to "cancel" the 6 in the bottom of the fraction means to divide by 6. To get an equivalent fraction you must divide all of the numerator by 6, not merely part of it.

You can cancel only when the numerator and denominator are written as products, not sums or differences.

The three mistakes shown have plagued algebra students for centuries. Be careful.

RECIPROCALS

The *reciprocal* of any real number is a second number such that the product of the two numbers is equal to 1.

The reciprocal of 4 is $\frac{1}{4}$ because $(4)\left(\frac{1}{4}\right) = 1$.

The reciprocal of $\frac{2}{3}$ is $\frac{3}{2}$ because $\left(\frac{2}{3}\right)\left(\frac{3}{2}\right) = 1$.

The reciprocal of x is $\frac{1}{x}$ because $(x)\left(\frac{1}{x}\right) = 1$.

To find the reciprocal of any number n we simply invert n to form the fraction $1 \div n$.

Here are a few more algebraic fraction reciprocals:

Quantity	Reciprocal
$\dfrac{x}{2}$	$\dfrac{2}{x}$
$\dfrac{x}{y}$	$\dfrac{y}{x}$
$\dfrac{1}{x}$	x
$\dfrac{x + y}{x - y}$	$\dfrac{x - y}{x + y}$
$3ab^2$	$\dfrac{1}{3ab^2}$
$\dfrac{2x^2 - y}{x + 4}$	$\dfrac{x + 4}{2x^2 - y}$
$-a$	$-\dfrac{1}{a}$ or $\dfrac{-1}{a}$

8 (a) $\dfrac{a - b}{b - a} = \dfrac{-(b - a)}{(b - a)} = \dfrac{(-1)(b - a)}{(b - a)} = -1$

(b) $-\dfrac{2(x - y)}{(y - x)} = \dfrac{-2(x - y)}{(y - x)} = \dfrac{(-2)(-1)(y - x)}{(y - x)} = (-2)(-1) = 2$

(c) $\dfrac{1 - x^2}{x + 1} = \dfrac{-(x^2 - 1)}{x + 1} = \dfrac{(-1)(x + 1)(x - 1)}{(x + 1)} = -(x - 1) = 1 - x$

(d) $-\dfrac{b - a}{xa - xb} = \dfrac{-(b - a)}{x(a - b)} = \dfrac{(-1)(-1)(a - b)}{x(a - b)} = \dfrac{(a - b)}{x(a - b)} = \dfrac{1}{x}$

Building Equivalent Fractions

Given any algebraic fraction, we can obtain other fractions equivalent to it by multiplying both numerator and denominator of the original fraction by the same non-zero number or algebraic quantity. For example,

$$\frac{3x}{y} = \frac{3x \cdot \boxed{2}}{y \cdot \boxed{2}} = \frac{6x}{2y}$$

The new fraction $\dfrac{6x}{2y}$ is equivalent to the original fraction $\dfrac{3x}{y}$.

In effect, we have multiplied the algebraic fraction by $\dfrac{2}{2}$ or 1, and multiplying any quantity by 1 does not change the value of that quantity. Using this procedure we can "build up" any fraction to produce an equivalent fraction.

Another example: $\dfrac{2x}{3ay^2} = \dfrac{?}{6a^2y^2}$

Here we are asked to build up the original fraction with denominator equal to $3ay^2$ to form a new and equivalent fraction with denominator equal to $6a^2y^2$. The first step in this build up is to find the multiplier:

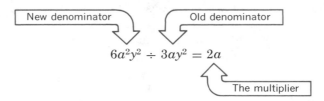

$$6a^2y^2 \div 3ay^2 = 2a$$

Second, use the multiplier $\boxed{2a}$ to build up the original fraction. Multiply both numerator and denominator of the original fraction by $\boxed{2a}$.

$$\frac{2x}{3ay^2} = \frac{2x \cdot \boxed{2a}}{3ay^2 \cdot \boxed{2a}} = \frac{4ax}{6a^2y^2}$$

The algebraic fractions $\dfrac{2x}{3ay^2}$ and $\dfrac{4ax}{6a^2y^2}$ are equivalent.

"Building up" a fraction is the reverse process to reducing to lowest terms.

Express each of the following fractions as an equivalent fraction with the denominator shown.

(a) $\dfrac{2}{x} = \dfrac{?}{3x}$ (b) $\dfrac{5}{3y} = \dfrac{?}{6xy^2}$ (c) $\dfrac{3x^2}{y^2} = \dfrac{?}{3x^2y^2}$

(d) $-\dfrac{2x}{3y} = \dfrac{?}{12xy^2}$ (e) $y = \dfrac{?}{x}$ (f) $-2x^2 = \dfrac{?}{3a}$

Check your work in **9**.

---

9   (a) $\dfrac{2}{x} = \dfrac{?}{3x}$ The multiplier is $\boxed{3}$ since $3x \div x = 3$.

Multiply the numerator and denominator of the original fraction by $\boxed{3}$.

$$\frac{2 \cdot \boxed{3}}{x \cdot \boxed{3}} = \frac{6}{3x}$$

(b) $\dfrac{5}{3y} = \dfrac{?}{6xy^2}$ The multiplier is $\boxed{2xy}$ since $6xy^2 \div 3y = 2xy$.

Multiply the numerator and denominator of the original fraction by $\boxed{2xy}$.

$$\frac{5 \cdot \boxed{2xy}}{3y \cdot \boxed{2xy}} = \frac{10xy}{6xy^2}$$

(c) $\dfrac{3x^2}{y^2} = \dfrac{?}{3x^2y^2}$ The multiplier is $\boxed{3x^2}$ since $3x^2y^2 \div y^2 = 3x^2$.

Multiply the numerator and denominator of the original fraction by $\boxed{3x^2}$.

$$\frac{3x^2 \cdot \boxed{3x^2}}{y^2 \cdot \boxed{3x^2}} = \frac{9x^4}{3x^2y^2}$$

(d) $-\dfrac{2x}{3y} = \dfrac{?}{12xy^2}$ Rewrite the fraction with the negative sign associated with the numerator.

$\dfrac{-2x}{3y} = \dfrac{?}{12xy^2}$ The multiplier is $\boxed{4xy}$ since $12xy^2 \div 3y = 4xy$.

Multiply numerator and denominator of the original fraction by $\boxed{4xy}$.

$$\frac{-2x \cdot \boxed{4xy}}{3y \cdot \boxed{4xy}} = \frac{-8x^2y}{12xy^2}$$

(e) $y = \dfrac{?}{x}$ First rewrite the left member of the equation as a fraction.

$\dfrac{y}{1} = \dfrac{?}{x}$ The multiplier is $\boxed{x}$ since $x \div 1 = x$.

Multiply numerator and denominator of the original fraction by $\boxed{x}$.

$$\frac{y \cdot \boxed{x}}{1 \cdot \boxed{x}} = \frac{xy}{x}$$

(f) $-2x^2 = \dfrac{?}{3a}$ Rewrite the left member of the equation as a fraction.

$\dfrac{-2x^2}{1} = \dfrac{?}{3a}$ The multiplier is $\boxed{3a}$ since $3a \div 1 = 3a$.

Multiply numerator and denominator of the original fraction by $\boxed{3a}$.

$$\frac{-2x^2 \cdot \boxed{3a}}{1 \cdot \boxed{3a}} = \frac{-6ax^2}{3a}$$

The process of building equivalent fractions is no different when the algebraic expressions that make up the fraction are binomials or trinomials or more complicated polynomials.

Change the following fractions to their equivalents as shown.

(a) $\dfrac{1}{2} = \dfrac{?}{4(x+1)}$ (b) $3x = \dfrac{?}{x+2}$

(c) $\dfrac{3x}{x+3} = \dfrac{?}{(x+3)(x-3)}$ (d) $x = \dfrac{?}{x^2+1}$

(e) $\dfrac{2}{x-1} = \dfrac{?}{x^2-1}$ (f) $-\dfrac{y}{y+2} = \dfrac{?}{(x+2)(y+2)}$

Check your work in **10**.

---

10 (a) $\dfrac{1}{2} = \dfrac{?}{4(x+1)}$ The multiplier is $\boxed{2(x+1)}$ since $4(x+1) \div 2 = 2(x+1)$

Multiply the numerator and denominator of the original fraction by $\boxed{2(x+1)}$.

$$\frac{1 \cdot \boxed{2(x+1)}}{2 \cdot \boxed{2(x+1)}} = \frac{2(x+1)}{4(x+1)}$$

(b) $3x = \dfrac{?}{x+2}$

$\dfrac{3x}{1} = \dfrac{?}{x+2}$ The multiplier is $\boxed{x+2}$ since $(x+2) \div 1 = x+2$.

$$\frac{3x \cdot \boxed{(x+2)}}{1 \cdot \boxed{(x+2)}} = \frac{3x(x+2)}{(x+2)}$$

(c) $\dfrac{3x}{x+3} = \dfrac{?}{(x+3)(x-3)}$ The multiplier is $\boxed{x-3}$ since $(x+3)(x-3) \div (x+3) = (x-3)$.

$$\frac{3x \cdot (x-3)}{(x+3) \cdot (x-3)} = \frac{3x(x-3)}{(x+3)(x-3)}$$

(d) $x = \dfrac{?}{x^2 + 1}$

$\dfrac{x}{1} = \dfrac{?}{x^2 + 1}$ The multiplier is $x^2 + 1$ since $(x^2 + 1) \div 1 = (x^2 + 1)$.

$$\frac{x \cdot (x^2 + 1)}{1 \cdot (x^2 + 1)} = \frac{x(x^2 + 1)}{(x^2 + 1)}$$

(e) $\dfrac{2}{x-1} = \dfrac{?}{x^2 - 1}$ Rewrite $x^2 - 1$ as $(x+1)(x-1)$.

$\dfrac{2}{x-1} = \dfrac{?}{(x+1)(x-1)}$ The multiplier is $x + 1$ since
$(x+1)(x-1) \div (x-1) = (x+1)$.

$$\frac{2 \cdot (x+1)}{(x-1) \cdot (x+1)} = \frac{2(x+1)}{x^2 - 1}$$

(f) $-\dfrac{y}{y+2} = \dfrac{?}{(x+2)(y+2)}$

$\dfrac{-y}{y+2} = \dfrac{?}{(x+2)(y+2)}$ The multiplier is $(x+2)$.

$$\frac{-y \cdot (x+2)}{(y+2) \cdot (x+2)} = \frac{-y(x+2)}{(y+2)(x+2)}$$

With some fractions, in order to find the multiplying factor needed to build up the equivalent fraction, it may be necessary to factor the denominator first. For example, in the problem

$\dfrac{x}{x+1} = \dfrac{?}{2x^2 + 5x + 3}$ Factor the denominator on the right.

$\dfrac{x}{x+1} = \dfrac{?}{(x+1)(2x+3)}$ The multiplier is $2x + 3$ since
$(x+1)(2x+3) \div (x+1) = 2x + 3$.

$$\frac{x \cdot (2x+3)}{(x+1) \cdot (2x+3)} = \frac{x(2x+3)}{(x+1)(2x+3)}$$

$$= \frac{x(2x+3)}{2x^2 + 5x + 3}$$

Now turn to **11** for a set of practice problems on working with algebraic fractions.

Unit 5
Fractions and Radicals

11   The answers are on page 558.

Algebraic Fractions

A. Reduce to lowest terms:

1. $\dfrac{2x^2}{x^3}$ _____

2. $\dfrac{a^5}{a^2}$ _____

3. $\dfrac{4x^3}{12x^4}$ _____

4. $\dfrac{15}{10y^2}$ _____

5. $\dfrac{5x}{20x^5}$ _____

6. $\dfrac{-2x}{6x^3}$ _____

7. $\dfrac{36}{-9y^3}$ _____

8. $\dfrac{a}{a^2b^2}$ _____

9. $\dfrac{x^2y^4}{xy^2}$ _____

10. $\dfrac{ab^3}{b}$ _____

11. $\dfrac{-a^2x}{ax}$ _____

12. $\dfrac{8ab^2}{4c}$ _____

13. $\dfrac{2xy}{14a}$ _____

14. $\dfrac{3a^2bc^2}{9ab}$ _____

15. $\dfrac{-p^2qt}{-p^2q^2}$ _____

16. $\dfrac{-4x^2y^2}{-2x^2y^2}$ _____

17. $\dfrac{8(x+2y)}{4(x+2y)}$ _____

18. $\dfrac{12(a+3)}{4(a+3)^2}$ _____

19. $\dfrac{2x^2-3x}{x^3}$ _____

20. $\dfrac{a^2+2ab}{3a}$ _____

21. $\dfrac{6x+18}{x+3}$ _____

22. $\dfrac{t-2}{4t-8}$ _____

23. $\dfrac{(x-y)^2}{x^2-y^2}$ _____

24. $\dfrac{a+b}{b+a}$ _____

25. $\dfrac{a^2-b^2}{a-b}$ _____

26. $\dfrac{x^2-x}{x^2+x-2}$ _____

27. $\dfrac{-2a}{4a^2-8a}$ _____

28. $\dfrac{x^2-3x}{x^2-4x+3}$ _____

29. $\dfrac{x-y}{y-x}$ _____

30. $\dfrac{x^2-y^2}{x^2+xy}$ _____

31. $\dfrac{a(a+b)}{2ab-a}$ _____

32. $\dfrac{pq(p+q^2)}{p^2q^2}$ _____

33. $\dfrac{4x^2+4x+1}{1-4x^2}$ _____

34. $\dfrac{8+2x}{x^2+8x+16}$ _____

35. $\dfrac{a^2+3a-4}{a^2-1}$ _____

36. $\dfrac{3a^2-a-4}{4+a-3a^2}$ _____

37. $\dfrac{2(x-1)(x^2+4x+4)}{(x^2-3x+2)(x+3)}$ _____

38. $\dfrac{-2x+4}{2-x}$ _____

39. $\dfrac{2x^2+2x+4}{2x^2+2x+6}$ _____

40. $\dfrac{2x^2-3x-2}{2x^2+7x+3}$ _____

B. Express each fraction as an equivalent fraction as shown.

1. $\dfrac{5}{2x}=\dfrac{?}{8x}$ _____

2. $ax=\dfrac{?}{bx}$ _____

3. $\dfrac{2a}{5}=\dfrac{?}{5x}$ _____

4. $4=\dfrac{?}{x}$ _____

5. $\dfrac{a}{b}=\dfrac{?}{4b^2}$ _____

6. $\dfrac{x^2}{y^2}=\dfrac{?}{3y^3}$ _____

7. $x=\dfrac{?}{y}$ _____

8. $2y=\dfrac{?}{y}$ _____

9. $\dfrac{2}{-3x}=\dfrac{?}{6x^2}$ _____

10. $\dfrac{4}{-5x} = \dfrac{?}{15x^3}$ _____

11. $\dfrac{-a}{b} = \dfrac{?}{2ab}$ _____

12. $-\dfrac{2x}{3y} = \dfrac{?}{12xy}$ _____

13. $\dfrac{-a}{bx} = \dfrac{?}{a^2bx^2}$ _____

14. $-xy = \dfrac{?}{2xy}$ _____

15. $-x = \dfrac{?}{y}$ _____

16. $x^2 = \dfrac{?}{2x^2y}$ _____

17. $\dfrac{1}{2} = \dfrac{?}{2(x+2)}$ _____

18. $\dfrac{3}{4} = \dfrac{?}{8(x+y)}$ _____

19. $5x = \dfrac{?}{x+1}$ _____

20. $\dfrac{-3x}{2} = \dfrac{?}{4(x+1)}$ _____

21. $\dfrac{2x}{x+y} = \dfrac{?}{(x+y)(x-y)}$ _____

22. $\dfrac{3}{x+3} = \dfrac{?}{(x+1)(x+3)}$ _____

23. $\dfrac{-1}{2x+1} = \dfrac{?}{(x+2)(2x+1)}$ _____

24. $\dfrac{8a}{-(a+b)} = \dfrac{?}{(2a+b)(a+b)}$ _____

25. $\dfrac{x}{x+2} = \dfrac{?}{x^2-4}$ _____

26. $\dfrac{2}{y} = \dfrac{?}{y+y^2}$ _____

27. $\dfrac{-4}{1-p} = \dfrac{?}{1-p^2}$ _____

28. $\dfrac{x+1}{x-1} = \dfrac{?}{1-x^2}$ _____

29. $\dfrac{x+1}{x^2-4} = \dfrac{?}{(x-2)(x^2+x-2)}$ _____

30. $\dfrac{-1}{a+b} = \dfrac{?}{a^2+2ab+b^2}$ _____

31. $\dfrac{2y}{y+2} = \dfrac{?}{y^2-y-6}$ _____

32. $\dfrac{x}{-(x+3)} = \dfrac{?}{x^3+4x^2+3x}$ _____

33. $\dfrac{-2x}{-x-1} = \dfrac{?}{x^2-1}$ _____

34. $\dfrac{-x+1}{-x-1} = \dfrac{?}{x^2+2x+1}$ _____

35. $\dfrac{3x^2}{2x-1} = \dfrac{?}{6x^2+x-2}$ _____

36. $(x+1)^2 = \dfrac{?}{(x+1)^2}$ _____

37. $x = \dfrac{?}{x^2+2x+1}$ _____

38. $-a = \dfrac{?}{a^2-a-6}$ _____

39. $\dfrac{1}{x} = \dfrac{?}{-x-x^2}$ _____

40. $\dfrac{y}{y+1} = \dfrac{?}{1-y^2}$ _____

When you have had the practice you need either return to the preview on page 335 or turn to **12** where you will learn to add and subtract algebraic fractions.

Date

Name

Course/Section

12 The operations of arithmetic—addition, subtraction, multiplication, division, finding powers and roots—are as important and useful with algebraic fractions as they are with the ordinary fractions of arithmetic. Most students, in this age of the electronic calculator, have learned to work with arithmetic fractions by first converting them to decimals by dividing, but algebraic fractions cannot be handled in this way. We will study the addition and subtraction of algebraic fractions in this section and their multiplication and division in the next section.

Adding algebraic fractions with like denominators is easy. The sum of two algebraic fractions with like denominators is a fraction whose numerator is the sum of their numerators and whose denominator is their common denominator. For example,

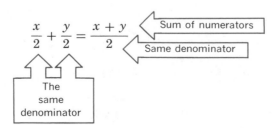

Another example:

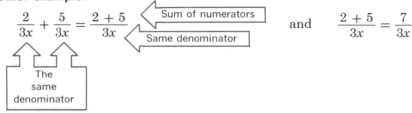   and $\dfrac{2+5}{3x} = \dfrac{7}{3x}$

Try it. Add $\dfrac{3}{xy} + \dfrac{4}{xy} = $ _____

Check your addition in **13**.

---

13 $\dfrac{3}{xy} + \dfrac{4}{xy} = \dfrac{3+4}{xy}$ ⟸ Sum of numerators / Same denominator   and $\dfrac{3+4}{xy} = \dfrac{7}{xy}$

Same denominator

This simple procedure for adding fractions is useful only when the denominators of the two fractions are the same.

Add the following algebraic fractions. No need for fancy calculations; you should be able to find the sum quickly by simply reading the problem.

(a) $\dfrac{1}{y} + \dfrac{2}{y}$ (b) $\dfrac{2}{ab^2} + \dfrac{2a}{ab^2}$ (c) $\dfrac{2x}{3} + \dfrac{8x}{3}$

(d) $\dfrac{2}{x+1} + \dfrac{3x}{x+1}$ (e) $\dfrac{a^2}{a+b} + \dfrac{b^2}{a+b}$ (f) $\dfrac{5}{2x^2} + \dfrac{3}{2x^2}$

Check your answers in **14**.

14   (a) $\dfrac{3}{y}$ (b) $\dfrac{2 + 2a}{ab^2}$ (c) $\dfrac{10x}{3}$ (d) $\dfrac{2 + 3x}{x + 1}$

       (e) $\dfrac{a^2 + b^2}{a + b}$ (f) $\dfrac{4}{x^2}$ reduced to lowest terms

Subtraction

Subtraction of algebraic fractions with like denominators is much the same as addition. For example,

$$\frac{5}{ax} - \frac{3}{ax} = \frac{5 - 3}{ax} = \frac{2}{ax}$$

However, in order to avoid the confusion that usually exists when the numerators are more complicated algebraic expressions, the super-student will rewrite every subtraction as an addition.

$$\frac{5}{ax} - \frac{3}{ax} = \frac{5}{ax} + \frac{-3}{ax} = \frac{5 + (-3)}{ax} = \frac{2}{ax}$$

Here the fraction $-\dfrac{3}{ax}$ is written as $\dfrac{-3}{ax}$. The negative sign is associated with the

numerator of the fraction. Notice also that we use parentheses to avoid any possibility of confusion.

The following example will show you why this slow and careful approach is needed.

$$\frac{x + 1}{a} - \frac{1 - 2x}{a} = \frac{x + 1}{a} + \frac{-(1 - 2x)}{a} = \frac{x + 1}{a} + \frac{-1 + 2x}{a}$$

$$= \frac{x + 1 - 1 + 2x}{a}$$

$$= \frac{3x}{a}$$

Even the most capable student may become confused by trying to do all of these algebraic operations in one or two steps.

Perform these subtractions as shown above.

(a) $\dfrac{7}{2x} - \dfrac{1}{2x}$ (b) $\dfrac{1}{5a} - \dfrac{x}{5a}$

(c) $\dfrac{x + 2}{y} - \dfrac{2}{y}$ (d) $\dfrac{a + b}{2x} - \dfrac{b - a}{2x}$

Check your work in 15.

---

15   (a) $\dfrac{7}{2x} - \dfrac{1}{2x} = \dfrac{7}{2x} + \dfrac{-1}{2x} = \dfrac{7 + (-1)}{2x} = \dfrac{7 - 1}{2x} = \dfrac{6}{2x} = \dfrac{3}{x}$

Always reduce your answer to lowest terms.

(b) $\dfrac{1}{5a} - \dfrac{x}{5a} = \dfrac{1}{5a} + \dfrac{-x}{5a} = \dfrac{1 + (-x)}{5a} = \dfrac{1 - x}{5a}$

(c) $\dfrac{x + 2}{y} - \dfrac{2}{y} = \dfrac{x + 2}{y} + \dfrac{-2}{y} = \dfrac{x + 2 - 2}{y} = \dfrac{x}{y}$

(d) $\dfrac{a + b}{2x} - \dfrac{b - a}{2x} = \dfrac{a + b}{2x} + \dfrac{-(b - a)}{2x} = \dfrac{a + b - (b - a)}{2x}$

$$= \frac{a + b - b + a}{2x}$$

$$= \frac{2a}{2x} = \frac{a}{x}$$

Notice that in problem (d) it was necessary to insert parentheses to avoid confusion.

In each problem we rewrote the problem as an addition before combining the fractions. The answer should always be reduced to lowest terms.

The following problem set provides practice in adding and subtracting algebraic fractions with like denominators.

1. $\dfrac{x}{2a} - \dfrac{2x}{2a} + \dfrac{3x}{2a}$ 	2. $\dfrac{2x - 1}{3x} + \dfrac{3x + 2}{3x}$

3. $\dfrac{a - 2b}{a + b} - \dfrac{2a + 3b}{a + b}$ 	4. $\dfrac{x - y}{x + 2} + \dfrac{2x - y}{x + 2} + \dfrac{x}{x + 2}$

5. $\dfrac{5p + 2q}{3p - q} + \dfrac{p - q}{3p - q}$ 	6. $\dfrac{7x - 2y}{x + y} - \dfrac{4x - 5y}{x + y}$

7. $\dfrac{a - 2b}{3} + \dfrac{5a - b}{3}$ 	8. $\dfrac{x + y}{2(x - y)} - \dfrac{x - y}{2(x - y)}$

9. $\dfrac{3x - 1}{x^2 + 3x + 2} - \dfrac{x - 5}{x^2 + 3x + 2}$ 	10. $\dfrac{2x + y}{x^2 - 1} - \dfrac{2 + y}{x^2 - 1}$

When you have completed these problems, check your answers in **16**.

---

16 1. $\dfrac{x}{a}$ 	2. $\dfrac{5x + 1}{3x}$ 	3. $\dfrac{-a - 5b}{a + b}$ 	4. $\dfrac{4x - 2y}{x + 2}$ 	5. $\dfrac{6p + q}{3p - q}$

6. 3 	7. $2a - b$ 	8. $\dfrac{y}{x - y}$ 	9. $\dfrac{2}{x + 1}$ 	10. $\dfrac{2}{x + 1}$

Note that in problems 6, 7, 8, 9, and 10 an extra step was needed to reduce the answer to lowest terms.

Fractions with Unlike Denominators

When the algebraic fractions to be added or subtracted have unlike denominators, the procedure is more difficult. To add algebraic fractions with unlike denominators, first build them up to form equivalent fractions with the same denominator, then add. For example,

$$\frac{x}{2} + \frac{2x}{3}$$

The new denominator is 6. Build up the two fractions to equivalent fractions with a denominator of 6.

$$\frac{x}{2} = \frac{?}{6} \qquad \frac{x \cdot 3}{2 \cdot 3} = \frac{3x}{6}$$

$$\frac{2x}{3} = \frac{?}{6} \qquad \frac{2x \cdot 2}{3 \cdot 2} = \frac{4x}{6}$$

Now add the new fractions 	$\dfrac{x}{2} + \dfrac{2x}{3} = \dfrac{3x}{6} + \dfrac{4x}{6} = \dfrac{7x}{6}$

For many problems the new denominators can be found from your knowledge of arithmetic or beginning algebra. Use as the new denominator the smallest number or simplest algebraic expression that is a multiple of both of the original denominators.

In the sum given, 6 is the smallest number evenly divisible by both denominators 2 and 3. The number 6 is the Least Common Denominator (LCD) of 2 and 3.

Find this sum:

$$\frac{2}{x} + \frac{1}{3x} = \underline{\hspace{4cm}}$$

Check your work in **17**.

MENTAL MATH QUIZ

Write down the answers to the following problems quickly. You should not need to do any paper-and-pencil calculations.

Average time: 2:55
Record time: 1:48

(These times are for students scoring at least 18 correct in a class in beginning algebra in a community college.)

1. $\dfrac{1}{x} + \dfrac{2}{x}$ 2. $\dfrac{2}{a} + \dfrac{3}{a}$ 3. $\dfrac{x}{4} + \dfrac{y}{4}$ 4. $\dfrac{a}{2} + \dfrac{2a}{2}$

5. $\dfrac{2}{3a} + \dfrac{5}{3a}$ 6. $\dfrac{x}{2} + \dfrac{x}{2}$ 7. $\dfrac{5}{7y} + \dfrac{1}{7y}$ 8. $\dfrac{3p}{q} + \dfrac{p}{q}$

9. $\dfrac{3}{x} - \dfrac{1}{x}$ 10. $\dfrac{5t}{4} - \dfrac{2t}{4}$ 11. $\dfrac{3}{xy} - \dfrac{1}{xy}$ 12. $\dfrac{x}{a} - \dfrac{y}{a}$

13. $\dfrac{b}{x} - \dfrac{2}{x}$ 14. $\dfrac{2}{x} - \dfrac{y}{x}$

15. $\dfrac{9}{3z} - \dfrac{2}{3z}$ 16. $\dfrac{4}{x-1} - \dfrac{1}{x-1}$

17. $\dfrac{11k}{t} + \dfrac{3k}{t} - \dfrac{8k}{t}$ 18. $\dfrac{1}{x^2} - \dfrac{5}{x^2} + \dfrac{7}{x^2}$

19. $\dfrac{x}{ab^2} + \dfrac{2x}{ab^2}$ 20. $\dfrac{xy}{ab} + \dfrac{x}{ab} - \dfrac{y}{ab}$

The correct answers are on page 559.

---

17 Build up the first fraction to an equivalent fraction with denominator equal to $3x$.

$$\frac{2}{x} = \frac{?}{3x} \qquad \frac{2 \cdot \boxed{3}}{x \cdot \boxed{3}} = \frac{6}{3x}$$

Add $\dfrac{2}{x} + \dfrac{1}{3x} = \dfrac{6}{3x} + \dfrac{1}{3x} = \dfrac{6+1}{3x} = \dfrac{7}{3x}$

Try these problems.

(a) $\dfrac{x}{2} + \dfrac{2x}{5}$ (b) $\dfrac{2}{a} + \dfrac{1}{2a}$ (c) $2 + \dfrac{1}{x}$ (d) $\dfrac{1}{x} + \dfrac{1}{y}$

Turn to **18** to check your answers.

18 (a) The Least Common Denominator is 10.

$$\frac{x}{2} = \frac{?}{10} \qquad \frac{x \cdot \boxed{5}}{2 \cdot \boxed{5}} = \frac{5x}{10} \qquad \frac{2x}{5} = \frac{?}{10} \qquad \frac{2x \cdot \boxed{2}}{5 \cdot \boxed{2}} = \frac{4x}{10}$$

$$\frac{x}{2} + \frac{2x}{5} = \frac{5x}{10} + \frac{4x}{10} = \frac{9x}{10}$$

(b) The least common denominator is $2a$.

$$\frac{2}{a} = \frac{?}{2a} \qquad \frac{2 \cdot \boxed{2}}{a \cdot \boxed{2}} = \frac{4}{2a}$$

$$\frac{2}{a} + \frac{1}{2a} = \frac{4}{2a} + \frac{1}{2a} = \frac{5}{2a}$$

(c) The least common denominator is x.

$$2 = \frac{?}{x} \qquad \frac{2}{1} = \frac{?}{x} \qquad \frac{2 \cdot \boxed{x}}{1 \cdot \boxed{x}} = \frac{2x}{x}$$

$$2 + \frac{1}{x} = \frac{2x}{x} + \frac{1}{x} = \frac{2x+1}{x}$$

(d) The least common denominator is xy.

$$\frac{1}{x} = \frac{?}{xy} \qquad \frac{1 \cdot \boxed{y}}{x \cdot \boxed{y}} = \frac{y}{xy}$$

$$\frac{1}{y} = \frac{?}{xy} \qquad \frac{1 \cdot \boxed{x}}{y \cdot \boxed{x}} = \frac{x}{xy}$$

$$\frac{1}{x} + \frac{1}{y} = \frac{y}{xy} + \frac{x}{xy} = \frac{y+x}{xy}$$

LCD of Algebraic Fractions

The least common denominator of two algebraic fractions is not always very obvious or easy to find. In many problems a special method must be followed. For example, in the sum

$$\frac{3}{2x} + \frac{2}{x^2 y}$$

the denominators are not the same and the least common denominator is not obvious to most students. To find the least common denominator follow these steps.

Step 1. Write each denominator as a product of its factors.

Example: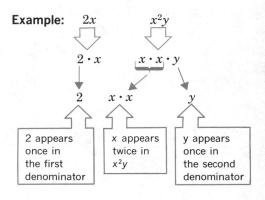

Step 2. Write each factor again, repeating it as many times as the most number of times it appears in either denominator.

Step 3. Multiply these factors to find the least common denominator.

$$LCD = 2 \cdot x \cdot x \cdot y = 2x^2y$$

Finally, use this number as the new denominator and build fractions equivalent to the original fractions and add them as before.

$$\frac{3}{2x} = \frac{?}{2x^2y} \qquad \frac{3 \cdot \boxed{xy}}{2x \cdot \boxed{xy}} = \frac{3xy}{2x^2y}$$

$$\frac{2}{x^2y} = \frac{?}{2x^2y} \qquad \frac{2 \cdot \boxed{2}}{x^2y \cdot \boxed{2}} = \frac{4}{2x^2y}$$

Add $\qquad \dfrac{3}{2x} + \dfrac{2}{x^2y} = \dfrac{3xy}{2x^2y} + \dfrac{4}{2x^2y} = \dfrac{3xy + 4}{2x^2y}$

The process can be long and tricky. Assume that the following expressions are denominators and practice finding the least common denominators of each pair.

(a) a^2b and ab^3
(b) $4xy$ and $6x^2z$
(c) $x(x + 2)$ and $(x + 2)^2$
(d) $x^2 - 1$ and $x(x - 1)$
(e) $x^2 + 4x + 3$ and $x^2 + 2x + 1$
(f) p^2 and pq and pq^2

Follow the three steps above to find each least common denominator, then turn to **19** to check your work.

19 (a)

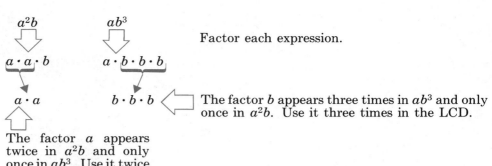

$$\text{LCD} = a \cdot a \cdot b \cdot b \cdot b = a^2b^3$$

(b)

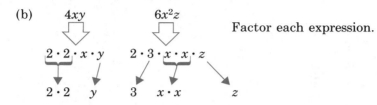

Write each factor as many times as it appears in that expression in which it appears most.

$$\text{LCD} = 2 \cdot 2 \cdot 3 \cdot x \cdot x \cdot y \cdot z = 12x^2yz$$

(c) $x(x + 2)$ $(x + 2)^2$

$x \cdot (x + 2)$ $(x + 2) \cdot (x + 2)$

$$\text{LCD} = x \cdot (x + 2) \cdot (x + 2) = x(x + 2)^2$$

(d) $x^2 - 1$ $x(x - 1)$

$(x + 1) \cdot (x - 1)$ $x \cdot (x - 1)$

$$\text{LCD} = (x + 1) \cdot (x - 1) \cdot x = x(x + 1)(x - 1)$$

(e) $x^2 + 4x + 3$ $x^2 + 2x + 1$

$(x + 1) \cdot (x + 3)$ $(x + 1) \cdot (x + 1)$

$$\text{LCD} = (x + 3)(x + 1)(x + 1)$$

(f) p^2 pq pq^2

$p \cdot p$ $p \cdot q$ $p \cdot q \cdot q$

$$\text{LCD} = p \cdot p \cdot q \cdot q = p^2q^2$$

Now apply your skill at finding the LCD to finding the following sums and differences.

(a) $\dfrac{a}{xy^2} + \dfrac{2}{2x^2y}$ (b) $\dfrac{1}{4ab} - \dfrac{1}{6a^2c}$

(c) $\dfrac{1}{x - 2} + \dfrac{2}{x^2 - 4}$ (d) $\dfrac{2}{x^2 - x - 6} + \dfrac{1}{x^2 + 4x + 4}$

Look for the answers in **20**.

20 (a) The LCD of xy^2 and $2x^2y$ is $2x^2y^2$.

$$\frac{a}{xy^2} = \frac{a \cdot 2x}{xy^2 \cdot 2x} = \frac{2ax}{2x^2y^2}$$

Build up the original fractions to equivalent fractions with the new denominator.

$$\frac{2}{2x^2y} = \frac{2 \cdot y}{2x^2y \cdot y} = \frac{2y}{2x^2y^2}$$

$$\frac{a}{xy^2} + \frac{2}{2x^2y} = \frac{2ax}{2x^2y^2} + \frac{2y}{2x^2y^2} = \frac{2ax + 2y}{2x^2y^2} = \frac{2(ax + y)}{2x^2y^2} = \frac{ax + y}{x^2y^2}$$

(b) The LCD of $4ab$ and $6a^2c$ is $12a^2bc$.

$$\frac{1}{4ab} = \frac{1 \cdot 3ac}{4ab \cdot 3ac} = \frac{3ac}{12a^2bc}$$

$$-\frac{1}{6a^2c} = \frac{-1}{6a^2c} = \frac{-1 \cdot 2b}{6a^2c \cdot 2b} = \frac{-2b}{12a^2bc}$$

$$\frac{1}{4ab} - \frac{1}{6a^2c} = \frac{3ac}{12a^2bc} + \frac{-2b}{12a^2bc} = \frac{3ac - 2b}{12a^2bc}$$

(c) The LCD of $x - 2$ and $x^2 - 4$ is $(x + 2)(x - 2)$.

$$\frac{1}{x - 2} = \frac{1 \cdot (x + 2)}{(x - 2) \cdot (x + 2)} = \frac{(x + 2)}{(x + 2)(x - 2)}$$

$$\frac{2}{x^2 - 4} = \frac{2}{(x - 2)(x + 2)}$$

$$\frac{1}{x - 2} + \frac{2}{x^2 - 4} = \frac{(x + 2)}{(x + 2)(x - 2)} + \frac{2}{(x + 2)(x - 2)}$$

$$= \frac{(x + 2) + 2}{(x + 2)(x - 2)} = \frac{x + 4}{(x + 2)(x - 2)}$$

(d) $x^2 - x - 6 = (x - 3)(x + 2)$ $\qquad$ $x^2 + 4x + 4 = (x + 2)(x + 2)$

The LCD of $(x - 3)(x + 2)$ and $(x + 2)(x + 2)$ is $(x - 3)(x + 2)(x + 2)$.

$$\frac{2}{x^2 - x - 6} = \frac{2 \cdot (x + 2)}{(x - 3)(x + 2) \cdot (x + 2)} = \frac{2(x + 2)}{(x - 3)(x + 2)(x + 2)}$$

$$\frac{1}{x^2 + 4x + 4} = \frac{1 \cdot (x - 3)}{(x + 2)(x + 2) \cdot (x - 3)} = \frac{(x - 3)}{(x + 2)(x + 2)(x - 3)}$$

$$\frac{2}{x^2 - x - 6} + \frac{1}{x^2 + 4x + 4} = \frac{2(x + 2)}{(x - 3)(x + 2)(x + 2)} + \frac{(x - 3)}{(x - 3)(x + 2)(x + 2)}$$

$$= \frac{2x + 4 + x - 3}{(x - 3)(x + 2)(x + 2)} = \frac{3x + 1}{(x - 3)(x + 2)(x + 2)}$$

Now you should be ready for a set of practice problems on the addition and subtraction of algebraic fractions. Turn to **21**.

Unit 5
Fractions
and Radicals

21   The answers are on page 559.

Adding and
Subtracting Algebraic
Fractions

A. Add:

1. $\dfrac{x}{7} + \dfrac{3x}{7}$ _____

2. $\dfrac{4}{x} + \dfrac{6}{x}$ _____

3. $\dfrac{a}{3x} + \dfrac{2a}{3x}$ _____

4. $\dfrac{2}{5y} + \dfrac{3a}{5y}$ _____

5. $\dfrac{y}{x} + \dfrac{y}{x}$ _____

6. $\dfrac{y}{x^2} + \dfrac{4}{x^2}$ _____

7. $\dfrac{y}{y-1} + \dfrac{3y}{y-1}$ _____

8. $\dfrac{2c+4}{5c} + \dfrac{4c+3}{5c}$ _____

9. $\dfrac{x-3y}{xy} + \dfrac{6x-y}{xy}$ _____

10. $\dfrac{x+4y}{x-y} + \dfrac{3x}{x-y}$ _____

11. $\dfrac{3-2t}{2t^2} + \dfrac{5t}{2t^2}$ _____

12. $\dfrac{9a+4}{4-y} + \dfrac{6a-9}{4-y}$ _____

13. $\dfrac{5t+3}{3(p+2)} + \dfrac{4t}{3(p+2)}$ _____

14. $\dfrac{a-x}{x-b} + \dfrac{a+x}{x-b}$ _____

15. $\dfrac{3a+b}{a^2+a+1} + \dfrac{3a-2b}{a^2+a+1}$ _____

16. $\dfrac{7y}{y-2x} + \dfrac{3x}{y-2x}$ _____

17. $\dfrac{7-w}{w-2} + \dfrac{3w}{w-2} + \dfrac{3-2w}{w-2}$ _____

18. $\dfrac{6-x}{x^2+5} + \dfrac{5x-2}{x^2+5}$ _____

19. $\dfrac{2x+7y}{x+y} + \dfrac{x-4y}{x+y} + \dfrac{-3x-3y}{x+y}$ _____

20. $\dfrac{a}{x} + \dfrac{b}{x} + \dfrac{ab}{x} + \dfrac{2a}{x}$ _____

21. $\dfrac{2}{y} + \dfrac{x}{y} + \dfrac{2x}{y}$ _____

22. $\dfrac{x}{x+1} + \dfrac{1}{x+1}$ _____

23. $\dfrac{3x}{y-1} + \dfrac{4}{y-1} + \dfrac{x}{y-1}$ _____

B. Subtract:

1. $\dfrac{5}{3b} - \dfrac{6}{3b}$ _____

2. $\dfrac{4x}{7} - \dfrac{8x}{7}$ _____

3. $\dfrac{9a}{x} - \dfrac{3a}{x}$ _____

4. $\dfrac{a-b}{a} - \dfrac{a+b}{a}$ _____

5. $\dfrac{2y-4}{y} - \dfrac{y-4}{y}$ _____

6. $\dfrac{a}{x} - \dfrac{b}{x}$ _____

7. $\dfrac{3p+q}{pq} - \dfrac{p-2q}{pq}$ _____

8. $\dfrac{5a-9}{3(a-2)} - \dfrac{4a}{3(a-2)}$ _____

9. $\dfrac{7a+4c}{a-c} - \dfrac{6a-4c}{a-c}$ _____

10. $\dfrac{5x}{x+2y} - \dfrac{4x-7}{x+2y}$ _____

11. $\dfrac{13-x}{x} - \dfrac{10+x}{x}$ _____

12. $\dfrac{3w}{wx} - \dfrac{5w-7}{wx}$ _____

13. $\dfrac{3x-2}{x+1} - \dfrac{x+4}{x+1}$ _____

14. $\dfrac{9a-3}{2(a+b)} - \dfrac{5a+7}{2(a+b)}$ _____

15. $\dfrac{7+x}{x^2+3x+2} - \dfrac{4+2x}{x^2+3x+2}$ _____

16. $\dfrac{3x-5y}{2x+2y} - \dfrac{x+7y}{2x+2y}$ _____

17. $\dfrac{5p-2q}{p+q} - \dfrac{p-q}{p+q}$ _____

18. $\dfrac{7y-15}{3y-4} - \dfrac{7y-6}{3y-4}$ _____

19. $\dfrac{x-4}{x+2} - \dfrac{4-x}{x+2} + \dfrac{2x}{x+2}$ _____

20. $\dfrac{a+2b}{a^2} - \dfrac{3a+4b}{a^2} + \dfrac{4a+b}{a^2}$ _____

21. $\dfrac{2}{x} - \dfrac{a}{x} + \dfrac{2a}{x}$ _____

22. $\dfrac{1}{y} - \dfrac{x}{y} - \dfrac{3x}{y}$ _____

23. $\dfrac{x}{x+1} - \dfrac{1}{x+1} - \dfrac{2x}{x+1}$ _____

24. $\dfrac{2y}{y+4} - \dfrac{1}{y+4} + \dfrac{3}{y+4}$ _____

C. Add:

1. $\dfrac{2}{a} + \dfrac{3}{b}$ _____

2. $\dfrac{3}{4} + \dfrac{5}{x}$ _____

3. $\dfrac{4}{p^2} + \dfrac{3}{p}$ _____

4. $\dfrac{2}{2x} + \dfrac{2}{x}$ _____

5. $\dfrac{b}{a} + \dfrac{a}{b}$ _____

6. $\dfrac{2}{x} + \dfrac{5}{2y}$ _____

7. $\dfrac{6x}{15x} + \dfrac{5}{x}$ _____

8. $\dfrac{12}{3a} + \dfrac{6}{3b}$ _____

9. $\dfrac{3}{x^2} + \dfrac{x}{2x}$ _____

10. $\dfrac{3t}{5t^2} + \dfrac{7}{t}$ _____

11. $\dfrac{5}{cd^2} + \dfrac{1}{c^2d}$ _____

12. $\dfrac{1}{6x^2} + \dfrac{2}{12x}$ _____

13. $\dfrac{11t}{st} + \dfrac{2s}{s^2t^2}$ _____

14. $\dfrac{5d}{6} + \dfrac{2d}{9} + \dfrac{d}{3}$ _____

15. $\dfrac{x}{13} + \dfrac{2x}{26}$ _____

16. $\dfrac{5a}{2b} + \dfrac{4a}{7bc}$ _____

17. $\dfrac{4}{r^2} + \dfrac{3}{r}$ _____

18. $\dfrac{4}{2abc} + \dfrac{3}{bc}$ _____

19. $\dfrac{3x-5}{8} + \dfrac{4x-3}{6}$ _____

20. $\dfrac{w+2}{3w} + \dfrac{w-2}{5w}$ _____

21. $\dfrac{1}{x} + \dfrac{1}{x-y}$ _____

22. $\dfrac{6}{x-4} + \dfrac{2}{3x+2}$ _____

23. $\dfrac{1}{a+b} + \dfrac{1}{a-b}$ _____

24. $\dfrac{2}{x(x-3)} + \dfrac{4}{x-3}$ _____

25. $\dfrac{2x}{x+y} + \dfrac{2y}{x-y}$ _____

26. $\dfrac{x-y}{x+y} + \dfrac{x+y}{x-y}$ _____

27. $\dfrac{1}{2x+6} + \dfrac{2}{x^2-9}$ _____

28. $4 + \dfrac{1}{2x}$ _____

29. $1 + \dfrac{1}{x}$ _____

30. $\dfrac{1}{1+x} + x + 1$ _____

31. $a + \dfrac{2a+4}{a-2}$ _____

32. $\dfrac{x}{x-3} + \dfrac{2x}{3-x}$ _____

33. $\dfrac{5}{a+3} + \dfrac{3}{2a+6}$ _____

Date

Name

Course/Section

34. $\dfrac{a+4}{4a} + \dfrac{6}{2a+2}$ _____

35. $\dfrac{2x}{x^2-5x+6} + \dfrac{x-1}{2x^2-7x+3}$ _____

36. $\dfrac{1}{2(x-2)} + \dfrac{2}{x^3-x^2-2x}$ _____

37. $\dfrac{x}{y} + \dfrac{x}{y^2} + \dfrac{x}{y^3}$ _____

38. $\dfrac{x}{2x} + \dfrac{3x+1}{4x-4} + \dfrac{1}{x^2-x}$ _____

39. $\dfrac{5a}{a^2+a-2} + \dfrac{4a}{a^2-a-6}$ _____

40. $\dfrac{x-1}{x^2+2x} + \dfrac{1}{x} + \dfrac{2}{x+2}$ _____

41. $\dfrac{1}{x} + \dfrac{2}{x+2} + \dfrac{3}{x+3}$ _____

42. $\dfrac{2}{y} + \dfrac{y}{y-1} + \dfrac{2y}{y-2}$ _____

43. $\dfrac{2}{x} + \dfrac{1}{x^2-5x}$ _____

44. $\dfrac{3}{4x-x^2} + \dfrac{1}{2x}$ _____

45. $\dfrac{1}{y} + \dfrac{1}{y-5} + \dfrac{1}{y^2-4y-5}$ _____

46. $\dfrac{2}{a} + \dfrac{1}{a^2} + \dfrac{1}{a^2-a}$ _____

47. $\dfrac{x+y}{x-y} + \dfrac{1}{x} + \dfrac{1}{y}$ _____

48. $\dfrac{x+a}{x-a} + \dfrac{a}{x} + \dfrac{x}{a}$ _____

49. $x + 2 + \dfrac{1}{x-1}$ _____

50. $2x + \dfrac{1}{x+3} + 3$ _____

D. Subtract:

1. $\dfrac{2a}{4} - \dfrac{a}{8}$ _____

2. $\dfrac{6x}{9} - \dfrac{4x}{12}$ _____

3. $\dfrac{xy}{5} - \dfrac{xy}{9}$ _____

4. $\dfrac{t}{4} - \dfrac{p}{5}$ _____

5. $\dfrac{x}{10} - \dfrac{y}{3}$ _____

6. $\dfrac{x}{a} - \dfrac{a}{y}$ _____

7. $\dfrac{3y}{7} - \dfrac{2y}{21}$ _____

8. $\dfrac{3}{x^2} - \dfrac{4}{x}$ _____

9. $\dfrac{7}{a^3} - \dfrac{4}{a}$ _____

10. $\dfrac{x+2}{6} - \dfrac{3x+4}{4}$ _____

11. $\dfrac{3a+5}{5a} - \dfrac{a-6}{3a}$ _____

12. $\dfrac{2n+1}{10n} - \dfrac{3n-1}{4n}$ _____

13. $6 - \dfrac{3b-5}{8}$ _____

14. $\dfrac{5c-2}{5} - 3c$ _____

15. $9 - \dfrac{x-2}{2x}$ _____

16. $\dfrac{y+4}{3} - \dfrac{2y-1}{5}$ _____

17. $\dfrac{9b-1}{b^2} - \dfrac{b+2}{b} - 7$ _____

18. $\dfrac{9}{a+1} - \dfrac{4}{5a+2}$ _____

19. $\dfrac{2}{d+4} - \dfrac{2}{d-4}$ _____

20. $\dfrac{6}{x^2-16} - \dfrac{3}{x-4}$ _____

21. $\dfrac{3y}{x+y} - \dfrac{5y}{x^2-y^2}$ _____

22. $\dfrac{7}{a^2-ab} - \dfrac{1}{ab-a^2}$ _____

23. $\dfrac{a+b}{a-b} - \dfrac{3a}{4b}$ _____

24. $\dfrac{3x+2}{x^2} + \dfrac{x-4}{x} - \dfrac{2}{3}$ _____

25. $\dfrac{7x}{x-y} - \dfrac{3}{2} + \dfrac{8y}{x+y}$ _____

26. $\dfrac{4t - 3}{5t - 5} - \dfrac{3t}{5} + \dfrac{t - 4}{t - 1}$ _____

27. $\dfrac{5q}{3q - 3} - \dfrac{2}{3q} + \dfrac{3}{q - 1}$ _____

28. $\dfrac{1}{x - y} - \dfrac{1}{x + y} - \dfrac{1}{x}$ _____

29. $x - \dfrac{1}{x + 1}$ _____

30. $\dfrac{5y}{y^2 - 9} - \dfrac{y + 1}{y + 3}$ _____

31. $\dfrac{2}{4x^2 + 6x + 2} - \dfrac{1}{2x^2 - x - 1}$ _____

32. $\dfrac{x - 2}{2x + 10} - \dfrac{2}{x^2 + 4x - 5}$ _____

33. $\dfrac{1}{x^2} - \dfrac{1}{x - 3}$ _____

34. $\dfrac{1}{y - 3} - \dfrac{1}{y - 1}$ _____

35. $\dfrac{x - y}{x + y} - \dfrac{1}{x}$ _____

36. $\dfrac{y}{x} - \dfrac{1}{x - y}$ _____

37. $x - \dfrac{x}{x - 2}$ _____

38. $\dfrac{2y}{y + 2} - y$ _____

39. $\dfrac{1}{x^2 - 8x + 12} - \dfrac{1}{x^2 - 5x - 6}$ _____

40. $\dfrac{2}{y^2 - y - 20} - \dfrac{1}{y^2 + y - 12}$ _____

When you have had the practice you need, either return to the preview on page 335 or turn to frame **22** where you will learn to multiply and divide algebraic fractions.

Date

Name

Course/Section

22 The multiplication of algebraic fractions is a simple, straightforward operation. The product of two fractions is another fraction whose numerator is the product of the original numerators and whose denominator is the product of the original denominators. For example, the product

$$\frac{1}{5} \cdot \frac{3}{2} = \frac{1 \cdot 3}{5 \cdot 2} = \frac{3}{10}$$
Multiply the numerators
Multiply the denominators

With algebraic fractions,

$$\frac{x}{3} \cdot \frac{2}{y} = \frac{2x}{3y}$$
Multiply the numerators
Multiply the denominators

Important Of course, we always reduce the product fraction to lowest terms.

$$\frac{2x^2}{5} \cdot \frac{10a}{6x} = \frac{2x^2 \cdot \overset{2}{\cancel{10}}a}{\cancel{5} \cdot \underset{3}{\cancel{6x}}} = \frac{2ax}{3}$$

Try it. Multiply:

(a) $\dfrac{1}{2} \cdot \dfrac{6x}{5}$ 　　　(b) $\dfrac{8x^3}{7} \cdot \dfrac{1}{4x}$ 　　　(c) $\dfrac{4a}{3y^3} \cdot \dfrac{3ay}{2a}$ 　　　(d) $\dfrac{12x^5}{5xy^3} \cdot \dfrac{15y}{8x^2}$

Check your multiplications in **23**.

23　(a) $\dfrac{1}{2} \cdot \dfrac{6x}{5} = \dfrac{\overset{3}{\cancel{6}}x}{\cancel{2} \cdot 5} = \dfrac{3x}{5}$ 　　　(b) $\dfrac{8x^3}{7} \cdot \dfrac{1}{4x} = \dfrac{\overset{2}{\cancel{8}}x^{\overset{2}{\cancel{3}}}}{7 \cdot \cancel{4x}} = \dfrac{2x^2}{7}$

　　(c) $\dfrac{4a}{3y^3} \cdot \dfrac{3ay}{2a} = \dfrac{\overset{2}{\cancel{4}}a \cdot \cancel{3}a\cancel{y}}{\cancel{3}y^{\overset{2}{\cancel{3}}} \cdot \cancel{2}\cancel{a}} = \dfrac{2a}{y^2}$ 　　(d) $\dfrac{\overset{3}{\cancel{12}}x^5}{\cancel{5}xy^3} \cdot \dfrac{\overset{3}{\cancel{15}}y}{\underset{2}{\cancel{8}}x^2} = \dfrac{9x^2}{2y^2}$

If one of the expressions to be multiplied is not a fraction, write it as a fraction with denominator equal to 1, and multiply as before.

Example:

$$4 \cdot \frac{2}{9} = \frac{4}{1} \cdot \frac{2}{9} = \frac{8}{9}$$

With algebraic fractions,

Example:

$$2 \cdot \frac{x}{3y} = \frac{2}{1} \cdot \frac{x}{3y} = \frac{2x}{3y}$$
Write as a fraction

Another example:

$$(2x^2y) \cdot \left(\frac{3y^2x}{2ax^3}\right) = \frac{\cancel{2}x^{\overset{}{\cancel{2}}}y}{1} \cdot \frac{3y^2\cancel{x}}{\cancel{2}a\cancel{x^3}} = \frac{3y^3}{a}$$

As usual we reduce the answer to lowest terms.

Notice that any dividing or "cancelling" is done before the actual multiplication. This will make the actual multiplication simpler.

Here are a few practice problems.

(a) $\quad 4 \cdot \dfrac{5x^2}{2}$
(b) $\quad 3a^2b \cdot \dfrac{7b^3}{6ab}$
(c) $\quad \dfrac{8x^3y^2}{3xy} \cdot 6x^2$

The answers are in **24**.

NEGATIVE FRACTIONS

When a fraction containing negative signs is written in the form $\dfrac{-a}{b}$ where the numerator is a negative term, the fraction is said to be in standard negative form. It will be very helpful to you to remember that

$$-\frac{a}{b} = \frac{-a}{b} \qquad \text{and} \qquad \frac{a}{-b} = \frac{-a}{b}$$

If a fraction has two negative signs, it can be simplified so that they are eliminated.

$$\frac{-a}{-b} = \frac{a}{b} \qquad \text{and} \qquad -\frac{a}{-b} \qquad \text{and} \qquad -\frac{-a}{b} = \frac{a}{b}$$

If a fraction contains three negative signs it can be rewritten this way:

$$-\frac{-a}{-b} = \frac{-a}{b}$$

Solve the following problems to get this idea firmly implanted in your mathematical muscles.

Write in standard negative form:

1. $\quad -\dfrac{2}{3}$
2. $\quad \dfrac{2x}{-3}$
3. $\quad -\dfrac{5x^2}{3y}$
4. $\quad \dfrac{x+2}{-3x}$

5. $\quad \dfrac{-3y}{-2a}$
6. $\quad -\dfrac{-4x^2}{5y}$
7. $\quad -\dfrac{pq}{-2a}$
8. $\quad -\dfrac{-3x^2}{-11y}$

9. $\quad -\dfrac{x+1}{-(x+2)}$
10. $\quad \dfrac{-2x}{-2x}$
11. $\quad -\dfrac{-pq}{pq}$
12. $\quad -\dfrac{-2t}{(t+1)}$

13. $\quad \dfrac{-1+x}{-x+1}$
14. $\quad -\dfrac{2(x-y)}{-(y-x)}$
15. $\quad -\dfrac{a^2-b^2}{b^2-a^2}$
16. $\quad -\dfrac{-2a^2bc}{-2a^2bc}$

Check your answers on page 561.

---

24   (a) $\quad 4 \cdot \dfrac{5x^2}{2} = \dfrac{\overset{2}{\cancel{4}}}{1} \cdot \dfrac{5x^2}{\cancel{2}} = \dfrac{2 \cdot 5x^2}{1} = 10x^2$

(b) $\quad 3a^2b \cdot \dfrac{7b^3}{6ab} = \dfrac{\cancel{3a^2b}}{1} \cdot \dfrac{7b^3}{\underset{2}{\cancel{6ab}}} = \dfrac{a \cdot 7b^3}{1 \cdot 2} = \dfrac{7ab^3}{2}$

(c) $\quad \dfrac{8x^3y^2}{3xy} \cdot 6x^2 = \dfrac{8x^3y^2}{\cancel{3xy}} \cdot \dfrac{\overset{2}{\cancel{6x^2}}}{1} = \dfrac{8x^3y \cdot 2x}{1} = 16x^4y$

After reducing to lowest terms by dividing out the common factors, simplify the answer by multiplying like quantities, and rearranging the factors. Notice in (b) for example, that we write the answer as $\dfrac{7ab^3}{2}$ rather than $\dfrac{a7b^3}{2}$ or $\dfrac{7b^3a}{2}$. When writing a multiplication of this sort, the convention is that the integer number part comes first and the variables follow in alphabetical order. No special reason for this, except that it keeps things neat and tidy.

Negative Signs on Fractions

If either of the terms of the fraction, or the fraction itself, are negative, rewrite the fraction so that the negative sign is attached to the numerator, and then multiply as before. For example,

$$-\frac{a}{b} \cdot \frac{x}{3} = \frac{-a}{b} \cdot \frac{x}{3} = \frac{(-a) \cdot x}{b \cdot 3} = \frac{-ax}{3b}$$

Rewrite the fraction

or $\quad \dfrac{2}{-x} \cdot \dfrac{x^2}{3y} = \dfrac{-2}{x} \cdot \dfrac{x^2}{3y} = \dfrac{(-2) \cdot x^2}{x \cdot 3y} = \dfrac{-2x}{3y}$

or $\quad -\dfrac{3b}{-a} \cdot \dfrac{a^3}{6b^2} = \dfrac{3b}{a} \cdot \dfrac{a^3}{6b^2} = \dfrac{3b \cdot a^3}{a \cdot 6b^2} = \dfrac{a^2}{2b}$

 If you find working with negative fractions to be troublesome, be certain to visit the box after frame **23** for help.

Another example:

$$(y - x) = (-1)(x - y)$$

$$-\frac{-3(y - x)}{x - y} = \frac{3(y - x)}{(x - y)} = \frac{-3(x - y)}{(x - y)} = -3$$

In this problem it was necessary to rewrite the numerator before we could reduce the fraction to lowest terms.

Practice with the following set of problems.

(a) $\quad -\dfrac{a}{5} \cdot \dfrac{x}{2}$

(b) $\quad \left(\dfrac{-2x}{3y}\right) \cdot \left(-\dfrac{3y^2}{4x^2}\right)$

(c) $\quad -2 \cdot \dfrac{x}{a}$

(d) $\quad 6 \cdot \left(\dfrac{-a^2}{2bc}\right)$

(e) $\quad \left(-\dfrac{6p^2}{5q}\right) \cdot \left(\dfrac{20q^2}{-2p^3}\right)$

(f) $\quad \left(\dfrac{(x-1)^2(x-2)}{-(x-3)}\right) \cdot \left(\dfrac{-(x-3)^2}{-(x-1)^3}\right)$

Check your work in **25**.

---

25 (a) $\dfrac{-ax}{10}$ (b) $\dfrac{y}{2x}$ (c) $\dfrac{-2x}{a}$

(d) $\dfrac{-3a^2}{bc}$ (e) $\dfrac{12q}{p}$ (f) $\dfrac{-(x-2)(x-3)}{(x-1)}$

Binomials and Trinomials in Fractions

When numerator or denominator are binomials or trinomials rather than monomials, it is usually helpful to factor them before multiplying. This will help you to find any common factors that may be eliminated before the multiplication is performed. For example,

$$\frac{3x - 6}{x^2 + x - 6} \cdot \frac{x^2 + 6x + 9}{x^2 - 1} = \frac{3(x - 2)}{(x - 2)(x + 3)} \cdot \frac{(x + 3)^2}{(x + 1)(x - 1)} = \frac{3(x + 3)}{(x + 1)(x - 1)}$$

The ability to factor polynomials, a skill you learned in Unit 3, is very important here. Always factor numerator and denominator before multiplying algebraic fractions.

 Be very careful when simplifying fractions by "cancelling." To "cancel" means to divide the entire numerator and the entire denominator by the factor being "cancelled."

Multiply these algebraic fractions as shown above.

(a) $\dfrac{3x - 6}{2x - 2} \cdot \dfrac{3x - 1}{6x - 12}$

(b) $\dfrac{3x + 9}{3y} \cdot \dfrac{4y}{2x + 6}$

(c) $\dfrac{x - 3}{4 - x^2} \cdot \dfrac{x^2 + 3x + 2}{x^2 + x - 12}$

(d) $\dfrac{x^2 - y^2}{x^2 + 3xy + 2y^2} \cdot \dfrac{2x + 4y}{x^2 - xy}$

Look for our answers in **26**.

26 (a) $\dfrac{3x - 6}{2x - 2} \cdot \dfrac{3x - 1}{6x - 12} = \dfrac{3(x - 2)}{2(x - 1)} \cdot \dfrac{3x - 1}{6(x - 2)}$ Factor all terms completely.

$$= \dfrac{\cancel{3}(\cancel{x - 2})(3x - 1)}{2(x - 1)\underset{2}{\cancel{6}}(\cancel{x - 2})}$$ Simplify by dividing out the common factors.

$$= \dfrac{3x - 1}{4(x - 1)}$$ Multiply.

(b) $\dfrac{3x + 9}{3y} \cdot \dfrac{4y}{2x + 6} = \dfrac{3(x + 3)}{3y} \cdot \dfrac{4y}{2(x + 3)}$ Factor all terms completely.

$$= \dfrac{\cancel{3}(\cancel{x + 3})(\cancel{4}y)}{(\cancel{3y})\cancel{2}(\cancel{x + 3})}$$ Simplify by dividing out the common factors.

$$= 2$$

(c) $\dfrac{x - 3}{4 - x^2} \cdot \dfrac{x^2 + 3x + 2}{x^2 + x - 12} = \dfrac{x - 3}{(2 - x)(2 + x)} \cdot \dfrac{(x + 2)(x + 1)}{(x + 4)(x - 3)}$ Factor all terms.

$$= \dfrac{(\cancel{x - 3})(\cancel{x + 2})(x + 1)}{(2 - x)(\cancel{2 + x})(x + 4)(\cancel{x - 3})}$$ Simplify by dividing out the common factors.

$$= \dfrac{x + 1}{(2 - x)(x + 4)}$$

(d) $\dfrac{x^2 - y^2}{x^2 + 3xy + 2y^2} \cdot \dfrac{2x + 4y}{x^2 - xy} = \dfrac{(x - y)(x + y)}{(x + 2y)(x + y)} \cdot \dfrac{2(x + 2y)}{x(x - y)}$ Factor all terms.

$$= \dfrac{(\cancel{x - y})(\cancel{x + y})2(\cancel{x + 2y})}{(\cancel{x + 2y})(\cancel{x + y})x(\cancel{x - y})}$$ Simplify by dividing out the common factors.

$$= \dfrac{2}{x}$$

Division

Division is the reverse operation to multiplication. The division $a \div b$ means that we are to find a number c such that $bc = a$.

$$a \div b = \frac{a}{b} = (a) \cdot \left(\frac{1}{b}\right)$$

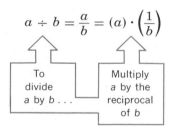

To divide a by b ...

Multiply a by the reciprocal of b

Reciprocals

The reciprocal of any fraction can be found by simply "inverting" the fraction. See page 345 for more on reciprocals if you need a review.

To divide arithmetic fractions:

Invert the divisor

$$\frac{2}{3} \div \frac{5}{6} = \frac{2}{3} \cdot \frac{6}{5} = \frac{2 \cdot \overset{2}{\cancel{6}}}{\cancel{3} \cdot 5} = \frac{4}{5}$$

Multiply

To divide algebraic fractions:

Invert

$$\frac{x}{2} \div \frac{a}{b} = \frac{x}{2} \cdot \frac{b}{a} = \frac{x \cdot b}{2 \cdot a} = \frac{bx}{2a}$$

Multiply

To divide by a fraction, multiply by its reciprocal.

Try it. Perform the following divisions using this "invert and multiply" rule.

(a) $\frac{1}{4} \div \frac{x}{y}$

(b) $2 \div \frac{x}{3}$

(c) $y \div \frac{1}{3}$

(d) $\frac{2}{x} \div 4$

(e) $\frac{ax}{b} \div \frac{x}{b}$

(f) $\frac{2a}{3x} \div \frac{4a}{6b}$

The step-by-step solutions are in **27**.

Where did the invert and multiply rule come from?

We are not really certain. Math historians do know that a ninth century Indian mathematician, Mahavira, used the rule, but it was probably an old idea even then.

27　　(a)　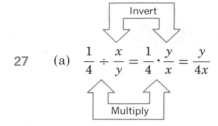

$$\frac{1}{4} \div \frac{x}{y} = \frac{1}{4} \cdot \frac{y}{x} = \frac{y}{4x}$$

(b)　$2 \div \dfrac{x}{3} = \dfrac{2}{1} \div \dfrac{x}{3} = \dfrac{2}{1} \cdot \dfrac{3}{x} = \dfrac{6}{x}$

Write
as a
fraction
first

(c)　$y \div \dfrac{1}{3} = \dfrac{y}{1} \div \dfrac{1}{3} = \dfrac{y}{1} \cdot \dfrac{3}{1} = 3y$

(d)　$\dfrac{2}{x} \div 4 = \dfrac{2}{x} \div \dfrac{4}{1} = \dfrac{2}{x} \cdot \dfrac{1}{4} = \dfrac{1}{2x}$　　　Always reduce the answer to lowest terms.

Write
as a
fraction
first

(e)　$\dfrac{ax}{b} \div \dfrac{x}{b} = \dfrac{a\cancel{x}}{\cancel{b}} \cdot \dfrac{\cancel{b}}{\cancel{x}} = a$

(f)　$\dfrac{2a}{3x} \div \dfrac{4a}{6b} = \dfrac{2\cancel{a}}{\cancel{3}x} \cdot \dfrac{\cancel{6}b}{\cancel{4}\cancel{a}} = \dfrac{b}{x}$

Negative Signs　　If either of the fractions in the division has a negative sign attached or if either the
on Fractions　　numerator or denominator is negative, write the fraction in standard negative form
　　　　　　　　　first before dividing.

Example:

Put the fraction in standard
negative form with the sign
on the numerator

$$-\frac{ax}{2} \div 2a^2 = \frac{-ax}{2} \div \frac{2a^2}{1}$$

$$= \frac{-ax}{2} \cdot \frac{1}{2a^2}$$　　　Invert the divisor and multiply.

$$= \frac{-\cancel{a}x}{4a^{\cancel{2}}}$$

$$= \frac{-x}{4a}$$　　　Multiply and reduce to lowest terms.

Another example:

$$3p \div \left(-\frac{p^2}{2}\right) = \frac{3p}{1} \cdot \left(-\frac{2}{p^2}\right)$$　　　Invert the divisor.

$$= \frac{3\cancel{p}}{1} \cdot \frac{-2}{p^{\cancel{2}}}$$　　　Put the fraction in standard negative form.

$$= \frac{-6}{p}$$

Try these few problems involving the division of fractions with negative signs.

(a) $\quad -\dfrac{2y}{3x} \div \dfrac{2x^2y}{xy}$

(b) $\quad \dfrac{5a^2b^2}{3} \div \dfrac{a^3}{-6b}$

(c) $\quad (-2x^3) \div \left(-\dfrac{4x^2}{y}\right)$

(d) $\quad \dfrac{4ab^2c}{-3} \div \dfrac{2a^2c}{b}$

Check your work in **28**.

28

(a) $\quad -\dfrac{2y}{3x} \div \dfrac{2x^2y}{xy} = \dfrac{-2y}{3x} \div \dfrac{2x^2y}{xy}$ Write in standard negative form.

$\quad = \dfrac{-2\cancel{y}}{3\cancel{x}} \cdot \dfrac{\cancel{x}y}{2x^2\cancel{y}}$ Invert the divisor and multiply.

$\quad = \dfrac{-y}{3x^2}$

(b) $\quad \dfrac{5a^2b^2}{3} \div \dfrac{a^3}{-6b} = \dfrac{5a^2b^2}{\cancel{3}} \cdot \dfrac{-\overset{2}{\cancel{6}}b}{a^3}$ Invert the divisor and multiply.

$\quad = \dfrac{-10b^3}{a}$

(c) $\quad (-2x^3) \div \left(-\dfrac{4x^2}{y}\right) = \dfrac{-2x^3}{1} \div \dfrac{-4x^2}{y}$ Put the fraction in standard negative form.

$\quad = \dfrac{-\cancel{2}x^{\cancel{3}}}{1} \cdot \dfrac{y}{-\underset{2}{\cancel{4}}\cancel{x^2}}$ Invert the divisor and multiply.

$\quad = \dfrac{xy}{2}$

(d) $\quad \dfrac{4ab^2c}{-3} \div \dfrac{2a^2c}{b} = \dfrac{-4ab^2c}{3} \div \dfrac{2a^2c}{b}$ Put in standard negative form.

$\quad = \dfrac{-\overset{2}{\cancel{4}}\cancel{a}b^2\cancel{c}}{3} \cdot \dfrac{b}{\cancel{2}a^{\cancel{2}}\cancel{c}}$ Invert and multiply.

$\quad = \dfrac{-2b^3}{3a}$

Fractions with Binomials or Trinomials

If either of the fractions to be divided contains binomial or trinomial expressions, invert the divisor as usual, then factor each expression before multiplying.

Example:

$\dfrac{3xy + x}{2y^2} \div \dfrac{3y + 1}{y} = \dfrac{3xy + x}{2y^2} \cdot \dfrac{y}{3y + 1}$ Invert the divisor and multiply.

$\quad = \dfrac{x(\cancel{3y + 1})}{2y^{\cancel{2}}} \cdot \dfrac{\cancel{y}}{\cancel{3y + 1}}$ Factor.

$\quad = \dfrac{x}{2y}$

Another example:

$\dfrac{x^2 + 2x - 8}{x^2 - 9} \div \dfrac{4x - 8}{2x + 6} = \dfrac{x^2 + 2x - 8}{x^2 - 9} \cdot \dfrac{2x + 6}{4x - 8}$ Invert the divisor.

$\quad = \dfrac{(x + 4)(\cancel{x - 2})}{(x - 3)(\cancel{x + 3})} \cdot \dfrac{\overset{}{\cancel{2}}(\cancel{x + 3})}{\underset{2}{\cancel{4}}(\cancel{x - 2})}$ Factor each expression.

$\quad = \dfrac{x + 4}{2(x - 3)}$

Test your understanding with the following problems.

(a) $\dfrac{x^2 - xy}{xy} \div \dfrac{3x - 3y}{y^2}$ (b) $\dfrac{9x^2 - 3x}{x + 1} \div \dfrac{3x - 1}{2xy^2 + 2y^2}$

(c) $\dfrac{x^2 - 4x - 5}{x^2 - 4} \div \dfrac{x^2 - 3x - 10}{2x^2 - 8}$ (d) $\left(x - \dfrac{1}{x}\right) \div \left(\dfrac{x - 1}{x + 1}\right)$

(e) $\dfrac{x^2 - 5x + 6}{x^2 - 2x - 15} \div \dfrac{x^2 + 5x - 14}{x^2 - 4x - 5}$

Check your work in **29**.

---

29 (a) $\dfrac{x^2 - xy}{xy} \div \dfrac{3x - 3y}{y^2} = \dfrac{x^2 - xy}{xy} \cdot \dfrac{y^2}{3x - 3y}$ Invert the divisor.

$= \dfrac{\cancel{x}(x - y)}{\cancel{x}y} \cdot \dfrac{y^{\cancel{2}}}{3(x - y)}$ Factor.

$= \dfrac{y}{3}$

(b) $\dfrac{9x^2 - 3x}{x + 1} \div \dfrac{3x - 1}{2xy^2 + 2y^2} = \dfrac{9x^2 - 3x}{x + 1} \cdot \dfrac{2xy^2 + 2y^2}{3x - 1}$ Invert the divisor.

$= \dfrac{3x(3x - 1)}{x + 1} \cdot \dfrac{2y^2(x + 1)}{(3x - 1)}$ Factor.

$= 6xy^2$

(c) $\dfrac{x^2 - 4x - 5}{x^2 - 4} \div \dfrac{x^2 - 3x - 10}{2x^2 - 8} = \dfrac{x^2 - 4x - 5}{x^2 - 4} \cdot \dfrac{2x^2 - 8}{x^2 - 3x - 10}$ Invert the divisor.

$= \dfrac{(x + 1)(x - 5)}{(x - 2)(x + 2)} \cdot \dfrac{2(x - 2)(x + 2)}{(x - 5)(x + 2)}$ Factor.

$= \dfrac{2(x + 1)}{x + 2}$

(d) $\left(x - \dfrac{1}{x}\right) \div \left(\dfrac{x - 1}{x + 1}\right) = \dfrac{x^2 - 1}{x} \cdot \dfrac{x + 1}{x - 1}$ Invert the divisor.

$= \dfrac{(x - 1)(x + 1)}{x} \cdot \dfrac{x + 1}{x - 1}$ Factor.

$= \dfrac{(x + 1)(x + 1)}{x} = \dfrac{x^2 + 2x + 1}{x}$ Simplify.

(e) $\dfrac{x^2 - 5x + 6}{x^2 - 2x - 15} \div \dfrac{x^2 + 5x - 14}{x^2 - 4x - 5} = \dfrac{x^2 - 5x + 6}{x^2 - 2x - 15} \cdot \dfrac{x^2 - 4x - 5}{x^2 + 5x - 14}$ Invert the divisor.

$= \dfrac{(x - 2)(x - 3)}{(x + 3)(x - 5)} \cdot \dfrac{(x + 1)(x - 5)}{(x - 2)(x + 7)}$ Factor.

$= \dfrac{(x - 3)(x + 1)}{(x + 3)(x + 7)}$ Multiply.

$= \dfrac{x^2 - 2x - 3}{x^2 + 10x + 21}$

Complex Fractions   A fraction whose numerator or denominator (or both) are themselves fractions is called a *complex fraction*. For example,

$\dfrac{\dfrac{1}{2}}{\dfrac{1}{4}}$ and $\dfrac{\dfrac{2a}{x}}{\dfrac{a^2}{x^2}}$

are complex fractions. Such fractions may look confusing, but they are not. They represent a division written as a fraction. Treat them the same as any other division of fractions.

Examples:

$$\frac{\frac{1}{2}}{\frac{1}{4}} = \frac{1}{2} \div \frac{1}{4} = \frac{1}{2} \cdot \frac{4}{1} = 2 \qquad \text{First, write it as a division, then invert the divisor and multiply.}$$

$$\frac{\frac{2a}{x}}{\frac{a^2}{x^2}} = \frac{2a}{x} \div \frac{a^2}{x^2} = \frac{2\cancel{a}}{\cancel{x}} \cdot \frac{x^{\cancel{2}}}{a^{\cancel{2}}} \qquad \text{First, write it as a division, then invert the divisor and multiply.}$$

$$= \frac{2x}{a} \qquad \text{Simplify by reducing to lowest terms.}$$

The numerator and denominator of a complex fraction may contain expressions that must be simplified before the division can be performed.

Example:

$$\frac{2 + \dfrac{1}{x}}{4 - \dfrac{1}{x^2}} = \frac{\dfrac{2}{1} + \dfrac{1}{x}}{\dfrac{4}{1} - \dfrac{1}{x^2}} \qquad \text{Rewrite the numerators and denominators so that they are algebraic fractions.}$$

$$= \frac{\dfrac{2 \cdot x}{1 \cdot x} + \dfrac{1}{x}}{\dfrac{4 \cdot x^2}{1 \cdot x^2} - \dfrac{1}{x^2}} \qquad \text{Build up to like fractions.}$$

$$= \frac{\dfrac{2x}{x} + \dfrac{1}{x}}{\dfrac{4x^2}{x^2} - \dfrac{1}{x^2}} = \frac{\dfrac{2x + 1}{x}}{\dfrac{4x^2 - 1}{x^2}} \qquad \text{The complex fraction now has fractions for both numerator and denominator.}$$

$$= \frac{2x + 1}{x} \div \frac{4x^2 - 1}{x^2} \qquad \text{Write as a division.}$$

$$= \frac{2x + 1}{x} \cdot \frac{x^2}{4x^2 - 1} \qquad \text{Invert and multiply.}$$

$$= \frac{\cancel{2x + 1}}{\cancel{x}} \cdot \frac{x^{\cancel{2}}}{\cancel{(2x + 1)}(2x - 1)} \qquad \text{Factor.}$$

$$= \frac{x}{2x - 1}$$

Example:

$$\frac{\dfrac{1}{x}}{1 - \dfrac{1}{x}} = \frac{\dfrac{1}{x}}{\dfrac{x - 1}{x}} \qquad \text{Rewrite the denominator as an algebraic fraction.}$$

$$= \frac{1}{x} \div \frac{x - 1}{x} \qquad \text{Write it as a division.}$$

$$= \frac{1}{\cancel{x}} \cdot \frac{\cancel{x}}{x - 1} \qquad \text{Invert the divisor and multiply.}$$

$$= \frac{1}{x - 1} \qquad \text{Simplify.}$$

Example:

$$\frac{1 - \dfrac{2}{x}}{\dfrac{1}{x} - \dfrac{1}{2}} = \frac{\dfrac{x-2}{x}}{\dfrac{2}{2x} - \dfrac{x}{2x}} = \frac{\dfrac{x-2}{x}}{\dfrac{2-x}{2x}}$$ Rewrite both numerator and denominator as algebraic fractions.

$$= \frac{x-2}{x} \div \frac{2-x}{2x}$$ Write it as a division.

$$= \frac{x-2}{x} \cdot \frac{2x}{2-x}$$ Invert the divisor and multiply.

$$= \frac{x-2}{x} \cdot \frac{2x}{(-1)(x-2)}$$ Factor: $2 - x = (-1)(x-2)$.

$$= \frac{2}{-1} = -2$$ Simplify.

Now for a set of practice problems on multiplying and dividing algebraic fractions, turn to 30.

Unit 5
Fractions
and Radicals

The answers are on page 561.

A. Multiply:

1. $\dfrac{3x}{2} \cdot \dfrac{1}{y}$ _____

2. $\dfrac{1}{2} \cdot \dfrac{4a^2}{3}$ _____

3. $\dfrac{5x}{y} \cdot \dfrac{1}{3}$ _____

4. $\dfrac{6ab}{2} \cdot \dfrac{1}{a}$ _____

5. $\dfrac{5}{2x} \cdot \dfrac{4x}{3}$ _____

6. $\dfrac{3y^2}{4} \cdot \dfrac{1}{2xy}$ _____

7. $\dfrac{2a^3}{5} \cdot \dfrac{3}{4a}$ _____

8. $\dfrac{7p}{4} \cdot \dfrac{2}{p^2}$ _____

9. $3 \cdot \dfrac{1}{x}$ _____

10. $5x \cdot \dfrac{x}{3}$ _____

11. $2y^2 \cdot \dfrac{2}{y^2}$ _____

12. $\dfrac{3}{ab} \cdot b^2$ _____

13. $\dfrac{18p^2q}{4pq^3} \cdot \dfrac{2p^2}{9p^2q}$ _____

14. $\dfrac{14abc^2}{2b^2c} \cdot \dfrac{24a^2b}{21ab^2}$ _____

15. $\dfrac{28x^2y}{7t} \cdot \dfrac{2t^2}{3xy^2}$ _____

16. $\dfrac{10x^3y}{xy^2} \cdot \dfrac{2xy^2}{5xy}$ _____

17. $\dfrac{21a^2c^2}{36abc} \cdot \dfrac{4b}{7c^2}$ _____

18. $\dfrac{16x^5y^2}{3xy^4} \cdot \dfrac{y}{8x^2}$ _____

19. $\dfrac{56at^4}{16a^2} \cdot \dfrac{2at}{14t^3}$ _____

20. $\dfrac{13xyz}{4x^2y} \cdot \dfrac{8x^2z^2}{2y^3}$ _____

21. $-2 \cdot \left(\dfrac{2}{x^2}\right)$ _____

22. $x^2y \cdot \left(-\dfrac{2}{x}\right)$ _____

23. $\dfrac{5}{-ab} \cdot \dfrac{a^2}{2}$ _____

24. $\dfrac{1}{y} \cdot \left(\dfrac{3x}{-2}\right)$ _____

25. $\dfrac{8x^2yz^2}{-3y} \cdot \dfrac{6y^3z}{4xz^4}$ _____

26. $\dfrac{32a^4p^5}{3ap^2} \cdot \dfrac{a}{-16p^3}$ _____

27. $-\dfrac{-2ax^3}{3a^3x} \cdot \dfrac{6x^2}{-a^2}$ _____

28. $-\dfrac{5x^5a}{-4xa^3} \cdot \left(-\dfrac{2a^3x}{2a^4}\right)$ _____

29. $\dfrac{3ax^2y}{4a^2} \cdot \dfrac{2y^3}{6xy} \cdot \dfrac{8x^2}{a^3y}$ _____

30. $\dfrac{16pq^2t}{3pt} \cdot \dfrac{p^2q}{2qt} \cdot \dfrac{3pq}{2t^4}$ _____

Date _____

Name _____

Course/Section _____

B. Multiply:

1. $\dfrac{3x-3}{2x+2} \cdot \dfrac{5x+5}{2x-2}$ _____

2. $\dfrac{4x+4}{x^2+x} \cdot \dfrac{x^2}{3x-3}$ _____

3. $\dfrac{5x+10}{2x+2} \cdot \dfrac{3x-3}{6x+12}$ _____

4. $\dfrac{x^3-4x^2}{2x+6} \cdot \dfrac{-2x-6}{x^2+5x}$ _____

5. $\dfrac{a^2 + ab}{a + 2b} \cdot \dfrac{2a + 4b}{ab + b^2}$ _____

6. $\dfrac{4p - 4}{p^2 + ap} \cdot \dfrac{2p - 2a}{p - 1}$ _____

7. $\dfrac{x^3 + x^2}{x^2 - 1} \cdot \dfrac{2x - 2}{x + 2}$ _____

8. $\dfrac{2x^2 + 2x - 24}{2x^2 - 6x} \cdot \dfrac{x^2 - 1}{x^2 + 3x - 4}$ _____

9. $\dfrac{x^2 + 2x - 3}{x^2 - 1} \cdot \dfrac{x^2 + x}{2x + 4}$ _____

10. $\dfrac{x^2 + 3x + 2}{x^2 - 4} \cdot \dfrac{x^2 - 3x + 2}{x^2 - 1}$ _____

11. $\dfrac{2x^2 - 10x}{x^2 - 6x + 5} \cdot \dfrac{x^2 + x - 2}{3x + 9}$ _____

12. $\dfrac{a^2b + ab^2}{a^2 - b^2} \cdot \dfrac{a^2 - 3ab + 2b^2}{2a^2 - 4ab}$ _____

13. $\dfrac{a^3 + 2a^2 + a}{a^2b + 3ab + 2b} \cdot \dfrac{a^2 - a}{a^3 - a}$ _____

14. $\dfrac{x^2 - 2xy + y^2}{y^2 - x^2} \cdot \dfrac{2x + 2y}{x^2 - xy}$ _____

15. $\dfrac{x^3y}{x + y} \cdot \dfrac{xy - y}{x^2 + x} \cdot \dfrac{x^2 + xy}{x^2y - xy}$ _____

16. $\dfrac{2}{x + 1} \cdot \dfrac{3x - 6}{x - 1} \cdot \dfrac{x^3 - x}{2x^2 - 4x}$ _____

17. $\dfrac{x^2y + 2xy}{y + 1} \cdot \dfrac{y^2 - 1}{x} \cdot \dfrac{1}{x^2 + 4x + 4}$ _____

18. $\dfrac{x^2 - 6x - 7}{x^2 + 8x + 12} \cdot \dfrac{x^2 + 5x - 6}{x^2 - 9x + 14}$ _____

19. $\dfrac{x^2 - 3x - 18}{x^2 - 8x + 7} \cdot \dfrac{x^2 - 9x + 14}{x^2 + 5x + 6}$ _____

20. $\dfrac{3x^3y - 21x^2y}{2x^2y^2 - 10xy^2} \cdot \dfrac{x^2 - 13x + 40}{x^2 - 3x - 28}$ _____

21. $\dfrac{2a^2b - 2ab}{3ab^3 - 3ab^2} \cdot \dfrac{b^2 - 3b + 2}{a^2 + a - 2}$ _____

C. Divide:

1. $\dfrac{a}{b} \div \dfrac{a}{b}$ ____

2. $\dfrac{x}{2} \div \dfrac{1}{x}$ ____

3. $\dfrac{x}{y} \div \dfrac{y}{x}$ ____

4. $\dfrac{a^2}{b} \div \dfrac{a}{b}$ ____

5. $\dfrac{x^2}{y} \div x^3$ ____

6. $2 \div \dfrac{1}{x}$ ____

7. $\dfrac{a}{b} \div 3$ ____

8. $4 \div \dfrac{2}{p}$ ____

9. $\dfrac{3}{2x} \div \dfrac{x}{6}$ ____

10. $\dfrac{p}{2} \div \dfrac{3p^2}{4}$ ____

11. $a^2 \div \dfrac{1}{a}$ ____

12. $\dfrac{1}{a} \div a^2$ ____

13. $\dfrac{1}{y} \div \dfrac{2}{y^2}$ ____

14. $\dfrac{2ax^2}{3b} \div \dfrac{a^2x}{3}$ ____

15. $\dfrac{3a}{2b} \div \dfrac{6c}{4d}$ ____

16. $\dfrac{x^2}{y^2} \div \dfrac{x}{y^2}$ ____

17. $\dfrac{24}{7ab^2} \div \dfrac{18a}{14b}$ ____

18. $\dfrac{x}{y^2} \div \dfrac{xy}{2y}$ ____

19. $2x^2 \div \dfrac{x}{y}$ ____

20. $18t^2 \div \dfrac{t^2}{2}$ ____

21. $-\dfrac{2a}{5x} \div \dfrac{a}{x^2}$ ____

22. $\dfrac{b^3}{a} \div \dfrac{ab}{-a^2}$ ____

23. $\dfrac{x^3y^2}{-x} \div \dfrac{y^3}{x}$ ____

24. $\dfrac{pq}{p^2} \div \dfrac{2q^2}{-p}$ ____

25. $\dfrac{a}{b} \div \dfrac{-c}{d}$ ____

26. $\dfrac{a}{-b} \div \left(-\dfrac{c}{d}\right)$ ____

27. $-\dfrac{a}{-b} \div \left(\dfrac{c}{-d}\right)$ ____

28. $\dfrac{a}{b} \div \left(\dfrac{-c}{-d}\right)$ _____

29. $\dfrac{12p^2q}{pq} \div \dfrac{3pq^2}{q}$ _____

30. $\dfrac{2ab}{x^2y} \div \dfrac{a^3b}{xy}$ _____

31. $\dfrac{3x}{y^3} \div \dfrac{xy^2}{y^2}$ _____

32. $\dfrac{21x^2y^3}{10x^2y} \div \dfrac{7xy^2}{2xy}$ _____

33. $\dfrac{2x}{y} \div \dfrac{x^2y}{2}$ _____

34. $\dfrac{x^2}{y^2} \div \dfrac{x}{y}$ _____

35. $\dfrac{x^2}{y^2} \div \dfrac{y}{x}$ _____

36. $\dfrac{36a^2b^3}{15ab^2} \div \dfrac{12a}{10b^2}$ _____

37. $\dfrac{x^2y}{xy^2} \div \dfrac{1}{x^2y^2}$ _____

38. $\dfrac{2xy^3}{x^2y} \div xy^2$ _____

39. $\dfrac{x^2y}{3x} \div 3x^3$ _____

40. $8x^2y^3 \div \dfrac{1}{2x}$ _____

41. $12ab^3c^2 \div \dfrac{3}{a^2b}$ _____

42. $\dfrac{1}{3x^2} \div \dfrac{1}{4x}$ _____

D. Divide:

1. $\dfrac{x+1}{x+2} \div \dfrac{x+3}{x+2}$ _____

2. $\dfrac{3x-3}{ax+2a} \div \dfrac{2x-2}{x+2}$ _____

3. $\dfrac{x^2-xy}{3x-3} \div \dfrac{xy-y^2}{6x+6}$ _____

4. $\dfrac{x^2-ax}{ax^2} \div \dfrac{2x-2a}{x-4}$ _____

5. $\dfrac{12x^2-3x}{x^2+x} \div \dfrac{4x^2-x}{2x+2}$ _____

6. $\dfrac{ax-ay}{bx+by} \div \dfrac{cx-cy}{dx+dy}$ _____

7. $\dfrac{x^2-1}{x^2-4} \div \dfrac{x-1}{x-4}$ _____

8. $\dfrac{1-a^2}{1-b^2} \div \dfrac{a+1}{b+1}$ _____

9. $\dfrac{x^2+3x+2}{x^2+x-2} \div \dfrac{x^2+2x+1}{x^2-1}$ _____

10. $\dfrac{2x^2-5x-3}{x^2+x-12} \div \dfrac{6x^2+3x}{2x^2+16x+32}$ _____

11. $\dfrac{x^2+8x+15}{2x^2+7x+3} \div \dfrac{x^2+x-20}{4x^2-1}$ _____

12. $\dfrac{p^2+6p-7}{p^2+3p-4} \div \dfrac{p^2+7p}{p^2+6p+8}$ _____

13. $\dfrac{a^2-ab}{a+b} \div \dfrac{b}{ab+b^2}$ _____

14. $\dfrac{x-2}{3x+6} \div \dfrac{x^2-2x}{x}$ _____

15. $\dfrac{x^2-8x+7}{x^2+1} \div (x-7)$ _____

16. $\dfrac{x^2-3x-10}{x^2+2} \div (x-5)$ _____

17. $\dfrac{x^2-3x+2}{x-4} \div (1-x)$ _____

18. $\dfrac{x^2-7x+10}{x-3} \div (2-x)$ _____

19. $\dfrac{1+\dfrac{1}{x}}{1-\dfrac{1}{x}}$ _____

20. $\dfrac{\dfrac{2a}{x}}{\dfrac{-a^2}{x}}$ _____

21. $\dfrac{x+\dfrac{1}{x}}{\dfrac{2}{x}}$ _____

22. $\dfrac{a-\dfrac{1}{a}}{1-\dfrac{1}{a}}$ _____

23. $\dfrac{a+\dfrac{a}{b}}{1+\dfrac{1}{b}}$ _____

24. $\dfrac{1-\dfrac{a}{b}}{1-\dfrac{b}{a}}$ _____

25. $\dfrac{\dfrac{1}{x}-\dfrac{1}{y}}{\dfrac{1}{x}+\dfrac{1}{y}}$ _____

26. $\dfrac{1-\dfrac{1}{x^2}}{1-\dfrac{1}{x}}$ _____

E. Brain Boosters

1. Divide: $\dfrac{x^4 - 3x^3 + 7x^2 - x}{x^4 - x^2} \div \dfrac{x^3 - 3x^2 + 7x - 1}{x^2 + x}$

2. Evaluate the following number without actually multiplying any arithmetic numbers. A whiz can do it in his or her head!

$$\frac{674321870}{(674321871)^2 - (674321870)(674321872)}$$

$\left(\textit{Hint:} \text{ This expression is of the form } \dfrac{A}{(A + 1)^2 - A(A + 2)}\right)$

3. An expression of the form

$$a + \cfrac{b}{c + \cfrac{d}{e + \cfrac{f}{g + \cdots}}}$$

is called a *continued fraction*. In general the numbers $a, b, c, d, e, \ldots$ may be any real numbers and the terms may continue indefinitely. A *simple continued fraction* is one in which the numerators are all equal to 1.

$$a + \cfrac{1}{a + \cfrac{1}{a + \cfrac{1}{a + \cfrac{1}{a + \cdots}}}}$$

(a) Find the value of $\quad 1 + \cfrac{1}{2 + \cfrac{1}{2}}$

(b) Show that $\quad \dfrac{43}{30} = 1 + \cfrac{1}{2 + \cfrac{1}{3 + \cfrac{1}{4}}}$

(c) Continued fractions are useful because they enable us to find rational or fractional approximations to numbers that have no finite decimal or fractional form. For example,

$$\sqrt{2} = 1 + \cfrac{1}{2 + \cfrac{1}{2 + \cfrac{1}{2 + \cdots}}} \qquad \text{where the three dots indicate that the 2's continue forever.}$$

$\sqrt{2} \cong 1 \qquad$ not a very good approximation

$\sqrt{2} \cong 1 + \dfrac{1}{2} \qquad$ a little better approximation

$\sqrt{2} \cong 1 + \cfrac{1}{2 + \cfrac{1}{2}} \qquad$ a good approximation

Date

Name

Course/Section

and so on. Calculate the first four approximations of $\sqrt{2}$ from the fraction above and compare them with the actual value of $\sqrt{2} \cong 1.41421356\ldots$

(d) Calculate the successive approximations of the following continued fraction. Put your answers in both fraction and decimal form. Do you recognize this number?

$$3 + \cfrac{1}{7 + \cfrac{1}{15 + \cfrac{1}{1 + \cfrac{1}{292 + \cfrac{1}{1 + \ldots}}}}}$$

(e) Find the successive approximations of the continued fraction

$$1 + \cfrac{1}{1 + \cfrac{1}{1 + \cfrac{1}{1 + \cfrac{1}{1 + \ldots}}}}$$
Do you notice anything interesting about this sequence of numbers?

---

4. Find the pattern in the following equations and write an equation showing the pattern.

$$\left(1 + \frac{1}{2}\right) \cdot 3 = 4\frac{1}{2} \quad \text{and} \quad \left(1 + \frac{1}{2}\right) + 3 = 4\frac{1}{2}$$

$$\left(1 + \frac{1}{3}\right) \cdot 4 = 5\frac{1}{3} \quad \text{and} \quad \left(1 + \frac{1}{3}\right) + 4 = 5\frac{1}{3}$$

$$\left(1 + \frac{1}{4}\right) \cdot 5 = 6\frac{1}{4} \quad \text{and} \quad \left(1 + \frac{1}{4}\right) + 5 = 6\frac{1}{4}$$

and so on.

---

5. A spaceship, moving at speed v_1 with respect to the earth, fires a rocket forward at speed v_2 with respect to the ship. Our intuition tells us that the speed of the rocket with respect to the earth is $v_1 + v_2$, but the Special Theory of Relativity tells us that the speed of the rocket with respect to the earth is

$$v = \frac{v_1 + v_2}{1 + \dfrac{v_1 v_2}{c^2}}$$

where c is the speed of light. Find v if $v_1 = \frac{1}{2}c$ and $v_2 = \frac{1}{4}c$.

Date

Name

Course/Section

When you have had the practice you need, either return to the preview on page 335 or turn to frame **31** and continue with the study of equations involving algebraic fractions.

31   In Unit 2 you learned to solve linear equations with integer coefficients such as

$$2x + 3 = 9$$

The simplest method of solving an equation containing fractions is to convert it to an equivalent equation with no fractions. For example, to solve

$$\frac{x}{3} + \frac{1}{2} = \frac{3}{2}$$

first multiply each term by 6 , the least common denominator of the denominators 2 and 3.

$$6 \left(\frac{x}{3}\right) + 6 \left(\frac{1}{2}\right) = 6 \left(\frac{3}{2}\right)$$
$$\frac{6x}{3} + \frac{6}{2} = \frac{18}{2}$$
$$2x + 3 = 9$$

The equation containing algebraic fractions has been rewritten as an equivalent equation with no fractions. This procedure is usually called "clearing the equation of fractions."

Solve this equation in the usual way.

$$2x + 3 \ -3 \ = 9 \ -3$$
$$2x = 6$$
$$x = 3$$

Check this answer by substituting it back into the *original* equation.

Check: $\dfrac{(3)}{3} + \dfrac{1}{2} = \dfrac{3}{2}$

$$\frac{3}{2} = \frac{3}{2}$$

Solve the following equation using this procedure.

$$\frac{2x}{3} - 1 = \frac{3}{2} + \frac{x}{2}$$

Check your work in **32**.

What do I do with an equation having decimal coefficients? Like this:
$0.5x + 0.6 = 1.6$

Multiply each term by 10 to get
$5x + 6 = 16$.
Clear decimals by multiplying by 10 or some multiple of 10.

32   To clear the equation of fractions, find the least common denominator and multiply each term in the equation by it. The LCD of 3 and 2 is 6. Multiply all terms by $\boxed{6}$.

$$\boxed{6}\left(\frac{2x}{3}\right) - \boxed{6}\,(1) = \boxed{6}\left(\frac{3}{2}\right) + \boxed{6}\left(\frac{x}{2}\right)$$

$$4x - 6 = 9 + 3x$$

Solve: $4x - 6 \;\boxed{-\,3x}\; = 9 + 3x \;\boxed{-\,3x}$

$$x - 6 = 9$$

$$x - 6 \;\boxed{+\,6} = 9 \;\boxed{+\,6}$$

$$x = 15$$

Check your answer by substituting it back into the original equation.

Check: $\dfrac{2(15)}{3} - 1 = \dfrac{3}{2} + \dfrac{(15)}{2}$

$$10 - 1 = \frac{3}{2} + \frac{15}{2}$$

$$9 = \frac{18}{2}$$

The equation to be solved may include algebraic fractions whose denominator contains the variable or an expression containing the variable. For example,

$$\frac{8}{x} + 2 = \frac{10}{3} \qquad \text{(Of course, this equation is meaningful only if } x \neq 0.\text{)}$$

To clear the equation of fractions, multiply each term by $\boxed{3x}$, the least common denominator of the two denominators x and 3.

$$\boxed{3x}\left(\frac{8}{x}\right) + \boxed{3x}\,(2) = \boxed{3x}\left(\frac{10}{3}\right)$$

$$24 + 6x = 10x$$
$$\text{or} \qquad\qquad 4x = 24$$
$$x = 6$$

Check the solution by substituting it back into the original equation.

Solve the following equations.

(a) $2 + \dfrac{3}{x} = \dfrac{7}{x}$

(b) $\dfrac{x+3}{5} - \dfrac{2-x}{3} = 9$

(c) $\dfrac{3}{x} + \dfrac{7}{3} = \dfrac{2x-1}{2x} + \dfrac{5}{2}$

(d) $\dfrac{2}{x-2} = \dfrac{7}{x+3}$

Check your solutions in **33**.

I was trying to solve the equation $\frac{1}{2x} + \frac{1}{3x} = \frac{1}{5}$ so I inverted everything to get $2x + 3x = 5$. Is that OK? — I did the same thing to both sides, I inverted the terms.

No, you didn't do the same thing to both sides, you did the same thing to all terms. Inverting $\frac{1}{5}$ gives 5, but inverting $\left(\frac{1}{2x} + \frac{1}{3x}\right)$ gives $\frac{6x}{5}$. It is much better to multiply all terms by the LCD of the denominators.

33 (a) $2 + \dfrac{3}{x} = \dfrac{7}{x}$ (x cannot equal zero or the fractions would be undefined.)

To clear the fractions, multiply each term by $\boxed{x}$.

$$\boxed{x}\,(2) + \boxed{x}\left(\dfrac{3}{x}\right) = \boxed{x}\left(\dfrac{7}{x}\right)$$

$$2x + 3 = 7$$
$$2x = 4$$
$$x = 2$$

Check: $2 + \dfrac{3}{(2)} = \dfrac{7}{(2)}$

$$2 + \dfrac{3}{2} = \dfrac{7}{2}$$

$$3\dfrac{1}{2} = 3\dfrac{1}{2}$$

(b) $\dfrac{x+3}{5} - \dfrac{2-x}{3} = 9$

Multiply each term by $\boxed{15}$, the least common denominator of 5 and 3.

$$\boxed{15}\left(\dfrac{x+3}{5}\right) - \boxed{15}\left(\dfrac{2-x}{3}\right) = \boxed{15}\,(9)$$

$$3(x+3) - 5(2-x) = 135$$
$$3x + 9 - 10 + 5x = 135$$
$$8x - 1 = 135$$
$$8x = 136$$
$$x = 17$$

Check: $\dfrac{(17)+3}{5} - \dfrac{2-(17)}{3} = 9$

$$4 - (-5) = 9$$
$$4 + 5 = 9$$

(c) $\dfrac{3}{x} + \dfrac{7}{3} = \dfrac{2x-1}{2x} + \dfrac{5}{2}$ ($x \neq 0$)

Multiply each term by $\boxed{6x}$, the least common denominator of x, 3, $2x$, and 2.

$$\boxed{6x}\left(\dfrac{3}{x}\right) + \boxed{6x}\left(\dfrac{7}{3}\right) = \boxed{6x}\left(\dfrac{2x-1}{2x}\right) + \boxed{6x}\left(\dfrac{5}{2}\right)$$

$$18 + 14x = 3(2x - 1) + 15x$$
$$18 + 14x = 6x - 3 + 15x$$
$$18 + 14x = 21x - 3$$
$$7x = 21$$
$$x = 3$$

Be certain to check your answer.

(d) $\dfrac{2}{x-2} = \dfrac{7}{x+3}$ (x cannot be equal to 2 or -3 since these values produce zero denominators for the fractions.)

Multiply each term by $\boxed{(x-2)(x+3)}$.

$$(x + 3)(x - 2)\left(\frac{2}{x - 2}\right) = (x + 3)(x - 2)\left(\frac{7}{x + 3}\right)$$

$$(x + 3)2 = (x - 2)(7)$$
$$2x + 6 = 7x - 14$$
$$5x = 20$$
$$x = 4$$

Again, be certain to check your answer.

Fractions with the Variable in the Denominator

When the equation to be solved includes algebraic fractions with the variable in the denominator, it is particularly important that you check the solution. Equations of this kind will be valid only for certain allowed values of the variable. If the solution is not one of these allowed values, the equation has no solution. In the example

$$\frac{8}{x} + 2 = \frac{10}{3}$$

any value of x will be a possible solution *except* $x = 0$. The fraction $\frac{8}{x}$ is not defined for $x = 0$.

The equation

$$4 + \frac{5}{x - 2} = \frac{x + 3}{x - 2}$$

is defined only for values of x other than $x = 2$. For $x = 2$, the denominator $x - 2$ is equal to 0 and the algebraic fractions $\frac{5}{x - 2}$ and $\frac{x - 3}{x - 2}$ are not real numbers.

To solve this equation, multiply each term by $(x - 2)$

$$(x - 2)(4) + (x - 2)\frac{5}{x - 2} = (x - 2)\frac{x + 3}{x - 2}$$

$$4x - 8 + 5 = x + 3$$
$$4x - 3 = x + 3$$
$$3x = 6$$
$$x = 2$$

But this is exactly the value of x for which this particular equation has no meaning. In such a situation we say that the equation has no solution.

Careful ▷ Always check your solution by substituting it back into the original equation and you will never be tripped up by equations having no solution.

Solve the following equations.

(a) $\dfrac{2x + 1}{x + 3} = \dfrac{3}{4}$ (b) $\dfrac{1}{x - 2} = \dfrac{5}{x^2 - x - 2}$

(c) $\dfrac{x}{x - 1} + \dfrac{2}{1 - x} = 2$ (d) $\dfrac{2}{x - 3} = \dfrac{5 - x}{x - 3} - 5$

When you have completed these and have checked your solutions, check your work in **34**.

---

34   (a) $\dfrac{2x + 1}{x + 3} = \dfrac{3}{4}$

Multiply by the least common denominator $4(x + 3)$.

$$4(x + 3)\left(\frac{2x + 1}{x + 3}\right) = 4(x + 3)\left(\frac{3}{4}\right)$$

$$4(2x + 1) = (x + 3)3 \quad \text{Remove parentheses and multiply.}$$
$$8x + 4 = 3x + 9$$
$$5x = 5$$
$$x = 1$$

Check: $\dfrac{2(1) + 1}{(1) + 3} = \dfrac{3}{4}$

$$\dfrac{3}{4} = \dfrac{3}{4}$$

(b) $\dfrac{1}{x - 2} = \dfrac{5}{x^2 - x - 2}$

Factor the denominator on the right:

$$\dfrac{1}{x - 2} = \dfrac{5}{(x - 2)(x + 1)}$$

Multiply by the least common denominator $(x - 2)(x + 1)$.

$$(x - 2)(x + 1) \left(\dfrac{1}{x - 2} \right) = (x - 2)(x + 1) \left(\dfrac{5}{(x - 2)(x + 1)} \right)$$

$$x + 1 = 5$$
$$x = 4$$

Check: $\dfrac{1}{(4) - 2} = \dfrac{5}{(4)^2 - (4) - 2}$

$$\dfrac{1}{2} = \dfrac{5}{10}$$

$$\dfrac{1}{2} = \dfrac{1}{2}$$

(c) $\dfrac{x}{x - 1} + \dfrac{2}{1 - x} = 2$

Note that $1 - x = -(x - 1)$. Rewrite the equation as

$$\dfrac{x}{x - 1} + \dfrac{2}{-(x - 1)} = 2$$

or $\quad \dfrac{x}{x - 1} + \dfrac{-2}{x - 1} = 2$

Multiply each term by $x - 1$.

$$x - 1 \left(\dfrac{x}{x - 1} \right) + x - 1 \left(\dfrac{-2}{x - 1} \right) = x - 1 \, (2)$$

$$x + (-2) = (x - 1)(2)$$
$$x - 2 = 2x - 2$$
$$-2 = x - 2$$
$$0 = x$$
$$\text{or} \quad x = 0$$

Check: $\dfrac{(0)}{(0) - 1} + \dfrac{2}{1 - (0)} = 2$

$$0 + 2 = 2$$
$$2 = 2$$

(d) $\dfrac{2}{x - 3} = \dfrac{5 - x}{x - 3} - 5$

Multiply by $x - 3$.

$$x - 3 \left(\dfrac{2}{x - 3} \right) = x - 3 \left(\dfrac{5 - x}{x - 3} \right) + x - 3 \, (-5)$$

$$2 = 5 - x + (x - 3)(-5)$$
$$2 = 5 - x - 5x + 15$$
$$2 = 20 - 6x$$
$$6x = 18$$
$$x = 3$$

Check: $\dfrac{2}{(3) - 3} = \dfrac{5 - (3)}{(3) - 3} - 5$

$$\dfrac{2}{0} = \dfrac{2}{0} - 5$$

Division by zero is not defined in arithmetic; therefore $x = 3$ is not a valid solution and the equation has no solution.

Solving Formulas
Containing Fractions

One of the most immediately useful skills of basic algebra is the ability to change a formula or equation to an equivalent form. In Unit 2 we solved linear formulas with integer coefficients. If the formula to be solved contains one or more fractions, the first step in solving it is to multiply each term by the least common denominator of the fractions. This will simplify the equation so that it can be solved easily. For example, solve the formula

$$h = \frac{A - 1}{R^2} \quad \text{for } A.$$

First, multiply both members of the equation by R^2 :

$$R^2\,(h) = R^2 \left(\frac{A - 1}{R^2} \right)$$

$$R^2 h = A - 1$$
$$R^2 h + 1 = A$$

or $\quad A = R^2 h + 1$

Solve the following formulas as shown.

(a) The formula relating principle P and interest rate r to the amount A accumulated in t years for an investment is

$$P = \frac{A}{1 + rt} \qquad \text{Solve for } r.$$

(b) In physics the formula relating power W to voltage V and resistance R in an electrical circuit is

$$W = \frac{V^2}{R} \qquad \text{Solve for } V.$$

(c) In electronics, the formula for the total resistance of resistors in parallel is

$$\frac{1}{R_1} + \frac{1}{R_2} = \frac{1}{R} \qquad \text{Solve for } R.$$

Our solutions are in **35**.

---

35 (a) $\quad P = \dfrac{A}{1 + rt} \qquad$ Solve for r.

Multiply each member of the equation by $(1 + rt)$.

$$(1 + rt)\,(P) = (1 + rt) \left(\frac{A}{1 + rt} \right)$$

$$P + Prt = A$$
$$Prt = A - P$$
$$r = \frac{A - P}{Pt}$$

(b) $\quad W = \dfrac{V^2}{R} \qquad$ Solve for V.

Multiply each member of the equation by R.

$$R\,(W) = R\left(\frac{V^2}{R}\right)$$
$$RW = V^2$$
$$V^2 = RW$$
$$V = \sqrt{RW}$$

(c) $\dfrac{1}{R_1} + \dfrac{1}{R_2} = \dfrac{1}{R}$

Multiply each term by RR_1R_2, the least common denominator of the three fractions.

$$RR_1R_2\left(\frac{1}{R_1}\right) + RR_1R_2\left(\frac{1}{R_2}\right) = RR_1R_2\left(\frac{1}{R}\right)$$
$$RR_2 + RR_1 = R_1R_2$$

Solve for R:

$$R(R_2 + R_1) = R_1R_2$$
$$R = \frac{R_1R_2}{(R_2 + R_1)}$$

Systems of Equations Containing Fractions

In Unit 4 you learned to solve systems of two simultaneous linear equations in two variables, where the coefficients were integers. To solve a system of equations in which algebraic fractions appear, first rewrite the equations in an equivalent form without fractions, then solve the system of equations in the usual way. For example, the system of equations

$$\frac{2x}{3} + y = 10$$

$$\frac{3x}{2} - \frac{y}{4} = \frac{5}{2}$$

becomes

$$3\left(\frac{2x}{3}\right) + 3\,(y) = 3\,(10) \qquad \text{Clear fractions by multiplying by } 3.$$

$$4\left(\frac{3x}{2}\right) + 4\left(\frac{-y}{4}\right) = 4\left(\frac{5}{2}\right) \qquad \text{Clear fractions by multiplying by } 4.$$

The system of equations is now

$$2x + 3y = 30$$
$$6x - y = 10$$

This system of equations can now be solved using the procedures explained in Unit 4. The solution is $x = 3$ and $y = 8$. Check this solution by substituting it back into the original system of equations.

Rewrite each of the following systems of equations as equivalent equations without algebraic fractions, then solve them.

(a) $\dfrac{x}{5} + \dfrac{y}{2} = \dfrac{1}{2}$ (b) $\dfrac{3x + 1}{2} + \dfrac{y - 1}{5} = 1$

$\quad\ \dfrac{x}{2} - \dfrac{y}{3} = 6$ $\dfrac{2x + 1}{3} + \dfrac{2y + 3}{5} = 0$

Check your work in **36**.

36 (a) $\dfrac{x}{5} + \dfrac{y}{2} = \dfrac{1}{2}$

$\dfrac{x}{2} - \dfrac{y}{3} = 6$

Multiply each term in the first equation by 10 , the least common denominator of 5 and 2.

$10\ \left(\dfrac{x}{5}\right) + 10\ \left(\dfrac{y}{2}\right) = 10\ \left(\dfrac{1}{2}\right) \longrightarrow 2x + 5y = 5$

Multiply each term in the second equation by 6 , the least common denominator of 2 and 3.

$6\ \left(\dfrac{x}{2}\right) + 6\ \left(\dfrac{-y}{3}\right) = 6\ (6) \longrightarrow 3x - 2y = 36$

Solve this equivalent system of equations: $2x + 5y = 5$
$3x - 2y = 36$

The solution is $x = 10$ and $y = -3$. Check it by substituting it back into the original system of equations.

(b) $\dfrac{3x + 1}{2} + \dfrac{y - 1}{5} = 1$

$\dfrac{2x + 1}{3} + \dfrac{2y + 3}{5} = 0$

Multiply each term in the first equation by 10 , the least common denominator of 2 and 5.

$10\ \left(\dfrac{3x + 1}{2}\right) + 10\ \left(\dfrac{y - 1}{5}\right) = 10\ (1) \longrightarrow 5(3x + 1) + 2(y - 1) = 10$

$15x + 5 + 2y - 2 = 10$
$15x + 2y = 7$

Multiply each term in the second equation by 15 , the least common denominator of 3 and 5.

$15\ \left(\dfrac{2x + 1}{3}\right) + 15\ \left(\dfrac{2y + 3}{5}\right) = 15\ (0) \longrightarrow 5(2x + 1) + 3(2y + 3) = 0$

$10x + 5 + 6y + 9 = 0$
$10x + 6y = -14$

Solve this equivalent system of equations: $15x + 2y = 7$
$10x + 6y = -14$

The solution is $x = 1$ and $y = -4$. Check it by substituting it back into the original system of equations.

Fractions with the
Variable in
the Denominator

Be especially careful when a system of equations includes algebraic fractions in which the variables appear in the denominators. For example,

$\dfrac{1}{x} + \dfrac{2}{y} = 3$

$\dfrac{3}{x} - \dfrac{2}{y} = -1$

To solve this system of equations, add them.

$\left(\dfrac{1}{x} + \dfrac{3}{x}\right) + \left(\dfrac{2}{y} - \dfrac{2}{y}\right) = (3 - 1)$

$\dfrac{4}{x} + 0 = 2$

$$\frac{4}{x} = 2 \qquad \text{Multiply both members of the equation by } x$$

$$4 = 2x$$

$$x = 2$$

Substitute this value of x back into the first equation to find y. $y = \frac{4}{5}$.

Notice that when the variable appears in the denominator, you must be especially careful to check that the solution obtained does not lead to a denominator of zero. Division by zero is not allowed, remember?

Word Problems

Word problems involving algebraic fractions should present no special problem for you. Solve them exactly as you would any other word problems:

1. Translate to an algebraic equation or system of equations.
2. Clear fractions and solve the equations.
3. Check your solution.

For example, solve this word problem:

"If one-half of a certain number is added to four times the number, the result is $\frac{3}{4}$. Find the number."

Look in **37** for our solution.

37   Translating the English sentence to an algebraic equation, we obtain

$$\frac{1}{2}(x) + 4(x) = \frac{3}{4}$$

or $$\frac{x}{2} + 4x = \frac{3}{4}$$

Multiply each term in the equation by $\boxed{4}$, the least common denominator of 2 and 4.

$$\boxed{4}\left(\frac{x}{2}\right) + \boxed{4}(4x) = \boxed{4}\left(\frac{3}{4}\right)$$

$$2x + 16x = 3$$

$$18x = 3$$

$$x = \frac{3}{18}$$

$$x = \frac{1}{6}$$

Check the solution by substituting it into the original equation.

Translate each of the following word problems into algebraic equations, then rewrite each equation in an equivalent form containing no algebraic fractions.

(a) Find two consecutive integers such that $\frac{1}{3}$ of the first integer plus $\frac{1}{2}$ of the second integer is equal to 8.

(b) The width of a rectangle is four feet more than one-third its length. If its perimeter is 32 feet, find the length and width.

(c) Eric's piggybank contains $4.40 in quarters, dimes, and nickles. He has three-fourths as many dimes as quarters and the number of nickles is six more than one-third the number of quarters. How many of each kind of coin does he have?

(d) Two pipes feed water into a storage tank. One of them acting alone will fill the tank in 8 hours and the second one, acting alone, will fill the tank in 12 hours. How long will it take to fill the tank if both inlet pipes are used?

(e) Carlos rows his boat 2 miles upstream and then returns downstream to his starting point. The trip up takes three times as long as the trip down. If Carlos rows 2 mph in still water, how fast is the stream moving?

Compare your answers with ours in **38**.

38   (a)   Let the integers be x and $x + 1$, then

$$\frac{1}{3}(x) + \frac{1}{2}(x + 1) = 8$$

or $$\frac{x}{3} + \frac{x + 1}{2} = 8$$

Clear fractions by multiplying all terms by $\boxed{6}$, the least common denominator of 3 and 2.

$$\boxed{6}\left(\frac{x}{3}\right) + \boxed{6}\left(\frac{x + 1}{2}\right) = \boxed{6}\,(8)$$

$$
\begin{aligned}
2x + 3(x + 1) &= 48 & &\text{Multiply.}\\
2x + 3x + 3 &= 48 & &\text{Collect like terms.}\\
5x &= 45\\
x &= 9
\end{aligned}
$$

The two consecutive integers are 9 and 10.

(b)   Let the width equal W and the length equal L, then

$$W = 4 + \frac{1}{3}(L)$$

or $\qquad W = 4 + \dfrac{L}{3}$

and $\qquad 2W + 2L = 32$

The system of equations to be solved is

$$W = 4 + \frac{L}{3}$$

$$2W + 2L = 32$$

Substitute the expression for W from the first equation into the second equation:

$$2\left(4 + \frac{L}{3}\right) + 2L = 32$$

$$8 + \frac{2L}{3} + 2L = 32 \qquad \text{Multiply.}$$

Multiply each term by $\boxed{3}$ to clear the fraction.

$$\boxed{3}\,(8) + \boxed{3}\,\left(\frac{2L}{3}\right) + \boxed{3}\,(2L) = \boxed{3}\,(32)$$

$$24 + 2L + 6L = 96$$
$$8L = 72$$
$$L = 9 \text{ ft}$$

Substitute this value of L back into the first equation to obtain the value of W.

$$W = 4 + \frac{1}{3}\,(9) = 7 \text{ ft}$$

(c) It is easiest to solve this one if we use three variables. Let Q be the number of quarters, D be the number of dimes, and N be the number of nickels. Then

$$25Q + 10D + 5N = 440$$

$$D = \frac{3}{4}\,(Q) \qquad \text{or} \qquad D = \frac{3Q}{4}$$

$$N = 6 + \frac{1}{3}\,(Q) \qquad \text{or} \qquad N = 6 + \frac{Q}{3}$$

Combine these equations by substituting these expressions for D and N in the first equation.

$$25Q + 10\left(\frac{3Q}{4}\right) + 5\left(6 + \frac{Q}{3}\right) = 440$$

or

$$25Q + \frac{30Q}{4} + 30 + \frac{5Q}{3} = 440$$

Clear fractions by multiplying by $\boxed{12}$, the least common denominator of 3 and 4.

$$\boxed{12}\,(25Q) + \boxed{12}\,\left(\frac{30Q}{4}\right) + \boxed{12}\,(30) + \boxed{12}\,\left(\frac{5Q}{3}\right) = \boxed{12}\,(440)$$

$$300Q + 90Q + 360 + 20Q = 5280$$
$$410Q = 4920$$
$$Q = 12$$

Substitute this value of Q into the original equations to find D and N.

$$D = \frac{3Q}{4} = \frac{3(12)}{4} \qquad\qquad N = 6 + \frac{Q}{3} = 6 + \frac{(12)}{3}$$

$$D = 9 \qquad\qquad\qquad\qquad N = 10$$

389 5-4 Equations Involving Algebraic Fractions

(d) One method of solving problems involving rates was shown in Unit 2. Now that you are able to work with algebraic fractions, you may want to use the following shortcut method.

Let x equal the number of hours needed to fill the tank when both pipes are used. Then,

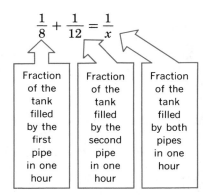

$$\frac{1}{8} + \frac{1}{12} = \frac{1}{x}$$

If the first pipe can fill the tank in 8 hours, it fills $\frac{1}{8}$ of the tank in 1 hour.

Fraction of the tank filled by the first pipe in one hour

Fraction of the tank filled by the second pipe in one hour

Fraction of the tank filled by both pipes in one hour

Clear fractions by multiplying each term in the equation by $24x$, the least common denominator of 8, 12, and x.

$$24x\left(\frac{1}{8}\right) + 24x\left(\frac{1}{12}\right) = 24x\left(\frac{1}{x}\right)$$
$$3x + 2x = 24$$
$$5x = 24$$
$$x = \frac{24}{5} \quad \text{or} \quad 4.8 \text{ hours}$$

(e) Let v be the speed of the stream, then $2 - v$ is the speed of the boat moving upstream and $2 + v$ is the speed of the boat moving downstream. Remember that for any object moving at a steady speed

Distance = speed · time

or

$$\text{Time} = \frac{\text{distance}}{\text{speed}}$$

For Carlo's boat, the time to travel upstream is $\dfrac{2}{2 - v}$ and the time to travel downstream is $\dfrac{2}{2 + v}$.

Then

$$\frac{2}{2 - v} = 3\left(\frac{2}{2 + v}\right) \quad \text{or} \quad \frac{2}{2 - v} = \frac{6}{2 + v}$$

To clear fractions multiply each term by $(2 - v)(2 + v)$, the least common denominator of the denominators $(2 - v)$ and $(2 + v)$.

$$(2 - v)(2 + v)\frac{2}{2 - v} = (2 - v)(2 + v)\frac{6}{2 + v}$$
$$2(2 + v) = 6(2 - v)$$
$$4 + 2v = 12 - 6v$$
$$8v = 8$$
$$v = 1 \text{ mph} \quad \text{The stream moves at 1 mph.}$$

Ratio

One variety of algebraic fraction problem that appears very often in science, business, technology, and many other applied areas involves the concepts of ratio and proportion. A *ratio* is a comparison of the sizes of two quantities found by dividing them. It is a single number usually written as a fraction.

For example, the steepness of a hill can be given as the ratio of its height to its horizontal extent.

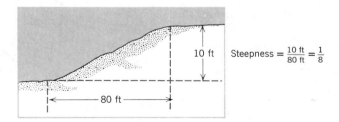

The ratio of the circumference of any circle to its diameter is a ratio used so often in mathematics that it has been given a special symbol: π.

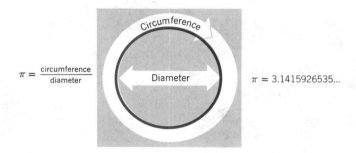

Notice that because a ratio is a division of like quantities, the units do not appear in the ratio number. A ratio can be written as a decimal, as a percent, or as a fraction.

Write the following ratios as fractions reduced to lowest terms.

(a) Sam takes 25 minutes to do a certain job and Barry takes one hour to do the same job. What is the ratio of their job times?

(b) What is the gear ratio of a bicycle using 24 teeth on the rear sprocket and 50 teeth on the front chainwheel?

(c) What is the won-to-lost ratio of a basketball team that wins 8 games and loses 18 in a given season?

(d) What is the pay-off ratio of a game that pays a winner $7 for every $2 he bets?

Check your answers in **39**.

39

(a) Job time ratio $= \dfrac{25 \text{ min}}{1 \text{ hr}} = \dfrac{25 \text{ min}}{60 \text{ min}} = \dfrac{25}{60} = \dfrac{5}{12}$

(b) Gear ratio $= \dfrac{\text{front}}{\text{rear}} = \dfrac{50 \text{ teeth}}{24 \text{ teeth}} = \dfrac{50}{24} = \dfrac{25}{12}$

(c) Won-to-lost ratio $= \dfrac{\text{wins}}{\text{losses}} = \dfrac{8}{18} = \dfrac{4}{9}$

(d) Pay-off ratio $= \dfrac{\text{pay-off}}{\text{bet}} = \dfrac{\$7}{\$2} = \dfrac{7}{2}$ or $3\dfrac{1}{2}$

A proportion is a statement that two ratios are equal. It can be a sentence in words, but most often it is in algebraic equation. For example, the question

"If the ratio of my height to his height is 7 to 8, and he is 6 ft tall, how tall am I?"

can be written as an equation.

$$\text{Ratio of our heights} = \frac{\text{my height}}{\text{his height}} = \frac{7}{8} = \frac{x}{6 \text{ ft}}$$

where x is my height.

The proportion equation is $\frac{7}{8} = \frac{x}{6}$.

Solve this equation for x.

Check your work in **40**.

40 $\frac{7}{8} = \frac{x}{6}$

Clear fractions by multiplying both terms by 24 , the least common denominator of 6 and 8.

$$24 \left(\frac{7}{8}\right) = 24 \left(\frac{x}{6}\right)$$

$$21 = 4x$$

$$x = \frac{21}{4} \qquad \text{or} \qquad 5\frac{1}{4} \text{ ft}$$

The Cross-Product Rule

Because proportions appear so often in practical applications of algebra, we normally use a shortcut method of solution known as the *cross-product* rule or procedure.

The equation $\frac{a}{b} = \frac{c}{d}$ is equivalent to the equation $ad = bc$.

The *cross products* of the proportion are equal.

The cross-product rule:

$$\frac{a}{b} \bowtie \frac{c}{d} \qquad ad = bc$$

For example, the proportion $\frac{5}{2} = \frac{x}{6}$ is equivalent to $5 \cdot 6 = 2 \cdot x$

$$\text{or} \quad 30 = 2x$$
$$x = 15$$

The proportion $\frac{2}{3} = \frac{x-2}{4}$ is equivalent to $2 \cdot 4 = 3(x-2)$ or $8 = 3x - 6$

Solving this equation,

$$3x = 14$$
$$x = 4\frac{2}{3}$$

Use the cross-product procedure to solve the following proportions.

(a) $\frac{1}{2} = \frac{x}{9}$ (b) $\frac{t}{7} = \frac{3}{4}$ (c) $\frac{3}{y} = \frac{2}{5}$

(d) $\frac{9}{16} = \frac{3}{b}$ (e) $\frac{6}{5} = \frac{2}{3-x}$ (f) $\frac{15}{12} = \frac{a-1}{a+3}$

(g) $\frac{2\frac{1}{2}}{3\frac{1}{2}} = \frac{Q}{2}$ (h) $\frac{1}{2} = \frac{x+2}{x+5}$

The solutions are in **41**.

How do you know that the cross product rule always works?

It's easy to prove. Suppose the proportion $\frac{a}{b} = \frac{c}{d}$ is true, then we can multiply both sides of this equation by the quantity bd and cancel factors, $\left(\frac{a}{b}\right)bd = \left(\frac{c}{d}\right)bd$ or $ad = cb$ — the cross products.

41   (a) $1 \cdot 9 = 2 \cdot x$ (b) $4t = 3 \cdot 7$ (c) $3 \cdot 5 = 2 \cdot y$
           $2x = 9$ $4t = 21$ $2y = 15$
           $x = 4.5$ $t = 5\frac{1}{4}$ $y = 7.5$

  (d) $9 \cdot b = 3 \cdot 16$ (e) $6(3-x) = 2 \cdot 5$ (f) $15(a+3) = 12(a-1)$
           $9b = 48$ $18 - 6x = 10$ $15a + 45 = 12a - 12$
           $b = 5\frac{1}{3}$ $-6x = -8$ $3a = -57$
                                $6x = 8$ $a = -19$
                                $x = 1\frac{1}{3}$

  (g) $2\frac{1}{2} \cdot 2 = 3\frac{1}{2} \cdot Q$ (h) $1(x+5) = 2(x+2)$
                                       $x + 5 = 2x + 4$
           $5 = 3\frac{1}{2} \cdot Q = \frac{7Q}{2}$ $1 = x$
                                      $x = 1$
           $10 = 7Q$

           $Q = \frac{10}{7}$

Use the ideas of ratio and proportion to solve the following problems.

(a) A telephone pole casts a shadow 9 ft long. A yardstick beside the pole casts a shadow 16 in. long. What is the height of the telephone pole?

(b) If five men can dig three ditches in one day, how many ditches will two men working at the same rate dig in one day?

Our solutions are in **42**. Check it when you are finished.

---

42 (a) Ratio of heights $= \dfrac{\text{height of pole}}{\text{height of stick}} = \dfrac{\text{h ft}}{\text{3 ft}}$

Ratio of shadow heights $= \dfrac{\text{shadow of pole}}{\text{shadow of stick}} = \dfrac{9 \text{ ft}}{16 \text{ in.}} = \dfrac{108 \text{ in.}}{16 \text{ in.}}$

The proportion is $\quad \dfrac{h}{3} = \dfrac{108}{16}$

Cross multiply:

$16 \cdot h = 3 \cdot 108$

$16h = 324$

$h = 20\dfrac{1}{4} \text{ ft}$

Notice that we set up ratios of like quantities and that top and bottom of each ratio are in the same units. Both fractions have pole information in the numerator and stick information in the denominator of the ratios.

(b) Diggers ratio $= \dfrac{5 \text{ men}}{2 \text{ men}} = \dfrac{5}{2}$

Work ratio $= \dfrac{3 \text{ ditches}}{D \text{ ditches}} = \dfrac{3}{D}$

The work corresponding to the first men is placed in the numerator of the fraction to match the first fraction.

$\dfrac{5}{2} = \dfrac{3}{D}$

$5 \cdot D = 3 \cdot 2$

$5D = 6$

$D = \dfrac{6}{5} = 1.2 \text{ ditches}$

Directly Proportional

One very useful way to describe events in the world around us is to look for patterns or relationships. In Unit 4 you learned to use graphs as a way to describe or illustrate a numerical relationship. When one variable increases or decreases in response to the increase or decrease in another variable, we say that the variables are related. When two variables are related so that their ratio is always a constant, we say that they are *directly proportional* or that one of the variables *varies directly* as the other. If the variables x and y are directly related then they satisfy the equation

$y = kx \quad$ where k is some constant number.

Graphically, direct variation produces a straight line graph.

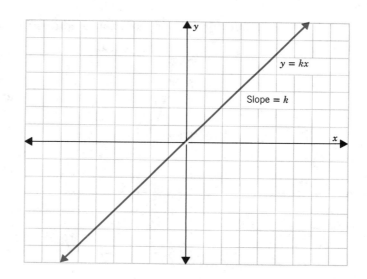

As an example of direct variation, consider the following: if you are paid for work at a fixed rate of $4 per hour, then your total monthly income I will be directly related to the number of hours N that you work. The equation describing the relationship is

$I = 4N$ As N increases, I increases. For $N = 10$, $I = 40$.
                                               For $N = 20$, $I = 80$.

Solve the following problem:

If y varies directly as x, and if $y = 4$ when $x = 2$, find the value of y when $x = 7$.

First set up the direct variation equation, then use the information given to determine the constant k, and finally use this equation to find y for $x = 7$.

Check your work in **43**.

---

43   If y varies directly as x, then $y = kx$.

Knowing that $y = 4$ when $x = 2$ allows us to determine the constant k. Substituting in the equation $y = kx$,

$(4) = k(2)$

or $2k = 4$
        $k = 2$

The equation relating y to x is $y = 2x$.

When $x = 7$, $y = 2(7) = 14$

Inversely Proportional   Two variables may be related so that one increases in response to a decrease in the other. We say that the variables are *inversely related* or *vary inversely* if they satisfy the equation

$y = \dfrac{k}{x}$ where k is some constant number

Graphically, inverse variation produces the graph curve shown.

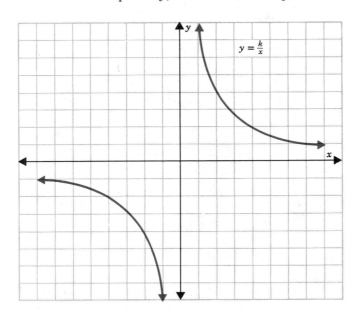

$y = \dfrac{k}{x}$

This curve is called a hyperbola.

The lower branch of the curve is for negative values of x and y.

For example, the pressure exerted on a surface by a fixed 100-lb force is inversely related to the area of the surface.

$$P = \frac{100}{A}$$

As A decreases, P increases. For $A = 10$, $P = 10$.
For $A = 5$, $P = 20$.
For $A = 2$, $P = 50$.

Solve the following inverse variation problem:

If y varies inversely as x, and if $y = 10$ when $x = 2$, find the value of y when $x = 5$.

First, set up the inverse variation equation, then solve the problem as you did the direct variation problem.

Check your work in **44**.

---

44   If y varies inversely as x, then $y = \dfrac{k}{x}$.

Knowing that $y = 10$ when $x = 2$ allows us to determine the constant k. Substituting in the equation

$$y = \frac{k}{x}$$

$$(10) = \frac{k}{(2)}$$

$$k = 20$$

The equation relating y to x is $y = \dfrac{20}{x}$.

When $x = 5$, $y = \dfrac{20}{(5)} = 4$

Solve the following practice problems on ratio, proportion, and variation.

(a)   Solve for x: $\dfrac{4}{x - 1} = \dfrac{8}{x}$

(b) The sum of two numbers is 20 and their ratio is $\frac{2}{3}$. Find the numbers.

(c) If 20 lb of potatoes costs $1.88, how much will a 25-lb bag cost?

(d) The gauge pressure on a scuba diver is directly proportional to the depth of the diver. If the pressure at 9 ft is 4 psi (pounds per square inch), what will be the gauge pressure at a depth of 100 ft? (Compare this with normal atmospheric pressure, which is about 15 psi.)

(e) When the temperature is kept constant, the pressure exerted by a gas on its container is inversely proportional to the volume of the container. If the pressure is 6 atmospheres when the volume is 2 liters, what will the pressure be when the volume is increased to 6 liters?

After you have solved these problems, check your answers in **45**.

---

45 (a) $\dfrac{4}{x-1} = \dfrac{8}{x}$ The cross-product rule gives

$$4x = (x-1)8$$
$$4x = 8x - 8$$
$$8 = 4x$$
$$x = 2$$

(b) Let the numbers be x and y, then the problem statement gives the equations

$$x + y = 20 \quad \text{and} \quad \frac{x}{y} = \frac{2}{3}$$

The second equation is equivalent to $3x = 2y$ using the cross-product rule. Solve the first equation for x:

$$x = 20 - y$$

and substitute this expression for x into the second equation.

$$3(20 - y) = 2y$$
$$60 - 3y = 2y$$
$$60 = 5y$$
$$y = 12 \quad \text{The numbers are 8 and 12.}$$

(c) $\dfrac{20}{25} = \dfrac{188}{x}$

Using the cross-product rule $20x = 25 \cdot 188$
$$20x = 4700$$
$$x = 235 \quad \text{or} \quad \$2.35$$

(d) For direct variation the pressure P and depth D are related by the equation

$$P = kD$$

Substituting $P = 4$ and $D = 9$ into this equation

$$(4) = k(9)$$

and solving for k

$$k = \frac{4}{9} \quad \text{The equation relating } P \text{ and } D \text{ is}$$

$$P = \frac{4D}{9}$$

At $D = 100$ ft, $P = \dfrac{4(100)}{9} = \dfrac{400}{9}$ or $P \cong 44$ psi

(e) For inverse variation the pressure P and volume V are related by the equation

$$P = \frac{k}{V}$$

Substituting $P = 6$ and $V = 2$ into this equation

$$(6) = \frac{k}{(2)} \quad \text{and solving for } k$$

$$k = 12$$

The equation relating P and V is

$$P = \frac{12}{V}$$

At $V = 6$ liters, $P = \frac{12}{(6)} = 2$ atmospheres

Now turn to **46** for a set of practice problems on equations involving algebraic fractions.

Unit 5
Fractions
and Radicals

Equations Involving
Algebraic Fractions

The answers are on page 563.

A. Solve the following equations.

1. $\dfrac{x}{4} + \dfrac{x}{2} = 6$ _____

2. $\dfrac{x}{3} + \dfrac{2x}{4} = \dfrac{5}{3}$ _____

3. $1 + \dfrac{a}{2} = \dfrac{5}{3}$ _____

4. $\dfrac{y}{3} - 2 = \dfrac{10}{4}$ _____

5. $\dfrac{1}{x} + \dfrac{1}{2x} = \dfrac{8}{5}$ _____

6. $1 - \dfrac{2}{x} = 3$ _____

7. $\dfrac{2}{a} - \dfrac{3}{2a} = \dfrac{1}{4}$ _____

8. $\dfrac{2}{3} - \dfrac{1}{p} = \dfrac{5}{3p}$ _____

9. $\dfrac{1}{x - 1} - \dfrac{2}{x} = 0$ _____

10. $\dfrac{x + 1}{2} + \dfrac{x - 1}{3} = \dfrac{1}{6}$ _____

11. $\dfrac{x - 1}{x - 2} = \dfrac{4}{3}$ _____

12. $\dfrac{1}{a + 1} = \dfrac{2}{a - 2}$ _____

13. $\dfrac{2}{y + 2} = \dfrac{1}{y}$ _____

14. $\dfrac{x - 2}{2x + 3} = \dfrac{1}{3}$ _____

15. $\dfrac{1}{x + 2} + \dfrac{1}{x + 3} = \dfrac{3}{x + 3}$ _____

16. $\dfrac{1}{x - 1} + \dfrac{1}{x + 1} = \dfrac{6}{x^2 - 1}$ _____

17. $\dfrac{t + 2}{4} - \dfrac{t - 1}{3} = 1$ _____

18. $\dfrac{2}{x^2 - 4} = \dfrac{3}{x - 2}$ _____

19. $\dfrac{x - 1}{2} - \dfrac{2x + 1}{3} = \dfrac{x + 2}{4} + \dfrac{4 - x}{2}$ _____

20. $\dfrac{x - 1}{x} + \dfrac{x + 1}{x - 1} = 2$ _____

21. $\dfrac{a - 1}{a - 2} = \dfrac{a + 1}{a + 3}$ _____

22. $\dfrac{2t}{t + 1} = 2 - \dfrac{3}{2t}$ _____

23. $\dfrac{2y - 1}{y + 1} = \dfrac{2y + 3}{y}$ _____

24. $\dfrac{x}{2x + 1} = \dfrac{x + 1}{2x}$ _____

B. Solve each of the following formulas for x.

1. $\dfrac{1}{a} + \dfrac{1}{x} = \dfrac{1}{b}$ _____

2. $\dfrac{ab}{x} = \dfrac{bc}{t}$ _____

3. $Q = \dfrac{ax - b}{r - x}$ _____

4. $t = \dfrac{nx}{A + Bx}$ _____

5. $\dfrac{a + x}{x} = \dfrac{a + 1}{2}$ _____

6. $\dfrac{a}{b} = \dfrac{x + a}{a}$ _____

Date _____

Name _____

Course/Section _____

7. $a = b + \dfrac{c}{x}$ _____

8. $\dfrac{A - x}{xr} = t$ _____

9. $L = \dfrac{2S}{x} - a$ _____

10. $\dfrac{x - a}{n - x} = d$ _____

C. Solve each of the following systems of equations.

1. $\dfrac{3x}{2} - y = 5$

   $x - \dfrac{y}{2} = \dfrac{7}{2}$ _____

2. $\dfrac{x}{3} + \dfrac{y}{2} = 7$

   $\dfrac{x}{2} - \dfrac{y}{5} = 1$ _____

3. $\dfrac{2x - 1}{2} - \dfrac{y + 1}{3} = \dfrac{1}{2}$

   $\dfrac{3x + 1}{5} + \dfrac{y - 1}{4} = 3$ _____

4. $\dfrac{1}{x + 1} = \dfrac{2}{y - 1}$

   $\dfrac{1}{2x + 3} = \dfrac{2}{y - 5}$ _____

5. $\dfrac{1}{3x + 1} = \dfrac{3}{8y - 1}$

   $\dfrac{2}{1 + x} = \dfrac{1}{y - \frac{3}{2}}$ _____

6. $\dfrac{2x + 3}{3} = \dfrac{y + 1}{2} + 1$

   $\dfrac{x - 1}{2} = \dfrac{2y}{4}$ _____

D. Solve.

1. If one-half of a number is added to $\frac{2}{3}$ of the number, the sum is equal to the original number minus 1. Find the number.

   _____

2. If 18 is divided by a certain number and if 30 is divided by twice that number, the sum of the two quotients is 11. What is the number?

   _____

3. If the reciprocal of a certain number is increased by $\frac{2}{3}$, the sum is equal to 2. What is the number?

   _____

4. The sum of two numbers is 15. Their ratio is 4. What are the two numbers?

   _____

5. A house and lot together cost $27,540. The lot costs one-eighth as much as the house. Find the cost of each.

   _____

6. Student tickets for the basketball game sell for $1.50 and general admission tickets

sell for $2.50. If the total ticket sales were $1922 and if $\frac{2}{5}$ as many student tickets were sold as general admission tickets, how many tickets were sold?

_____

7. Maria is twice as old as Rudy. Two years from now Rudy will be two-thirds as old as she will be. How old are they now?

_____

8. Tom has four times as much money as Steve. If Tom gives Steve $1, Steve will then have one-third as much money as Tom will have. How much does each boy have?

_____

9. Change a $10 bill into quarters, dimes, and nickels so that the number of nickels is one-third the number of dimes and the number of quarters is one-fifth the number of dimes. How many of each kind of coin are there in the change?

_____

10. Find three consecutive even integers such that the first is three-fifths of the last.

_____

11. In the annual Riverside Regatta, boats race three miles up the mighty Roubidoux River and return three miles downstream. If the river current is 5 mph and it takes three times as long to travel up the river as to return, what is the maximum speed of the winning boat in still water?

_____

12. Mike can plow and seed a certain lawn in three days. His brother Dan can do the same job in 4 days. How long would it take them to do the job if they worked together?

_____

13. After their annual New Year's Eve poker game, the Porter family found that Uncle Dave had won all 340 markers. He had one-fourth as many raisins as peanuts and two-thirds as many walnuts as raisins. How many of each kind of poker "chip" did cardsharp Dave have?

_____

14. What number divided by 8 is the same as 5 divided by 3?

_____

Date
_____

Name
_____

Course/Section
_____

15. If six cans of soup cost $1.12, how much would five cans cost?

_____

16. Solve:

(a) $\dfrac{7}{17} = \dfrac{x}{442}$ _____

(b) $\dfrac{a}{54} = \dfrac{7}{6}$ _____

(c) $\dfrac{6}{J} = \dfrac{1}{7}$ _____

(d) $\dfrac{y}{3} = \dfrac{1008}{112}$ _____

17. Solve:

(a) $\dfrac{3}{4} = \dfrac{6}{x-2}$ _____

(b) $-\dfrac{3}{5} = \dfrac{x-1}{x+3}$ _____

(c) $\dfrac{2x}{x+1} = \dfrac{11}{6}$ _____

(d) $\dfrac{2x+1}{1+x} = 3$ _____

18. The Zerox Copycat machine produces 270 copies of a booklet in $8\frac{1}{2}$ minutes. How much time, to the nearest minute, is needed to produce 1000 copies?

_____

19. Gary O., the clean-up batter on the Wiley Warriors softball team, managed to get 88 hits in his last 110 times at bat. At this rate, how many hits will he get in 300 times at bat?

_____

20. In a scale model clipper ship, 10 ft on the ship is represented by $1\frac{1}{2}$ in. on the model. How high would the mast on the model be if the actual mast is 75 ft high?

_____

21. If y varies directly with x, and if y equals 14 when x is 3, find y when x equals 12.

_____

22. If y varies inversely as x, and if y equals 16 when x is 2, find y when x equals 8.

_____

23. The stretch of a spring is directly proportional to the force pulling it. If a certain spring stretches 2 in. when a 5-lb weight is hung on it, how much force must be applied to it to stretch it 7 in.?

_____

24. Write equations that reflect each of the following statements:

(a) Supply varies directly with demand.

_____

(b) The distance traveled by a freely falling object is directly proportional to the square of the time of fall.

_____

(c)   The unit cost of a natural resource varies inversely with the amount available.

_____

(d)   The area of a triangle is directly proportional to its height.

_____

(e)   Running speed is inversely proportional to hill steepness.

_____

(f)   The rate of heat flow through a furnace wall is directly proportional to the temperature difference between the two sides of the wall.

_____

E. Brain Boosters

1.   The gravitational force of attraction F between two stars varies inversely with D^2, where D is the distance between the stars. Describe how F changes when

(a) D increases,

_____

(b) D doubles in size so that the stars are twice as far apart

_____

(c) D decreases to one-half so that the stars are only one-half as far apart.

_____

2.   Here are a few problems from the Rhind Papyrus, an ancient Egyptian manuscript more than 4000 years old.

(a)   Problem 24 "A quantity and its seventh, added together, become 19. What is the quantity?"

_____

(b)   Problem 25 "A quantity and its half added together become 16. What is the quantity?"

_____

(c)   Problem 26 "A quantity and its fourth added together become 15. What is the quantity?"

_____

Date

_____

Name

_____

Course/Section

(d) Problem 27 "A quantity and its fifth added together become 21. What is the quantity?"

_____

(e) Problem 28 "A quantity and its two-thirds are added together and from the sum one-third of the sum is subtracted and 10 remains. What is the quantity?"

_____

(f) Problem 35 "A quantity, its two-third, its one-half, and its three-seventh added together becomes 33. What is the quantity?"

_____

3. Solve. $\dfrac{x+5}{x-7} - 5 = \dfrac{4x-40}{13-x}$

_____

4. Simplify. $\left(x + \dfrac{1}{x}\right)\left(y + \dfrac{1}{y}\right) + \left(x - \dfrac{1}{x}\right)\left(y - \dfrac{1}{y}\right)$

_____

5. Simplify.

(a) $\dfrac{a}{a-b} + \dfrac{b}{b-a}$ _____

(b) $\dfrac{x^2}{x-y} + \dfrac{y^2}{y-x}$ _____

(c) $\dfrac{p+q}{p-q} - \dfrac{p-q}{p+q}$ _____

(d) $\dfrac{1 - \dfrac{x(1-y)}{x+y}}{1 + \dfrac{1-y}{x+y}}$ _____

6. In the odd-ball relay, the first runner on the team runs half the total distance plus half of a mile. The second runner runs one-third of the remaining distance plus one-third of a mile. The third and last runner on the team finishes after having run one-fourth of the remaining distance plus one-fourth of a mile. How long was the relay race?

_____

When you have had the practice you need either return to the preview on page 335 or turn to frame **47** and continue with the study of radical expressions.

47 The solution of most equations you have met so far in this book has involved only integers, such as 2, -3, or 152, fractions, such as $\frac{1}{2}$ or $-\frac{3}{4}$, or decimal numbers, such as 4.2, -2.7, or 1.333. . . . Because each such number can be written as a fraction or the ratio of two numbers, we call these *rational* numbers. But there are problems and equations whose solution involves another very different class of real numbers known as *irrational* numbers.

For example, the equation

$$x^2 - 2 = 0$$

cannot be factored using any integer, fraction, or decimal numbers. The values of x that make this equation a true statement are

$$x = \sqrt{2} \quad \text{and} \quad x = -\sqrt{2}$$

The number $\sqrt{2}$ is an irrational number and is defined as that number whose square is equal to 2.

$$\sqrt{2} \cdot \sqrt{2} = 2$$

The number $\sqrt{2}$ is a real number and can be written, at least approximately, as a decimal

$$\sqrt{2} = 1.4142136 \ldots$$

The decimal number continues endlessly.
It never stops and never repeats in a pattern.

Because irrational numbers are real numbers, all of the operations of arithmetic—addition, subtraction, multiplication, and division—can be used with them. In order to be able to work with radical numbers in algebra and to use them in the solution of algebra equations, you must first be able to do arithmetic with them. To check your need for a review of the arithmetic of square roots, try the following quiz.

1. Use the table of square roots on page 533 to find the value of the following numbers. Round to three decimal places.

  (a) $\sqrt{6}$ (b) $\sqrt{11}$ (c) $-\sqrt{30}$

2. Calculate each of the following and round to three decimal places.

  (a) $\sqrt{2} + \sqrt{3}$ (b) $1 - \sqrt{3}$ (c) $2\sqrt{5}$

  (d) $\sqrt{\dfrac{2}{3}}$ (e) $\dfrac{3 + \sqrt{3}}{2}$ (f) $\sqrt{3} \cdot \sqrt{5}$

Turn to **48** to check your answers.

---

48 Answers to the square root arithmetic review quiz:

1. (a) 2.449 (b) 3.317 (c) -5.477
2. (a) 3.146 (b) -0.732 (c) 4.472
   (d) 0.816 (e) 2.366 (f) 3.873

If you answered any of these problems incorrectly, turn to the arithmetic review on page 527 at the back of this book.

If you were able to answer all of these correctly, turn to frame **49** and continue.

---

49 For any number n greater than zero, we define $\sqrt{n}$ to be a positive number such that $\sqrt{n} \cdot \sqrt{n} = n$

Radical Sign
Radicand

The symbol $\sqrt{}$ is called a *radical symbol* or radical sign and the number written under the radical symbol is called the *radicand*. For example, $\sqrt{4} = 2$ and $\sqrt{49} = 7$. In general, the square root produces an irrational number.

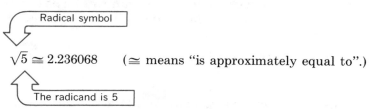

$\sqrt{5} \cong 2.236068$ ($\cong$ means "is approximately equal to".)

Radical
Radical Expression

The expression $\sqrt{5}$ is called a *radical expression* or simply a *radical*.

The radicand of a real number can *never* be negative since there is no real number whose square is negative.

Every positive number has two square roots, one positive and the other negative. For example, the square roots of 25 are 5 and -5 since

$5 \cdot 5 = 25$

and

$(-5)(-5) = 25$

The radical sign is used to name the positive square root.

$\sqrt{25} = 5$

The negative square root of 25 is written as $-\sqrt{25}$ or -5.

Use the radical symbol to name the two square roots of 2.

Check your answer in **50**.

---

50 The square roots of 2 are $\sqrt{2}$ and $-\sqrt{2}$ or, numerically, $1.4142 \ldots$ and $-1.4142 \ldots$ since by definition,

$\sqrt{2} \cdot \sqrt{2} = 2$ or $(1.4142 \ldots)(1.4142 \ldots) = 2$

and

$(-\sqrt{2})(-\sqrt{2}) = 2$ or $(-1.4142\ldots)(-1.4142\ldots) = 2$

Similarly, the square roots of any number n are $\sqrt{n}$ and $-\sqrt{n}$.

Of course $\sqrt{0} = 0$.

Principal Square Root

The symbol $\sqrt{}$ means that we are naming the *positive* square root only. For any positive real number n, $\sqrt{n}$ is the *principal* square root of n.

With algebraic expressions, if the radicand is a perfect square, the square root is easy to find. For example, if all variables in the following are positive,

$\sqrt{a^2} = a$ since $a \cdot a = a^2$
$-\sqrt{a^2} = -a$ since $(-a)(-a) = a^2$
$\sqrt{x^2 y^2} = xy$ since $(xy)(xy) = x^2 y^2$
$\sqrt{4p^2} = 2p$ since $(2p)(2p) = 4p^2$

Find each of the following square roots. (Assume that all variables are positive.)

(a) $\sqrt{x^2}$ (b) $\sqrt{a^4}$ (c) $-\sqrt{9c^2}$ (d) $\sqrt{16p^2 q^4}$

(e) $-\sqrt{\dfrac{9t^2}{m^2}}$ (f) $\sqrt{(x+y)^2}$ (g) $\sqrt{64a^{10}}$ (h) $-\sqrt{25(g-h)^2}$

Look in **51** for our answers.

---

51 (a) x (b) a^2 (c) $-3c$ (d) $4pq^2$

(e) $-\dfrac{3t}{m}$ (f) $x+y$ (g) $8a^5$ (h) $-5(g-h)$

Even Power of a Variable

You should remember from your study of exponents that when the exponent is an even integer, the expression is a perfect square.

$$a^4 = a^2 \cdot a^2$$

Even integer

Simplifying Radical Expressions

The radical expression $\sqrt{4x}$ may be simplified by factoring the radicand.

$$\sqrt{4x} = \sqrt{4 \cdot x}$$
$$= \sqrt{4} \cdot \sqrt{x}$$
$$= 2 \cdot \sqrt{x} \quad \text{or} \quad 2\sqrt{x}$$

Another example:

$$\sqrt{x^3} = \sqrt{x^2 \cdot x}$$
$$= \sqrt{x^2} \cdot \sqrt{x}$$
$$= x\sqrt{x}$$

The general rule is that the square root of a product is equal to the product of the square roots of its factors.

$$\sqrt{xy} = \sqrt{x} \cdot \sqrt{y}$$

To simplify a radical expression,

1. Factor it into two parts, one of which is a perfect square.
2. Find the square root of the perfect square factor.
3. Rewrite the radical expression using the new square root.

For example,

$$\sqrt{12a^3} = \sqrt{4a^2 \cdot 3a}$$

A perfect square

$$= \sqrt{4a^2} \cdot \sqrt{3a}$$
$$= 2a\sqrt{3a} \qquad \text{since } \sqrt{4a^2} = 2a$$

This process of simplifying radical expressions can be used to find square roots of numbers that are beyond the range of the table of square roots. For example,

$$\sqrt{1008} = \sqrt{2 \cdot 2 \cdot 2 \cdot 2 \cdot 3 \cdot 3 \cdot 7} \qquad \text{Factor the radicand completely.}$$
$$= \sqrt{2^4 \cdot 3^2 \cdot 7}$$
$$= \sqrt{2^4} \cdot \sqrt{3^2} \cdot \sqrt{7}$$
$$= 2^2 \cdot 3 \cdot \sqrt{7}$$
$$= 12\sqrt{7}$$

From the table of square roots $\sqrt{7} \cong 2.6458$ so that

$12\sqrt{7} \cong 31.7490$

or

$\sqrt{1008} \cong 31.749$ rounded to three decimal digits.

Simplify the following radical expressions.

(a) $\sqrt{12}$ (b) $-\sqrt{819}$ (c) $\sqrt{x^6}$ (d) $\sqrt{20t^7}$ (e) $\sqrt{8x^2y^3}$

Check your work in **52**.

---

52   (a) $\sqrt{12} = \sqrt{4 \cdot 3}$ (b) $-\sqrt{819} = -\sqrt{9 \cdot 91}$

$\qquad\qquad\quad = \sqrt{4} \cdot \sqrt{3}$ $= -\sqrt{9} \cdot \sqrt{91}$

$\qquad\qquad\quad = 2\sqrt{3}$ $= -3\sqrt{91}$

(c) $\sqrt{x^6} = x^3$ (d) $\sqrt{20t^7} = \sqrt{4t^6 \cdot 5t}$ An odd power of
An even power of a variable is a $= \sqrt{4t^6} \cdot \sqrt{5t}$ a variable is
perfect square. Its square root $= 2t^3\sqrt{5t}$ never a perfect
is found by dividing the expo-   square. Sim-
nent by 2.   plify it by fac-
      (e) $\sqrt{8x^2y^3} = \sqrt{4x^2y^2 \cdot 2y}$ toring out the
$\qquad\qquad\qquad\quad = \sqrt{4x^2y^2} \cdot \sqrt{2y}$ next lower even
$\qquad\qquad\qquad\quad = 2xy\sqrt{2y}$ power.

Use the following set of problems as practice in simplifying radical expressions.

1.   (a) $\sqrt{18}$ (b) $-\sqrt{40}$ (c) $\sqrt{500}$

      (d) $\sqrt{270}$ (e) $-\sqrt{72}$ (f) $\sqrt{1000}$

      (g) $\sqrt{60}$ (h) $-\sqrt{392}$ (i) $\sqrt{116}$

      (j) $\sqrt{171}$ (k) $\sqrt{750}$ (l) $-\sqrt{396}$

2.   (a) $\sqrt{x^3}$ (b) $\sqrt{x^5}$ (c) $\sqrt{a^8}$

      (d) $-\sqrt{x^9}$ (e) $\sqrt{4y^3}$ (f) $-\sqrt{9x^7}$

      (g) $\sqrt{20a^5}$ (h) $\sqrt{24p^{11}}$ (i) $\sqrt{ax^2}$

      (j) $\sqrt{p^2q}$ (k) $\sqrt{x^2y^5}$ (l) $\sqrt{m^3n^4}$

      (m) $-\sqrt{k^3p^3}$ (n) $\sqrt{x^3y^5}$ (o) $\sqrt{x^2yz^3}$

      (p) $-\sqrt{a^5b^2c^3}$ (q) $\sqrt{xy^3}$ (r) $-\sqrt{a^4c}$

      (s) $\sqrt{x^5y^6z^7}$ (t) $\sqrt{a^6b^7c^8}$

3.   (a) $\sqrt{2x^2y}$ (b) $\sqrt{5x^2y^2}$ (c) $-\sqrt{20p^3q}$

      (d) $2\sqrt{x^3}$ (e) $3\sqrt{xy^2}$ (f) $-4\sqrt{2a^2b^3}$

      (g) $3p\sqrt{p^2q}$ (h) $y\sqrt{25y^3}$ (i) $\frac{1}{2}\sqrt{28x^3y^2z}$

      (j) $-\dfrac{a}{b}\sqrt{b^3c}$ (k) $\sqrt{(x+y)^3}$ (l) $\sqrt{(a-b)^5}$

      (m) $\sqrt{x^2y+x^3}$ (n) $\sqrt{a^2bc^2-a^3bc}$ (o) $\sqrt{x^4y-x^4y^3}$

      (p) $\sqrt{4pq^2+8p^2q^2}$ (q) $\sqrt{x^2+2xy+y^2}$ (r) $\sqrt{2a^2-4a+2}$

      (s) $\sqrt{\dfrac{1}{xy} \cdot x^4y^3}$ (t) $\dfrac{p^2}{q}\sqrt{q^2p}$

Check your answers in **53**.

53

1. (a) $3\sqrt{2}$ (b) $-2\sqrt{10}$ (c) $10\sqrt{5}$
   (d) $3\sqrt{30}$ (e) $-6\sqrt{2}$ (f) $10\sqrt{10}$
   (g) $2\sqrt{15}$ (h) $-14\sqrt{2}$ (i) $2\sqrt{29}$
   (j) $3\sqrt{19}$ (k) $5\sqrt{30}$ (l) $-6\sqrt{11}$

2. (a) $x\sqrt{x}$ (b) $x^2\sqrt{x}$ (c) a^4
   (d) $-x^4\sqrt{x}$ (e) $2y\sqrt{y}$ (f) $-3x^3\sqrt{x}$
   (g) $2a^2\sqrt{5a}$ (h) $2p^5\sqrt{6p}$ (i) $x\sqrt{a}$
   (j) $p\sqrt{q}$ (k) $xy^2\sqrt{y}$ (l) $mn^2\sqrt{m}$
   (m) $-kp\sqrt{kp}$ (n) $xy^2\sqrt{xy}$ (o) $xz\sqrt{yz}$
   (p) $-a^2bc\sqrt{ac}$ (q) $y\sqrt{xy}$ (r) $-a^2\sqrt{c}$
   (s) $x^2y^3z^3\sqrt{xz}$ (t) $a^3b^3c^4\sqrt{b}$

3. (a) $x\sqrt{2y}$ (b) $xy\sqrt{5}$ (c) $-2p\sqrt{5pq}$
   (d) $2x\sqrt{x}$ (e) $3y\sqrt{x}$ (f) $-4ab\sqrt{2b}$
   (g) $3p^2\sqrt{q}$ (h) $5y^2\sqrt{y}$ (i) $xy\sqrt{7xz}$
   (j) $-a\sqrt{bc}$ (k) $(x+y)\sqrt{x+y}$ (l) $(a-b)^2\sqrt{a-b}$
   (m) $x\sqrt{y+x}$ (n) $a\sqrt{bc^2-abc}$ (o) $x^2\sqrt{y-y^3}$
   (p) $2q\sqrt{p+2p^2}$ (q) $(x+y)$ (r) $(a-1)\sqrt{2}$
   (s) $xy\sqrt{x}$ (t) $p^2\sqrt{p}$

Adding and Subtracting Radical Expressions

If you remember that

$$2a + 3a = 5a$$

Like Radicands

then it is easy to add radical expressions with the same radicand. For example,

$$2\sqrt{x} + 3\sqrt{x} = 5\sqrt{x}$$

2 + 3 = 5

Same radicand

Here we factor out $\sqrt{x}$ from each term to get

$$(2 + 3)\sqrt{x} = 5\sqrt{x}$$

If no numerical coefficient is written in front of the radical sign, the coefficient is understood to be 1.

$$\sqrt{a} = 1 \cdot \sqrt{a}$$

For example, in the sum

$$3\sqrt{a} + \sqrt{a} = 3\sqrt{a} + 1 \cdot \sqrt{a} = 4\sqrt{a}$$

The coefficient is 1

Try it. Add or subtract to simplify the following radical expressions.

(a) $2\sqrt{6} + 3\sqrt{6}$ (b) $3\sqrt{x} - \sqrt{x}$

(c) $\sqrt{pq} + 2\sqrt{pq}$ (d) $2\sqrt{y} - a\sqrt{y}$

Check your answers in **54**.

Where did the square root sign $\sqrt{\ }$ come from?

Square roots entered algebra as solutions or *roots* of equations. About 1525 someone began to abbreviate the word root, or *radix* in Latin, using the symbol R, or the letter *r* in handwriting. Soon *r* led to $\sqrt{\ }$ then to $\sqrt{\ }$ and finally to $\sqrt{\ }$.

54   (a) $2\sqrt{6} + 3\sqrt{6} = (2 + 3)\sqrt{6}$
$$= 5\sqrt{6}$$

(b) $3\sqrt{x} - \sqrt{x} = 3\sqrt{x} - 1\sqrt{x}$
$$= (3 - 1)\sqrt{x}$$
$$= 2\sqrt{x}$$

(c) $\sqrt{pq} + 2\sqrt{pq} = 1\sqrt{pq} + 2\sqrt{pq}$
$$= (1 + 2)\sqrt{pq}$$
$$= 3\sqrt{pq}$$

(d) $2\sqrt{y} - a\sqrt{y} = (2 - a)\sqrt{y}$

Unlike Radicands

If the radicands are different, it may be possible to rewrite the expressions so that they can be combined. For example,

$$\sqrt{2} + \sqrt{8} = \sqrt{2} + 2\sqrt{2} \qquad \text{since} \qquad \sqrt{8} = \sqrt{4 \cdot 2} = \sqrt{4} \cdot \sqrt{2} = 2\sqrt{2}$$
$$= 1\sqrt{2} + 2\sqrt{2} \qquad \text{The radicands are now the same and the expressions can}$$
$$= (1 + 2)\sqrt{2} \qquad \text{be added.}$$
$$= 3\sqrt{2}$$

Another example: to find the difference $3\sqrt{a^3} - a\sqrt{4a}$, first factor out all perfect squares from the radicands.

$$3\sqrt{a^3} - a\sqrt{4a} = 3a\sqrt{a} - 2a\sqrt{a} \qquad \text{since} \qquad \sqrt{a^3} = \sqrt{a^2 \cdot a} = \sqrt{a^2} \cdot \sqrt{a} = a\sqrt{a}$$
$$\text{and} \qquad \sqrt{4a} = \sqrt{4 \cdot a} = \sqrt{4} \cdot \sqrt{a} = 2\sqrt{a}$$
$$= (3a - 2a)\sqrt{a}$$
$$= a\sqrt{a}$$

Of course not all radical sums can be simplified in this way. The sum $\sqrt{2} + \sqrt{3}$ for example cannot be simplified further, but it can be evaluated approximately using the table of square roots on page 533.

$$\sqrt{2} + \sqrt{3} \cong 1.4142 + 1.7321 = 3.1463 \qquad \text{rounded to four decimal digits.}$$

Simplify each of the following radical expressions by adding or subtracting as shown. If the expression cannot be simplified, say so, and evaluate it numerically using the table of square roots.

(a) $3\sqrt{5} - \sqrt{5}$

(b) $a\sqrt{3} - b\sqrt{3}$

(c) $\sqrt{12} - 5\sqrt{3}$

(d) $\sqrt{12} + \sqrt{27}$

(e) $2\sqrt{x} - 4\sqrt{x}$

(f) $\sqrt{3} - \sqrt{5}$

(g) $\sqrt{a^3x} + \sqrt{ax^3}$

(h) $\sqrt{pq} + \sqrt{pq^3}$

(i) $3\sqrt{3y} - \sqrt{12y}$

(j) $2\sqrt{a} + 2\sqrt{9a}$

(k) $\sqrt{6} + \sqrt{2}$

(l) $\sqrt{32a} - \sqrt{50a} + \sqrt{18a}$

(m) $\sqrt{b^3} + \sqrt{b^2} + \sqrt{b}$

(n) $\sqrt{8x} - \sqrt{2x}$

When you have completed these problems, turn to **55** to check your work.

Who dreamed up radical numbers?

Hindu mathematicians used radical numbers over 1000 years ago. They knew, for example, that $\sqrt{3} + \sqrt{12} = 3\sqrt{3}$ and that $\sqrt{a} + \sqrt{b} = \sqrt{a + b + 2\sqrt{ab}}$.

---

55 (a) $3\sqrt{5} - \sqrt{5} = 3\sqrt{5} - 1\sqrt{5}$
$$= (3 - 1)\sqrt{5}$$
$$= 2\sqrt{5}$$

(b) $a\sqrt{3} - b\sqrt{3} = (a - b)\sqrt{3}$

(c) $\sqrt{12} - 5\sqrt{3} = 2\sqrt{3} - 5\sqrt{3} \qquad \text{where} \qquad \sqrt{12} = \sqrt{4 \cdot 3} = \sqrt{4} \cdot \sqrt{3} = 2\sqrt{3}$
$$= -3\sqrt{3}$$

(d) $\sqrt{12} + \sqrt{27} = 2\sqrt{3} + 3\sqrt{3} \qquad \text{where} \qquad \sqrt{27} = \sqrt{9 \cdot 3} = \sqrt{9} \cdot \sqrt{3} = 3\sqrt{3}$
$$= 5\sqrt{3}$$

(e) $2\sqrt{x} - 4\sqrt{x} = (2 - 4)\sqrt{x}$
$$= -2\sqrt{x}$$

(f) $\sqrt{3} - \sqrt{5} \qquad$ cannot be simplified further.
$$\sqrt{3} - \sqrt{5} \cong 1.7321 - 2.2361 = -0.5040$$

(g) $\sqrt{a^3x} + \sqrt{ax^3} = a\sqrt{ax} + x\sqrt{ax}$
$$= (a + x)\sqrt{ax}$$

(h) $\sqrt{pq} + \sqrt{pq^3} = 1\sqrt{pq} + q\sqrt{pq}$
$$= (1 + q)\sqrt{pq}$$

(i) $3\sqrt{3y} - \sqrt{12y} = 3\sqrt{3y} - 2\sqrt{3y}$ since $\sqrt{12y} = \sqrt{4 \cdot 3y} = \sqrt{4} \cdot \sqrt{3y} = 2\sqrt{3y}$
$$= (3 - 2)\sqrt{3y}$$
$$= \sqrt{3y}$$

(j) $2\sqrt{a} + 2\sqrt{9a} = 2\sqrt{a} + 6\sqrt{a}$ since $\sqrt{9a} = \sqrt{9} \cdot \sqrt{a} = 3\sqrt{a}$
$$= (2 + 6)\sqrt{a}$$
$$= 8\sqrt{a}$$

(k) $\sqrt{6} + \sqrt{2}$ cannot be simplified further
$$\sqrt{6} + \sqrt{2} \cong 2.4495 + 1.4142 = 3.8637$$

(l) $\sqrt{32a} - \sqrt{50a} + \sqrt{18a} = 4\sqrt{2a} - 5\sqrt{2a} + 3\sqrt{2a}$
$$= (4 - 5 + 3)\sqrt{2a}$$
$$= 2\sqrt{2a}$$

(m) $\sqrt{b^3} + \sqrt{b^2} + \sqrt{b} = b\sqrt{b} + b + \sqrt{b}$ since $\sqrt{b^3} = \sqrt{b^2 \cdot b} = b\sqrt{b}$
$$= (b + 1)\sqrt{b} + b$$

(n) $\sqrt{8x} - \sqrt{2x} = 2\sqrt{2x} - 1\sqrt{2x}$
$$= (2 - 1)\sqrt{2x}$$
$$= \sqrt{2x}$$

Radical in the Numerator of a Fraction

If a radical is included in the numerator of a fraction, the fraction can be added to or subtracted from other fractions in the usual way. For example,

$$\frac{1}{2} + \frac{\sqrt{5}}{2} = \frac{1 + \sqrt{5}}{2}$$

    Add numerators

    Same denominator

or $\dfrac{1}{3} - \dfrac{\sqrt{5}}{6} = \dfrac{1}{3} \cdot \dfrac{2}{2} - \dfrac{\sqrt{5}}{6}$ Build up the fractions so that they have the same denominators.

$$= \frac{2}{6} - \frac{\sqrt{5}}{6}$$ Combine the like fractions.

$$= \frac{2 - \sqrt{5}}{6}$$

Use this procedure to combine the following fractions.

(a) $\dfrac{1}{4} + \dfrac{\sqrt{3}}{2}$ (b) $\dfrac{1}{3} - \dfrac{\sqrt{5}}{2}$

(c) $\dfrac{1}{x} + \dfrac{\sqrt{2}}{y}$ (d) $\dfrac{a}{b} - \dfrac{\sqrt{x}}{a}$

Check your work in **56**.

56 (a) $\dfrac{1}{4} + \dfrac{\sqrt{3}}{2} = \dfrac{1}{4} + \dfrac{\sqrt{3} \cdot 2}{2 \cdot 2}$

Build up the second fraction to an equivalent fraction with the same denominator as the first fraction.

$= \dfrac{1}{4} + \dfrac{2\sqrt{3}}{4}$

Now add the like fractions.

$= \dfrac{1 + 2\sqrt{3}}{4}$

(b) $\dfrac{1}{3} - \dfrac{\sqrt{5}}{2} = \dfrac{1 \cdot 2}{3 \cdot 2} - \dfrac{\sqrt{5} \cdot 3}{2 \cdot 3}$

Build up both fractions to equivalent fractions with a denominator of 6.

$= \dfrac{2}{6} - \dfrac{3\sqrt{5}}{6}$

Add the like fractions.

$= \dfrac{2 - 3\sqrt{5}}{6}$

(c) $\dfrac{1}{x} + \dfrac{\sqrt{2}}{y} = \dfrac{1 \cdot y}{x \cdot y} + \dfrac{\sqrt{2} \cdot x}{y \cdot x}$

Build up both fractions to equivalent fractions with a denominator of xy.

$= \dfrac{y}{xy} + \dfrac{x\sqrt{2}}{xy}$

$= \dfrac{y + x\sqrt{2}}{xy}$

(d) $\dfrac{a}{b} - \dfrac{\sqrt{x}}{a} = \dfrac{a \cdot a}{b \cdot a} - \dfrac{\sqrt{x} \cdot b}{a \cdot b}$

Build up both fractions to equivalent fractions with a denominator of ab.

$= \dfrac{a^2}{ab} - \dfrac{b\sqrt{x}}{ab}$

$= \dfrac{a^2 - b\sqrt{x}}{ab}$

Multiplying Radical Expressions

As we have already shown, any radical expression can be written as the product of the square root of its factors.

$$\sqrt{ab} = \sqrt{a} \cdot \sqrt{b}$$

Of course the reverse of this is also true. The product of two square roots is equal to the square root of the product of their radicands. For example,

$$\sqrt{2} \cdot \sqrt{5} = \sqrt{2 \cdot 5}$$
$$= \sqrt{10}$$

413 **5-5 Radical Expressions**

or
$$\sqrt{3} \cdot 2\sqrt{6} = 2\sqrt{3 \cdot 6}$$
$$= 2\sqrt{3 \cdot 3 \cdot 2}$$
$$= 2\sqrt{3 \cdot 3} \cdot \sqrt{2}$$
$$= 2\sqrt{9} \cdot \sqrt{2}$$
$$= 2 \cdot 3\sqrt{2}$$
$$= 6\sqrt{2}$$

When the radicands are equal, the product is equal to the radicand itself.

$$\sqrt{2} \cdot \sqrt{2} = \sqrt{2 \cdot 2}$$
$$= \sqrt{4}$$
$$= 2$$

Multiply the following radicals.

(a) $\sqrt{3} \cdot \sqrt{2}$ (b) $\sqrt{5} \cdot \sqrt{6}$ (c) $\sqrt{a} \cdot \sqrt{3}$ (d) $\sqrt{2x} \cdot \sqrt{3y}$

(e) $\sqrt{3a} \cdot \sqrt{3b}$ (f) $\sqrt{2x} \cdot \sqrt{5x}$ (g) $2\sqrt{x} \cdot 3\sqrt{y}$ (h) $\sqrt{2} \cdot \sqrt{3} \cdot \sqrt{5}$

Check your work in **57**.

Can the radical $\sqrt{x^2 + y^2}$ be simplified to $x + y$?

No. It is already in its simplest form. Remember that $\sqrt{x^2 + y^2}$ means $\sqrt{(x^2 + y^2)}$ —it is the square root of the entire quantity $x^2 + y^2$. $\sqrt{x^2 + y^2}$ is <u>not</u> equal to $\sqrt{x^2} + \sqrt{y^2}$.

57 (a) $\sqrt{3} \cdot \sqrt{2} = \sqrt{3 \cdot 2}$ Multiply the radicands.
$$= \sqrt{6}$$

(b) $\sqrt{5} \cdot \sqrt{6} = \sqrt{5 \cdot 6}$
$$= \sqrt{30}$$

(c) $\sqrt{a} \cdot \sqrt{3} = \sqrt{a \cdot 3}$
$$= \sqrt{3a}$$

(d) $\sqrt{2x} \cdot \sqrt{3y} = \sqrt{2x \cdot 3y}$ Rewrite the radicand with a single numerical coeffi-
$$= \sqrt{6xy}$$ cient.

(e) $\sqrt{3a} \cdot \sqrt{3b} = \sqrt{3a \cdot 3b}$
$$= \sqrt{9ab}$$ Simplify by factoring out all perfect squares.
$$= \sqrt{9} \cdot \sqrt{ab}$$
$$= 3\sqrt{ab}$$

414   **Fractions and Radicals**

(f) $\sqrt{2x} \cdot \sqrt{5x} = \sqrt{2x \cdot 5x}$

$\phantom{(f)\quad \sqrt{2x} \cdot \sqrt{5x}} = \sqrt{10x^2}$ Simplify by factoring out all perfect squares.

$\phantom{(f)\quad \sqrt{2x} \cdot \sqrt{5x}} = \sqrt{10} \cdot \sqrt{x^2}$

$\phantom{(f)\quad \sqrt{2x} \cdot \sqrt{5x}} = x\sqrt{10}$

(g) $2\sqrt{x} \cdot 3\sqrt{y} = 2 \cdot 3\sqrt{x \cdot y}$

$\phantom{(g)\quad 2\sqrt{x} \cdot 3\sqrt{y}} = 6\sqrt{xy}$

(h) $\sqrt{2} \cdot \sqrt{3} \cdot \sqrt{5} = \sqrt{2 \cdot 3 \cdot 5}$

$\phantom{(h)\quad \sqrt{2} \cdot \sqrt{3} \cdot \sqrt{5}} = \sqrt{30}$

Radical in
a Binomial
Expression

If the radical is included in a binomial expression the multiplication follows the normal procedure for multiplying binomials. For example,

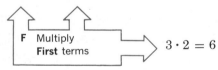

$$\boxed{\sqrt{2}}\,(2 + \sqrt{3}) = \boxed{\sqrt{2}} \cdot 2 + \boxed{\sqrt{2}} \cdot \sqrt{3}$$ Multiply each term in the binomial by $\boxed{\sqrt{2}}$.

$$= 2\sqrt{2} + \sqrt{2 \cdot 3}$$

$$= 2\sqrt{2} + \sqrt{6}$$

and

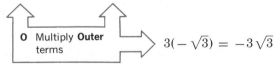

$$(3 + \sqrt{2})(2 - \sqrt{3}) = \boxed{3}\,(2 - \sqrt{3}) + \boxed{\sqrt{2}}\,(2 - \sqrt{3})$$

$$= \boxed{3} \cdot 2 - \boxed{3} \cdot \sqrt{3} + \boxed{\sqrt{2}} \cdot 2 - \boxed{\sqrt{2}} \cdot \sqrt{3}$$

$$= 6 - 3\sqrt{3} + 2\sqrt{2} - \sqrt{6}$$

The following step-by-step outline shows this multiplication of binomials.

Step 1. $(3 + \sqrt{2})(2 - \sqrt{3})$

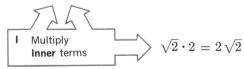

  F Multiply **First** terms $3 \cdot 2 = 6$

Step 2. $(3 + \sqrt{2})(2 - \sqrt{3})$

  O Multiply **Outer** terms $3(-\sqrt{3}) = -3\sqrt{3}$

Step 3. $(3 + \sqrt{2})(2 - \sqrt{3})$

  I Multiply **Inner** terms $\sqrt{2} \cdot 2 = 2\sqrt{2}$

Step 4. $(3 + \sqrt{2})(2 - \sqrt{3})$

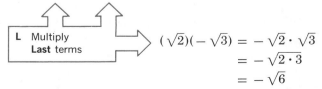

  L Multiply **Last** terms $(\sqrt{2})(-\sqrt{3}) = -\sqrt{2} \cdot \sqrt{3}$

$$\phantom{(\sqrt{2})(-\sqrt{3})} = -\sqrt{2 \cdot 3}$$

$$\phantom{(\sqrt{2})(-\sqrt{3})} = -\sqrt{6}$$

Finally, add the products.

$(3 + \sqrt{2})(2 - \sqrt{3}) = 6 - 3\sqrt{3} + 2\sqrt{2} - \sqrt{6}$

Multiply the following radical expressions.

(a) $2(1 - \sqrt{5})$ (b) $\sqrt{2}(1 + \sqrt{2})$

(c) $\sqrt{2}(\sqrt{3} + \sqrt{5})$ (d) $(1 + \sqrt{x})(1 - \sqrt{x})$

(e) $(1 + \sqrt{y})^2$ (f) $(2\sqrt{a} - \sqrt{b})(\sqrt{a} + 2\sqrt{b})$

Check your answers in **58**.

HIGHER ROOTS

The radical symbol can be used to denote other roots of a number in addition to the square root. For example,

$\sqrt[3]{8} = 2$ because $2 \cdot 2 \cdot 2 = 2^3 = 8$

Cube root   Cube

The small 3 before the radical sign is called the *root index*. If no index number appears, the radical symbol means square root as usual.

In general, cube roots or higher roots of integers are irrational numbers. For example,

$\sqrt[3]{2} \cong 1.259921$ because $(1.259921)^3 \cong 2$

$\sqrt[3]{3} \cong 1.4422496$ because $(1.4422496)^3 \cong 3$

and so on.

In advanced mathematics, even higher roots may be useful.

$\sqrt[4]{81} = 3$ because $3 \cdot 3 \cdot 3 \cdot 3 = 3^4 = 81$

Root index   Exponent

so that

$\sqrt[4]{2} \cong 1.1892071$

$\sqrt[4]{3} \cong 1.3160740$

and similarly,

$\sqrt[5]{2} \cong 1.1486984$ $\sqrt[6]{2} \cong 1.1224620 \ldots$

In basic algebra the radical symbol is used only to represent square roots. Higher roots will appear only in more advanced algebra courses.

---

58   (a)   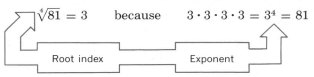

$\boxed{2}\,(1 - \sqrt{5}) = \boxed{2} \cdot 1 - \boxed{2} \cdot \sqrt{5}$ Multiply each term in the parentheses by $\boxed{2}$

$= 2 - 2\sqrt{5}$

(b) $\boxed{\sqrt{2}}\,(1 + \sqrt{2}) = \boxed{\sqrt{2}} \cdot 1 + \boxed{\sqrt{2}} \cdot \sqrt{2}$ Multiply each term in the parentheses by $\boxed{\sqrt{2}}$.

$= \sqrt{2} + \sqrt{2 \cdot 2}$

$= \sqrt{2} + \sqrt{4}$

$= \sqrt{2} + 2$

(c) $\sqrt{2}(\sqrt{3} + \sqrt{5}) = \sqrt{2} \cdot \sqrt{3} + \sqrt{2} \cdot \sqrt{5}$ Multiply each term in the parentheses
$$= \sqrt{2 \cdot 3} + \sqrt{2 \cdot 5} \quad\quad \text{by } \sqrt{2}.$$
$$= \sqrt{6} + \sqrt{10}$$

(d) $(1 + \sqrt{x})(1 - \sqrt{x}) = 1 \cdot 1 + 1 \cdot (-\sqrt{x}) + \sqrt{x} \cdot 1 + \sqrt{x}(-\sqrt{x})$
$$= 1 - \sqrt{x} + \sqrt{x} - \sqrt{x \cdot x}$$
$$= 1 - \sqrt{x^2}$$
$$= 1 - x \quad \text{Notice that the product of the sum } (1 + \sqrt{x}) \text{ and}$$

difference $(1 - \sqrt{x})$ of two terms is equal to the difference of their squares.

Multiplying Radical Expressions

The procedures learned in Unit 3 for multiplying polynomials are still valid when the terms contain radicals.

(e) $(1 + \sqrt{y})^2 = (1 + \sqrt{y})(1 + \sqrt{y})$
$$= 1 \cdot 1 + 1 \cdot \sqrt{y} + \sqrt{y} \cdot 1 + \sqrt{y} \cdot \sqrt{y}$$
$$= 1 + \sqrt{y} + \sqrt{y} + \sqrt{y \cdot y}$$
$$= 1 + 2\sqrt{y} + \sqrt{y^2}$$
$$= 1 + 2\sqrt{y} + y$$

(f) $(2\sqrt{a} - \sqrt{b})(\sqrt{a} + 2\sqrt{b}) = 2\sqrt{a} \cdot \sqrt{a} + 2\sqrt{a} \cdot 2\sqrt{b} - \sqrt{b} \cdot \sqrt{a} - \sqrt{b} \cdot 2\sqrt{b}$
$$= 2\sqrt{a^2} + 4\sqrt{ab} - \sqrt{ab} - 2\sqrt{b^2}$$
$$= 2a + 3\sqrt{ab} - 2b$$
$$= 2a - 2b + 3\sqrt{ab}$$

Dividing Radical Expressions

The quotient of two square roots is equal to the square root of the quotient of their radicals. For example,

$$\sqrt{\frac{3}{4}} = \frac{\sqrt{3}}{\sqrt{4}} = \frac{\sqrt{3}}{2}$$

Or, in general,

$$\sqrt{\frac{a}{b}} = \frac{\sqrt{a}}{\sqrt{b}} \quad\quad \text{for any real number } b \text{ not equal to zero.}$$

To find the numerical value of $\sqrt{\frac{2}{3}}$ the most obvious procedure is to divide $\sqrt{2}$ by $\sqrt{3}$.

$$\sqrt{\frac{2}{3}} = \frac{\sqrt{2}}{\sqrt{3}} \cong \frac{1.4142}{1.7321} \cong 0.8165 \text{ using the table of square roots.}$$

But it is much easier and less time-consuming to simplify the fraction first before doing the arithmetic.

$$\frac{\sqrt{2}}{\sqrt{3}} = \frac{\sqrt{2}}{\sqrt{3}} \cdot \frac{\sqrt{3}}{\sqrt{3}} \quad\quad$$ Rewrite the fraction so that the denominator is a perfect square. In this case we multiply both numerator and denominator of the fraction by $\sqrt{3}$.

$$= \frac{\sqrt{2 \cdot 3}}{\sqrt{3 \cdot 3}}$$

$$= \frac{\sqrt{6}}{3}$$

$$\cong \frac{2.4495}{3} \quad\quad \text{Evaluate using the table of square roots.}$$

$$\cong 0.8165$$

By simplifying first, we have only one square root to find and a much easier division to perform.

Rationalizing the Denominator

This process of rewriting a radical fraction so that there is no radical in the denominator is called *rationalizing the denominator* of the fraction. It is a useful convention in mathematics that we do not work with fractions that have a radical in their denominator. Instead, we rationalize the denominator and work with a simpler fraction. These examples will help you to understand the process.

Example:

$$\sqrt{\frac{1}{2}} = \frac{1}{\sqrt{2}} \qquad \text{Rationalize the denominator to remove the radical.}$$

$$= \frac{1}{\sqrt{2}} \cdot \frac{\sqrt{2}}{\sqrt{2}} \qquad \text{Multiply both numerator and denominator of the fraction by } \sqrt{2}.$$

$$= \frac{\sqrt{2}}{\sqrt{2 \cdot 2}}$$

$$= \frac{\sqrt{2}}{\sqrt{4}}$$

$$= \frac{\sqrt{2}}{2}$$

Another example:

$$\frac{2}{\sqrt{2a}} = \frac{2}{\sqrt{2a}} \cdot \frac{\sqrt{2a}}{\sqrt{2a}} \qquad \text{Multiply both numerator and denominator of the fraction by } \sqrt{2a}.$$

$$= \frac{2\sqrt{2a}}{2a}$$

$$= \frac{\sqrt{2a}}{a}$$

Rationalizing the denominator is a procedure where we build up the fraction to an equivalent fraction whose denominator includes no radicals. If the denominator is a simple radical as shown above, then multiply by that radical to rationalize the denominator.

Examples:

To rationalize $\dfrac{1}{\sqrt{2}}$ multiply by $\sqrt{2}$,

to rationalize $\dfrac{1}{\sqrt{x}}$ multiply by $\sqrt{x}$,

to rationalize $\dfrac{1}{\sqrt{3abc}}$ multiply by $\sqrt{3abc}$,

and so on.

Try it. Simplify each of the following fractions by rationalizing the denominator.

(a) $\dfrac{2}{\sqrt{x}}$ (b) $\sqrt{\dfrac{4a}{b}}$ (c) $\dfrac{\sqrt{8}}{\sqrt{2p}}$

(d) $\sqrt{\dfrac{2}{a}}$ (e) $\dfrac{\sqrt{12}}{\sqrt{3t}}$ (f) $\dfrac{2\sqrt{3}}{\sqrt{8}}$

(g) $\dfrac{a\sqrt{2}}{\sqrt{a}}$ (h) $\sqrt{\dfrac{x^3}{xy}}$ (i) $\dfrac{\sqrt{2x}}{\sqrt{3y^2}}$

Our step-by-step answers are in **59**.

FINDING THE SQUARE ROOT OF A DECIMAL NUMBER USING THE TABLE OF SQUARE ROOTS

To find the square root of any decimal number using the table of square roots, follow these steps:

Step 1. Rewrite the number as a fraction whose denominator is a power of ten.

Example A: $\sqrt{0.3} = \sqrt{\dfrac{3}{10}}$

Example B: $\sqrt{2.4} = \sqrt{\dfrac{24}{10}}$

Example C: $\sqrt{0.002} = \sqrt{\dfrac{2}{1000}}$

Step 2. Rationalize the denominator of the fraction obtained and write it in its simplest radical form.

Example A: $\sqrt{\dfrac{3}{10}} = \dfrac{\sqrt{3}}{\sqrt{10}} \cdot \dfrac{\sqrt{10}}{\sqrt{10}} = \dfrac{\sqrt{30}}{10}$

Example B: $\sqrt{\dfrac{24}{10}} = \dfrac{\sqrt{24}}{\sqrt{10}} \cdot \dfrac{\sqrt{10}}{\sqrt{10}} = \dfrac{\sqrt{240}}{10} = \dfrac{4\sqrt{15}}{10} = \dfrac{2\sqrt{15}}{5}$

Example C: $\sqrt{\dfrac{2}{1000}} = \dfrac{\sqrt{2}}{\sqrt{1000}} = \dfrac{\sqrt{2}}{10\sqrt{10}}$

$= \dfrac{\sqrt{2}}{10\sqrt{10}} \cdot \dfrac{\sqrt{10}}{\sqrt{10}} = \dfrac{\sqrt{20}}{100} = \dfrac{\sqrt{5}}{50}$

Step 3. Use the table of square roots to evaluate the final expression.

Example A: $\sqrt{0.3} = \dfrac{\sqrt{30}}{10} \cong \dfrac{5.4772}{10} \cong 0.548$

Example B: $\sqrt{2.4} = \dfrac{2\sqrt{15}}{5} \cong \dfrac{2 \cdot 3.8730}{5} \cong 1.549$

Example C: $\sqrt{0.002} = \dfrac{\sqrt{5}}{50} \cong \dfrac{2.2361}{50} \cong 0.0447$

Step 4. Check your answers.

Example A: $0.548 \cdot 0.548 = 0.30030 \cong 0.3$

Example B: $1.549 \cdot 1.549 = 2.39940 \cong 2.4$

Example C: $0.0447 \cdot 0.0447 = 0.001998 \cong 0.002$

The checks show that the answers are approximately correct.

59 (a) $\dfrac{2}{\sqrt{x}} = \dfrac{2}{\sqrt{x}} \cdot \dfrac{\sqrt{x}}{\sqrt{x}}$ Multiply numerator and denominator of the fraction by $\sqrt{x}$.

$= \dfrac{2\sqrt{x}}{\sqrt{x \cdot x}}$

$= \dfrac{2\sqrt{x}}{x}$

(b) $\sqrt{\dfrac{4a}{b}} = \dfrac{\sqrt{4a}}{\sqrt{b}}$

$= \dfrac{\sqrt{4a} \cdot \sqrt{b}}{\sqrt{b} \cdot \sqrt{b}}$ Multiply numerator and denominator by $\sqrt{b}$.

$= \dfrac{\sqrt{4a \cdot b}}{\sqrt{b \cdot b}}$

$= \dfrac{2\sqrt{ab}}{b}$ $\sqrt{4ab} = \sqrt{4} \cdot \sqrt{ab} = 2\sqrt{ab}$

(c) $\dfrac{\sqrt{8}}{\sqrt{2p}} = \dfrac{\sqrt{8}}{\sqrt{2p}} \cdot \dfrac{\sqrt{2p}}{\sqrt{2p}}$ Multiply numerator and denominator by $\sqrt{2p}$.

$= \dfrac{\sqrt{8 \cdot 2p}}{\sqrt{2p \cdot 2p}}$

$= \dfrac{\sqrt{16p}}{2p}$ Remember that $\sqrt{16p} = \sqrt{16} \cdot \sqrt{p} = 4\sqrt{p}$ and $\sqrt{2p \cdot 2p} = 2p$

$= \dfrac{4\sqrt{p}}{2p}$

$= \dfrac{2\sqrt{p}}{p}$

(d) $\sqrt{\dfrac{2}{a}} = \dfrac{\sqrt{2}}{\sqrt{a}}$

$= \dfrac{\sqrt{2} \cdot \sqrt{a}}{\sqrt{a} \cdot \sqrt{a}}$ Multiply numerator and denominator by $\sqrt{a}$.

$= \dfrac{\sqrt{2 \cdot a}}{\sqrt{a \cdot a}}$

$= \dfrac{\sqrt{2a}}{a}$

(e) $\dfrac{\sqrt{12}}{\sqrt{3t}} = \dfrac{\sqrt{12}}{\sqrt{3t}} \cdot \dfrac{\sqrt{3t}}{\sqrt{3t}}$ Multiply numerator and denominator by $\sqrt{3t}$.

$= \dfrac{\sqrt{12 \cdot 3t}}{\sqrt{3t \cdot 3t}}$

$= \dfrac{\sqrt{36t}}{3t}$ $\sqrt{36t} = \sqrt{36} \cdot \sqrt{t} = 6\sqrt{t}$

$= \dfrac{6\sqrt{t}}{3t}$

$= \dfrac{2\sqrt{t}}{t}$

(f) $\dfrac{2\sqrt{3}}{\sqrt{8}} = \dfrac{2\sqrt{3}}{\sqrt{8}} \cdot \dfrac{\sqrt{8}}{\sqrt{8}}$ Multiply numerator and denominator by $\sqrt{8}$.

$$= \frac{2\sqrt{24}}{8} \qquad 2 \cdot \sqrt{3} \cdot \sqrt{8} = 2\sqrt{3 \cdot 8} = 2\sqrt{24}$$

$$\sqrt{24} = \sqrt{4 \cdot 6} = 2\sqrt{6}$$

$$= \frac{2 \cdot 2\sqrt{6}}{8}$$

$$= \frac{\sqrt{6}}{2}$$

(g) $\dfrac{a\sqrt{2}}{\sqrt{a}} = \dfrac{a\sqrt{2} \cdot \boxed{\sqrt{a}}}{\sqrt{a} \cdot \boxed{\sqrt{a}}}$

$$= \frac{a\sqrt{2a}}{a}$$

$$= \sqrt{2a}$$

(h) $\sqrt{\dfrac{x^3}{xy}} = \dfrac{\sqrt{x^3}}{\sqrt{xy}}$

$= \dfrac{\sqrt{x^3} \cdot \boxed{\sqrt{xy}}}{\sqrt{xy} \cdot \boxed{\sqrt{xy}}}$ Multiply numerator and denominator by $\boxed{\sqrt{xy}}$.

$= \dfrac{\sqrt{x^4 y}}{xy}$ $\sqrt{x^3} \cdot \sqrt{xy} = \sqrt{x^3 \cdot xy} = \sqrt{x^4 y}$

$= \dfrac{x^2 \sqrt{y}}{xy}$ $\sqrt{x^4 y} = \sqrt{x^4} \cdot \sqrt{y} = x^2\sqrt{y}$

$= \dfrac{x\sqrt{y}}{y}$

(i) $\dfrac{\sqrt{2x}}{\sqrt{3y^2}} = \dfrac{\sqrt{2x}}{y\sqrt{3}}$ Simplifying the denominator.

$= \dfrac{\sqrt{2x} \cdot \boxed{\sqrt{3}}}{y\sqrt{3} \cdot \boxed{\sqrt{3}}}$ Multiply numerator and denominator by $\boxed{\sqrt{3}}$.

$= \dfrac{\sqrt{6x}}{3y}$

Radical in the Denominator of a Fraction

If the denominator of the fraction is a radical expression of the form $a + \sqrt{b}$, then the denominator can be rationalized by multiplying both numerator and denominator by the radical expression $a - \sqrt{b}$. For example, to rationalize $\dfrac{1}{2 + \sqrt{2}}$ multiply by $\boxed{2 - \sqrt{2}}$.

$$\frac{1}{2 + \sqrt{2}} = \frac{1}{2 + \sqrt{2}} \cdot \boxed{\frac{2 - \sqrt{2}}{2 - \sqrt{2}}}$$

$$= \frac{2 - \sqrt{2}}{(2 + \sqrt{2})(2 - \sqrt{2})}$$

But $(2 + \sqrt{2})(2 - \sqrt{2}) = 4 - 2 = 2$

Therefore the original fraction is $\dfrac{2 - \sqrt{2}}{2}$.

Notice that the product $(2 + \sqrt{2})(2 - \sqrt{2})$ is equal to a rational number.

$$(2 + \sqrt{2})(2 - \sqrt{2}) = 2 \cdot 2 + 2(-\sqrt{2}) + \sqrt{2} \cdot 2 + \sqrt{2}(-\sqrt{2})$$

$$= 4 - 2\sqrt{2} + 2\sqrt{2} - \sqrt{2} \cdot \sqrt{2}$$

$$= 4 - \sqrt{2 \cdot 2} = 4 - 2 = 2$$

This last product should remind you of a similar product you studied in Unit 3.

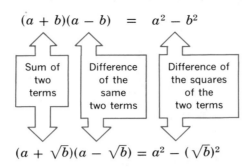

$$(a + \sqrt{b})(a - \sqrt{b}) = a^2 - (\sqrt{b})^2$$

Any radical expression of the form $(a + b)$, where a or b may contain radicals, can be rationalized by multiplying by another radical expression $(a - b)$. The expression $(a - b)$ is called the *conjugate* of $(a + b)$.

The conjugate of $\sqrt{2} + 1$ is $\sqrt{2} - 1$ and $(\sqrt{2} + 1)(\sqrt{2} - 1) = (\sqrt{2})^2 - 1^2$
$$= 2 - 1 = 1$$

The conjugate of $\sqrt{5} + \sqrt{2}$ is $\sqrt{5} - \sqrt{2}$ and $(\sqrt{5} + \sqrt{2})(\sqrt{5} - \sqrt{2}) = (\sqrt{5})^2 - (\sqrt{2})^2$
$$= 5 - 2 = 3$$

The conjugate of $-3\sqrt{5} - 2\sqrt{3}$ is $-3\sqrt{5} + 2\sqrt{3}$ and
$$(-3\sqrt{5} - 2\sqrt{3})(-3\sqrt{5} + 2\sqrt{3}) = (-3\sqrt{5})^2 - (2\sqrt{3})^2$$
$$= 9 \cdot 5 - 4 \cdot 3$$
$$= 45 - 12 = 33$$

For each of the following fractions, multiply by the conjugate of the denominator to rationalize the fraction.

(a) $\dfrac{3}{\sqrt{3} - 1}$ (b) $\dfrac{7}{2\sqrt{2} + 1}$ (c) $\dfrac{1 + \sqrt{x}}{2 + \sqrt{x}}$ (d) $\dfrac{2\sqrt{2}}{3\sqrt{2} - 2\sqrt{3}}$

Check your work in **60**.

---

60   (a)   The conjugate of $\sqrt{3} - 1$ is $\sqrt{3} + 1$.

$$\frac{3}{\sqrt{3} - 1} = \frac{3}{(\sqrt{3} - 1)} \cdot \frac{(\sqrt{3} + 1)}{(\sqrt{3} + 1)}$$ Multiply both numerator and denominator of the fraction by the conjugate.

$$= \frac{3(\sqrt{3} + 1)}{(\sqrt{3})^2 - 1^2}$$

$$= \frac{3\sqrt{3} + 3}{3 - 1}$$

$$= \frac{3\sqrt{3} + 3}{2}$$

(b)   The conjugate of $2\sqrt{2} + 1$ is $2\sqrt{2} - 1$.

$$\frac{7}{2\sqrt{2} + 1} = \frac{7}{(2\sqrt{2} + 1)} \cdot \frac{(2\sqrt{2} - 1)}{(2\sqrt{2} - 1)}$$ Multiply both numerator and denominator by the conjugate.

$$= \frac{7(2\sqrt{2} - 1)}{(2\sqrt{2})^2 - 1^2}$$

$$= \frac{7(2\sqrt{2} - 1)}{8 - 1}$$

$$= \frac{7(2\sqrt{2} - 1)}{7}$$

$$= 2\sqrt{2} - 1$$

(c) The conjugate of $2 + \sqrt{x}$ is $2 - \sqrt{x}$.

$$\frac{1 + \sqrt{x}}{2 + \sqrt{x}} = \frac{(1 + \sqrt{x}) \cdot (2 - \sqrt{x})}{(2 + \sqrt{x}) \cdot (2 - \sqrt{x})}$$

$$= \frac{2 + \sqrt{x} - x}{2^2 - (\sqrt{x})^2}$$

$$= \frac{2 + \sqrt{x} - x}{4 - x}$$

(d) The conjugate of $3\sqrt{2} - 2\sqrt{3}$ is $3\sqrt{2} + 2\sqrt{3}$.

$$\frac{2\sqrt{2}}{3\sqrt{2} - 2\sqrt{3}} = \frac{2\sqrt{2}}{(3\sqrt{2} - 2\sqrt{3})} \cdot \frac{(3\sqrt{2} + 2\sqrt{3})}{(3\sqrt{2} + 2\sqrt{3})}$$

$$= \frac{2\sqrt{2}(3\sqrt{2} + 2\sqrt{3})}{(3\sqrt{2})^2 - (2\sqrt{3})^2}$$

$$= \frac{6\sqrt{2}\sqrt{2} + 4\sqrt{2}\sqrt{3}}{18 - 12}$$

$$= \frac{12 + 4\sqrt{6}}{6}$$

$$= \frac{6 + 2\sqrt{6}}{3}$$

Simplest Radical Form

The purpose of learning to rationalize the denominator of a fraction is to enable you to write a radical expression in its simplest form. If an algebraic expression includes radicals we can rewrite it in its *simplest radical form* by

1. Rewriting any fractions that appear under the radical sign.

    Example: $\sqrt{\dfrac{2x}{3}} = \dfrac{\sqrt{2x}}{\sqrt{3}}$

2. Rationalizing the denominator so that no radicals appear in the denominator.

    Example: $\dfrac{4}{\sqrt{2a}} = \dfrac{2\sqrt{2a}}{a}$

3. Factoring out any perfect squares that appear under the radical sign.

    Example: $\sqrt{8x^3 y} = \sqrt{4x^2} \cdot \sqrt{2xy} = 2x\sqrt{2xy}$

Write each of the following radical expressions in its simplest radical form.

(a) $\sqrt{\dfrac{4\pi R^2}{3}}$ (b) $\sqrt{\dfrac{9x^3}{2x - 1}}$

Check your work in **61**.

---

61 (a) First, rewrite the fraction as a quotient of radicals.

$$\sqrt{\frac{4\pi R^2}{3}} = \frac{\sqrt{4\pi R^2}}{\sqrt{3}}$$

Second, rationalize the denominator.

$$= \frac{\sqrt{4\pi R^2}}{\sqrt{3}} \cdot \frac{\sqrt{3}}{\sqrt{3}}$$

$$= \frac{\sqrt{4\pi R^2 \cdot 3}}{3}$$

Third, factor out all perfect squares.

$$= \frac{\sqrt{4R^2} \cdot \sqrt{3\pi}}{3}$$

$$= \frac{2R\sqrt{3\pi}}{3} \qquad \text{This expression is in its simplest radical form.}$$

(b) First, rewrite the fraction as a quotient of radical expressions.

$$\sqrt{\frac{9x^3}{2x - 1}} = \frac{\sqrt{9x^3}}{\sqrt{2x - 1}}$$

Second, rationalize the denominator.

$$= \frac{\sqrt{9x^3} \cdot \sqrt{2x - 1}}{\sqrt{2x - 1} \cdot \sqrt{2x - 1}}$$

$$= \frac{\sqrt{9x^3(2x - 1)}}{2x - 1}$$

Third, factor out all perfect squares under the radical sign.

$$= \frac{\sqrt{9x^2} \cdot \sqrt{x(2x - 1)}}{2x - 1}$$

$$= \frac{3x\sqrt{2x^2 - x}}{2x - 1} \qquad \text{This expression is in its simplest radical form.}$$

Fractional Exponents

It is possible to use fractional exponents to represent square roots or other radicals. By definition, for any positive real number a,

$$\sqrt{a} = a^{1/2}$$

For example, $\qquad 2^{1/2} = \sqrt{2}$

$$3^{1/2} = \sqrt{3}$$

or $\qquad 4^{1/2} = \sqrt{4} = 2$

and so on.

In general, if a is a real number, $a \geq 0$, and n is any positive integer greater than 1,

$$\boxed{a^{1/n} = \sqrt[n]{a}}$$

Examples:

$$9^{1/2} = \sqrt{9} = 3 \qquad \text{since} \qquad 3 \cdot 3 = 3^2 = 9$$

$$8^{1/3} = \sqrt[3]{8} = 2 \qquad \text{since} \qquad 2 \cdot 2 \cdot 2 = 2^3 = 8$$

$$625^{1/4} = \sqrt[4]{625} = 5 \qquad \text{since} \qquad 5 \cdot 5 \cdot 5 \cdot 5 = 5^4 = 625$$

Find the value of each of the following:

(a) $\sqrt[4]{16}$ $\qquad$ (b) $\sqrt[3]{27}$ $\qquad$ (c) $32^{1/5}$ $\qquad$ (d) $256^{1/4}$

Check your work in **62**.

62

(a) $\sqrt[4]{16} = 2$ since $2 \cdot 2 \cdot 2 \cdot 2 = 2^4 = 16$

(b) $\sqrt[3]{27} = 3$ since $3 \cdot 3 \cdot 3 = 3^3 = 27$

(c) $32^{1/5} = 2$ since $2 \cdot 2 \cdot 2 \cdot 2 \cdot 2 = 2^5 = 32$

(d) $256^{1/4} = 4$ since $4 \cdot 4 \cdot 4 \cdot 4 = 4^4 = 256$

In general, if $a \geq 0$ and if m and n are positive integers, with n greater than 1, there are two equivalent ways of writing a number with a fractional exponent.

$$\boxed{a^{m/n} = (\sqrt[n]{a})^m \qquad \text{and} \qquad a^{m/n} = \sqrt[n]{a^m}}$$

Examples:

$$8^{2/3} = (\sqrt[3]{8})^2 = (2)^2 = 4 \qquad \text{or} \qquad 8^{2/3} = \sqrt[3]{8^2} = \sqrt[3]{64} = 4$$

$$4^{5/2} = (\sqrt{4})^5 = (2)^5 = 32 \qquad \text{or} \qquad 4^{5/2} = \sqrt{4^5} = \sqrt{1024} = 32$$

Of course, the fractional exponent can always be reduced to lowest terms. For example,

$$5^{3/6} = 5^{1/2} = \sqrt{5}$$

and

$$8^{2/6} = 8^{1/3} = \sqrt[3]{8} = 2$$

Very often it is easier to work with radical expressions if we write them in terms of fractional exponents and then use the basic rules for working with exponents presented in Unit 1.

Rule 1 $x^m \cdot x^n = x^{m+n}$

Rule 2 $(x^n)^m = x^{nm}$

Rule 3 $\dfrac{x^m}{x^n} = x^{m-n}$

Rule 4 $x^{-n} = \dfrac{1}{x^n}$

Here are some examples of each of these rules applied to fractional exponents.

Examples: **Rule 1** $x^m \cdot x^n = x^{m+n}$

$$x^{1/4} \cdot x^{1/2} = x^{1/4 + 1/2} = x^{3/4}$$

$$a^3 \cdot a^{-1/2} \cdot a^{2/3} = a^{3 - 1/2 + 2/3} = a^{19/6}$$

$$y^{1/2} \cdot y^{1/3} = y^{1/2 + 1/3} = y^{5/6}$$

Examples: **Rule 2** $(x^n)^m = x^{nm}$

$$(x^2)^{1/3} = x^{2(1/3)} = x^{2/3}$$

$$(xy^3)^{1/2} = (x^1)^{1/2} \cdot (y^3)^{1/2} = x^{1/2}y^{3/2}$$

$$(a^{-2}b^3)^{1/4} = a^{-2(1/4)} \cdot b^{3(1/4)} = a^{-1/2}b^{3/4}$$

$$(4x^2)^{3/2} = 4^{3/2} \cdot x^{2(3/2)} = (4^{1/2})^3 \cdot x^3 = (\sqrt{4})^3 x^3 = 2^3 x^3 = 8x^3$$

Examples: **Rule 3** $\dfrac{x^m}{x^n} = x^{m-n}$

$$\frac{x^{1/2}}{x^{1/3}} = x^{1/2 - 1/3} = x^{1/6}$$

$$\frac{a^{3/4}}{a^{-2/3}} = a^{3/4 - (-2/3)} = a^{3/4 + 2/3} = a^{17/12}$$

$$\frac{x^3}{x^{1/3}} = x^{3 - 1/3} = x^{8/3}$$

Examples: Rule 4 $x^{-n} = \dfrac{1}{x^n}$

$x^{-2/3} = \dfrac{1}{x^{2/3}}$

$(ab)^{-1/2} = \dfrac{1}{a^{1/2}b^{1/2}}$

$(4x^2)^{-3/2} = \dfrac{1}{4^{3/2}x^{2(3/2)}} = \dfrac{1}{8x^3}$

We can use these rules in combination to simplify expressions involving radicals and fractional exponents.

Examples:

$(a^2b^3)^{1/2}b^{1/3} = a^{2 \cdot 1/2} \cdot b^{3 \cdot 1/2} \cdot b^{1/3} = a^1 \cdot b^{3/2 + 1/3} = ab^{11/6}$

$(x^{-2}y)^{-1/2} = x^{-2(-1/2)} \cdot y^{-1/2} = x^1 \cdot y^{-1/2} = \dfrac{x}{y^{1/2}}$ or $xy^{-1/2}$

$(8x^2)^{2/3}(x^{-1/4})^2 = 8^{2/3} \cdot x^{2(2/3)} \cdot x^{-1/4(2)} = 4 \cdot x^{4/3} \cdot x^{-1/2} = 4 \cdot x^{4/3 - 1/2} = 4x^{5/6}$

$\left(\dfrac{3}{4x}\right)^{-1/2} = \dfrac{3^{-1/2}}{4^{-1/2}x^{-1/2}} = 3^{-1/2} \cdot 4^{1/2} \cdot x^{1/2} = \dfrac{2x^{1/2}}{3^{1/2}}$ or $2 \cdot 3^{-1/2}x^{1/2}$

Try it. Use these rules to simplify each of the following expressions. Write the answer with positive exponents only.

(a) $16^{3/4}$

(b) $32^{-3/5}$

(c) $\left(\dfrac{4}{9}\right)^{3/2}$

(d) $\left(\dfrac{8}{27}\right)^{4/3}$

(e) $x^{3/6}$

(f) $y^{8/4}$

(g) $a^{-1/2}$

(h) $b^{-3/4}$

(i) $y^{-1/2} \cdot y^{2/3}$

(j) $x^{3/4} \cdot x^{-1/2} \cdot x^{1/3}$

(k) $\left(\dfrac{x^2}{y^{-2/3}}\right)^{1/2}$

(l) $\left(\dfrac{x^{2/3}}{y^{-1/4}}\right)^{1/3}$

(m) $(9x)^{-3/2}$

(n) $(8y^2)^{-2/3}$

(o) $(x^{-3}y^4)^{1/2}$

(p) $(p^2t^{-3})^{1/3}$

(q) $\left(\dfrac{x}{y}\right)^{-1/3}$

(r) $\left(\dfrac{a^2b^{-1/3}}{c^{-1/4}}\right)^{-2}$

(s) $\dfrac{x^{-2/3} \cdot x^{3/2}}{x^{1/3}}$

(t) $\dfrac{x^{1/4} \cdot x^{-1/2}}{x^{2/3}}$

Check your answers in **63**.

---

63   (a) 8

(b) $\dfrac{1}{8}$

(c) $\dfrac{8}{27}$

(d) $\dfrac{16}{81}$

(e) $x^{1/2}$

(f) y^2

(g) $\dfrac{1}{a^{1/2}}$

(h) $\dfrac{1}{b^{3/4}}$

(i) $y^{1/6}$

(j) $x^{7/12}$

(k) $xy^{1/3}$

(l) $x^{2/9}y^{1/12}$

(m) $\dfrac{1}{27x^{3/2}}$

(n) $\dfrac{1}{4y^{4/3}}$

(o) $\dfrac{y^2}{x^{3/2}}$

(p) $\dfrac{p^{2/3}}{t}$

(q) $\dfrac{y^{1/3}}{x^{1/3}}$

(r) $\dfrac{b^{2/3}c^{1/2}}{a^4}$

(s) $x^{1/2}$

(t) $\dfrac{1}{x^{11/12}}$

Now turn to frame **64** for a set of practice problems on radical expressions.

Unit 5
Fractions and Radicals

PROBLEM SET 5-5

64

Radical Expressions

The answers are on page 563.

A. Simplify each of the following radicals.

1. $\sqrt{28}$ ___

2. $\sqrt{27}$ ___

3. $\sqrt{45}$ ___

4. $\sqrt{48}$ ___

5. $-\sqrt{63}$ ___

6. $\sqrt{125}$ ___

7. $\sqrt{200}$ ___

8. $-\sqrt{98}$ ___

9. $\sqrt{108}$ ___

10. $\sqrt{320}$ ___

11. $-\sqrt{44}$ ___

12. $\sqrt{117}$ ___

13. $-\sqrt{50}$ ___

14. $\sqrt{700}$ ___

15. $\sqrt{567}$ ___

16. $\sqrt{p^3}$ ___

17. $-\sqrt{x^5}$ ___

18. $\sqrt{n^2}$ ___

19. $\sqrt{4x^2}$ ___

20. $\sqrt{4y}$ ___

21. $-\sqrt{9ab}$ ___

22. $\sqrt{\dfrac{4x}{a^2}}$ ___

23. $\sqrt{20x^3}$ ___

24. $\sqrt{24y^5}$ ___

25. $-\sqrt{2a^2b}$ ___

26. $\sqrt{12p^6}$ ___

27. $\sqrt{45a^4b}$ ___

28. $\sqrt{x^3y^2z}$ ___

29. $-\sqrt{ab^2c^3}$ ___

30. $\sqrt{36p^2q^3}$ ___

31. $-2\sqrt{4x}$ ___

32. $3a\sqrt{9a^3}$ ___

33. $-4\sqrt{3a^5}$ ___

34. $\dfrac{3a}{b}\sqrt{ab^2}$ ___

35. $\dfrac{2x}{y}\sqrt{4x^2y^3}$ ___

36. $\sqrt{18(x+y)^2}$ ___

37. $\sqrt{8(a-b)^3}$ ___

38. $\sqrt{4x^2+4x+1}$ ___

39. $\sqrt{x^3+x^2y}$ ___

40. $-\sqrt{2x^2-x^3}$ ___

B. Add or subtract as shown.

1. $2\sqrt{5}+3\sqrt{5}$ ___

2. $5\sqrt{3}-3\sqrt{3}$ ___

3. $2\sqrt{6}-5\sqrt{6}$ ___

4. $3\sqrt{5}-\sqrt{5}$ ___

5. $7\sqrt{6}-2\sqrt{6}$ ___

6. $\sqrt{3}-2\sqrt{3}$ ___

7. $2\sqrt{x}-\sqrt{x}$ ___

8. $3\sqrt{a}-4\sqrt{a}$ ___

9. $\sqrt{pq}+2\sqrt{pq}$ ___

10. $\sqrt{y}-3\sqrt{y}$ ___

11. $-\sqrt{x}-2\sqrt{x}$ ___

12. $2\sqrt{2a}+\sqrt{2a}$ ___

13. $a\sqrt{b}-b\sqrt{b}$ ___

14. $k\sqrt{x}-\sqrt{x}$ ___

15. $2\sqrt{3t}+\sqrt{3t}$ ___

16. $\sqrt{2}+\sqrt{18}$ ___

17. $2\sqrt{3}+\sqrt{12}$ ___

18. $\sqrt{20}-\sqrt{5}$ ___

Date ___

Name ___

Course/Section ___

427 Problem Set 5-5

Copyright © 1982 by John Wiley & Sons, Inc.

19. $\sqrt{4x} + \sqrt{9x}$ _____

20. $3\sqrt{a} - \sqrt{4a}$ _____

21. $8\sqrt{2x} - 2\sqrt{8x}$ _____

22. $\sqrt{ab} + \sqrt{a^3b}$ _____

23. $\sqrt{2x} - \sqrt{50x}$ _____

24. $\sqrt{4x^3} - a\sqrt{x}$ _____

25. $2\sqrt{3p} + \sqrt{12p}$ _____

26. $\sqrt{3x^2} + \sqrt{3y^2}$ _____

27. $\sqrt{2p^2} - \sqrt{8p^2}$ _____

28. $\sqrt{2} + 2\sqrt{18} - \sqrt{50}$ _____

29. $\sqrt{4x^3} - x\sqrt{9x} + x\sqrt{x}$ _____

30. $\sqrt{ab} + \sqrt{4a^3b} + \sqrt{9ab^3}$ _____

31. $\dfrac{1}{3} - \dfrac{\sqrt{3}}{3}$ _____

32. $\dfrac{2}{a} + \dfrac{\sqrt{2}}{a}$ _____

33. $\dfrac{\sqrt{3}}{x} - \dfrac{\sqrt{2}}{x}$ _____

34. $\dfrac{1}{5} - \dfrac{\sqrt{2}}{2}$ _____

35. $\dfrac{3}{a} - \dfrac{\sqrt{5}}{b}$ _____

36. $\dfrac{2}{x} + \dfrac{\sqrt{x}}{3}$ _____

37. $\dfrac{\sqrt{3}}{5} - \dfrac{1}{3}$ _____

38. $\dfrac{2\sqrt{3}}{3} - \dfrac{\sqrt{2}}{2}$ _____

39. $\dfrac{2\sqrt{3}}{3} - 2$ _____

40. $\dfrac{2\sqrt{5}}{3} + 2$ _____

C. Multiply and simplify.

1. $\sqrt{3} \cdot \sqrt{5}$ _____

2. $\sqrt{2} \cdot \sqrt{6}$ _____

3. $\sqrt{3} \cdot \sqrt{8}$ _____

4. $\sqrt{2} \cdot \sqrt{7}$ _____

5. $\sqrt{3} \cdot \sqrt{6}$ _____

6. $\sqrt{2} \cdot \sqrt{8}$ _____

7. $\sqrt{5} \cdot \sqrt{7}$ _____

8. $\sqrt{6} \cdot \sqrt{5}$ _____

9. $\sqrt{2} \cdot \sqrt{3} \cdot \sqrt{5}$ _____

10. $\sqrt{3} \cdot \sqrt{6} \cdot \sqrt{10}$ _____

11. $\sqrt{5} \cdot \sqrt{10} \cdot \sqrt{3}$ _____

12. $\sqrt{3} \cdot \sqrt{10} \cdot \sqrt{8}$ _____

13. $\sqrt{x} \cdot \sqrt{y}$ _____

14. $\sqrt{a} \cdot \sqrt{a}$ _____

15. $\sqrt{b} \cdot \sqrt{bc}$ _____

16. $\sqrt{2x} \cdot \sqrt{xy}$ _____

17. $\sqrt{2a} \cdot \sqrt{3a}$ _____

18. $\sqrt{3ab} \cdot \sqrt{2ab}$ _____

19. $\sqrt{2pq} \cdot \sqrt{6pq^2}$ _____

20. $\sqrt{3a^2b} \cdot \sqrt{6b^2}$ _____

21. $\sqrt{8y} \cdot \sqrt{4y^3}$ _____

22. $\sqrt{12t} \cdot \sqrt{32t}$ _____

23. $\sqrt{5k^2} \cdot \sqrt{10k}$ _____

24. $2\sqrt{2a} \cdot 3\sqrt{3a}$ _____

25. $2\sqrt{x} \cdot 3\sqrt{x} \cdot \sqrt{x}$ _____

26. $a\sqrt{5} \cdot a\sqrt{3}$ _____

27. $x\sqrt{y} \cdot y\sqrt{x} \cdot \sqrt{xy}$ _____

28. $x\sqrt{x} \cdot x^2\sqrt{x} \cdot \sqrt{x^3}$ _____

29. $2(1 + \sqrt{6})$ _____

30. $3(\sqrt{2} + \sqrt{3})$ _____

31. $5(\sqrt{3} - 1)$ _____

32. $2(\sqrt{8} + \sqrt{6})$ _____

33. $\sqrt{2}(2 + \sqrt{2})$ _____

34. $\sqrt{3}(\sqrt{2} + \sqrt{3})$ _____

35. $\sqrt{3}(\sqrt{2} - \sqrt{6})$ _____

36. $\sqrt{5}(2 + \sqrt{10})$ _____

37. $\sqrt{3}(\sqrt{3} - \sqrt{5})$ _____

38. $\sqrt{2}(\sqrt{2} - \sqrt{6})$ _____

39. $\sqrt{x}(\sqrt{x} - \sqrt{y})$ _____

40. $a\sqrt{b}(b\sqrt{a} - a\sqrt{b})$ _____

41. $(1 + \sqrt{2})(1 + \sqrt{2})$ _____

42. $(1 + \sqrt{2})(1 - \sqrt{2})$ _____

43. $(2 + \sqrt{2})^2$ _____

44. $(\sqrt{2} + \sqrt{3})^2$ _____

45. $(3 - 2\sqrt{3})(2 + \sqrt{3})$ _____

46. $(\sqrt{5} - \sqrt{2})(1 + \sqrt{2})$ _____

47. $(a + \sqrt{b})(a - \sqrt{b})$ _____

48. $(x + \sqrt{y})^2$ _____

49. $(2 + 3\sqrt{x})^2$ _____

50. $(3\sqrt{5} - 2\sqrt{3})(\sqrt{5} + \sqrt{3})$ _____

D. Write each expression in the simplest radical form.

1. $\dfrac{1}{\sqrt{6}}$ _____

2. $\dfrac{1}{\sqrt{5}}$ _____

3. $\dfrac{4}{\sqrt{y}}$ _____

4. $\dfrac{1}{\sqrt{2x^2}}$ _____

5. $\dfrac{1}{\sqrt{2x^2y}}$ _____

6. $\dfrac{2}{\sqrt{2}}$ _____

7. $\sqrt{\dfrac{7}{8}}$ _____

8. $\sqrt{\dfrac{4}{3}}$ _____

9. $\sqrt{\dfrac{5}{3}}$ _____

10. $\sqrt{\dfrac{1}{x}}$ _____

11. $\sqrt{\dfrac{8}{p^2q}}$ _____

12. $\dfrac{2\sqrt{3}}{\sqrt{8}}$ _____

13. $\sqrt{2\tfrac{1}{2}}$ _____

14. $\sqrt{1\tfrac{1}{5}}$ _____

15. $2\sqrt{3\tfrac{1}{2}}$ _____

16. $3\sqrt{1\tfrac{1}{3}}$ _____

17. $\dfrac{6x}{\sqrt{3}}$ _____

18. $\sqrt{\dfrac{12}{2t}}$ _____

19. $\dfrac{a\sqrt{3}}{\sqrt{a}}$ _____

20. $\dfrac{x\sqrt{6}}{\sqrt{3x}}$ _____

21. $\sqrt{\dfrac{4x^2}{y}}$ _____

22. $\sqrt{\dfrac{4c}{ab}}$ _____

23. $\sqrt{\dfrac{75}{4x}}$ _____

24. $\sqrt{\dfrac{27t}{2p^3}}$ _____

25. $\dfrac{2}{1 - \sqrt{2}}$ _____

26. $\dfrac{4}{1 + \sqrt{3}}$ _____

27. $\dfrac{6}{2 + \sqrt{2}}$ _____

Date _____

Name _____

Course/Section _____

28. $\dfrac{10}{\sqrt{6} - 1}$ _____

29. $\dfrac{a}{1 + \sqrt{a}}$ _____

30. $\dfrac{\sqrt{x}}{1 + \sqrt{x}}$ _____

31. $\dfrac{1 + \sqrt{x}}{1 - \sqrt{x}}$ _____

32. $\dfrac{\sqrt{x} + \sqrt{y}}{\sqrt{x} - \sqrt{y}}$ _____

E. Simplify. Write each of these expressions using a radical sign.

1. $7^{1/2}$ _____

2. $6^{1/3}$ _____

3. $12^{1/2}$ _____

4. $16^{1/3}$ _____

5. $4x^{1/2}$ _____

6. $(17y^2)^{1/3}$ _____

7. $(1\frac{1}{4})^{1/2}$ _____

8. $(2\frac{1}{8})^{1/3}$ _____

9. $5^{-1/2}$ _____

10. $6^{-1/3}$ _____

11. $2^{2/3}$ _____

12. $3^{3/4}$ _____

13. $8^{3/2}$ _____

14. $9^{2/3}$ _____

15. $x^{1/2}y^{1/2}$ _____

16. $a^{1/3}b^{2/3}$ _____

17. $x^{4/3}y^{1/3}$ _____

18. $p^{3/4}q^{1/4}$ _____

19. $(ab^2)^{1/2}$ _____

20. $(x^2y^3)^{1/3}$ _____

Simplify. Write with positive exponents only.

21. $\sqrt{3}$ _____

22. $\sqrt[3]{11}$ _____

23. $\sqrt{\dfrac{2}{3}}$ _____

24. $\sqrt{\dfrac{7}{5}}$ _____

25. $\sqrt{xy}$ _____

26. $\sqrt{ab^3}$ _____

27. $\sqrt{\dfrac{p}{q}}$ _____

28. $\sqrt{\dfrac{a}{b^3}}$ _____

29. $\sqrt[4]{3t}$ _____

30. $\sqrt[3]{2x^2}$ _____

31. $x^{8/6}$ _____

32. $y^{3/9}$ _____

33. $x^{-1/3}$ _____

34. $t^{-1/4}$ _____

35. $3y^{-1/2}$ _____

36. $5x^{-2/3}$ _____

37. $\dfrac{1}{x^{-2/3}}$ _____

38. $\dfrac{2}{x^{-3/4}}$ _____

39. $\dfrac{3}{y^{-2/3}x^{-2}}$ _____

40. $\dfrac{1}{a^{-3}b^{-3/4}}$ _____

41. $\dfrac{x^{3/2}}{x^{-1/2}}$ _____

42. $\dfrac{y^{2/5}}{y^{-2/3}}$ _____

43. $\dfrac{a^{3/4}}{a^{-1/4}}$ _____

44. $\dfrac{x^{-2}}{x^{-1/3}}$ _____

45. $(x^{-1/2})^5$ _____

46. $(x^{2/3})^{-6}$ _____

47. $(xy^2)^{-3/5}$ _____

48. $(x^3y^{-2})^{1/4}$ _____

49. $x^{1/2} \cdot x^{3/5} \cdot x^{-2}$ _____

50. $y^3 \cdot y^{2/3} \cdot y^{-1/6}$ _____

51. $(a^4b^{-2/3})^{-2}$ _____

52. $(x^{-2}y^5)^{-1/3}$ _____

53. $\dfrac{x^{-3/2} \cdot x^{1/3} \cdot x^{1/4}}{x^{-1/2}}$ _____

54. $\dfrac{y^{-2} \cdot y^{-1/3} \cdot y^{1/4}}{y^{-4}}$ _____

55. $\dfrac{a^{1/2} \cdot a^{2/5} \cdot a^{-1/5}}{a^{-3/5}}$ _____

56. $\dfrac{x^{3/4} \cdot x^{-1/3} \cdot x^{1/2}}{x^{-2/3}}$ _____

57. $\left(\dfrac{2x^{-2}y^{3/4}}{z^{1/2}}\right)^2$ _____

58. $\left(\dfrac{3a^{1/4}b^{-3}}{c^{-2/3}}\right)^{-3}$ _____

59. $\left(\dfrac{a^{-1/2}}{b^{-1/4}c^{2/3}}\right)^{-3}$ _____

60. $\left(\dfrac{x^{-1/2}}{y^{-4}z^{2/3}}\right)^2$ _____

F. Brain Boosters

1. Show that   (a) $\sqrt{2\dfrac{2}{3}} = 2\sqrt{\dfrac{2}{3}}$ (b) $\sqrt{3\dfrac{3}{8}} = 3\sqrt{\dfrac{3}{8}}$

   (c)   In general $\sqrt{a + \dfrac{a}{a^2 - 1}} = a\sqrt{\dfrac{a}{a^2 - 1}}$ for all $a > 1$.

2. Show that   (a) $\sqrt{4\dfrac{4}{3}} = 2\sqrt{\dfrac{4}{3}}$ (b) $\sqrt{9\dfrac{9}{8}} = 3\sqrt{\dfrac{9}{8}}$

   (c)   In general $\sqrt{a^2 + b} = a\sqrt{b}$ for $b = \dfrac{a^2}{a^2 - 1}$ when $a > 1$

3. Show that   (a) $\sqrt{19 + 6\sqrt{2}} = 1 + 3\sqrt{2}$

   (b) $\sqrt{75 + 12\sqrt{21}} = 3\sqrt{7} + 2\sqrt{3}$

   (*Hint:* Square both sides of the equation and simplify.)

4. A rocketship zooms past the earth moving with speed v with respect to the earth. A scientist on the rocketship measures its length as L_0, but a scientist on the earth measures the length of the moving rocketship as L. According to Einstein's special theory of relativity

   $$L = L_0\sqrt{1 - \dfrac{v^2}{c^2}}$$ where c is the speed of light.

   For any value of v greater than zero and less than c, L is less than L_0. The rocketship shrinks when it moves!

   (a)   Find L if L_0 is 100 ft and $v = 0.1c$. (The ship is moving at about 67 million mph.)

   _____

   (b)   Find L if $v = 0.8c$. (The speed is about 530 million mph.)

   _____

   (c)   Find L if $v = c$. (The speed is about 670 million mph.)

   _____

5. How far can you see from an airplane on a clear day? It depends on the altitude of the airplane. If we take account of the curvature of the earth, then the view in miles on either side of the plane can be calculated from the formula

   $V \cong 1.22\sqrt{A}$ where V is the view in miles,

   and A is the altitude in feet. For example, at 5000 ft

   $V \cong 1.22\sqrt{5000} \cong 86.3$ miles

   Calculate how far you can see from a commercial jetliner cruising at 35,000 ft.

   _____

Date _____

Name _____

Course/Section _____

6. Suppose you want to mail a pencil 19 cm long in a box 10 cm by 8 cm by 14 cm.

   (a) Can it be packed in the box without bending or breaking?

   _____

   (b) Can it be packed in a cubical box 11 cm on each edge?

   _____

   (_Hint:_ the diagonal of a rectangular box with edges a, b, and c is $\sqrt{a^2 + b^2 + c^2}$.)

7. In order to escape from the gravitational pull of the earth, a satellite must be launched with a speed of at least $v = \sqrt{2gR}$ where g is a constant equal to 32 ft/sec^2 and R is the radius of the earth, about $21{,}000{,}000$ ft. What is the minimum escape speed of a satellite?

   _____

8. Use the fact that $(x^{1/2})^2 = x$ to factor the expression $(x - y)$.

   _____

   (_Hint:_ Think of the quantity $(x - y)$ as the difference of two squares.)

When you have had the practice you need, turn to **65** for a self-test over the work of Unit 5.

Unit 5
Fractions
and Radicals

Self-Test 5

65

1. Reduce to lowest terms $\dfrac{8x + 2y}{12x + 3y}$ = _____

2. Reduce to lowest terms $-\dfrac{3x^2 + x - 2}{x^2 - 1}$ = _____

3. $-\dfrac{2a^2}{3ab} = \dfrac{?}{6a^2b^2}$ = _____

4. $\dfrac{x^2}{x + 1} + \dfrac{4}{x + 1}$ = _____

5. $\dfrac{x - y}{y - 1} - \dfrac{2x - y}{y - 1}$ = _____

6. $\dfrac{a}{3} + \dfrac{2a}{5}$ = _____

7. $\dfrac{2}{2xy} - \dfrac{1}{3x^2}$ = _____

8. $\dfrac{x}{x + 1} + \dfrac{2x}{x^2 - 1}$ = _____

9. $\left(\dfrac{6a^3b}{5}\right)\left(\dfrac{a}{3a^2b}\right)$ = _____

10. $\left(\dfrac{2x + 8}{3x + 3}\right) \div \left(\dfrac{4x + 16}{6x - 6}\right)$ = _____

11. Simplify $\dfrac{\frac{2x^2}{3}}{\frac{6xy}{3y^2}}$ = _____

12. Simplify $\sqrt{20xy^3}$ = _____

13. Simplify $\sqrt{3x^2} \cdot \sqrt{6xy^2}$ = _____

14. Simplify $\sqrt{\dfrac{a^4}{ab}}$ = _____

15. Simplify $\dfrac{6x}{\sqrt{3x}}$ = _____

Date _____

16. Simplify $\dfrac{\sqrt{y}}{\sqrt{y} + 1}$ = _____

Name _____

17. Multiply $(2 + \sqrt{x})(3 - \sqrt{x})$ = _____

Course/Section _____

18. $5\sqrt{2x} - \sqrt{8x}$ = _____

19. Simplify $(x\sqrt{a})(x\sqrt{b})$ = _____

20. Solve $\dfrac{x+9}{3} = \dfrac{x-6}{2}$ = _____

21. Solve $\dfrac{2x+3}{x-4} = \dfrac{2}{3}$ = _____

22. If three-fourths of a number is subtracted from one-sixth of the number, the result is 7. Find the number.

    _____

23. A runner covers 6 miles in the same time that a bicyclist covers 15 miles. The cyclist moves at a speed 12 miles per hour faster than the runner. Find the speed of each person.

    _____

24. Solve $\dfrac{8}{x} = \dfrac{12}{x+7}$

    _____

25. If a college baseball player hits 9 home runs in the first 12 games of the season, how many can he be expected to hit in the entire 28 game season at this rate?

    _____

26. Simplify. Write with positive exponents.

    (a) $\dfrac{1}{x^{-3/5}}$ = _____

    (b) $\dfrac{2x^{-4/3}}{y^{-2/5}z^3}$ = _____

    (c) $(x^{-1/2} \cdot y^{2/3})^{-4/3}$ = _____

    (d) $\dfrac{x^{1/3} \cdot x^{-2/3} \cdot x^{-1/2}}{x^{-1/4}}$ = _____

    (e) $\left(\dfrac{2x^{-3}y^{3/2}}{z^{-1/4}}\right)^{-3}$ = _____

Answers are on page 565.

Unit 5
Fractions
and Radicals

A. Reduce to lowest terms.

1. $\dfrac{3x^4}{x^2}$ ____

2. $\dfrac{y}{y^2}$ ____

3. $\dfrac{6t}{2t^3}$ ____

4. $\dfrac{12x^3}{9x^4}$ ____

5. $\dfrac{-5x^3}{x^2}$ ____

6. $\dfrac{a^2b^3}{a^3b^2}$ ____

7. $\dfrac{y^2}{-2y^5}$ ____

8. $\dfrac{3a}{6xy}$ ____

9. $\dfrac{xy^3}{-x^2y}$ ____

10. $\dfrac{x^2}{2}$ ____

11. $\dfrac{a^2b^2}{a}$ ____

12. $\dfrac{-6pq^3}{-2p^4q}$ ____

13. $\dfrac{x^{14}}{x^9}$ ____

14. $\dfrac{18a^5b^7c^6}{12a^9b^4c^8}$ ____

15. $\dfrac{4xyz^3}{6xy^2}$ ____

16. $\dfrac{12t^2}{-10t^2}$ ____

17. $\dfrac{a-b}{b-a}$ ____

18. $\dfrac{x^2-y^2}{x-y}$ ____

19. $\dfrac{x^2-y^2}{x+y}$ ____

20. $\dfrac{t^2-1}{t-1}$ ____

21. $\dfrac{p+q}{q+p}$ ____

22. $\dfrac{(a-b)^3}{a-b}$ ____

23. $\dfrac{3x+x^2}{2x}$ ____

24. $\dfrac{a^2x-ax^3}{a^2x^2}$ ____

25. $\dfrac{-2x}{6x-4x^2}$ ____

26. $\dfrac{x^2y^3}{xy^2-x^2y^2}$ ____

27. $\dfrac{2x^3-3x}{6x^4}$ ____

28. $\dfrac{4a^3}{4a-a^2}$ ____

29. $\dfrac{x^2+2x+1}{x+1}$ ____

30. $\dfrac{a+2}{a^2+4a+4}$ ____

31. $\dfrac{a-2}{a^2-5a+6}$ ____

32. $\dfrac{3p^2-3p}{1-p}$ ____

33. $\dfrac{x^2-1}{x^2-4x+3}$ ____

34. $\dfrac{2x-6}{x^2+2x-15}$ ____

35. $\dfrac{4+9x^2+12x}{4-9x^2}$ ____

36. $\dfrac{(x+y)^2(2y^2-3y+1)}{(x^2-y^2)(y-1)}$ ____

37. $\dfrac{a^2-2a-3}{a^2-a-6}$ ____

38. $\dfrac{2t^2+t-1}{3-5t-2t^2}$ ____

39. $\dfrac{(a^2+ab)(a^2+ab-2b^2)}{(a^2-b^2)(a+2b)}$ ____

40. $\dfrac{2xy}{x-y}$ ____

Date

Name

Course/Section

B. Express each fraction as an equivalent fraction as shown.

1. $\dfrac{2}{3a} = \dfrac{?}{6a}$ ____

2. $2x = \dfrac{?}{x}$ ____

3. $\dfrac{3p}{2} = \dfrac{?}{4q}$ _____

4. $\dfrac{3}{2y} = \dfrac{?}{y}$ _____

5. $\dfrac{x}{y} = \dfrac{?}{3y}$ _____

6. $\dfrac{4}{2x} = \dfrac{?}{3x}$ _____

7. $2 = \dfrac{?}{x}$ _____

8. $\dfrac{x}{y} = \dfrac{?}{2xy}$ _____

9. $\dfrac{a^2}{b^2} = \dfrac{?}{b^3}$ _____

10. $p = \dfrac{?}{x}$ _____

11. $p = \dfrac{?}{-x}$ _____

12. $-p = \dfrac{?}{x}$ _____

13. $-p = \dfrac{?}{-x}$ _____

14. $\dfrac{1}{2} = \dfrac{?}{3x^2 y}$ _____

15. $\dfrac{a}{b} = \dfrac{?}{2ab(b-1)}$ _____

16. $\dfrac{1}{2x} = \dfrac{?}{y+1}$ _____

17. $\dfrac{y}{x+y} = \dfrac{?}{2(x+y)}$ _____

18. $\dfrac{y}{x+y} = \dfrac{?}{x^2 - y^2}$ _____

19. $\dfrac{y}{x+y} = \dfrac{?}{x^2 + 2xy + y^2}$ _____

20. $\dfrac{a}{b} = \dfrac{?}{c}$ _____

21. $\dfrac{a-b}{b-a} = \dfrac{?}{a}$ _____

22. $\dfrac{2}{1-x^2} = \dfrac{?}{1-x}$ _____

23. $\dfrac{3}{x} = \dfrac{?}{x-x^3}$ _____

24. $\dfrac{-2}{x^2} = \dfrac{?}{x(x+1)}$ _____

25. $\dfrac{-3a}{2b^2} = \dfrac{?}{4ab^5}$ _____

26. $\dfrac{1}{t} = \dfrac{?}{-2t^2 - t}$ _____

27. $x - y = \dfrac{?}{y-x}$ _____

28. $a = \dfrac{?}{a+x}$ _____

29. $\dfrac{x^2}{x+1} = \dfrac{?}{x^2 + 2x + 1}$ _____

30. $\dfrac{a}{a-1} = \dfrac{?}{a^2 - 1}$ _____

Unit 5
Fractions
and Radicals

A. Add:

1. $\dfrac{a}{3} + \dfrac{a}{3}$ _____

2. $\dfrac{2a}{b} + \dfrac{a}{b}$ _____

3. $\dfrac{2}{x+1} + \dfrac{x}{x+1}$ _____

4. $\dfrac{3a+1}{2a} + \dfrac{a+2}{2a}$ _____

5. $\dfrac{x+y}{x-y} + \dfrac{x-2y}{x-y}$ _____

6. $\dfrac{2a+1}{a-1} + \dfrac{1-a}{a-1}$ _____

7. $\dfrac{1}{x^2} + \dfrac{2}{x^2}$ _____

8. $\dfrac{1+y}{y^2} + \dfrac{y-1}{y^2}$ _____

9. $\dfrac{a}{a^2-1} + \dfrac{1}{a^2-1}$ _____

10. $\dfrac{1}{x} + \dfrac{1}{y}$ _____

11. $\dfrac{1}{2} + \dfrac{2}{x}$ _____

12. $\dfrac{2}{a} + \dfrac{1}{2a}$ _____

13. $\dfrac{1}{x^2} + \dfrac{1}{x}$ _____

14. $\dfrac{2}{3t^2} + \dfrac{2}{t}$ _____

15. $\dfrac{1}{2a^2} + \dfrac{2}{3a}$ _____

16. $\dfrac{3}{2x} + \dfrac{1}{2y}$ _____

17. $\dfrac{1}{pqt} + \dfrac{t}{p^2q}$ _____

18. $\dfrac{1}{a} + \dfrac{1}{a-b}$ _____

19. $\dfrac{1}{a+b} + \dfrac{1}{a-b}$ _____

20. $\dfrac{x}{x-1} + \dfrac{1}{x-2}$ _____

21. $2 + \dfrac{1}{a}$ _____

22. $y + \dfrac{2}{y}$ _____

23. $\dfrac{p}{q^2} + \dfrac{p}{q} + \dfrac{1}{q^3}$ _____

24. $(a+b) + \dfrac{1}{a-b}$ _____

25. $\dfrac{1}{x^2-1} + \dfrac{2}{x+1}$ _____

26. $\dfrac{2a}{a^2+2a+1} + \dfrac{1}{a+1}$ _____

27. $\dfrac{4x}{x^2-3x+2} + \dfrac{3x}{x^2+x-2}$ _____

28. $\dfrac{1}{x} + \dfrac{1}{x^2+x-6}$ _____

29. $\dfrac{1}{x} + \dfrac{1}{x^2} + \dfrac{1}{2x}$ _____

30. $\dfrac{1}{x} + \dfrac{1}{y} + \dfrac{1}{z}$ _____

31. $\dfrac{1}{x+1} + \dfrac{1}{x} + \dfrac{1}{x-1}$ _____

32. $\dfrac{2}{a} + \dfrac{3}{ab} + \dfrac{1}{b}$ _____

Date

Name

Course/Section

B. Subtract:

1. $\dfrac{3}{a} - \dfrac{5}{a}$ _____

2. $\dfrac{a}{x} - \dfrac{a^2}{x}$ _____

3. $\dfrac{2 + p}{p} - \dfrac{p - 1}{p}$ _____

4. $\dfrac{x + 3}{2x + 1} - \dfrac{3x + 4}{2x + 1}$ _____

5. $\dfrac{1 - x}{x} - \dfrac{x - 1}{2x}$ _____

6. $\dfrac{x}{2} - \dfrac{x}{4}$ _____

7. $\dfrac{x}{2a} - \dfrac{1}{3a}$ _____

8. $\dfrac{1}{4x} - \dfrac{1}{3x}$ _____

9. $1 - \dfrac{2}{p}$ _____

10. $3 - \dfrac{1}{x}$ _____

11. $a - \dfrac{1}{2a}$ _____

12. $\dfrac{3}{t} - t^2$ _____

13. $\dfrac{2}{x^2} - \dfrac{1}{x}$ _____

14. $\dfrac{d^2 - 1}{d^3} - \dfrac{d - 1}{d^2}$ _____

15. $\dfrac{2}{ax^2} - \dfrac{x}{ax}$ _____

16. $\dfrac{2x}{x - 1} - \dfrac{1}{x^2 - 1}$ _____

17. $\dfrac{1}{a^2 - b^2} - \dfrac{1}{a + b}$ _____

18. $\dfrac{x}{x + 2} - \dfrac{2}{2x + 1}$ _____

19. $\dfrac{2}{n} - \dfrac{1}{n + 1}$ _____

20. $\dfrac{3}{2y} - \dfrac{y}{y + 1}$ _____

21. $\dfrac{x}{3} - \dfrac{3}{x}$ _____

22. $\dfrac{1}{x - y} - \dfrac{3}{2y}$ _____

23. $\dfrac{2x}{x^2 - 4x + 3} - \dfrac{1}{x^2 + x - 2}$ _____

24. $\dfrac{x}{x^2 - 4} - \dfrac{2x}{x^2 - 4x + 4}$ _____

25. $t - \dfrac{2}{t - 2}$ _____

26. $\dfrac{b}{b + 1} - b$ _____

27. $2 - \dfrac{1}{1 - x} - \dfrac{x}{1 + x}$ _____

28. $1 - \dfrac{1}{x} - \dfrac{1}{x^2}$ _____

29. $\dfrac{1}{x + 1} + \dfrac{1}{y + 1} - \dfrac{1}{xy}$ _____

30. $\dfrac{1}{a} + \dfrac{1}{b} - \dfrac{1}{ab}$ _____

Unit 5
Fractions
and Radicals

A. Multiply:

1. $\dfrac{2t}{3} \cdot \dfrac{1}{2}$ _____

2. $\dfrac{1}{4} \cdot \dfrac{6x}{y}$ _____

3. $\dfrac{ab}{3} \cdot \dfrac{a}{b}$ _____

4. $2x \cdot \dfrac{3}{x}$ _____

5. $ab \cdot \dfrac{4}{b}$ _____

6. $2y^2 \cdot \dfrac{1}{3y}$ _____

7. $b^2 \cdot \dfrac{4}{3ab}$ _____

8. $\dfrac{xy}{z} \cdot \dfrac{z}{x}$ _____

9. $\dfrac{y^2}{4} \cdot \dfrac{12}{5y}$ _____

10. $\dfrac{21x^2y^3}{9xy^4} \cdot \dfrac{3y^5}{7x^4y}$ _____

11. $\dfrac{12a^3bc^2}{3b^2c} \cdot \dfrac{6ab^3c^4}{8a^2c}$ _____

12. $\dfrac{36m^3np}{20p^3n} \cdot \dfrac{15n^4p^4}{27m^5}$ _____

13. $\dfrac{3}{x^2 - 1} \cdot \dfrac{2x + 2}{9}$ _____

14. $\dfrac{3x - 3}{6x} \cdot \dfrac{4}{(x - 1)^2}$ _____

15. $\dfrac{4x - 6}{4x + 8} \cdot \dfrac{6x + 12}{5x - 15}$ _____

16. $\dfrac{a^2 - 4}{4a^2 - 9} \cdot \dfrac{2a - 3}{a + 2}$ _____

17. $\dfrac{4t + 12}{2t - 8} \cdot \dfrac{t^2 - 4}{2t^2 - 18}$ _____

18. $\dfrac{4x + 6}{2x + 4} \cdot \dfrac{x + 1}{2x^2 + 3x}$ _____

19. $\dfrac{3xy}{x - 1} \cdot \dfrac{x^3 - x^2}{6x^2}$ _____

20. $\dfrac{6x}{x^2 + 4x + 4} \cdot \dfrac{2x + 4}{24x^2}$ _____

21. $\dfrac{1}{x} \cdot \dfrac{x + 1}{x - 1} \cdot \dfrac{x^2 + 1}{x^2 - 1}$ _____

22. $\dfrac{2}{ab} \cdot \dfrac{a^2}{b^2} \cdot \dfrac{3ab^2}{4a^2}$ _____

23. $\dfrac{p - q}{p + q} \cdot \dfrac{p^2 - q^2}{p} \cdot \dfrac{q}{p - q}$ _____

24. $\dfrac{m - n}{m + n} \cdot \dfrac{m^2 - n^2}{m^2 - 2mn + n^2}$ _____

25. $\left(1 + \dfrac{1}{x}\right)\left(1 - \dfrac{1}{x}\right)$ _____

26. $\left(a + \dfrac{2}{a}\right)\left(\dfrac{a^2}{a + 1}\right)$ _____

27. $\left(\dfrac{p + q}{p}\right)\left(\dfrac{1}{p - q}\right)$ _____

28. $\dfrac{x^2 + 5x + 6}{x^2 - 7x + 12} \cdot \dfrac{x^2 - 5x + 4}{x^2 + x - 2}$ _____

29. $\dfrac{4t^2 - 2t}{t^2 + 3t + 2} \cdot \dfrac{t^2 + t - 2}{2t^2 - 3t + 1}$ _____

30. $\dfrac{x^2 - 3x + 2}{x^2 - 6x + 5} \cdot \dfrac{x^2 - 3x - 10}{2 + x - x^2}$ _____

Date _____

Name _____

Course/Section _____

B. Divide:

1. $\dfrac{2x}{y} \div \dfrac{x}{2}$ _____

2. $\dfrac{a}{b} \div \dfrac{2}{b}$ _____

3. $\dfrac{x^2}{3} \div \dfrac{1}{x}$ _____

4. $3 \div \dfrac{1}{x}$ _____

5. $x \div \dfrac{2}{a}$ _____

6. $a \div \dfrac{1}{2}$ _____

7. $\dfrac{2p}{q} \div \dfrac{2q}{p}$ _____

8. $\dfrac{3b}{5c} \div \dfrac{18b}{15c}$ _____

9. $\dfrac{6xy}{5z} \div \dfrac{24y^2}{20xz}$ _____

10. $3abc \div \dfrac{a}{b}$ _____

11. $4pq^2 \div \dfrac{2p}{3q}$ _____

12. $\dfrac{12x}{4y} \div 2x^2y$ _____

13. $\dfrac{-x}{2y} \div \dfrac{2y^2}{x}$ _____

14. $\dfrac{ab^2}{3c} \div \dfrac{3b}{-ac}$ _____

15. $\dfrac{5x^2}{-3xy^2} \div \dfrac{-15x^3}{9y}$ _____

16. $\dfrac{x}{x+y} \div \dfrac{y}{x+y}$ _____

17. $\dfrac{1}{x-y} \div \dfrac{1}{y-x}$ _____

18. $\dfrac{1-x}{1-y} \div \dfrac{1-x^2}{1-y^2}$ _____

19. $\left(1 + \dfrac{2}{x}\right) \div \left(x + \dfrac{1}{2}\right)$ _____

20. $\dfrac{1}{x} \div \left(1 + \dfrac{1}{x}\right)$ _____

21. $(a - 1) \div \dfrac{1+a}{1-a}$ _____

22. $\dfrac{x^2 - 4x + 3}{x^2 - 3x - 4} \div \dfrac{x^2 - x - 6}{x^2 + 3x + 2}$ _____

23. $\dfrac{x^2 - 6x + 5}{x^2 - 5x + 4} \div \dfrac{x^2 - 8x + 15}{x^2 + x - 12}$ _____

24. $\dfrac{2x^2 - 12}{x^2 - 5x - 6} \div \dfrac{x^3 - 3x^2 + 2x}{x^2 - 1}$ _____

25. $\left(x + \dfrac{1}{x}\right) \div \left(x - \dfrac{1}{x}\right)$ _____

26. $\left(x - \dfrac{1}{x}\right) \div \left(1 - \dfrac{1}{x^2}\right)$ _____

27. $(x^2 - 4) \div \left(\dfrac{x^2 - x - 2}{x + 1}\right)$ _____

28. $\dfrac{30x^2y^4z^3}{4x^4y} \div \dfrac{10xy^3}{6z^4}$ _____

29. $\dfrac{63x^9y^8}{18x^{11}y^5} \div \dfrac{21x^7y^4}{36x^3y^{12}}$ _____

30. $\dfrac{36ab^7c^{10}}{20a^6b^6} \div \dfrac{9ab^9c^{11}}{-15a^7b^6c^8}$ _____

Unit 5
Fractions
and Radicals

A. Solve the following equations.

1. $\dfrac{3x}{2} = 12$ _____

2. $\dfrac{2x}{7} = 12$ _____

3. $\dfrac{x-3}{4} = 2$ _____

4. $\dfrac{3y-1}{4} = 2$ _____

5. $\dfrac{3-2a}{5} = 3$ _____

6. $\dfrac{p+2}{4} = \dfrac{5}{2}$ _____

7. $\dfrac{x}{3} + \dfrac{x}{7} = 10$ _____

8. $\dfrac{2a}{5} = 3 + \dfrac{a}{4}$ _____

9. $\dfrac{y}{4} + \dfrac{y}{3} + \dfrac{y}{2} = 13$ _____

10. $1 = \dfrac{3x}{4} - \dfrac{2x}{3}$ _____

11. $\dfrac{a+2}{2} + \dfrac{a-2}{5} = 9$ _____

12. $\dfrac{x+1}{2} - \dfrac{x-3}{3} = 3$ _____

13. $\dfrac{x}{3} + \dfrac{x}{5} = 16$ _____

14. $\dfrac{t}{3} + 3 = \dfrac{t}{5} - 7$ _____

15. $\dfrac{1}{x-1} = \dfrac{3}{x-3}$ _____

16. $\dfrac{1}{x+2} + \dfrac{1}{x+1} = \dfrac{2}{x}$ _____

17. $\dfrac{x+1}{x-3} = \dfrac{x-2}{x+2}$ _____

18. $\dfrac{x}{1+x} = 1 + \dfrac{1}{x}$ _____

19. $\dfrac{1-a}{1+a} = \dfrac{2}{3}$ _____

20. $\dfrac{x-1}{x} = \dfrac{x-3}{x-1} - \dfrac{5}{x^2-x}$ _____

B. Solve each of the following formulas for A.

1. $\dfrac{1}{A} - \dfrac{1}{B} = C$ ____

2. $\dfrac{AB}{C} = \dfrac{1}{B}$ ____

3. $B = \dfrac{1-AC}{1-A}$ ____

4. $\dfrac{P}{Q} = \dfrac{A+B}{A}$ ____

5. $\dfrac{A-B}{2A} = T$ ____

6. $x = y - \dfrac{2}{Ax}$ ____

7. $\dfrac{A+1}{A+2} = B$ ____

8. $1 + \dfrac{B^2}{A} = 2B$ ____

9. $\dfrac{B}{A+1} = \dfrac{C}{A+2}$ ____

10. $\dfrac{1+Ax^2}{Ax} = B$ ____

11. $\dfrac{1}{A} + 1 = B$ ____

12. $\dfrac{1}{A} + \dfrac{1}{2A} = x$ ____

C. Solve each of the following systems of equations.

1. $\dfrac{x}{2} + \dfrac{y}{6} = 3$

$\dfrac{5y}{6} - \dfrac{x}{2} = 6$ _____

2. $\dfrac{x}{3} + \dfrac{y}{2} = 9$

$\dfrac{x}{2} + \dfrac{y}{5} = 8$ _____

3. $\dfrac{x}{4} - \dfrac{y}{2} = 5$

$\dfrac{y}{6} = \dfrac{x}{2} - 5$ _____

4. $\dfrac{x}{2} - \dfrac{y}{5} = -1$

$\dfrac{x}{8} + \dfrac{y}{2} = -3$ _____

5. $\dfrac{x}{7} + \dfrac{y}{3} = 1$

$\dfrac{x}{5} - \dfrac{2y}{3} = -2$ _____

6. $\dfrac{2x - 1}{3} + \dfrac{y + 1}{2} = 4$

$\dfrac{3x + 1}{7} - \dfrac{3y + 5}{4} = -4$ _____

D. Solve:

1. If one-half of a certain number is 6 more than the number itself, find the number.

   _____

2. Write 63 as the sum of two numbers such that the smaller is three-fourths of the larger.

   _____

3. When the reciprocal of a certain number is decreased by two, the result is three. Find the number.

   _____

4. On five mathematics quizzes, Tom scored exactly the same on the first two exams and ten points higher on the last three. His average was 86. What was his score on the first quiz?

   _____

5. It is possible to have $5 in quarters, dimes, and nickles so that there are exactly 40 coins and the number of quarters is equal to the number of nickles. Find the number of each coin needed.

   _____

Unit 5
Fractions
and Radicals

A. Simplify each of the following radicals.

1. $\sqrt{a^3}$ _____

2. $\sqrt{4x^5}$ _____

3. $-\sqrt{2x^2y^7}$ _____

4. $\sqrt{a^2bc^6}$ _____

5. $\sqrt{18x^6y^7}$ _____

6. $\sqrt{45xy^2z^3}$ _____

7. $\sqrt{p^3 - 2p^2}$ _____

8. $-\sqrt{40 - 8x^2}$ _____

9. $\dfrac{2x}{y}\sqrt{xy^2}$ _____

10. $-2\sqrt{8x^2}$ _____

11. $\sqrt{a^2b - b^2a^2}$ _____

12. $\sqrt{4 - x^2}$ _____

13. $\sqrt{x^2 + 2x + 4}$ _____

14. $\sqrt{2(x + 3)^2}$ _____

15. $\sqrt{20(p - q)^3}$ _____

16. $\sqrt{50(a - b)^2}$ _____

B. Multiply and simplify.

1. $\sqrt{p} \cdot \sqrt{q^2}$ _____

2. $\sqrt{a} \cdot \sqrt{ax}$ _____

3. $\sqrt{2x} \cdot \sqrt{2y}$ _____

4. $\sqrt{3p} \cdot \sqrt{5p}$ _____

5. $\sqrt{6ab^2} \cdot \sqrt{2ab}$ _____

6. $\sqrt{5t} \cdot \sqrt{20t^2}$ _____

7. $\sqrt{4a^2bc} \cdot \sqrt{3abc^3}$ _____

8. $2\sqrt{3x} \cdot 5\sqrt{2x}$ _____

9. $2\sqrt{x} \cdot 4\sqrt{xy} \cdot \sqrt{y}$ _____

10. $p\sqrt{p^3} \cdot p^2\sqrt{p} \cdot \sqrt{p}$ _____

11. $\sqrt{x}(\sqrt{x} + \sqrt{y})$ _____

12. $2\sqrt{y}(3\sqrt{y} - 2\sqrt{z})$ _____

13. $(a + \sqrt{b})^2$ _____

14. $(\sqrt{p} + \sqrt{q})^2$ _____

15. $(\sqrt{x} + \sqrt{y})(\sqrt{x} - \sqrt{y})$ _____

16. $(1 + 2\sqrt{t})^2$ _____

17. $(2\sqrt{x} - 3\sqrt{y})(\sqrt{x} + 2\sqrt{y})$ _____

18. $(1 + \sqrt{a})(2 + \sqrt{a})$ _____

19. $(m\sqrt{m} + 1)(m\sqrt{m} - 2)$ _____

20. $(\sqrt{ab} + \sqrt{b})(\sqrt{a} - \sqrt{ab})$ _____

C. Add or subtract as shown.

1. $2\sqrt{p} + \sqrt{p}$ _____

2. $3\sqrt{5x} + 2\sqrt{5x}$ _____

3. $\sqrt{a^2b} + \sqrt{a^2b^3}$ _____

4. $3\sqrt{x} - \sqrt{x}$ _____

5. $2\sqrt{a} - \sqrt{16a}$ _____

6. $\sqrt{pq^2} - \sqrt{p^3}$ _____

7. $\sqrt{3t} + \sqrt{12t}$ _____

8. $\sqrt{5a} - \sqrt{20a}$ _____

9. $\sqrt{3x^2} - \sqrt{12x^2}$ _____

10. $a\sqrt{bc} + b\sqrt{b^3c}$ _____

11. $\sqrt{xy^2} + \sqrt{4x} - \sqrt{9x^3y^2}$ _____

12. $q\sqrt{q} + q^2\sqrt{q}$ _____

Date _____

Name _____

Course/Section _____

D. Write each expression in its simplest radical form.

1. $\dfrac{1}{\sqrt{2}}$ ____

2. $\dfrac{2}{\sqrt{x}}$ ____

3. $\dfrac{1}{\sqrt{4a}}$ ____

4. $\dfrac{1}{\sqrt{3x^2y}}$ ____

5. $\dfrac{a^2}{\sqrt{a}}$ ____

6. $\dfrac{ab}{\sqrt{a^2b}}$ ____

7. $\sqrt{\dfrac{x}{4}}$ ____

8. $\sqrt{\dfrac{8}{y}}$ ____

9. $\sqrt{\dfrac{12}{pq^2t}}$ ____

10. $\sqrt{\dfrac{18}{3yz^2}}$ ____

11. $a\sqrt{\dfrac{5}{a}}$ ____

12. $\sqrt{\dfrac{xy^2}{2}}$ ____

13. $\sqrt{\dfrac{50a}{b^3}}$ ____

14. $\sqrt{\dfrac{2x}{yz}}$ ____

15. $\dfrac{2\sqrt{b}}{\sqrt{ab^2}}$ ____

16. $\dfrac{3\sqrt{xy}}{2\sqrt{x^2y}}$ ____

17. $\dfrac{1}{1+\sqrt{2}}$ ____

18. $\dfrac{1}{2-\sqrt{2}}$ ____

19. $\dfrac{a}{a+\sqrt{b}}$ ____

20. $\dfrac{x}{\sqrt{x}+\sqrt{y}}$ ____

21. $\dfrac{1+\sqrt{2}}{1-\sqrt{2}}$ ____

22. $\dfrac{a+\sqrt{a}}{b+\sqrt{b}}$ ____

23. $\dfrac{\sqrt{8x}}{\sqrt{x}-1}$ ____

24. $\dfrac{\sqrt{p^3}}{1+\sqrt{p^3}}$ ____

E. Simplify. Write using a radical sign.

1. $9^{1/3}$ ____

2. $27^{1/2}$ ____

3. $6^{-1/2}$ ____

4. $7^{2/3}$ ____

5. $a^{2/3}b^{1/3}$ ____

6. $(xy^3)^{1/2}$ ____

Simplify. Write with positive exponents only.

7. $\sqrt{10}$ ____

8. $\sqrt[3]{5}$ ____

9. $\sqrt{\dfrac{3}{4}}$ ____

10. $\sqrt[3]{ax^2}$ ____

11. $x^{2/6}$ ____

12. $y^{-1/2}$ ____

13. $4x^{-1/2}$ ____

14. $5a^{-3/2}$ ____

15. $\dfrac{1}{x^{-1/3}}$ ____

16. $\dfrac{4}{y^{-3/4}}$ ____

17. $\dfrac{2}{x^3y^{-5/2}}$ ____

18. $\dfrac{x^{-1/2}}{x^{-3/4}x^2}$ ____

19. $(x^2y^{-3})^{-2/3}$ ____

20. $\left(\dfrac{3x^{-2}y^{2/3}}{z^{-1/3}}\right)^2$ ____

21. $\left(\dfrac{2a^2b^{-3/4}}{4c^{-1/2}}\right)^{-3}$ ____

Quadratic Equations

Objective	Sample Problems	Where To Go for Help	
		Page	Frame
Upon successful completion of this program you will be able to:			
1. Solve quadratic equations by factoring.	Solve by factoring:		
	(a) $x^2 = 3$ _____	447	**1**
	(b) $x^2 - 4 = 0$ _____		
	(c) $4x^2 = 25$ _____		
	(d) $(x - 1)(x + 2) = 0$ _____	455	**7**
	(e) $x^2 = 3x$ _____		
	(f) $(2x - 1)(2x + 3) = 0$ _____		
	(g) $x^2 - 8x + 15 = 0$ _____		
	(h) $x(x + 3) = 4$ _____		
	(i) $x^2 - 5x + 2 = 2x - x^2 - 3$ _____		
2. Solve quadratic equations by completing the square.	Solve each equation by completing the square:		
	(a) $x^2 + 4x = 12$ _____	467	**14**
	(b) $x^2 - 3 = 2x$ _____		
	(c) $9x^2 + 6x - 8 = 0$ _____		
3. Graph quadratic equations.	Graph each equation and find its roots:		
	(a) $y = x^2$ _____	489	**27**
	(b) $y = x^2 - 2$ _____		
	(c) $y = 2x^2 - 3x + 1$ _____		
	(d) $y = x^2 - x - 12$ _____		
4. Solve quadratic equations using the quadratic formula.	Solve each equation using the quadratic formula:		
	(a) $x^2 + 2x - 2 = 0$ _____	475	**21**
	(b) $x^2 - 5x = 3$ _____		
	(c) $2x^2 + 4 + 7x = 0$ _____		
	(d) $x^2 - x - 1 = 0$ _____		
	(e) $4x^2 = 7$ _____		
	(f) $2x^2 = 3x$ _____		

Date _____

Name _____

Course/Section _____

PREVIEW 6

Answers to Sample Problems

1. (a) $\sqrt{3}, -\sqrt{3}$ (b) $2, -2$ (c) $\dfrac{5}{2}, \dfrac{-5}{2}$ (d) $1, -2$

   (e) $0, 3$ (f) $\dfrac{1}{2}, \dfrac{-3}{2}$ (g) $3, 5$ (h) $1, -4$

   (i) $1, \dfrac{5}{2}$

2. (a) $2, -6$ (b) $3, -1$ (c) $\dfrac{2}{3}, \dfrac{-4}{3}$

3. (a) Root at $x = 0$ (b) Roots at $x = \sqrt{2}, -\sqrt{2}$

   (c) Roots at $x = 1, \dfrac{1}{2}$ (d) Roots at $x = 4, -3$

4. (a) $-1 + \sqrt{3}, -1 - \sqrt{3}$ (b) $\dfrac{5 + \sqrt{37}}{2}, \dfrac{5 - \sqrt{37}}{2}$

   (c) $\dfrac{-7 + \sqrt{17}}{4}, \dfrac{-7 - \sqrt{17}}{4}$ (d) $\dfrac{1 + \sqrt{5}}{2}, \dfrac{1 - \sqrt{5}}{2}$

   (e) $\dfrac{\sqrt{7}}{2}, \dfrac{-\sqrt{7}}{2}$ (f) $0, \dfrac{3}{2}$

    If you are certain that you can work all of these problems correctly, turn to page 513 for a self-test. If you want help with any of these objectives or if you cannot work one of the sample problems, turn to the page indicated. Super-students will want to learn all of this and will turn to frame 1 and begin work there.

(Answers to these problems are at the bottom of this page.)

6 Quadratic Equations

6-1 INTRODUCTION

1 Quadratic equations are not "new Math." Consider the following imaginary but very reasonable problem from about 2000 B.C. Amyr, Babylon's leading bean merchant, has a bit of money to invest in the new bean crop and he is fairly certain he can sell 60 sacks of beans over the coming holiday. Sad experience has taught him that the difference between his purchasing price from the farmer and the selling price to his customers should not be more than three bags per shekel. What price, in bags of beans per shekel, can he afford to pay the farmer if he wants a return of 10 shekels profit on his holiday's work?

Amyr's problem, in modern notation, looks like this:

$$\frac{60}{x-3} - \frac{60}{x} = 10$$ where x is Amyr's purchase price from the farmer in bags per shekel.

This equation can be put in the form

$x^2 - 3x - 18 = 0$ . . . a quadratic equation.

Solving equations of this kind was an everyday chore for Babylonian merchants and the oldest recorded solution of a quadratic equation appears on a papyrus document dating from about 1650 B.C. In this unit we will examine quadratic equations, consider several ways of solving them and, by the end of the chapter, show you how to solve Amyr's problem by a method he would have envied.

Quadratic Equation A *quadratic equation* is an equation that, when written in its simplest form, contains the second power, but no higher powers, of the variable. If the variable is x, a quadratic equation will have an x^2 term and it may have an x term or terms containing only

         Copyright © 1982 by John Wiley & Sons, Inc.

numbers, but it will *not* contain terms involving x^3 or x^4 or any other power of x. For example,

$$x^2 + 2x - 1 = 0 \qquad \text{is a quadratic equation}$$

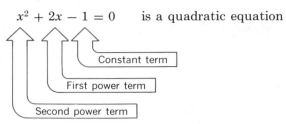

Constant term

First power term

Second power term

Complete and Incomplete Quadratics

The equations $2a^2 - 3a = 5$ and $3y^2 = 1 - y$ are also quadratic equations. If all three kinds of terms are present in the equation it is called a *complete* quadratic equation. The equations above are all complete quadratics. If either the first power term or the constant term are missing, the equation is called an *incomplete* quadratic. The equations

$$x^2 - 2 = 0$$
$$2x^2 + 3x = 0$$
$$4a^2 = 1$$

and

$$3y^2 = 5y \qquad \text{are all incomplete quadratic equations.}$$

Of course *every* quadratic equation must contain a second power of x^2 term—that is in fact what makes it a quadratic.

Which of the following are quadratic equations? Label each quadratic equation, QE.

1. $a^2 - 1 = 0$
2. $x - 1 = 0$
3. $p^2 - p + 1 = 0$
4. $y^3 - y + 1 = 0$
5. $2x^2 + 1 = x + 2$
6. $x(x + 1) = 3x^2 + 4$
7. $m(m + 2)(m + 3) = m - 2$
8. $2t^2 + 5 = 3$
9. $(x + 2)(x - 2) = x^2 + x$
10. $3y - 8 = y^2$

Check your answers in **2**.

---

2 1. QE $a^2 - 1 = 0$ is a quadratic equation.
   2. $x - 1 = 0$ is *not* a quadratic equation; it has no x^2 term.
   3. QE $p^2 - p + 1 = 0$ is a quadratic equation.
   4. $y^3 - y + 1 = 0$ is *not* a quadratic equation; it contains a y^3 term.
   5. QE $2x^2 + 1 = x + 2$ is a quadratic equation.
   6. QE $x(x + 1) = 3x^2 + 4$ can be rewritten as $x^2 + x = 3x^2 + 4$ or $2x^2 - x + 4 = 0$, which is a complete quadratic equation.
   7. $m(m + 2)(m + 3) = m - 2$ can be rewritten as $m^3 + 5m^2 + 5m + 2 = 0$. Because it contains a m^3 term, it is *not* a quadratic equation.
   8. QE $2t^2 + 5 = 3$ is a quadratic equation.
   9. $(x + 2)(x - 2) = x^2 + x$ can be rewritten as $x^2 - 4 = x^2 + x$ or $x = -4$. This is not a quadratic equation. It does not contain an x^2 term.
   10. QE $3y - 8 = y^2$ is a quadratic equation.

Standard Form of a Quadratic Equation

A quadratic equation is said to be in *standard form* when it is written in the form

$$ax^2 + bx + c = 0$$

Notice that all non-zero terms are written in the left member of the equation with zero as the right member of the equation. The x^2 term appears first on the left in the sum, the x term appears second, and the constant term c appears last in the sum.

The letters a, b, and c can represent any real numbers, except of course a cannot equal zero.

As you will see, the first step in many methods of solving quadratic equations is to rewrite the equation in standard quadratic form. It is important that you be able to translate any quadratic equation into standard form quickly and accurately.

In order to rewrite a quadratic equation into standard form, you may need to perform any or all of the following operations.

1. Arrange terms in order.

    Example: $2 - x + 3x^2 = 0$ becomes $3x^2 - x + 2 = 0$

    (a) If the coefficient of the x^2 term is negative we usually rewrite the equation changing the sign of each term so that the x^2 term becomes positive.

    $-x^2 + 3x - 5 = 0$ is usually written $x^2 - 3x + 5 = 0$ (multiply by -1)

    (b) If all terms have a common factor, divide by that factor to simplify the equation.

    $2x^2 + 2x + 4 = 0$ is written $x^2 + x + 2 = 0$ (divide by 2)

2. Collect like terms and combine them.

    Example: $x^2 - x + 3 = 2 + 3x$
    $x^2 - \underbrace{x - 3x}_{} + \underbrace{3 - 2}_{} = 0$ Add $-2 - 3x$ to both sides of the equation.
    $x^2 - 4x + 1 = 0$

3. Remove any parentheses.

    Example: $2(x - 1) = x^2$ Multiply the terms in the parentheses by 2.
    $2x - 2 = x^2$ Collect like terms on the right and rewrite
    $x^2 - 2x + 2 = 0$

4. Clear all fractions.

    Example: $\dfrac{x}{2} - x^2 = 1$ Multiply each term by 2.
    $x - 2x^2 = 2$ Now rewrite as:
    $2x^2 - x + 2 = 0$

5. Remove all radical signs.

    Example: $\sqrt{x - 1} = x$
    Square both sides to get
    $x - 1 = x^2$ or $x^2 - x + 1 = 0$

Rewrite each of the following quadratic equations in standard form.

1. $2 - x^2 = x$
2. $1 + x = 2 + x^2$
3. $x^2 - 2x + 3 = 2x^2 + x - 5$
4. $(x + 1)^2 = 2x + 3$
5. $x(x + 1) - 12 = 3(x - 4)$
6. $\dfrac{2x}{3} + 1 = x(2 - x)$
7. $\dfrac{x + 1}{x} = \dfrac{x - 1}{2}$
8. $\sqrt{2x} = 2x + 1$
9. $2(x - 4) - x(1 - 2x) = \dfrac{x - 2}{3}$
10. $\sqrt{2x^2 - 1} = 2x - 3$

Check your answers in **3**.

3 1. $2 - x^2 = x$ Rewrite as $-x^2 - x + 2 = 0$
    or $x^2 + x - 2 = 0$

2. $1 + x = 2 + x^2$
    Collect terms on the left: $1 + x - 2 - x^2 = 0$
    Combine like terms: $-x^2 + x - 1 = 0$
    or $x^2 - x + 1 = 0$

3. $x^2 - 2x + 3 = 2x^2 + x - 5$
    Collect terms on the left: $x^2 - 2x + 3 - 2x^2 - x + 5 = 0$
    Combine like terms: $-x^2 - 3x + 8 = 0$
    or $x^2 + 3x - 8 = 0$

4. $(x + 1)^2 = 2x + 3$
    Square to remove the parentheses: $x^2 + 2x + 1 = 2x + 3$
    Collect terms on the left: $x^2 + 2x + 1 - 2x - 3 = 0$
    Combine like terms: $x^2 - 2 = 0.$

5. $x(x + 1) - 12 = 3(x - 4)$
    Remove the parentheses by multiplying: $x^2 + x - 12 = 3x - 12$
    Collect terms on the left: $x^2 + x - 12 - 3x + 12 = 0$
    Combine like terms: $x^2 - 2x = 0$

6. $\dfrac{2x}{3} + 1 = x(2 - x)$

    To clear the fraction multiply each term of the equation by 3:
    $2x + 3 = 3x(2 - x)$
    Remove the parentheses by multiplying: $2x + 3 = 6x - 3x^2$
    Collect terms: $2x + 3 - 6x + 3x^2 = 0$
    Combine like terms: $3x^2 - 4x + 3 = 0$

7. $\dfrac{x + 1}{x} = \dfrac{x - 1}{2}$

    To clear the fractions multiply both sides of the equation by $2x$:
    $\dfrac{2x(x + 1)}{x} = \dfrac{2x(x - 1)}{2}$ or $2(x + 1) = x(x - 1)$
    Remove parentheses by multiplying: $2x + 2 = x^2 - x$
    Collect terms on the left: $2x + 2 - x^2 + x = 0$
    Combine like terms and rearrange: $-x^2 + 3x + 2 = 0$
    Multiply each term by -1: $x^2 - 3x - 2 = 0$

8. $\sqrt{2x} = 2x + 1$
    Square both members of the equation to remove the radical sign:
    $(\sqrt{2x})^2 = (2x + 1)^2$ or $2x = (2x + 1)^2$
    Remove the parentheses by multiplying: $2x = 4x^2 + 4x + 1$
    Collect terms: $2x - 4x^2 - 4x - 1 = 0$
    Combine like terms: $-4x^2 - 2x - 1 = 0$
    Multiply each term by -1: $4x^2 + 2x + 1 = 0$

9. $2(x - 4) - x(1 - 2x) = \dfrac{x - 2}{3}$

    Clear the fraction by multiplying each term by 3:
    $6(x - 4) - 3x(1 - 2x) = x - 2$
    Multiply to remove the parentheses: $6x - 24 - 3x + 6x^2 = x - 2$
    Collect terms on the left: $6x - 24 - 3x + 6x^2 - x + 2 = 0$
    Combine like terms: $6x^2 + 2x - 22 = 0$
    Simplify by dividing each term by 2: $3x^2 + x - 11 = 0$

10. $\sqrt{2x^2 - 1} = 2x - 3$
    Remove the radical by squaring both members of the equation:
    $(\sqrt{2x^2 - 1})^2 = (2x - 3)^2$ or $2x^2 - 1 = (2x - 3)^2$

450   **Quadratic Equations**

Remove the parentheses by squaring the quantity on the right:
$2x^2 - 1 = 4x^2 - 12x + 9$
Collect terms: $\quad 2x^2 - 1 - 4x^2 + 12x - 9 = 0$
Combine like terms: $\quad -2x^2 + 12x - 10 = 0$
Multiply each term by -1: $\quad 2x^2 - 12x + 10 = 0$
Simplify by dividing each term by 2: $\quad x^2 - 6x + 5 = 0$

Solution of a
Quadratic
Equation

The solution of a linear equation is a single number that when substituted for the variable makes the linear equation a true statement. For example, the solution of the linear equation $3x + 1 = 7$ is $x = 2$ because $3(2) + 1 = 7$ is a true statement.

The solution of a quadratic equation is in general a pair of numbers, each of which makes the equation a true statement. For example, the solution of the quadratic equation

$x^2 - 5x + 6 = 0 \quad$ consists of the numbers 2 and 3 because

$x = 2$ gives $(2)^2 - 5(2) + 6 = 0 \quad$ or $\quad 4 - 10 + 6 = 0 \quad$ which is a true statement.

and

$x = 3$ gives $(3)^2 - 5(3) + 6 = 0 \quad$ or $\quad 9 - 15 + 6 = 0 \quad$ which is also a true statement.

Every quadratic equation has two roots, either a pair of positive or negative real numbers including zero, or a pair of irrational numbers expressed as radical expressions. The two roots may be equal.

Show by substituting that each of the following equations is satisfied by both members of the solution given.

1. $x^2 + x = 0 \quad$ for $\quad x = 0$ and $x = -1$
2. $2x^2 - 5x - 12 = 0 \quad$ for $\quad x = 4$ and $x = -\dfrac{3}{2}$
3. $4x^2 - 1 = 0 \quad$ for $\quad x = \dfrac{1}{2}$ and $x = -\dfrac{1}{2}$
4. $x^2 - 4x + 4 = 0 \quad$ for $\quad x = 2$
5. $x^2 + 2x - 1 = 0 \quad$ for $\quad x = -1 + \sqrt{2}$ and $x = -1 - \sqrt{2}$

Check your work in **4**.

---

4 1. For $x = 0$ $\quad$ $x^2 + x = 0$ gives $(0)^2 + (0) = 0$
or $0 = 0$
For $x = -1$ $\quad$ $x^2 + x = 0$ gives $(-1)^2 + (-1) = 0$
or $1 - 1 = 0$

2. For $x = 4$ $\quad$ $2x^2 - 5x - 12 = 0$ gives $2(4)^2 - 5(4) - 12 = 0$
or $32 - 20 - 12 = 0$
For $x = -\dfrac{3}{2}$ $\quad$ $2x^2 - 5x - 12 = 0$ gives $2\left(-\dfrac{3}{2}\right)^2 - 5\left(-\dfrac{3}{2}\right) - 12 = 0$

or $\dfrac{9}{2} + \dfrac{15}{2} - 12 = 0$

or $\dfrac{24}{2} - 12 = 0$

3. For $x = \dfrac{1}{2}$ $\quad$ $4x^2 - 1 = 0$ gives $4\left(\dfrac{1}{2}\right)^2 - 1 = 0$

or $4\left(\dfrac{1}{4}\right) - 1 = 0$ or $1 - 1 = 0$

For $x = -\dfrac{1}{2}$ $4x^2 - 1 = 0$ gives $4\left(-\dfrac{1}{2}\right)^2 - 1 = 0$

or $4\left(\dfrac{1}{4}\right) - 1 = 0$ or $1 - 1 = 0$

4. For $x = 2$ $x^2 - 4x + 4 = 0$ gives $(2)^2 - 4(2) + 4 = 0$
or $4 - 8 + 4 = 0$

5. For $x = -1 + \sqrt{2}$ $x^2 + 2x - 1 = 0$
gives $(-1 + \sqrt{2})^2 + 2(-1 + \sqrt{2}) - 1 = 0$
or $1 - 2\sqrt{2} + 2 - 2 + 2\sqrt{2} - 1 = 0$
$1 + 2 - 2 - 1 = 0$
$3 - 3 = 0$

For $x = -1 - \sqrt{2}$ $x^2 + 2x - 1 = 0$
gives $(-1 - \sqrt{2})^2 + 2(-1 - \sqrt{2}) - 1 = 0$
or $1 + 2\sqrt{2} + 2 - 2 - 2\sqrt{2} - 1 = 0$
$1 + 2 - 2 - 1 = 0$
$3 - 3 = 0$

The kind of quadratic equation most easy to solve is one that contains only the x^2 and constant terms. For example, the equation

$x^2 - 4 = 0$

can be solved most simply by rewriting it as $x^2 = 4$ and taking the square root of both members of the equation. Since the two roots of 4 are $+2$ and -2, the two roots of the equation $x^2 - 4 = 0$ are $x = 2$ and $x = -2$.

Another example: the equation $x^2 - 5 = 0$ can be rewritten as

$x^2 = 5$

The two square roots of 5 are $\sqrt{5}$ and $-\sqrt{5}$; therefore, the roots of the equation $x^2 = 5$ are $x = \sqrt{5}$ and $x = -\sqrt{5}$.

Use this method to solve each of the following quadratic equations.

(a) $x^2 - 9 = 0$ (b) $x^2 - 2 = 0$ (c) $2x^2 - 6 = 0$
(d) $x^2 = 0$ (e) $3x^2 - 16 = 0$

Check your work in **5**.

---

5   (a) $x^2 - 9 = 0$
Rewrite this equation as $x^2 = 9$
The roots are $x = \sqrt{9}$ and $x = -\sqrt{9}$ or $x = 3$ and $x = -3$.

(b) $x^2 - 2 = 0$
Rewrite this equation as $x^2 = 2$
The roots of this equation are $x = \sqrt{2}$ and $x = -\sqrt{2}$.

(c) $2x^2 - 6 = 0$
Rewrite this equation as $2x^2 = 6$ or $x^2 = 3$.
The roots of this equation are $x = \sqrt{3}$ and $x = -\sqrt{3}$.

(d) $x^2 = 0$
The roots of this equation are both equal to $x = 0$.

(e) $3x^2 - 16 = 0$
Rewrite this equation as $3x^2 = 16$ or $x^2 = \dfrac{16}{3}$.

The roots of this equation are $x = \sqrt{\dfrac{16}{3}} = \dfrac{4}{\sqrt{3}} = \dfrac{4\sqrt{3}}{3}$ and

$$x = -\sqrt{\frac{16}{3}} = \frac{-4\sqrt{3}}{3}.$$

You should notice that certain quadratic equations of this kind do not have a solution in real numbers. For example, the equation

$x^2 + 2 = 0$ can be rewritten as $x^2 = -2$

and there is no real number value for x that can make this equation a true statement. There is no real number x whose square is equal to -2.

6-2 THE PYTHAGOREAN THEOREM

Right Triangle

Hypotenuse

One very useful application of this simplest kind of quadratic equation involves finding the lengths of the sides of a right triangle. In a *right triangle,* the two shorter sides, or *legs,* of the triangle are perpendicular and, therefore, meet at a *right* or 90° angle. The longest side of the triangle is known as the *hypotenuse.*

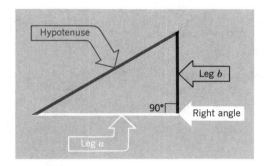

A very old and very useful theorem of geometry, known as the *Pythagorean theorem,* states that the sum of the squares of the lengths of the two legs of a right triangle is equal to the square of the length of the hypotenuse.

$$\boxed{a^2 + b^2 = c^2}$$

If we know the lengths of any two of the three sides of a right triangle, this rule enables us to find the length of the third side. For example, if we are told that the legs of a right triangle are $a = 2$ ft and $b = 1$ ft in length, then we can find the hypotenuse using the Pythagorean theorem.

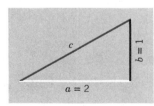

$c^2 = a^2 + b^2$
$c^2 = (2)^2 + (1)^2$
$c^2 = 4 + 1$
$c^2 = 5$

Solving this quadratic equation we have $c = \sqrt{5}$ and $c = -\sqrt{5}$.

The root $c = -\sqrt{5}$ is not a physically meaningful answer—we cannot have a line of negative length—so we disregard this part of the solution. The length of the hypotenuse of this triangle is $c = \sqrt{5} \cong 2.23$ ft.

It is important that you realize that the Pythagorean theorem is true *only* for right triangles, triangles that have two sides meeting perpendicular to each other.

Use the Pythagorean theorem to solve the following problems.

(a) 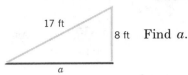 8 ft Find a.

(b) 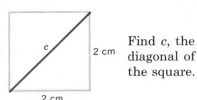 Find c, the diagonal of the square.

(c) 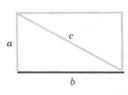 Find b if c is 6 in. and a is 4 in.

Compare your answers with those in **6.**

---

6 (a) $b = 8$ ft, $c = 17$ ft   Substituting these values into the equation
$a^2 + b^2 = c^2$ we find
$a^2 + (8)^2 = (17)^2$
or $a^2 + 64 = 289$

Rewrite this equation as $a^2 = 289 - 64$ or $a^2 = \sqrt{225}$. The solution to this quadratic equation is $a = \sqrt{225} = 15$ ft. We disregard the other root $a = -15$ ft because it is physically meaningless.

(b) $a = 2$ cm, $b = 2$ cm   Substituting these values into the equation
$a^2 + b^2 = c^2$ we find
$(2)^2 + (2)^2 = c^2$
or $c^2 = 4 + 4$
$c^2 = 8$

The solution to this quadratic equation is $c = 2\sqrt{2}$ and $c = -2\sqrt{2}$. Discard the negative root. The diagonal of the square is $2\sqrt{2}$ cm or about 2.83 cm.

(c) $a = 4$ in.; $c = 6$ in.   Substitute these values into the equation
$a^2 + b^2 = c^2$
$(4)^2 + b^2 = (6)^2$
or $16 + b^2 = 36$
$b^2 = 20$

The solution to this quadratic equation is $b = 2\sqrt{5}$ and $b = -2\sqrt{5}$. Discard the negative root. The length of the side b is $2\sqrt{5}$ inches or about 4.47 inches.

The following set of problems will provide practice in solving simple quadratic equations.

1. Solve: (a) $x^2 - 121 = 0$ (b) $y^2 = 10$
             (c) $x^2 + 4 = 12$ (d) $p^2 - 2 = 62$
             (e) $a^2 - \dfrac{1}{2} = 0$ (f) $x^2 - \dfrac{3}{4} = 0$
             (g) $2x^2 - 3 = 0$ (h) $5t^2 - 1 = 0$
             (i) $x^2 + 4 = 0$ (j) $3q^2 + 2 = 6$
             (k) $p^2 - 100 = 100$ (l) $13x^2 - 468 = 0$

2. Solve using the Pythagorean theorem.

    (a)   Find a.   (b)   Find c.
    $b = 45$ $c = 53$ c 15
              a 112

(c) 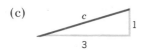 Find c. (d) 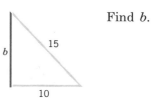 Find b.

(e) 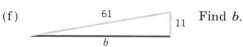 Find c. (f) Find b.

(g) Find the diagonal of a rectangle with sides 40 cm and 9 cm.

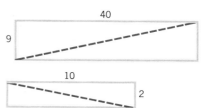

(h) Find the diagonal of a rectangle with sides 10 in. and 2 in.

(i) In a certain right triangle the hypotenuse is 20 ft long. If the longest leg is three times the length of the shorter, find the length of the two legs.

(j) The area A of a circle is given by the formula $A = \pi R^2$, where R is the radius of the circle. Find the radius of a circle whose area is 60 cm^2. Use $\pi \cong 3.14$. Round your answer to two decimal digits.

Check your answers in **7**.

7 1. (a) $11, -11$ (b) $\sqrt{10}, -\sqrt{10}$ (c) $2\sqrt{2}, -2\sqrt{2}$ (d) $8, -8$

(e) $\dfrac{\sqrt{2}}{2}, \dfrac{-\sqrt{2}}{2}$ (f) $\dfrac{\sqrt{3}}{2}, \dfrac{-\sqrt{3}}{2}$ (g) $\dfrac{\sqrt{6}}{2}, \dfrac{-\sqrt{6}}{2}$ (h) $\dfrac{\sqrt{5}}{5}, \dfrac{-\sqrt{5}}{5}$

(i) No solution in real numbers (j) $\dfrac{2\sqrt{3}}{3}, \dfrac{-2\sqrt{3}}{3}$ (k) $10\sqrt{2}, -10\sqrt{2}$ (l) $6, -6$

2. (a) 28 (b) 113 (c) $\sqrt{10}$ (d) $5\sqrt{5}$

(e) 101 (f) 60 (g) 41 cm (h) $2\sqrt{26}$ in. or about 10.2 in.

(i) The legs are $2\sqrt{10}$ ft and $6\sqrt{10}$ ft or about 6.32 ft and 18.97 ft.

(j) The radius is about 4.37 cm.

The following property of real numbers, known as the *zero-factor property,* is very useful in solving certain kinds of quadratic equations.

Zero-Factor Property

> If the product of two or more factors is equal to zero, then at least one of the factors must equal zero.

In symbols, if $ab = 0$, then either $a = 0$
or $b = 0$
or both a and $b = 0$

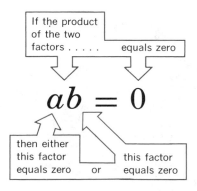

For example, suppose we want to solve the quadratic equation

$x^2 - 2x = 0$

First, factor the left member of this equation:

$x(x - 2) = 0$

Second, use the zero-factor property to write this as a pair of equations, each of which leads to one of the roots of the original equation:

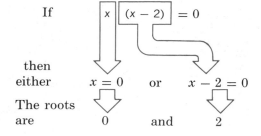

then
either $\quad x = 0 \quad$ or $\quad x - 2 = 0$

The roots
are $\qquad 0 \qquad$ and $\qquad 2$

Check this solution by substituting both values into the original equation to see if they satisfy it.

For x equal to 0 $\quad (0)^2 - 2(0) = 0$ gives $0 - 0 = 0$
For x equal to 2 $\quad (2)^2 - 2(2) = 0$ gives $4 - 4 = 0$

Use the zero-factor property to solve the equation $x^2 = -4x$.

Check your work in **8**.

8   Solve $x^2 = -4x$

Step 1.   Write the equation in standard form: $x^2 + 4x = 0$

Step 2.   Factor the left member of the equation: $x(x + 4) = 0$

Step 3.   Use the zero-factor property to write two new equations:
Either $x = 0$ or $(x + 4) = 0$

Step 4.   Solve these two equations to find the roots of the original quadratic equation:
The equations are $x = 0$ and $x = -4$, so the roots are 0 and -4.

Step 5.   Check both roots by substituting them back into the original quadratic equation:
For $x = 0$ $(0)^2 = -4(0)$ or $0 = 0$
For $x = -4$ $(-4)^2 = -4(-4)$ or $16 = 16$

This procedure for solving quadratic equations by using the zero-factor property has been used by mathematicians for more than three hundred years. It was first outlined by Thomas Harriot, an English mathematician who is also famous for having visited the American colonies in 1585 to survey the area now known as North Carolina.

Use the zero-factor property to solve the equation

$(x + 1)(x - 3) = 0$

Choose one of the following answers and go to the frame indicated.

(a)   The roots are given by $x = 1$ and $x = 3$.   Go to **9**
(b)   The roots are given by $x = -1$ and $x = 3$.   Go to **10**
(c)   I need more help. Please explain it again.   Go to **11**

---

9   This answer is *not* correct.

Do it this way: If $(x + 1)$ $(x - 3)$ $= 0$

then either $(x + 1) = 0$ or $(x - 3) = 0$

If $(x + 1) = 0$ then $x + \underbrace{1 - 1}_{0} = 0 - 1$ adding -1 to both
sides of the equation

or $x = -1$.
The first root is -1.

457   **6-3 Solving Quadratic Equations by Factoring**

Be careful here. Solve each of the zero-factor equations just as you would any linear equation.

Now return to the bottom of frame **8** and continue.

---

10   Right you are.

If $(x + 1)$ $(x - 3) = 0$

then either $x + 1 = 0$ or $x - 3 = 0$
$x = -1$ or $x = 3$

The roots are -1 and 3.

Be certain to check your answers by substituting them back into the original equation.

Solve each of the following equations using the zero-factor property.

(a) $(x + 2)(x - 2) = 0$ (b) $(2x - 1)(x + 2) = 0$
(c) $(3x + 2)(2x - 5) = 0$ (d) $2a(3a - 1) = 0$
(e) $y(y - 5) = 0$ (f) $(4 - x)(x - 1) = 0$
(g) $(2 - x)(x + 3) = 0$ (h) $(2t - 1)(3 + t) = 0$
(i) $3(p - 4)(2p + 1) = 0$ (j) $5y(5y - 2) = 0$

Check your answers in **12**.

---

11   Let's review.

If the product $(x + 1)(x - 3)$ is equal to zero, then one of the two factors $(x + 1)$ or $(x - 3)$ *must* equal zero.

If $(x + 1)$ $(x - 3) = 0$

then either $x + 1 = 0$ or $x - 3 = 0$
$x + 1 - 1 = 0 - 1$ or $x - 3 + 3 = 0 + 3$

$x = -1$ $x = 3$

In other words, if the equation $(x + 1)(x - 3) = 0$ is true, then either x must equal -1 or x must equal 3. This means that the roots of the equation are -1 and 3.

Now hop back to frame **10** and continue.

---

12   (a) $x = 2, x = -2$ (b) $x = \dfrac{1}{2}, x = -2$

(c) $x = -\dfrac{2}{3}, x = \dfrac{5}{2}$ (d) $a = 0, a = \dfrac{1}{3}$

(e) $y = 0, y = 5$ (f) $x = 1, x = 4$

(g) $x = 2, x = -3$ (h) $t = \dfrac{1}{2}, t = -3$

(i) $p = 4, p = -\dfrac{1}{2}$ (j) $y = 0, y = \dfrac{2}{5}$

Important > Any equation written in the form $x^2 = mx$ should be solved using the zero-factor property.

First, write it in standard form. $x^2 - mx = 0$

Second, factor it. $x(x - m) = 0$

Third, solve it using the zero-factor property. $x = 0, x = m$

Of course most of the quadratic equations you will encounter must be factored before the zero-factor rule can be applied. For example, to solve the equation

$x^2 - x = 6$

first, write it in standard form. $x^2 - x - 6 = 0$

then factor it, $(x + 2)(x - 3) = 0$

and finally solve it using the zero-factor property. $x + 2 = 0$ or $x - 3 = 0$

The roots are -2 and 3.

Not every quadratic equation can be factored in this way, and there is no easy way to tell which quadratic equations are factorable and which are not. But the first steps in solving *any* quadratic equation are

1. Rewrite it in standard form.
2. Factor it if possible.

If you need to review the process of factoring trinomials, return to page 221, frame **23**. Otherwise, turn to **13** for a set of problems on solving quadratic equations by factoring.

Unit 6
Quadratic
Equations

13   The answers are on page 566.

Solving Quadratic
Equations by
Factoring

A. Solve:

1. $x^2 = 1$ _____

2. $x^2 = 6$ _____

3. $x^2 - 18 = 0$ _____

4. $12 - x^2 = 0$ _____

5. $2x^2 - 10 = 0$ _____

6. $3x^2 - 24 = 0$ _____

7. $2x^2 + 1 = 9$ _____

8. $5x^2 + 3 = 18$ _____

9. $2x^2 + 3 = 0$ _____

10. $4x^2 - 7 = 0$ _____

11. $3x^2 - 4 = 0$ _____

12. $x^2 + \dfrac{1}{2} = 0$ _____

13. $x^2 = \dfrac{2}{3}$ _____

14. $x^2 = \dfrac{1}{2}$ _____

15. $2x^2 = \dfrac{1}{3}$ _____

16. $4x^2 = 8$ _____

17. $4 - x^2 = 0$ _____

18. $3 - 2x^2 = 0$ _____

B. Solve using the Pythagorean theorem:

1.

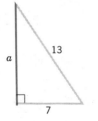

_____

2.

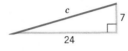

_____

3.

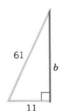

_____

4.

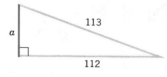

_____

5.

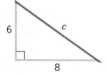

_____

6.

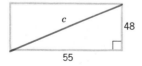

_____

7.

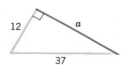

_____

8.

_____

9.

_____

10.

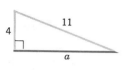

_____

11.

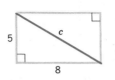

_____

12.

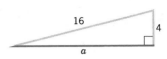

_____

13.

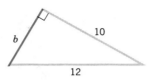

_____

14.

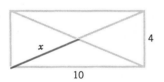

_____

15.

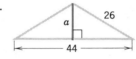

_____

16.

_____

C. Solve by factoring:

1. $(x + 1)(x - 3) = 0$ _____

2. $(y - 8)(y + 2) = 0$ _____

3. $(a - 4)(a - 3) = 0$ _____

4. $2(x - 2)(2x - 5) = 0$ _____

5. $(2x + 1)(3x - 1) = 0$ _____

6. $(7 - p)(3 + p) = 0$ _____

7. $a^2 - 3a - 18 = 0$ _____

8. $a^2 - 5a - 6 = 0$ _____

9. $2x^2 - 5x - 3 = 0$ _____

10. $2x^2 - x - 3 = 0$ _____

11. $3x^2 - 5x + 2 = 0$ _____

12. $3x^2 + 8x - 3 = 0$ _____

13. $6x^2 + 5x + 1 = 0$ _____

14. $6p^2 + 13p + 6 = 0$ _____

15. $25q^2 - 5q - 2 = 0$ _____

16. $9x^2 - 9x - 10 = 0$ _____

17. $4x(x + 2) = 5$ _____

18. $x(6x - 5) = 4$ _____

19. $4x(x + 3) + 12 = 3$ _____

20. $x(2x - 7) - 10 = 5$ _____

21. $x^2 = 7 - 6x$ _____

22. $2x^2 = 7x + 4$ _____

23. $a^2 - 3a + 8 = 6$ _____

24. $9q^2 = 9(q - 1) + 7$ _____

25. $(x + 1)(x - 6) = 18$ _____

26. $(x + 2)(2x + 1) = 6x + 5$ _____

27. $(x + 3)^2 = 1$ _____

28. $(x - 5)^2 - 1 = 0$ _____

29. $x^2 - 10x + 25 = 4(x - 2)$ _____

30. $2x^2 - x - 10 = 2(x + 5)$ _____

31. $1 = 6q - 9q^2$ _____

32. $3 = 4x^2 + x$ _____

33. $3t = t^2 - 4$ _____

34. $y^2 = 14 + 5y$ _____

35. $a^2 = a$ _____

36. $(q + 1)(q - 1) + 1 = 0$ _____

37. $(x + 2)(x - 2) = 3x$ _____

38. $(x - 1)(x + 2) = 10$ _____

39. $(2x - 1)(x + 2) = x^2 - 2$ _____

40. $(x - 3)(2x + 5) = x^2 + x$ _____

D. Solve by factoring:

1. $x^2 - \dfrac{1}{2} = 0$ _____

2. $2x^2 = \dfrac{25}{2}$ _____

3. $\dfrac{1}{2}x^2 = \dfrac{3}{4}$ _____

4. $\dfrac{2}{3}x^2 - \dfrac{1}{6} = 0$ _____

5. $\dfrac{a^2}{6} + \dfrac{a}{3} = \dfrac{1}{2}$ _____

6. $\dfrac{x^2}{3} + x = \dfrac{4}{3}$ _____

7. $\dfrac{x^2}{4} + 1 = x$ _____

8. $\dfrac{x^2}{2} - 3x + 4 = 0$ _____

Date _____

Name _____

Course/Section _____

9. $x^2 - 2 = \dfrac{x}{6}$ _____

10. $x^2 = 10 + \dfrac{3x}{2}$ _____

11. $x + \dfrac{2}{x} = 3$ _____

12. $x = 8 - \dfrac{15}{x}$ _____

13. $a - \dfrac{10}{a - 3} = 0$ _____

14. $x = \dfrac{32}{x - 4}$ _____

15. $\dfrac{x^2 - 5}{4} = 10 - x$ _____

16. $x = \dfrac{6}{x - 1}$ _____

17. $\dfrac{1}{2} + \dfrac{1}{2y} = \dfrac{1}{y^2}$ _____

18. $x + 1 = \dfrac{1}{1 - x}$ _____

19. $\dfrac{3}{x + 1} = 2 - \dfrac{4}{2x}$ _____

20. $\dfrac{7}{a - 3} - \dfrac{1}{2} = \dfrac{3}{a - 4}$ _____

21. $(x + 2)^2 = 9$ _____

22. $(x - 3)^2 = 16$ _____

23. $x(3x + 2) = x + 2$ _____

24. $5x(x + 1) = 8 - x$ _____

25. $2x - 5 + \dfrac{3}{x} = 0$ _____

26. $3x + 2 - \dfrac{8}{x} = 0$ _____

27. $10x - 13 = \dfrac{3}{x}$ _____

28. $7x - 11 = \dfrac{6}{x}$ _____

29. $5 - x = \dfrac{x + 4}{2x}$ _____

30. $2x + 1 = \dfrac{x + 2}{6x}$ _____

E. Brain Boosters

1. Solve for x. $(x - 1)(x + 2)(x + 1) = 0$ _____

2. Solve for x. $(x - a)^2 + (x - b)^2 = a^2 + b^2$ _____

3. Solve for x. $x - 3 = \dfrac{x^3 - 27}{x^2 + 8}$ _____

4. Solve for x: $\dfrac{a}{b + x} + \dfrac{b}{a + x} = 1$ _____

5. Solve for x:

(a) _____

(b) _____

Date _____

Name _____

Course/Section _____

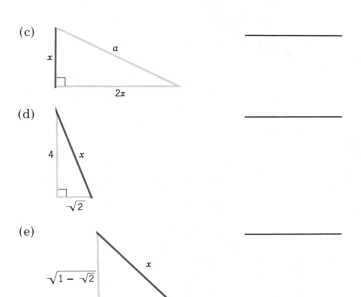

(c)

(d)

(e)

6. A farmer plants 600 corn plants in equal rows, then to make room for an irrigation ditch he takes out five plants from each row and with these he makes six more rows. How many plants did he originally have in a row?

$\left(\textit{Hint:}\text{ Let }x\text{ equal the original number in a row, then }\dfrac{600}{x}\text{ is the original number of rows.}\right)$

7. Given this problem 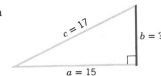 Eric found b this way:

$b = \sqrt{2 \cdot 17 + 2 \cdot 15} = \sqrt{64} = 8$, which is correct. He claims this method is always correct if the hypotenuse c is 2 more than side a. Show that if $c = a + 2$ then $b = \sqrt{2a + 2c}$.

---

8. The most interesting kind of right triangle is one in which the length of each of the sides is an integer. Any set of integers a, b, and c such that $a^2 + b^2 = c^2$ is called a Pythagorean triplet. Babylonian clay tablets reveal that mathematicians knew about such numbers and knew rules for finding them more than 2500 years ago. The oldest and simplest such triplet known has sides of length 3, 4, and 5: $3^2 + 4^2 = 5^2$. Ancient Egyptians used the 3–4–5 triangle to make right angles for surveying fields and constructing buildings. Here are a few others:

a	b	c
3	4	5
5	12	13
15	8	17
7	24	25
35	12	37
9	40	41
11	60	61
45	28	53

To find Pythagorean triples, choose any two positive odd integers m and n, such that m is greater than n, and m and n have no common factors, then

$a = mn$

$b = \dfrac{m^2 - n^2}{2}$

$c = \dfrac{m^2 + n^2}{2}$

For example, if $m = 5, n = 1$ $\quad a = 5 \cdot 1 = 5$

$$b = \frac{5^2 - 1^2}{2} = 12$$

$$c = \frac{5^2 + 1^2}{2} = 13$$

Notice that the hypotenuse c is always an odd number and that one of the two legs, b, is even and the other, a, is odd. Also, notice that (a) the even leg is always evenly divisible by 4, (b) in every triplet one of the sides is evenly divisible by 3, and (c) one of the sides is always evenly divisible by 5.

(a) Find all such Pythagorean triplets with sides less than 100. (There are sixteen triplets with all sides less than 100, not counting those that are multiples of smaller ones.)

————————————————————

(b) Show that in every triplet, $b + c$ is always a perfect square.

————————————————————

(c) Show that $a + c$ is always equal to twice a perfect square.

————————————————————

When you have had the practice you need either return to the preview on page 445 or continue in **14** where you will learn to solve more difficult quadratic equations.

———

Date

Name

Course/Section

14   The quickest and simplest way to solve a quadratic equation is to factor it as shown in the last section. However, this is only possible with equations that can be factored—rewritten as the zero product of two factors—and not every quadratic equation can be factored in this way. For example, the equation

$$x^2 - 2x - 1 = 0$$

cannot be written as the product of factors with integer coefficients.

In this section we will show you two methods that can be used to solve *any* quadratic equation.

First, you should recall from Unit 3 that

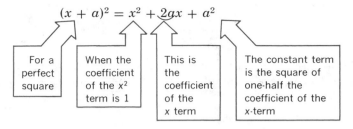

For example,

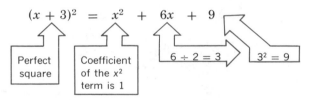

or

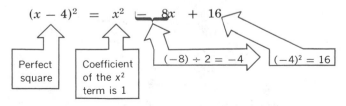

Use this procedure to find the missing last term for each of the following squares.

(a) $(x + 5)^2 = x^2 + 10x + \boxed{}$

(b) $(x - 3)^2 = x^2 - 6x + \boxed{}$

(c) $(x + 1)^2 = x^2 + 2x + \boxed{}$

(d) $(x - 2)^2 = x^2 - 4x + \boxed{}$

Check your work in **15**.

---

15   (a) $(x + 5)^2 = x^2 + 10x + \boxed{25}$

$$10 \div 2 = 5 \qquad 5^2 = 25$$

Step 1. Find the coefficient of the x term: 10

Step 2. Divide this number by 2: $10 \div 2 = 5$

Step 3. Square to find the constant term: $5^2 = 25$

(b) $(x - 3)^2 = x^2 - 6x + \boxed{9}$

$(-6) \div 2 = -3$ $(-3)^2 = 9$

(c) $(x + 1)^2 = x^2 + 2x + \boxed{1}$

$2 \div 2 = 1$ $1^2 = 1$

(d) $(x - 2)^2 = x^2 - 4x + \boxed{4}$

$(-4) \div 2 = -2$ $(-2)^2 = 4$

What number must be added to each of the following expressions in order to make it a perfect square? Complete each of the following equations.

(a) $x^2 - 10x + \square = ($ $)^2$

(b) $x^2 - 3x + \square = ($ $)^2$

(c) $x^2 + 12x + \square = ($ $)^2$

(d) $x^2 + x + \square = ($ $)^2$

Check your answers in **16**.

---

16 (a) $x^2 - 10x + 25 = (x - 5)^2$

$(-10) \div 2 = -5$ $(-5)^2 = 25$

(b) $x^2 - 3x + \dfrac{9}{4} = \left(x - \dfrac{3}{2}\right)^2$

$(-3) \div 2 = -\frac{3}{2}$ $(-\frac{3}{2})^2 = \frac{9}{4})$

(c) $x^2 + 12x + 36 = (x + 6)^2$

$12 \div 2 = 6$ $6^2 = 36$

(d) $x^2 + x + \dfrac{1}{4} = \left(x + \dfrac{1}{2}\right)^2$

$1 \div 2 = \frac{1}{2}$ $(\frac{1}{2})^2 = \frac{1}{4}$

Solving a Quadratic Equation by Completing the Square

We can use this process of completing the square to solve quadratic equations. For example, to solve

$$x^2 + 6x + 4 = 0$$

follow these steps.

Step 1. Rewrite the equation as $x^2 + 6x = -4$.

Put all terms containing the variable on the left of the equation and all constant terms on the right.

Step 2. Check to be certain that the coefficient of the x^2 term is 1. If the coefficient of x^2 is not 1, divide each term by the coefficient of x^2 and continue.

Step 3. Complete the square on the left.

$$x^2 + 6x + \boxed{9} = -4 + \boxed{9} \qquad \text{Add } \boxed{9} \text{ to both sides of the equation.}$$
$$6 \div 2 = 3 \rightarrow 3^2 = 9$$

$$x^2 + 6x + 9 = 5$$

Take one-half of the coefficient of the x term, square it, and add this square to *both* sides of the equation.

Step 4. Rewrite the left member of the equation as a perfect square.

$$(x + 3)^2 = 5$$

Step 5. Take the square root of both members of the equation and set up two equations.

$$x + 3 = \sqrt{5} \qquad \text{and} \qquad x + 3 = -\sqrt{5}$$

Remember that every real number has two square roots.

Step 6. Solve these two equations to get the two roots of the original equation.

$$x = -3 + \sqrt{5} \qquad \text{and} \qquad x = -3 - \sqrt{5}$$

The roots are $-3 + \sqrt{5}$ and $-3 - \sqrt{5}$ or $-3 \pm \sqrt{5}$.

Step 7. Check the solution by substituting both roots back into the original equation.

For $x = -3 + \sqrt{5} \qquad x^2 + 6x + 4 = 0$
gives $(-3 + \sqrt{5})^2 + 6(-3 + \sqrt{5}) + 4 = 0$
or $\quad 9 - 6\sqrt{5} + 5 - 18 + 6\sqrt{5} + 4 = 0$
$$18 - 18 = 0$$

For $x = -3 - \sqrt{5} \qquad x^2 + 6x + 4 = 0$
gives $(-3 - \sqrt{5})^2 + 6(-3 - \sqrt{5}) + 4 = 0$
or $\quad 9 + 6\sqrt{5} + 5 - 18 - 6\sqrt{5} + 4 = 0$
$$18 - 18 = 0$$

Your turn. Use this method to solve the equation

$$x^2 - 2x - 1 = 0$$

Our step-by-step answer is in **17**.

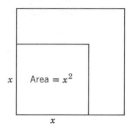
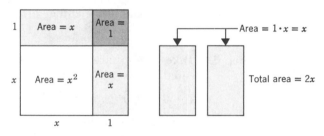
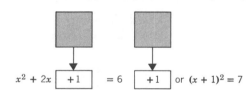
17

Step 1. $x^2 - 2x - 1 = 0$ should be rewritten as $x^2 - 2x = 1$

Step 2. Check to be certain that the coefficient of x^2 is 1.
(In other words, the first term must be x^2 and not $2x^2$ or $3x^2$ or any other term.)

Step 3. Complete the square.

$$x^2 - 2x + \boxed{1} = 1 + \boxed{1} \qquad \text{Add 1 to each side of the equation.}$$
$$(-2) \div 2 = -1 \leftarrow (-1)^2 = 1$$

Step 4. Write the left member of the equation as a perfect square.

$$(x - 1)^2 = 2$$

Step 5. Solve by taking the square root of both members of the equation.

$$x - 1 = \sqrt{2} \quad \text{and} \quad x - 1 = -\sqrt{2}$$

Step 6. Solve these two equations to get the two roots of the original equation.

$$x = 1 + \sqrt{2} \quad \text{and} \quad x = 1 - \sqrt{2} \qquad \text{The roots are } 1 + \sqrt{2} \text{ and } 1 - \sqrt{2}$$
$$\text{or } 1 \pm \sqrt{2}.$$

Step 7. Check the solution by substituting the roots back into the original equation.

For $x = 1 + \sqrt{2}$ $\qquad x^2 - 2x - 1 = 0$

$$\text{gives } (1 + \sqrt{2})^2 - 2(1 + \sqrt{2}) - 1 = 0$$
$$\text{or} \quad 1 + 2\sqrt{2} + 2 - 2 - 2\sqrt{2} - 1 = 0$$
$$3 - 3 = 0$$

For $x = 1 - \sqrt{2}$ $\qquad x^2 - 2x - 1 = 0$

$$\text{gives } (1 - \sqrt{2})^2 - 2(1 - \sqrt{2}) - 1 = 0$$
$$\text{or} \quad 1 - 2\sqrt{2} + 2 - 2 + 2\sqrt{2} - 1 = 0$$
$$3 - 3 = 0$$

In general, any quadratic equation that can be put in the form

$$(mx + q)^2 = p$$

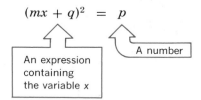

will have the solution $mx + q = \pm\sqrt{p}$ $\qquad$ or

$$x = \frac{-q \pm \sqrt{p}}{m}$$

Example:

$$(2x - 3)^2 = 5$$
$$2x - 3 = \pm\sqrt{5}$$
$$2x = 3 \pm \sqrt{5}$$
$$x = \frac{3 \pm \sqrt{5}}{2}$$

Practice this part of the process by solving the following equations.

(a) $(x - 2)^2 = 3$ $\qquad$ (b) $(2x - 1)^2 = 5$ $\qquad$ (c) $(x + 1)^2 = \frac{1}{2}$

Check your work in **18**.

---

18 (a) $(x - 2)^2 = 3$ $\qquad$ Taking the square root of both members of this equation leads to the following two equations.

$$x - 2 = \sqrt{3} \quad \text{and} \quad x - 2 = -\sqrt{3}$$

Adding 2 to each side of the equation,

$$x = 2 + \sqrt{3} \quad \text{and} \quad x = 2 - \sqrt{3} \qquad \text{These are the roots of the equation.}$$

(b) $(2x - 1)^2 = 5$ leads to the pair of equations

$$2x - 1 = \sqrt{5} \quad \text{and} \quad 2x - 1 = -\sqrt{5}$$

Add 1 to each side of the equation.

$$2x = 1 + \sqrt{5} \quad \text{and} \quad 2x = 1 - \sqrt{5}$$

Divide by 2 on each side of the equation.

$$x = \frac{1 + \sqrt{5}}{2} \quad \text{and} \quad x = \frac{1 - \sqrt{5}}{2} \qquad \text{These are the roots of the equation.}$$

(c) $(x + 1)^2 = \dfrac{1}{2}$ Take square roots of both members of the equation.

$$x + 1 = \sqrt{\dfrac{1}{2}} \qquad \text{and} \qquad x + 1 = -\sqrt{\dfrac{1}{2}}$$

or

$$x + 1 = \dfrac{\sqrt{2}}{2} \qquad \text{and} \qquad x + 1 = \dfrac{-\sqrt{2}}{2}$$

Add -1 to both members of the equation.

$$x = -1 + \dfrac{\sqrt{2}}{2} \qquad \text{and} \qquad x = -1 - \dfrac{\sqrt{2}}{2}$$

or

$$x = \dfrac{-2 + \sqrt{2}}{2} \qquad \text{and} \qquad x = \dfrac{-2 - \sqrt{2}}{2}$$

Now use this procedure of completing the square to solve the following quadratic equations.

(a) $x^2 - 4x = 1$ (b) $x^2 + 3x + 1 = 0$

(c) $x^2 - \dfrac{1}{9} = \dfrac{2}{3}x$ (d) $x^2 - \dfrac{3}{2}x + \dfrac{1}{2} = 0$

When you have solved all of these equations, check your work against ours in **19**.

19   (a) $x^2 - 4x = 1$

Complete the square on the left.

$$x^2 - 4x + \boxed{4} \;=\; 1 \;+\; \boxed{4}$$

$$(-4) \div 2 = -2 \longrightarrow (-2)^2 = 4$$

Write the left member as a perfect square.

$x^2 - 4x + 4 = (x - 2)^2$ so that the equation is now

$(x - 2)^2 = 5$

Take the square root of both members of the equation to find the two roots.

$x - 2 = \sqrt{5} \qquad \text{and} \qquad x - 2 = -\sqrt{5}$

The roots of the equation are $x = 2 + \sqrt{5} \qquad$ and $\qquad x = 2 - \sqrt{5}$

(b) $x^2 + 3x + 1 = 0$

Rewrite the equation so that the terms containing the variable are on the left and the constant term is on the right.

$x^2 + 3x = -1$

Complete the square.

$$x^2 + 3x + \boxed{\frac{9}{4}} = -1 + \boxed{\frac{9}{4}} \qquad \left(-1 + \frac{9}{4} = \frac{-4}{4} + \frac{9}{4} = \frac{5}{4}\right)$$

$$3 \div 2 = \frac{3}{2} \rightarrow \left(\frac{3}{2}\right)^2 = \frac{9}{4}$$

Write the left member of the equation as a perfect square.

$$\left(x + \frac{3}{2}\right)^2 = \frac{5}{4}$$

Take the square root of both members of the equation.

$$x + \frac{3}{2} = \sqrt{\frac{5}{4}} = \frac{\sqrt{5}}{2} \qquad \text{and} \qquad x + \frac{3}{2} = -\sqrt{\frac{5}{4}} = \frac{-\sqrt{5}}{2}$$

The roots are $x = \dfrac{-3 + \sqrt{5}}{2}$ and $x = \dfrac{-3 - \sqrt{5}}{2}$.

(c) Rewrite $x^2 - \dfrac{1}{9} = \dfrac{2}{3}x$ as $x^2 - \dfrac{2}{3}x = \dfrac{1}{9}$

Complete the square to get $x^2 - \dfrac{2}{3}x + \dfrac{1}{9} = \dfrac{2}{9}$ or $\left(x - \dfrac{1}{3}\right)^2 = \dfrac{2}{9}$.

The roots of the equation are $x = \dfrac{1 + \sqrt{2}}{3}$ and $x = \dfrac{1 - \sqrt{2}}{3}$.

(d) Rewrite $x^2 - \dfrac{3}{2}x + \dfrac{1}{2} = 0$ as $x^2 - \dfrac{3}{2}x = -\dfrac{1}{2}$.

Complete the square to get $x^2 - \dfrac{3}{2}x + \dfrac{9}{16} = -\dfrac{1}{2} + \dfrac{9}{16} = \dfrac{1}{16}$

or $\left(x - \dfrac{3}{4}\right)^2 = \dfrac{1}{16}$

The roots are $x - \dfrac{3}{4} = \sqrt{\dfrac{1}{16}}$ and $x - \dfrac{3}{4} = -\sqrt{\dfrac{1}{16}}$

or $x = \dfrac{3}{4} + \dfrac{1}{4}$ and $x = \dfrac{3}{4} - \dfrac{1}{4}$

$x = 1$ and $x = \dfrac{1}{2}$

Coefficient of x^2 Term not Equal to 1

So far we have solved only equations in which the coefficient of the x^2 term is equal to 1. To solve a quadratic equation in which the coefficient of the x^2 term is *not* equal to 1, first divide every term of the equation by the coefficient of the x^2 term, then continue as before. For example, to solve the equation

$2x^2 + 4x - 1 = 0$

The coefficient of the x^2 term is 2

Divide each term by 2

$$\frac{2x^2}{2} + \frac{4x}{2} - \frac{1}{2} = 0 \qquad \text{or} \qquad x^2 + 2x - \frac{1}{2} = 0$$

Now solve by completing the square:

$$(x + 1)^2 = \frac{3}{2} \qquad \text{or} \qquad x = -1 \pm \frac{\sqrt{6}}{2}$$

Not very difficult, is it? Solve each of these equations by completing the square.

(a) $3x^2 - 6x - 12 = 0$ (b) $2x^2 - 5x + \frac{1}{8} = 0$

Check your work in **20**.

A GRAPHICAL WAY TO SOLVE QUADRATIC EQUATIONS

Algebra methods for solving quadratic equations have been used for only a few hundred years. Before algebra was developed, geometric methods were used. Here is a graphical method that was taught to college students in the early nineteenth century. It was discovered by Thomas Carlyle, the famous Scottish writer and historian.

First, put the quadratic equation in the form $x^2 + gx + h = 0$, where the coefficient of x^2 is equal to 1.

Second, plot the points $A = (0, 1)$ and $B = (-g, h)$ on a coordinate graph.

Third, draw a circle using the line AB as a diameter. This circle will cut the x axis exactly at the roots of the equation $x^2 + gx + h = 0$.

For example, to solve $x^2 - 6x + 5 = 0$

$g = -6, h = 5$ so that $A = (0, 1)$ and $B = (6, 5)$

Plot these points on a coordinate system and draw a circle with the line AB as a diameter.

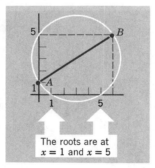

The roots are at
$x = 1$ and $x = 5$

---

20   (a) $3x^2 - 6x - 12 = 0$

The coefficient of the
x^2 term is 3

Divide each term of the equation by 3.

$$\frac{3x^2}{3} - \frac{6x}{3} - \frac{12}{3} = 0 \qquad \text{or} \qquad x^2 - 2x - 4 = 0$$

474   **Quadratic Equations**

The equation has been rewritten with an x^2 coefficient of 1 and can now be solved by completing the square. The roots of this equation are

$$x = 1 + \sqrt{5} \quad \text{and} \quad x = 1 - \sqrt{5}$$

(b) $\quad 2x^2 - 5x + \dfrac{1}{8} = 0$

> The coefficient of the x^2 term is 2

Divide each term of the equation by 2.

$$\frac{2x^2}{2} - \frac{5x}{2} + \left[\frac{1}{8} \div 2\right] = 0 \quad \text{or} \quad x^2 - \frac{5x}{2} + \frac{1}{16} = 0$$

Solve by completing the square. The roots of this equation are

$$x = \frac{5 + 2\sqrt{6}}{4} \quad \text{and} \quad x = \frac{5 - 2\sqrt{6}}{4}$$

The following set of problems will provide you with practice in solving quadratic equations by completing the square.

1. $x^2 + 8x - 2 = 0$ 2. $x^2 - 11 = 2x$

3. $x^2 - x - \dfrac{1}{2} = 0$ 4. $y^2 - 6y = 7$

5. $x^2 - 3x = 9$ 6. $p^2 - 5p + 5 = 0$

7. $3a + 9 = 2a^2$ 8. $4x^2 - 4x - 5 = 0$

9. $1 - x = 4x^2$ 10. $2x^2 = 4 - 2x$

The answers are in **21**.

---

21   1. $-4 + 3\sqrt{2}, \; -4 - 3\sqrt{2}$ 2. $1 + 2\sqrt{3}, \; 1 - 2\sqrt{3}$

3. $\dfrac{1 + \sqrt{3}}{2}, \dfrac{1 - \sqrt{3}}{2}$ 4. $7, \; -1$

5. $\dfrac{3 + 3\sqrt{5}}{2}, \dfrac{3 - 3\sqrt{5}}{2}$ 6. $\dfrac{5 + \sqrt{5}}{2}, \dfrac{5 - \sqrt{5}}{2}$

7. $3, \; -\dfrac{3}{2}$ 8. $\dfrac{1 + \sqrt{6}}{2}, \dfrac{1 - \sqrt{6}}{2}$

9. $\dfrac{-1 + \sqrt{17}}{8}, \dfrac{-1 - \sqrt{17}}{8}$ 10. $1, \; -2$

Notice that in problems 4 and 10, you could have solved the quadratic equation by factoring if you were clever enough to see that

$y^2 - 6y = 7 \quad$ is equivalent to $\quad (y - 7)(y + 1) = 0$

and $\quad 2x^2 = 4 - 2x \quad$ is equivalent to $\quad (x - 1)(x + 2) = 0.$

6-5 THE QUADRATIC FORMULA

For more than twenty centuries mathematicians dreamed of being able to solve any polynomial equation exactly using no geometry, no factoring, no tedious and time-comsuming "completing the square." They searched for formulas or procedures that could be used to calculate the solution of *any* polynomial equation from its numerical coefficients. About 150 years ago the brilliant young mathematicians Évariste Galois of France and Niels Abel of Norway were able to prove that this search had been in vain: it is not possible to find a formula of this kind for *every* equation. Happily, however, we do have an easy-to-use formula that enables us to calculate the roots of any quadratic equation.

If we write any quadratic equation in standard form as

$$ax^2 + bx + c = 0$$

where the coefficients a, b, and c are any real numbers, with a not equal to zero, of course, then the *Quadratic Formula* provides a quick and dependable way to calculate the two roots of the equation in terms of the coefficients a, b, and c.

The roots of $ax^2 + bx + c = 0$ are

$$x = \frac{-b + \sqrt{b^2 - 4ac}}{2a} \quad \text{and} \quad x = \frac{-b - \sqrt{b^2 - 4ac}}{2a}$$

To save time we usually write this as

$$\boxed{x = \frac{-b \pm \sqrt{b^2 - 4ac}}{2a}}$$ the quadratic formula.

(For a derivation of this formula look on page 477.)

In order to use the quadratic formula, first rewrite the equation in standard form. Then identify the numbers a, b, and c. For example, the equation

$$3x - 4 = x^2 \quad \text{is} \quad x^2 - 3x + 4 = 0 \quad \text{in standard form}$$

and the coefficients a, b, and c can be identified quickly.

$$x^2 \quad - 3x \quad + 4 = 0$$

| $a = 1$ | $b = -3$ | $c = 4$ |

Notice that the coefficient of the x term is -3. Think of the equation as

$$x^2 + (-3)x + 4 = 0 \quad \text{if you get confused with negative coefficients.}$$

If the equation to be solved contains fractions, it is best to clear all fractions before rewriting the equation in standard form. For example, in the equation

$$x^2 - \frac{1}{2} = \frac{2}{3}x$$

multiply each term by 6, the least common multiple of 2 and 3, to get

$$6x^2 - 3 = 4x$$

Now rewrite the equation in standard form.

$$6x^2 - 4x - 3 = 0$$

Finally, identify a, b, and c.

$$6x^2 \quad - 4x \quad - 3 = 0$$

| $a = 6$ | $b = -4$ | $c = -3$ |

Practice this first step in using the quadratic formula by rewriting each of the following equations in standard form and identifying a, b, and c.

Original Equation	Standard Form	a	b	c
$2x^2 + 1 - 3x = 0$				
$x - x^2 = 1$				
$4x = x^2$				
$y^2 - 1 = \dfrac{3}{4}y$				
$\dfrac{x^2}{2} = 1 - \dfrac{2}{3}x$				
$3p^2 - 5 = 0$				
$4x^2 + x = 1$				
$-2x^2 + x - 3 = 0$				
$-t = 4 - 6t^2$				
$2x^2 - x = \dfrac{1}{4}$				

Check your work in **22**.

DERIVING THE QUADRATIC FORMULA

There are many ways to show that the quadratic formula does give the solution to the quadratic equation. Here is one demonstration.

Start with $ax^2 + bx + c = 0$.

Multiply each term by $4a$: $\qquad 4a^2x^2 + 4abx + 4ac = 0$

Rewrite this as $\qquad\qquad\quad 4a^2x^2 + 4abx = -4ac$

Add b^2 to each side of the equation

$$4a^2x^2 + 4abx + b^2 = b^2 - 4ac$$

The left member of this equation is a perfect square. The equation becomes

$(2ax + b)^2 = b^2 - 4ac$

Take the square root of both members of this equation to get

$2ax + b = \sqrt{b^2 - 4ac} \qquad$ or $\qquad 2ax = -b + \sqrt{b^2 - 4ac}$

$\qquad\qquad\qquad\qquad\qquad$ or $\qquad x = \dfrac{-b + \sqrt{b^2 - 4ac}}{2a}$

and

$2ax + b = -\sqrt{b^2 - 4ac} \quad$ or $\qquad 2ax = -b - \sqrt{b^2 - 4ac}$

$\qquad\qquad\qquad\qquad\qquad$ or $\qquad x = \dfrac{-b - \sqrt{b^2 - 4ac}}{2a}$

The solution of the original equation is

$x = \dfrac{-b \pm \sqrt{b^2 - 4ac}}{2a} \qquad$ the quadratic formula.

Original Equation	Standard Form	a	b	c
$2x^2 + 1 - 3x = 0$	$2x^2 - 3x + 1 = 0$	2	-3	1
$x - x^2 = 1$	$x^2 - x + 1 = 0$	1	-1	1
$4x = x^2$	$x^2 - 4x = 0$	1	-4	0
$y^2 - 1 = \dfrac{3}{4}y$	$4y^2 - 3y - 4 = 0$	4	-3	-4
$\dfrac{x^2}{2} = 1 - \dfrac{2}{3}x$	$3x^2 + 4x - 6 = 0$	3	4	-6
$3p^2 - 5 = 0$	$3p^2 - 5 = 0$	3	0	-5
$4x^2 + x = 1$	$4x^2 + x - 1 = 0$	4	1	-1
$-2x^2 + x - 3 = 0$	$2x^2 - x + 3 = 0$	2	-1	3
$-t = 4 - 6t^2$	$6t^2 - t - 4 = 0$	6	-1	-4
$2x^2 - x = \dfrac{1}{4}$	$8x^2 - 4x - 1 = 0$	8	-4	-1

Don't forget that in standard quadratic form the equation is written so that a is a positive integer.

Once the numbers a, b, and c have been identified, the second step is to substitute them into the quadratic formula. Practice this second step in using the quadratic formula by completing the following table.

Use the given values of a, b, and c to calculate the value of each of the expressions shown.

a	b	c	$-b$	b^2	$4ac$	$b^2 - 4ac$	$\sqrt{b^2 - 4ac}$	$2a$
2	4	1						
1	-2	-1						
2	-1	-1						
1	0	-1						
1	-2	0						
1	1	-1						
12	5	-2						
3	2	-10						
4	-12	9						
2	-5	1						

Compare your answers with ours in **23**.

23

a	b	c	$-b$	b^2	$4ac$	$b^2 - 4ac$	$\sqrt{b^2 - 4ac}$	$2a$
2	4	1	-4	16	8	8	$\sqrt{8} = 2\sqrt{2}$	4
1	-2	-1	2	4	-4	8	$\sqrt{8} = 2\sqrt{2}$	2
2	-1	-1	1	1	-8	9	3	4
1	0	-1	0	0	-4	4	2	2
1	-2	0	2	4	0	4	2	2
1	1	-1	-1	1	-4	5	$\sqrt{5}$	2
12	5	-2	-5	25	-96	121	11	24
3	2	-10	-2	4	-120	124	$2\sqrt{31}$	6
4	-12	9	12	144	144	0	0	8
2	-5	1	5	25	8	17	$\sqrt{17}$	4

Now that you have had some practice in identifying the numbers a, b, and c for any quadratic equation and in using these numbers to calculate the terms in the quadratic formula, you should be ready to use this formula to solve a quadratic equation. Follow these steps.

Step 1. Write the equation in standard form, clearing all fractions.

Example: $3x + 2x^2 = 1$ becomes $2x^2 + 3x - 1 = 0$

Step 2. Identify a, b, and c.

Example: $2x^2 + 3x - 1 = 0$

$a = 2$ $b = 3$ $c = -1$

Step 3. Substitute these values for a, b, and c into the quadratic formula.

Example: $x = \dfrac{-b \pm \sqrt{b^2 - 4ac}}{2a}$

$$x = \dfrac{-(3) \pm \sqrt{(3)^2 - 4(2)(-1)}}{2(2)}$$

$$x = \dfrac{-3 \pm \sqrt{9 + 8}}{4}$$

$$x = \dfrac{-3 \pm \sqrt{17}}{4}$$

The two roots of the original quadratic equation are

$$x = \dfrac{-3 + \sqrt{17}}{4} \qquad \text{and} \qquad x = \dfrac{-3 - \sqrt{17}}{4}$$

Follow this same three step procedure to solve the following quadratic equation.

$6x = 2x^2 + 1$

Our answer is in **24**.

Why can't I solve equations by factoring like this—
$(x + 1)(x + 2) = 2$
gives $x + 1 = 2$ or $x = 1$
for the first root
and $x + 2 = 2$ or $x = 0$
for the second root.
OK?

No. That is incorrect. The zero-factor rule only works if the right member of the equation is equal to zero. Solve that equation like
This: $(x + 1)(x + 2) = 2$
$x^2 + 3x + 2 = 2$
$x^2 + 3x = 0$
$x(x + 3) = 0$.
The roots are $x = 0$ and $x = -3$.

24 **Step 1.** Write the equation in standard form.

$6x = 2x^2 + 1$ becomes $2x^2 - 6x + 1 = 0$ in standard form.

Step 2. Identify a, b, and c.

$$2x^2 \quad - 6x \quad + 1 = 0$$

$a = 2$ $b = -6$ $c = 1$

Step 3. Substitute these numbers into the quadratic formula.

$$x = \dfrac{-(-6) \pm \sqrt{(-6)^2 - 4(2)(1)}}{2(2)}$$

$$x = \dfrac{6 \pm \sqrt{36 - 8}}{4}$$

$$x = \dfrac{6 \pm \sqrt{28}}{4} = \dfrac{6 \pm 2\sqrt{7}}{4} \qquad \text{or} \qquad x = \dfrac{3 \pm \sqrt{7}}{2}$$

The two roots of the original equation are

$$x = \dfrac{3 + \sqrt{7}}{2} \qquad \text{and} \qquad x = \dfrac{3 - \sqrt{7}}{2}$$

Solve each of the following quadratic equations using the quadratic formula.

(a) $x^2 - 4x = 21$ (b) $5x - 1 = 2x^2$

(c) $3x^2 = x$ (d) $5x^2 = 2(1 - x)$

When you have solved these, turn to **25**.

EQUATIONS CONTAINING RADICALS

How do you solve an equation such as $\sqrt{x} = 3$?

First, remember that x must be positive in value, otherwise the equation will not have any real number solution.

Second, to solve this equation we cannot use the addition, subtraction, multiplication, or division rules that enabled us to solve linear equations. Instead we use a new rule:

if $a = b$ then $a^2 = b^2$

Applying this rule, $(\sqrt{x})^2 = (3)^2$ or $x = 9$

Third, always check your answer.

$\sqrt{9} = 3$

   $3 = 3$

This kind of equation can also lead to quadratic equations. For example, to solve $x + \sqrt{x + 2} = 4$

first, isolate the radical on the left.

$$\sqrt{x + 2} = 4 - x$$

Squaring, we get $(\sqrt{x + 2})^2 = (4 - x)^2$

$$x + 2 = 16 - 8x + x^2$$

or $x^2 - 9x + 14 = 0$

Solve this equation in the usual way. The roots are $x = 2$ and $x = 7$.

Check: $x = 2$ is a valid solution, since $\sqrt{(2) + 2} = 4 - 2$

$$\sqrt{4} = 2$$

But $x = 7$ is *not* a valid solution, since $\sqrt{(7) + 2} \neq 4 - 7$

$$\sqrt{9} \neq -3$$

Remember that $\sqrt{9}$ refers to the positive root: $\sqrt{9} = 3$.

The procedure of squaring both members of the equation may sometimes produce a solution that does not satisfy the equation. Beware! Check your answers!

Here are a few practice problems in radical equations.

(a) $\sqrt{x - 3} = x - 5$ (b) $\sqrt{x + 2} = x$

(c) $\sqrt{2x - 1} = x - 2$ (d) $\sqrt{x - 3} = x - 9$

(e) $4 + \sqrt{11 - 2x} = x$ (f) $\sqrt{2(x - 1)(1 - 2x)} = 2x - 1$

(g) $3 + \sqrt{2x - 6} = x$ (h) $\sqrt{x + 4} = x + 1$

(*Hint:* In problems (e) and (g) isolate the radical on the left before squaring.)

The answers are on page 567.

25 (a) Put $x^2 - 4x = 21$ in standard form.

$$x^2 \quad - 4x \quad - 21 = 0$$

$a = 1$ $b = -4$ $c = -21$

Substitute these values into the quadratic formula.

$$x = \frac{-(-4) \pm \sqrt{(-4)^2 - 4(1)(-21)}}{2(1)} = \frac{4 \pm \sqrt{16 + 84}}{2}$$

$$x = \frac{4 \pm \sqrt{100}}{2} = \frac{4 \pm 10}{2}$$

The roots of the original quadratic equation are

$$x = \frac{4 + 10}{2} \quad \text{and} \quad x = \frac{4 - 10}{2} \quad \text{or} \quad x = 7 \quad \text{and} \quad x = -3$$

We could have solved this equation by writing it in the form $(x - 7)(x + 3) = 0$ and using the zero-factor rule.

(b) Put $5x - 1 = 2x^2$ in standard form

$$2x^2 \quad - 5x \quad + 1 = 0$$

$a = 2$ $b = -5$ $c = 1$

Substitute these numbers into the quadratic formula.

$$x = \frac{-(-5) \pm \sqrt{(-5)^2 - 4(2)(1)}}{2(2)} = \frac{5 \pm \sqrt{17}}{4}$$

The roots of the original equation are

$$x = \frac{5 + \sqrt{17}}{4} \quad \text{and} \quad x = \frac{5 - \sqrt{17}}{4}$$

(c) Put $3x^2 = x$ in standard form

$$3x^2 \quad - x \quad + 0 = 0$$

$a = 3$ $b = -1$ $c = 0$

Substitute these numbers into the quadratic formula.

$$x = \frac{-(-1) \pm \sqrt{(-1)^2 - 4(3)(0)}}{2(3)} = \frac{1 \pm 1}{6}$$

The roots are $x = \dfrac{1 + 1}{6} = \dfrac{1}{3}$ and $x = \dfrac{1 - 1}{6} = 0.$

We could have solved this equation by factoring it.

$3x^2 - x + 0 = 0$ can be written as $x(3x - 1) = 0.$

(d) Put the equation $5x^2 = 2(1 - x)$ in standard form.

$$5x^2 \quad + 2x \quad - 2 = 0$$

$a = 5$ $b = 2$ $c = -2$

Substitute these values into the quadratic formula.

$$x = \frac{-(2) \pm \sqrt{(2)^2 - 4(5)(-2)}}{2(5)} = \frac{-2 \pm \sqrt{44}}{10}$$

$$x = \frac{-2 \pm 2\sqrt{11}}{10} = \frac{-1 \pm \sqrt{11}}{5}$$

The roots of the original equation are

$$x = \frac{-1 + \sqrt{11}}{5} \qquad \text{and} \qquad x = \frac{-1 - \sqrt{11}}{5}$$

Notice that when the root is a radical expression we always write it in its simplest radical form, as described in Unit 5, page 423.

For all of these problems, the solutions should be checked by substituting them back into the original quadratic equations.

SUMMARY Many students worry about which of the methods we have shown should be used in solving a given quadratic equation. Follow these guidelines:

1. If the quadratic equation can be written in the form $ax^2 + c = 0$, solve it for x^2 and find the square roots.
2. If the quadratic equation can be readily factored, then solve it by factoring and using the zero-factor property.
    If the quadratic equation cannot be factored, or if you simply do not see how to factor it, but you do recognize it immediately as an incomplete square, then solve it by completing the square. Most students would *never* try to solve it by completing the square.
3. If you are not able to factor the quadratic equation quickly, then use the quadratic formula to find the solution. If the coefficient numbers are very very large, try to solve the equation by factoring it or completing the square before using the quadratic formula.

Remember, the quadratic formula—if used correctly—will always give you the correct solution to the equation.

For some quadratic equations the roots found by using the quadratic formula may involve the square root of a negative number. In the quadratic formula if $b^2 < 4ac$, then the number under the radical sign will be negative. For example, the roots of the quadratic equation

$$x^2 - 2x + 4 = 0$$

are $\quad x = 1 + \sqrt{-3} \quad$ and $\quad x = 1 - \sqrt{-3}.$

Imaginary Numbers Because there is no real number whose square is -3, the number $\sqrt{-3}$ is not a real number. A number such as $\sqrt{-3}$ is called an *imaginary number* and numbers such as $1 + \sqrt{-3}$, formed by combining real and imaginary numbers, are known as *complex numbers*. The equation

$x^2 - 2x + 4 = 0$ has no solution in real numbers, but it does have the complex number solution shown above.

Complex numbers are important in mathematics and have very practical applications in science and engineering, but they are not usually studied in a basic algebra course. In this book we will not consider quadratic equations with complex number solutions.

Now turn to **26** for a set of practice problems on the solution of quadratic equations by completing the square and using the quadratic formula.

PROBLEM SET 6-2

26

Solving Quadratic
Equations

The answers are on page 567.

A. Solve by completing the square.

1. $x^2 + 4 = 4x$ _____

2. $x^2 = 6x + 19$ _____

3. $x^2 - 5 + 3x = 0$ _____

4. $x^2 - 2x = 6$ _____

5. $x^2 + 3 - 4x = 0$ _____

6. $x^2 = 9 + 6x$ _____

7. $2 - x^2 = 2x$ _____

8. $x - x^2 + 8 = 0$ _____

9. $5(1 - x) = x^2$ _____

10. $x(1 + 2x) - 6 = 0$ _____

11. $3x^2 + 5x = 2$ _____

12. $3x - 2x^2 + 4 = 0$ _____

13. $6x^2 + 2 = 7x$ _____

14. $2x^2 - 4x - 5 = 0$ _____

15. $2x^2 - 20x - 5 = 0$ _____

16. $3x^2 - 4x = 2$ _____

17. $(y + 1)(y - 3) = 1$ _____

18. $(x + 3)(x + 1) - 2 = 0$ _____

19. $x(x + 5) = x - 3$ _____

20. $2x + 1 = x(5 - x)$ _____

B. Solve using the quadratic formula.

1. $x^2 - 7x + 12 = 0$ _____

2. $5x^2 - 7x + 2 = 0$ _____

3. $2x^2 + x = 4$ _____

4. $x^2 - x = 11$ _____

5. $2x^2 + x - 5 = 0$ _____

6. $x^2 - 7x = 0$ _____

7. $x^2 - 6 = 0$ _____

8. $x^2 + 3x = 5$ _____

9. $5x^2 - 1 = 2x$ _____

10. $x^2 - 10 - 9x = 0$ _____

11. $2x^2 + 3 = 2x - 9x$ _____

12. $x^2 - 4x = -3$ _____

13. $5(x^2 + 2) = 7(x + 3)$ _____

14. $x^2 = 2(1 - x)$ _____

15. $5y^2 + 4y = 1$ _____

16. $x^2 - 5 = 3x$ _____

17. $x^2 - 8 = 8x$ _____

18. $5y^2 + 2y + 4 = 5 + y$ _____

19. $5x^2 - 2x = x^2 + 3x - 1$ _____

20. $2x^2 - x + 1 = 3x + 4$ _____

21. $p^2 = 3p - 2$ _____

22. $q^2 = 3q + 7$ _____

23. $2t^2 - 4t + 1 = 0$ _____

24. $1 = 5x - 3x^2$ _____

25. $(x + 2)^2 = 4$ _____

26. $(2x - 1)^2 = 3$ _____

27. $2x^2 + 3x - 2 = 0$ _____

28. $3x^2 + 5x - 1 = 0$ _____

Date _____

Name _____

Course/Section _____

485   **Problem Set 6-2**

29. $(x - 1)^2 = 3 - 4x$ _____

30. $(2x + 1)^2 = 5 - x$ _____

31. $x^2 = 2(x + 3)$ _____

32. $5(x + 1) = 2x^2$ _____

33. $\dfrac{x}{x - 1} = \dfrac{1}{x - 2}$ _____

34. $\dfrac{x + 1}{x + 2} = \dfrac{2x - 1}{x - 3}$ _____

35. $2y(y - 1) = (y + 2)^2$ _____

36. $9(y - 1)(y + 2) = 8$ _____

37. $x - 1 = \dfrac{1}{x}$ _____

38. $\dfrac{2}{x - 2} = x + 3$ _____

39. $3(x - 1)^2 - 2(x - 2)^2 = 0$ _____

40. $2x^2 + 3x + 5 = 3x^2 + 4x + 1$ _____

C.   Brain Boosters

1. Solve: (a) $(x - 1)(x - 2) + (x - 1)(x - 3) + (x - 2)(x - 3) = 0$

_____

(b) $\left(x - \dfrac{3}{4}\right)\left(x - \dfrac{6}{8}\right) + \left(x - \dfrac{3}{4}\right)\left(x - \dfrac{1}{2}\right) = 0$

_____

2. Solve: $x^3 + 8x^2 + 16x - 1 = (x + 3)^3$ _____

3. Solve by completing the square:

(a) $x^2 - 4x - 2597 = 0$ _____

(b) $x^2 + 102x + 2597 = 0$ _____

4. Solve using the quadratic formula:

(a) $x^2 - 2x(1 + \sqrt{2}) + 2\sqrt{2} = 0$ _____

(b) $x^2 + 6x\sqrt{2} + 5 = 0$ _____

5. Use the quadratic formula to show that for any quadratic equation $ax^2 + bx + c = 0$ the following statements are true.

(a)   The sum of the two roots is equal to $-\dfrac{b}{a}$. _____

(b)   The product of the two roots is equal to $\dfrac{c}{a}$. _____

(c)   The two roots are equal when $b = 2\sqrt{ac}$. _____

6. (a) Solve: $\quad x = 1 + \cfrac{1}{1 + \cfrac{1}{1 + \cfrac{1}{1 + \cdots}}}$ $\qquad \left(\textit{Hint: } x = 1 + \dfrac{1}{x}\right)$

(b) Solve: $\quad x = 1 + \cfrac{1}{2 + \cfrac{1}{2 + \cfrac{1}{2 + \cdots}}}$ $\qquad \left(\textit{Hint: } x = 1 + \dfrac{1}{1 + x}\right)$

---

7. (a) Find the value of $x = \sqrt{2 + \sqrt{2 + \sqrt{2 + \cdots}}}$

(*Hint:* First square both sides.)

(b) Find the value of $x = \sqrt{12 + \sqrt{12 + \sqrt{12 + \cdots}}}$

(c) Find the value of $x = \sqrt{1 - \sqrt{1 - \sqrt{1 - \cdots}}}$

(d) Solve for x in terms of n. $\quad x = \sqrt{n + \sqrt{n + \sqrt{n + \cdots}}}$

(e) Solve for x in terms of n. $\quad x = \sqrt{n - \sqrt{n - \sqrt{n - \cdots}}}$

8.

Of all the rectangles that can be drawn, which is the most visually pleasing? This is a question asked by Greek and Egyptian sculptors and architects 2500 years ago, and the answer they found, the "divine proportion" as it has been called since 1509, was used in the design of ancient buildings such as the Parthenon, churches, temples, gardens, courtyards, in painting and sculpture. It appears in an incredible variety of natural and man-made objects, from seashells to books.

We will define the "most pleasing" rectangle as one that can be divided into a square plus another rectangle exactly similar to the original.

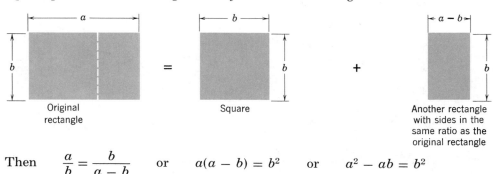

Original rectangle $\qquad$ Square $\qquad$ Another rectangle with sides in the same ratio as the original rectangle

Then $\quad \dfrac{a}{b} = \dfrac{b}{a-b} \qquad$ or $\qquad a(a-b) = b^2 \qquad$ or $\qquad a^2 - ab = b^2$

Divide each term by b^2:

$\dfrac{a^2}{b^2} - \dfrac{ab}{b^2} = \dfrac{b^2}{b^2} \qquad$ or $\qquad \left(\dfrac{a}{b}\right)^2 - \dfrac{a}{b} = 1$

$\qquad$

$\dfrac{a}{b}$ is the ratio of the sides of the golden rectangle. Call this ratio G. Then the equation above is $G^2 - G = 1$.

(a) Solve this equation to find the exact value of the golden ratio G. What is the value of G to two decimal digits?

_____

(b) Compare this answer with the answers to problems 6(a) and 7(c) on page 487.

(c) Show that

$$G^2 = 1 + G, \ G^3 = 1 + 2G, \ G^4 = 2 + 3G, \ G^5 = 3 + 5G, \text{ and } \dfrac{1}{G} = G - 1$$

Can you find a pattern in this set of equations?

(d) Find the ratio of the area of the square to the area of the smaller rectangle in the drawing on page 487.

_____

9. In the cartoon on page 447, the equation in the center frame is a quadratic. Solve it for X in terms of A and Y.

_____

When you have had the practice you need either return to the preview on page 445 or continue in **27** where you will learn to graph quadratic equations.

27   As you learned in Unit 4, a graph enables us to visualize an algebraic equation—that is, to see it. The graph of a linear equation in two variables is a straight line. The graph of a quadratic equation in two variables of the form

$$y = ax^2 + bx + c$$

is not a straight line but a curve.

We may find the graph of the equation

$$y = x^2 - 4$$

as follows.

First, select values for x and find the corresponding values for y.

For $x = 0$, $y = (0)^2 - 4$ or $y = -4$,
for $x = 1$, $y = (1)^2 - 4$ or $y = -3$,
for $x = 2$, $y = (2)^2 - 4$ or $y = 0$,
for $x = -1$, $y = (-1)^2 - 4$ or $y = -3$,

and so on.

Second, arrange these ordered pairs (x, y) in a table of values.

x	y
-3	5
-2	0
-1	-3
0	-4
1	-3
2	0
3	5

Finally, plot these number pairs as points on a rectangular coordinate system and draw a smooth curve through them.

You try it. Plot the points given in the table above on this coordinate axis.

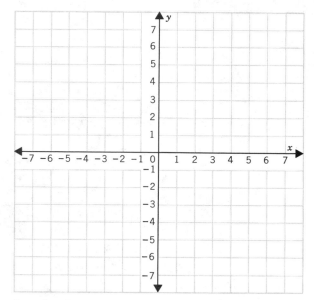

Check your graph in **28**.

28

x	y
-3	5
-2	0
-1	-3
0	-4
1	-3
2	0
3	5

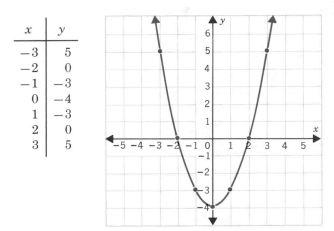

$y = x^2 - 4.$

We have drawn in only a few of the infinitely many points that satisfy this equation. The graph is the set of *all* of the ordered pairs of numbers that satisfy the equation, and the smooth curve is our best approximation to the actual complete graph.

The graph of any quadratic equation in two variables of the form

$$y = ax^2 + bx + c$$

Parabola

is called a *parabola*.

This name was given to the curve by the Greek mathematician Appolonius in the third century B.C. The word "parabola" means *to put beside* or *to put parallel,* and the parabola is the curve obtained when a plane surface is made to cut a cone parallel to the side of the cone.

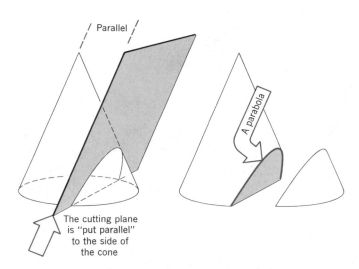

Parallel

A parabola

The cutting plane is "put parallel" to the side of the cone

The path of any projectile, from baseballs to bullets, is a parabola if air friction can be ignored.

Oops!

Easy out
$y = \dfrac{-x^2}{16} + \dfrac{x}{2} + 3$

Hit it Gary!

Warren H.

EAT AT JOE'S

Automobile headlights and spotlights have parabolic cross-sections. If light rays start from a point in the curve and hit a parabola-shaped mirror, they will emerge from the mirror in a parallel beam of light.

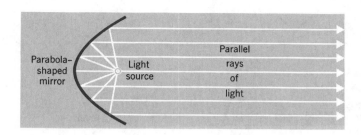

There are several characteristics of this parabola that are true for the graph of any quadratic equation.

First, it is always a cup-shaped curve. For the equation $y = x^2 - 4$, as you have seen, the cup is open at the top, but it could open downward

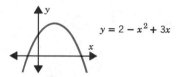

$$y = 2 - x^2 + 3x$$

or even open to the sides.

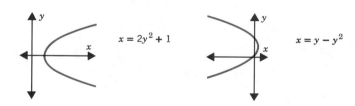

$$x = 2y^2 + 1 \qquad x = y - y^2$$

Vertex

The turning point of the curve, where the variable y is either a maximum or a minimum, is called the *vertex* of the parabola.

Second, the curve is always symmetric. It can be divided into two parts each with the same shape. The part of the graph on the left is the mirror image of the part of the graph on the right.

Symmetry

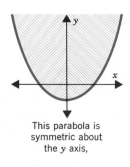

This parabola is symmetric about the y axis,

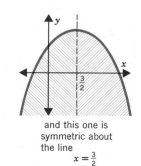

and this one is symmetric about the line $x = \frac{3}{2}$

When you draw a parabola, be certain that you draw a smooth curve similar to ours above. Here are some examples of what *not* to draw.

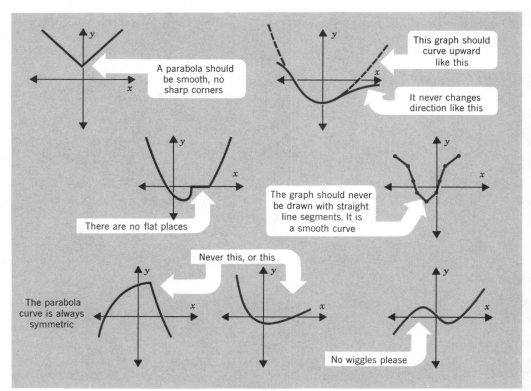

If you get any of the shapes shown above on your graph it is probably a signal that you have calculated or plotted a point incorrectly.

If we want to solve the quadratic equation

$$ax^2 + bx + c = 0$$

we can do so by graphing the equation

$$y = ax^2 + bx + c$$

and finding the roots of the quadratic equation from the graph. The quadratic equation has a root wherever $y = 0$, that is, wherever the parabola intersects the x axis.

Try it. To solve the quadratic equation $x^2 - 6x + 8 = 0$ graph the equation $y = x^2 - 6x + 8$ and find the points where the curve cuts the x axis.

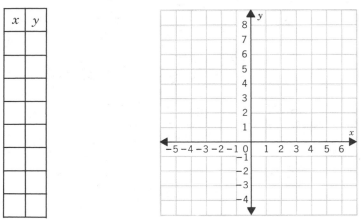

Check your work in **30**.

29

x	y
0	8
1	3
2	0
3	−1
4	0
5	3
6	8

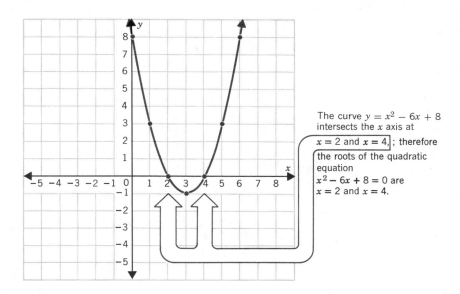

The curve $y = x^2 - 6x + 8$ intersects the x axis at $x = 2$ and $x = 4$; therefore the roots of the quadratic equation $x^2 - 6x + 8 = 0$ are $x = 2$ and $x = 4$.

Once the roots have been found, they should be substituted back into the original quadratic equation to be certain that they satisfy the equation.

For $x = 2$, $y = x^2 - 6x + 8$ gives $(0) = (2)^2 - 6(2) + 8$
$$\text{or} \quad 0 = 4 - 12 + 8$$
$$0 = 0$$

For $x = 4$, $y = x^2 - 6x + 8$ gives $(0) = (4)^2 - 6(4) + 8$
$$\text{or} \quad 0 = 16 - 24 + 8$$
$$0 = 0$$

There are three possible situations that can occur when a quadratic equation is graphed.

Three Possible Situations

1. The graph of $y = ax^2 + bx + c$ can touch the x axis in only one point.

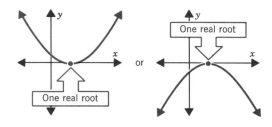

In this case the quadratic equation $ax^2 + bx + c = 0$ has only *one* root.

2. The graph of $y = ax^2 + bx + c$ can cut the x axis at two points.

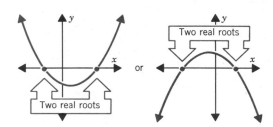

In this case the quadratic equation $ax^2 + bx + c = 0$ has *two* real roots.

3. The graph of $y = ax^2 + bx + c$ can be entirely above or entirely below the x axis and not cut the axis at all.

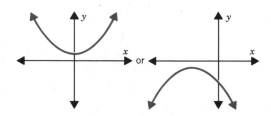

In this case the quadratic equation $ax^2 + bx + c = 0$ has *no* real roots at all.

Solve each of the following quadratic equations by graphing. If necessary, round your answer to one decimal digit.

(a) $x^2 - 2x - 3 = 0$

(b) $2x^2 + x - 6 = 0$

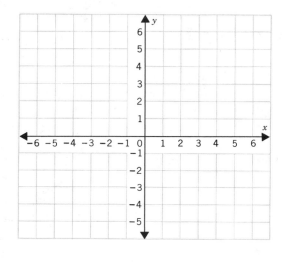

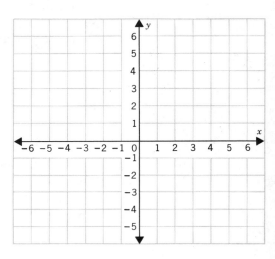

(c) $x^2 - x + 2 = 0$

(d) $x^2 + 2x - 2 = 0$

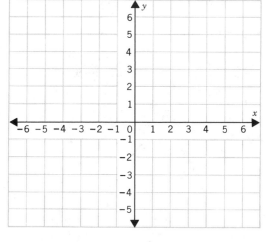

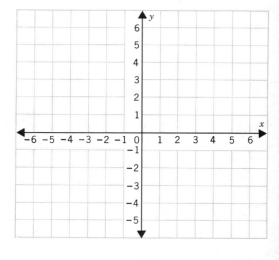

(e) $x^2 - 4x + 4 = 0$

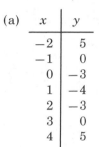

Check your graphs and answers in **30**.

---

30 (a)

x	y
-2	5
-1	0
0	-3
1	-4
2	-3
3	0
4	5

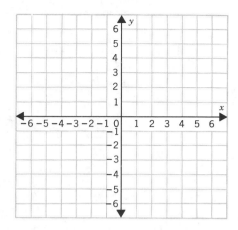

$y = x^2 - 2x - 3$

The roots of this equation are $x = -1$ and $x = 3$.

Check:

$x = -1$
$(0) = (-1)^2 - 2(-1) - 3$
$0 = 1 + 2 - 3$

$x = 3$
$(0) = (3)^2 - 2(3) - 3$
$0 = 9 - 6 - 3$

(b)

x	y
-3	9
-2	0
-1	-5
0	-6
1	-3
2	4
3	15

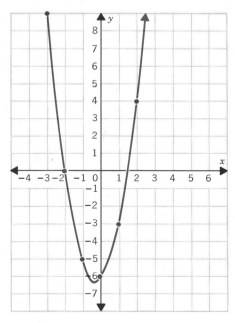

$y = 2x^2 + x - 6$

The roots of this equation are $x = -2$ and $x = \dfrac{3}{2}$.

(c)

x	y
-2	8
-1	4
0	2
1	2
2	4
3	8

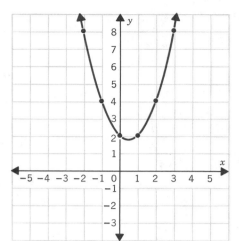

$y = x^2 - x + 2$

This equation has no real roots.

(d)

x	y
-4	6
-3	1
-2	-2
-1	-3
0	-2
1	1
2	6

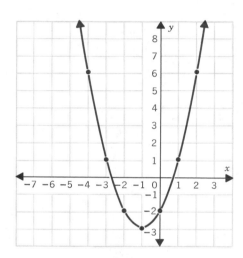

$y = x^2 + 2x - 2$

The roots are approximately -2.7 and 0.7.

The values of the roots found from the graph do not satisfy the equation exactly. The exact roots are $-1 \pm \sqrt{3}$.

(e)

x	y
0	4
1	1
2	0
3	1
4	4

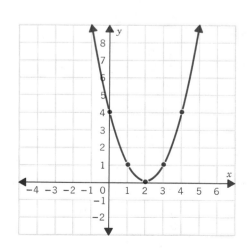

$y = x^2 - 4x + 4$

This equation has one root at $x = 2$

Always check your solution by substituting *both* roots back into the original equation.

In getting the table of values for plotting the graph it may help you to know that the vertex of the parabola for the equation

$y = ax^2 + bx + c$ will always be at $x = \dfrac{-b}{2a}$.

If you first calculate this point and find the y value for it and for integer values of x on either side of it, you will save time and effort.

Example: $y = 3x^2 - 12x + 7$

The vertex is at $x = \dfrac{-b}{2a} = \dfrac{-(-12)}{2(3)} = 2$

Find y at $x = 2$ and at values around $x = 2$, such as $x = -1$, 0, 1, 2, 3, 4, ...

Write each of the following quadratic equations as an equation in two variables, then draw the graph, and finally find the roots, rounding to the nearest one-half unit if necessary.

(a) $(x - 2)^2 = 0$

(b) $x^2 = 0$

(c) $-2x^2 = 0$

(d) $x^2 - 2 = 0$

(e) $x^2 - 3x + 2 = 0$

(f) $2 - x + 2x^2 = 0$

(g) $(x + 3)^2 - 1 = 0$

(h) $-x^2 + x = 0$

(i) $\dfrac{x^2}{4} - x - 2 = 0$

(j) $3 - 2x^2 - 4x = 0$

Check your work in **31**.

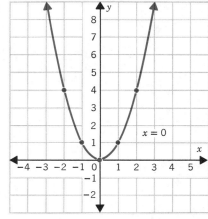

31 (a)
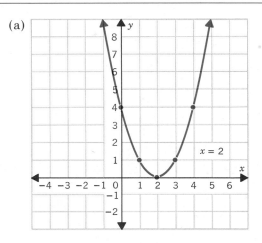
     (b)

Copyright © 1982 by John Wiley & Sons, Inc.

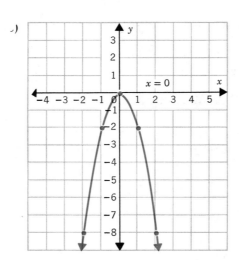

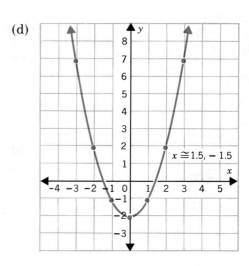

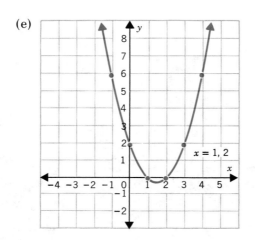

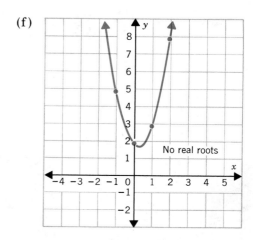

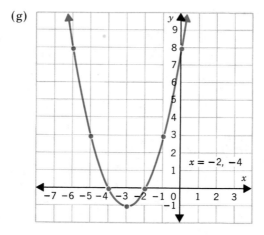

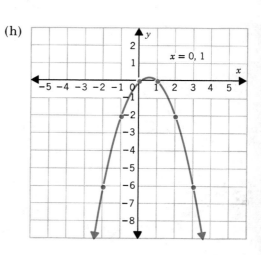

498   **Quadratic Equations**

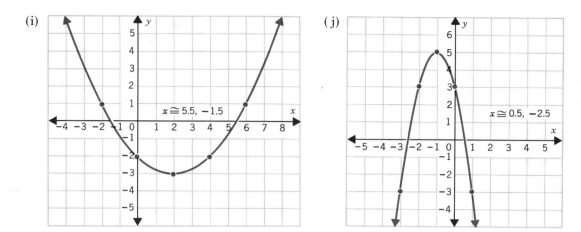

(i) $x \cong 5.5, -1.5$

(j) $x \cong 0.5, -2.5$

Now turn to **32** to learn how to solve word problems involving quadratic equations.

6-7 WORD PROBLEMS INVOLVING QUADRATIC EQUATIONS

32   Quadratic equations appear in a wide variety of mathematical, scientific, and technical problems and arise from many different word problems. If you are able to use the techniques for solving quadratic equations already discussed in this unit, your main difficulty in solving word problems will be to translate them from English sentences and phrases to algebraic equations. You may want to pause here and review the explanation of translating English to algebra given in Unit 1 (pages 59 to 67) or the sections on word problems in Unit 2 (pages 153 to 177) and Unit 4 (pages 313 to 321).

In this section of Unit 6 we will study several common kinds of word problems that involve quadratic equations, describe procedures to be used to solve them, and point out some ways in which these kinds of word problems are different from other word problems you already worked.

Problems Involving Integers

As explained in Unit 2, problems of this kind can usually be solved by translating the English phrases and sentences directly to algebraic expressions and equations. For example, to solve the problem

> "The square of a positive integer is equal to three more than twice the integer. Find the integer."

follow these steps.

Step 1.   Translate the problem directly to an algebraic equation.

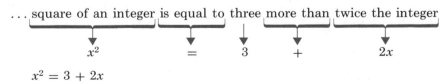

... square of an integer is equal to three more than twice the integer

$$x^2 \qquad = \qquad 3 \quad + \qquad 2x$$

$$x^2 = 3 + 2x$$

By this time in your study of algebra you know that not every word problem can be translated as easily and directly as this one. Careful, imaginative work is needed to translate most problems.

Step 2. Rewrite the quadratic equation in standard form and solve it using the methods you have already learned.

$$x^2 - 2x - 3 = 0$$

This equation can be factored as

$$(x - 3)(x + 1) = 0$$

Using the zero-factor rule, either $(x - 3) = 0$ or $(x + 1) = 0$.

The roots of the equation are $x = 3$ and $x = -1$.

Step 3. Check both roots to see if they satisfy the quadratic equation and to determine if both are reasonable answers to the word problem.

Check:

$x = 3$	$x = -1$
$(3)^2 = 3 + 2(3)$	$(-1)^2 = 3 + 2(-1)$
$9 = 3 + 6$	$1 = 3 - 2$
$9 = 9$	$1 = 1$

The word problem asks for a *positive* integer; therefore the root $x = -1$ is not appropriate even though it is a root of the equation. Discard this part of the answer. The answer to the word problem is $x = 3$.

Important ⟩ Because a quadratic equation has two roots, it is very important that you check *both* roots to see if they both are reasonable answers to the word problem. Discard any root that is not appropriate.

Your turn. Solve this problem.

If three times the square of a certain integer is increased by the integer itself, the sum is equal to 14. Find the integer.

Check your work in **33**.

EXTRANEOUS ROOTS

The act of solving an equation may itself cause part of the correct solution to be lost or may create new and incorrect solutions. For example, a very careless student might try to solve the equation

$$x^2 = 2x$$

by dividing both sides of the equation by x to get the root $x = 2$. This is *not* the correct way to solve this equation. The value $x = 0$ is also a root and substituting zero for x makes the equation a true statement. Solving the equation by dividing both sides of the equation by x is not legal because $x = 0$ is a possible solution and division by zero is not allowed.

This equation should be solved by rewriting it in standard form,

$$x^2 - 2x = 0$$

factoring, $x(x - 2) = 0$

and using the zero factor rule to get the roots

$$x = 0 \quad \text{and} \quad x = 2$$

Solving an equation can cause new, false, or *extraneous,* roots to appear. For example, if we solve this equation

$$\frac{1}{x - 2} = \frac{4}{x^2 - 4} + 1$$

by multiplying each term by the least common denominator $(x - 2)(x + 2)$ we get

$(x + 2) = 4 + (x - 2)(x + 2)$

or $x^2 - x - 2 = 0$

The roots of this equation are $x = 2$ and $x = -1$. If you check these roots in the original equation, you will find that $x = 2$ is not a correct solution. Substituting $x = 2$ into the original equation requires you to divide by zero, which of course is impossible.

Always check to be certain the roots you find are correct.

---

33   **Step 1.**   Translate the problem from English to algebra.

Three times the square . . . increased by . . . the integer . . . is equal to 14

$3 \quad \times \quad x^2 \qquad\qquad + \qquad\qquad x \qquad\qquad = \qquad 14$

$3x^2 + x = 14$

Step 2.   Rewrite in standard form and solve.

$3x^2 + x - 14 = 0$

$(3x + 7)(x - 2) = 0$ The equation can be factored.

Use the zero-factor rule.

$3x + 7 = 0 \qquad \text{or} \qquad x - 2 = 0$

The roots are $x = -\dfrac{7}{3}$ and $x = 2$.

Step 3.   Check both roots to see if they are correct and if they fit the problem.

In this case both roots are correct solutions to the equation, but the word problem asks you to find an integer and only the second root is an integer. Discard the first root. The correct answer to the word problem is $x = 2$.

In step 2, choose the method of solving the quadratic equation that seems easiest for you and most appropriate for the equation. If the problem asks for an integer solution the equation will be factorable, but if you do not immediately see how to factor it, use the quadratic formula.

Translating a Word Problem

Very often a direct translation of the word problem to algebra produces several equations in two variables. If so, solve these equations as a system of equations. If you want to review this process, return to Unit 4, page 295.

Lets look at an example. Solve this problem:

The sum of two numbers is 12 and their product is 32.
Find the two numbers.

Try it, then check your work in **34**

---

34   **Step 1.**   Translate the English sentence.

. . . sum of two numbers is 12 . . .

$x + y \qquad\qquad = 12$

$$\ldots \underbrace{\text{their product}}_{\downarrow} \ \underbrace{\text{is}}_{\downarrow} \ \underbrace{32}_{\downarrow}$$

$$xy \qquad = \quad 32$$

Step 2. Solve this system of two equations by solving the first equation for y and substituting the expression for y into the second equation.

The first equation becomes $\qquad y = 12 - x$
and the second equation becomes $\qquad x(12 - x) = 32$.
Rewrite in standard form: $\qquad x^2 - 12x + 32 = 0$.
Factor: $\qquad (x - 8)(x - 4) = 0$.
The roots are $\qquad x = 8$ and $x = 4$.

Step 3. Check both roots by substituting into the original quadratic equation, then check again to be certain that the roots are appropriate for the original word problem.

If $x + y = 12$ then $(8) + y = 12$ or $y = 4$

The two numbers asked for in the word problem are 8 and 4. Notice that both roots lead to the same solution of the word problem.

Use this procedure to solve the following problems.

(a)   When the square of a certain positive number is added to five times that number, the sum is equal to 36. Find the number.
(b)   Find two consecutive integers such that the sum of their squares is equal to 41.
(c)   The difference between two positive numbers is 3. The sum of their squares is 89. Find the numbers.

Work carefully. Our solutions are in **35**.

---

35   (a)   **Translate:**

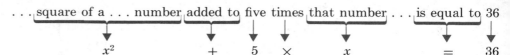

$$x^2 + 5x = 36$$

Solve:

$$x^2 + 5x - 36 = 0$$

Factor it.

$$(x + 9)(x - 4) = 0$$

Either $x + 9 = 0 \qquad$ or $\qquad x - 4 = 0$

The roots are $x = -9 \qquad$ and $\qquad x = 4$

Check:

Because the problem asks for a positive number, only $x = 4$ is a valid solution to the word problem.

(b)   **Translate:**   Let the two consecutive integers be n and $n + 1$, then the sum of their squares is

$$n^2 \qquad + \qquad (n + 1)^2$$

and the word problem produces the equation

$$n^2 + (n + 1)^2 = 41$$

Solve:

$$n^2 + n^2 + 2n + 1 = 41$$

$$2n^2 + 2n - 40 = 0 \quad \text{or} \quad n^2 + n - 20 = 0$$

This equation can be factored as

$$(n + 5)(n - 4) = 0$$

Use the zero-factor rule to solve this equation. The roots are $n = -5$ and $n = 4$.

For $n = -5$ the two consecutive integers are -5 and -4. For $n = 4$ the two consecutive integers are 4 and 5. Both pairs of integers satisfy the original word problem.

(c) **Translate:**

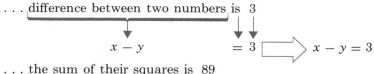

Solve: To solve these equations, combine them to form a single equation. The first equation is equivalent to $x = 3 + y$. Substitute this expression for x into the second equation to get

$$(3 + y)^2 + y^2 = 89$$

Multiply to remove parentheses: $\quad 9 + 6y + y^2 + y^2 = 89$
Rewrite in standard form: $\qquad\qquad 2y^2 + 6y - 80 = 0$
$$\text{or} \quad y^2 + 3y - 40 = 0$$

Factor: $\quad (y - 5)(y + 8) = 0$

Use the zero-factor property to get the equations

$$y - 5 = 0 \quad \text{and} \quad y + 8 = 0$$

The roots are $y = 5 \quad$ and $\quad y = -8$.

Check: The problem calls for a positive number solution; therefore the second root does not fit the problem. Discard this part of the solution.

Substitute $y = 5$ back into the original equation to get

$$x - y = 3 \quad x - (5) = 3$$
$$x = 8$$

The two numbers are 8 and 5.

Geometric Problems

A second variety of word problem whose solution may require solving a quadratic equation involves the dimensions, perimeter, and area of geometric figures. For example, consider the following problem:

The length of a rectangular rug is two feet more than its width. If its area is $11\frac{1}{4}$ square feet, find the dimensions of the rug.

To solve this problem, first represent the dimensions of the rectangle by algebra variable symbols.

$x =$ length of the rectangle
$y =$ width of the rectangle

Now proceed as before.

Step 1. Translate the word problem to one or more algebraic equations.

length . . . is two feet more than its width

$$x = 2 + y \qquad\Longrightarrow\qquad x = 2 + y$$

. . . area is $11\frac{1}{4}$ square feet

$$xy = 11\frac{1}{4} \qquad\Longrightarrow\qquad xy = 11\frac{1}{4}$$

(The area of a rectangle is equal to the product of its length and width.)

Step 2. Solve. Combine these two equations to get

$$(2 + y)(y) = 11\frac{1}{4}$$

You should be able to solve this equation. Try it and check your work in **36**.

---

36 To solve this equation, first rewrite it as $\quad 2y + y^2 = \dfrac{45}{4}$

or $\quad 8y + 4y^2 = 45$

In standard form, it appears as $4y^2 + 8y - 45 = 0$

Factor: $\quad (2y - 5)(2y + 9) = 0$

Use the zero-factor rule to get the equations

$2y - 5 = 0 \qquad$ and $\qquad 2y + 9 = 0$

The roots are $\quad y = \dfrac{5}{2} \qquad$ and $\qquad y = -\dfrac{9}{2}$

If you find this equation difficult to factor, you can solve it by using the quadratic formula.

Finally, check the answer. Since the width of a rectangle cannot be a negative number, the root $y = -\frac{9}{2}$ must be discarded as not appropriate for the problem.

Substitute the root $y = \frac{5}{2}$ back into the original equations to find $x = \frac{9}{2}$. The dimensions of the rug are $2\frac{1}{2}$ ft by $4\frac{1}{2}$ ft.

Solve the following problems using this same procedure.

(a) The perimeter of a rectangle is 40 in. and its area is 64 sq in. Find the dimensions of the rectangle.
(b) The length of a rectangular flower garden is three times its width. If the width is increased by one meter and the length is increased by two meters, the new rectangle has twice the area as the original. Find the dimensions of the original garden.
(c) In a certain right triangle, the difference in length of the two legs is 1 in. The length of the hypotenuse is 4 in. What are the lengths of the sides of the triangle? (Remember, for any right triangle if the sides are a and b and the hypotenuse is c, then $a^2 + b^2 = c^2$.)

Compare your work with ours in **37**.

---

504 **Quadratic Equations**

37 (a) **Translate:** Let $x =$ length, $y =$ width,
then the perimeter is $2x + 2y = 40$ or $x + y = 20$,
and the area is $xy = 64$.

Solve: Combine these equations: $y = 20 - x$
The second equation becomes $x(20 - x) = 64$
Rewrite this equation in standard form: $x^2 - 20x + 64 = 0$
Factor: $(x - 4)(x - 16) = 0$
The roots of this equation are $x = 4$ and $x = 16$.

Check: Substituting back into the original equations, we find that the dimensions of the rectangle are 4 in. by 16 in.

(b) **Translate:** Let $x =$ length, $y =$ width,
then the original area was xy and we know that $x = 3y$.
The new length is $x + 2$.
The new width is $y + 1$.
The new area is $(x + 2)(y + 1)$.
So that $(x + 2)(y + 1) = 2xy$.

Solve: Combine the two equations $x = 3y$ and $(x + 2)(y + 1) = 2xy$.

Substitute x from the first equation into the second equation

$(3y + 2)(y + 1) = 2(3y)y$

Rewrite in standard form: $3y^2 - 5y - 2 = 0$

Solve either by factoring or by using the quadratic formula.
Factor: $(3y + 1)(y - 2) = 0$

The roots are $y = -\dfrac{1}{3}$ and $y = 2$

Check: Check both roots to be certain they are appropriate for the original word problem. The solution

$y = -\frac{1}{3}$ is impossible, since the width cannot be a negative number.

Substitute $y = 2$ meters into the original equation to get $x = 6$ meters.
The dimensions of the garden are 2 meters by 6 meters.

(c) **Translate:** By the Pythagorean theorem $a^2 + b^2 = 4^2$
From the problem $a - b = 1$

Solve: To combine these two equations solve the second equation for b:
$b = a - 1$
and substitute this expression for b into the first equation.

$a^2 + (a - 1)^2 = 4^2$

Rewrite in standard form: $a^2 + a^2 - 2a + 1 = 16$
$$2a^2 - 2a - 15 = 0$$

Solve by using the quadratic formula.

$$a = \frac{-(-2) \pm \sqrt{(-2)^2 - 4(2)(-15)}}{2(2)}$$

$$a = \frac{2 \pm \sqrt{4 + 120}}{4} = \frac{2 \pm 2\sqrt{31}}{4} = \frac{1 \pm \sqrt{31}}{2}$$

The roots are $a = \dfrac{1 + \sqrt{31}}{2}$ and $a = \dfrac{1 - \sqrt{31}}{2}$

Check: The root $a = \dfrac{1 - \sqrt{31}}{2} \cong \dfrac{1 - 5.568}{2} \cong -2.284$

The length of the side of a triangle cannot be a negative number, so we must discard this value for a.

The root $a = \dfrac{1 + \sqrt{31}}{2} \cong \dfrac{1 + 5.568}{2} \cong 3.284$ is a reasonable answer.

Substituting this root back into the equation $a - b = 1$, we find that

$$b = a - 1 = \frac{1 + \sqrt{31}}{2} - 1 = \frac{-1 + \sqrt{31}}{2}$$

$$b \cong \frac{-1 + 5.568}{2} \cong 2.284$$

The sides of the triangle are roughly 3.28 in., 2.28 in., and 4 in.

SUMMARY   In general, to solve any word problems that lead to quadratic equations, follow the procedures described in Unit 4 for motion, work, or rate problems. These procedures will help you translate the word problem to a quadratic equation or a set of simultaneous equations that may be combined to produce a quadratic equation. Solve this equation as shown in the examples above, then check both roots carefully to be certain that they are valid solutions to the quadratic equation and appropriate answers to the original word problem. Always keep in mind that the value of many physical quantities such as the dimensions of geometric figures, areas, times, weights, and so on, cannot be negative numbers. Disregard any roots that are not reasonable or appropriate.

Now for a set of practice word problems involving quadratic equations turn to **38**.

Unit 6
Quadratic
Equations

38

The answers are on page 568.

A. **Graph each of the following equations and find its roots. Round the roots to the nearest tenth.**

1. $y = 1 - x^2$ _____

2. $y = -x^2$ _____

3. $3x^2 = y$ _____

4. $y = x^2 + 1$ _____

5. $y = x^2 - 3x$ _____

6. $2x^2 - x = y$ _____

7. $y = 3 - x^2 + 4x$ _____

8. $y = 1 - 6x + x^2$ _____

9. $y = 4x - 2x^2 + 1$ _____

10. $x^2 - 6x + 8 = y$ _____

11. $x^2 - 4x + 3 = y$ _____

12. $y = 1 + x(x - 2)$ _____

13. $y = x^2 - 2x + 5$ _____

14. $(2x - 3)^2 - 2 = y$ _____

15. $x^2 + x - 6 = y$ _____

16. $-x^2 - x + 12 = y$ _____

17. $y = x(x - 3) - 4$ _____

18. $y = 4x^2 - 9$ _____

19. $y = x^2 - 2x - 4$ _____

20. $y = 5 - x^2$ _____

B. **Solve:**

1. The square of a number exceeds the number by 30. Find the number.

_____

2. The square of a number exceeds four times the number by 20. Find the number.

_____

3. Twice the square of a certain positive integer decreased by the integer itself is equal to 15. Find the integer.

_____

4. Find two consecutive even integers such that the square of the smaller is 10 more than the larger.

_____

5. The sum of the squares of two positive integers is 841. The two integers differ by 1. Find the two integers.

_____

Date

Name

Course/Section

507   **Problem Set 6-3**

6. The sum of two numbers is 6. The sum of their squares is 40. Find the numbers.

   _____

7. The sum of a number and its reciprocal is $3\frac{1}{2}$. Find the number.

   _____

8. Find three consecutive odd integers such that the square of the largest is equal to 5 less than twice the product of the other two.

   _____

9. The product of two positive consecutive even integers is 80. Find the integers.

   _____

10. The sum of the reciprocals of two consecutive even integers is $\frac{5}{12}$. Find the integers.

    _____

11. The width of a certain rectangle is one-half its length. The area of the rectangle is 20 sq ft. Find the dimensions of the rectangle.

    _____

12. The perimeter of a rectangle is 68 in. and its area is 273 sq in. Find the dimensions of the rectangle.

    _____

13. The length of a certain rectangular garden is 4 ft longer than its width. If each dimension were increased by 2 ft, its area would double. Find the present dimensions of the garden.

    _____

14. Mary's flower garden is a rectangle measuring 10 ft by 16 ft. She wants to double its area by increasing each dimension by the same amount. By how much should she increase the length of each side, to the nearest one-half foot?

    _____

15. The length of a certain rectangle is six ft more than twice its width. If its area is 56 sq ft, find its dimensions.

    _____

16. A rectangular picture frame mat has outside dimensions of 22 in. by 30 in. The border is a strip of the same width on all sides of the mat. How wide must the border be if it contains one-third of the area of the mat? (Round your answer to one decimal digit.)

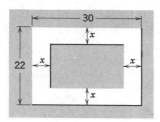

---

17. Find x for the triangle shown.

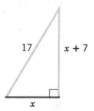

---

18. Find x.

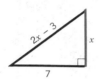

---

19. Find x.

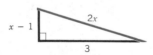

---

20. Find x.

---

21. One leg of a right triangle is twice as long as the other. If six inches is added to each leg, the area of the new triangle is five times the area of the original triangle. What are the lengths of the legs of the original triangle? Round to one decimal digit. (The area of a right triangle is equal to $\frac{1}{2}ab$, where a and b are the lengths of the legs of the triangle.)

---

509   **Problem Set 6-3**

22. In the diagram, the side of the larger square is 4 meters more than the side of the smaller square. The sum of the areas of the two squares is 136 square meters. Find the length of the sides of the smaller square.

_____

23. The length of a rectangular garden is 30 ft. If we add to it a square area with sides equal to the width of the rectangle, the total area becomes 500 sq ft. What was the width of the rectangle?

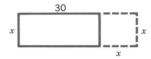

_____

24. The hypotenuse of a right triangle is 25 in. Find the lengths of the legs of the triangle if one leg is 17 in. longer than the other.

_____

25. The lengths of the legs of a certain right triangle sum to 49 inches and the length of the hypotenuse is 41 inches. What are the dimensions of the triangle?

_____

26. The legs of a certain right triangle differ by 1 inch. The hypotenuse is 29 inches long. What are the dimensions of the triangle?

_____

27. George and Dan each ran 8 miles. George ran at a speed 2 mph faster than Dan and arrived 10 minutes sooner. How fast did they each run? Round to one decimal digit. (_Hint:_ Let 10 minutes $= \frac{1}{6}$ hr.)

_____

28. If Joe and Alice work together, they can paint a room in 4 hours. If they work alone, Alice needs 6 hours more than Joe to paint the room. How many hours would each person need to paint the room working alone?

_____

29. If both the inlet and outlet pipes of a water storage tank are open it takes five hours for the tank to fill. The outlet pipe empties the tank in four hours more than the time it takes the inlet pipe to fill it. How long does it take the inlet pipe to fill the tank when the outlet pipe is closed? Round to one decimal digit.

_____

30. During a boat race the contestants travel 24 miles upstream and return 24 miles downstream. The winning boat can travel 10 mph in still water and takes one hour longer to do the upstream half than to do the downstream half. What is the speed of the current?

_____

31. On a vacation trip the return half of the journey took one hour longer than the first half. If each half of the trip was 300 miles in length, and if we drove 10 mph faster on the first half of the trip, how fast did we drive on the return half?

_____

32. Solve the bean merchant's problem given on page 447.

_____

C. Brain Boosters

1. Find the area of the largest rectangle that can be enclosed by 200 ft of fence. (*Hint:* Write the area as a quadratic equation with one side as x. Graph this equation to find the maximum value of the area.)

_____

2. The sum of the n consecutive integers $1 + 2 + 3 + \cdots + n = \dfrac{n(n + 1)}{2}$. Find n if the sum is 153.

_____

3. Here is a problem that appeared in a mathematics textbook published in 1556.

   "Find a square from which if $15\frac{3}{4}$ is subtracted the result is its own root."

   (*Hint:* Solve $x^2 - 15\frac{3}{4} = x$)

_____

4. In 1543 Girolamo Cardano, professor of medicine at Milan University, wrote a very famous mathematics book entitled *Ars Magna*. In chapter 37, Cardano solved this problem:

   "Divide 10 into two numbers whose product is 40."

   (In other words, the two numbers add up to 10 and their product is 40.) Can you solve Cardano's problem?

_____

5. (a) Solve $\dfrac{2}{x - 2} + \dfrac{3}{x - 3} + \dfrac{4}{x - 4} = 0$

_____

Date

_____

Name

_____

Course/Section

(b)   Solve $(x - 2)(x - 3) + (x - 3)(x - 1) + (x - 1)(x - 2) = 0$

When you have had the practice you need, turn to **39** for a self-test on the work of Unit 6.

Unit 6
Quadratic
Equations

39 Solve by factoring.

1. $2x^2 = 3x$ _____

2. $x^2 - 5x - 14 = 0$ _____

3. $44 = x^2 - 7x$ _____

4. $12x + 36 + x^2 = 0$ _____

5. $5 = 6x^2 + 13x$ _____

6. $2x^2 + 17x + 21 = 0$ _____

Solve by completing the square.

7. $x^2 + 4x - 7 = 0$ _____

8. $x^2 + 8x + 4 = 0$ _____

9. $2x^2 - 6x - 2 = 0$ _____

10. $2x^2 - 1 = 7x$ _____

Solve using the quadratic formula.

11. $x^2 + 2x - 63 = 0$ _____

12. $x^2 = 7x$ _____

13. $2x^2 - 6x + 1 = 0$ _____

14. $x^2 = 2(x + 5)$ _____

15. $5x^2 = 1 + 8x$ _____

16. $\dfrac{1}{x} + \dfrac{2}{x + 2} = 3$ _____

Graph each equation and find its roots.

17. $y = -4x^2$ _____

18. $y = x^2 + 3$ _____

19. $y = 2x^2 + 4x - 16$ _____

20. $y = x^2 - x - 1$ _____

Date _____

21. $y = -x^2 + 4x + 5$ _____

Name _____

22. Find two integers whose sum is 21 and whose product is 104.

Course/Section _____

_____

513 **Self-Test 6**

23. Find two consecutive positive even integers whose product is 168.

   _____

24. The sum of a number and its reciprocal is $\frac{12}{5}$. Find the number.

   _____

25. The length of a certain rectangle is 4 ft more than twice its width. Its area is 24 sq ft. Find the dimensions of the rectangle. Round to one decimal digit.

   _____

Answers are on page 571.

Unit 6 Quadratic Equations

Solving Quadratic
Equations by
Factoring

A. Solve:

1. $x^2 = 9$ _____

2. $p^2 - 24 = 0$ _____

3. $18 - y^2 = 0$ _____

4. $2x^2 - 40 = 0$ _____

5. $3y^2 - 96 = 0$ _____

6. $5a^2 - 3 = 1$ _____

7. $2x^2 - 9 = 1$ _____

8. $4p^2 - 3 = 1$ _____

9. $2x^2 - \dfrac{1}{4} = 0$ _____

10. $3y^2 - 12 = 0$ _____

11. $4x^2 = 1$ _____

12. $q^2 - 12 = 0$ _____

13. $2a^2 = 5$ _____

14. $3y^2 = \dfrac{1}{4}$ _____

15. $8 - x^2 = 3$ _____

16. $1 - x^2 = -17$ _____

B. Solve using the Pythagorean theorem.

1.

_____

2.

_____

3.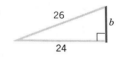

_____

4.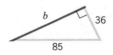

_____

5.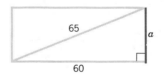

_____

6.

_____

7.

_____

8.

_____

9.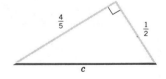

_____

C. Solve by factoring.

1. $(x - 7)(x + 2) = 0$ _____

2. $(2x - 6)(x - 8) = 0$ _____

3. $(2y - 3)(3y - 1) = 0$ _____

4. $x^2 - 14x + 48 = 0$ _____

5. $x^2 + 4x - 5 = 0$ _____

6. $x^2 - 5x - 14 = 0$ _____

7. $a^2 - 36 = 0$ _____

8. $10p = 25 + p^2$ _____

9. $x^2 - 5x - 1 = 35$ _____

10. $x^2 - 6x = 7$ _____

11. $y^2 + 6y - 16 = 0$ _____

12. $x^2 - 11x + 10 = 0$ _____

13. $a^2 - 3a - 1 = a + 20$ _____

14. $t^2 + 16 = 9t - 4$ _____

15. $y^2 - 3y = 18$ _____

16. $6x^2 - 7x - 24 = 0$ _____

17. $3p^2 - 4p = 15$ _____

18. $5x^2 + 29x - 6 = 0$ _____

19. $(x - 2)(x - 3) = 2(2x - 7)$ _____

20. $2x^2 + 11x = 6$ _____

21. $3q^2 = 2(4 - 5q)$ _____

22. $t - 1 = \dfrac{110}{t}$ _____

23. $x(x + 10) + 9 = 0$ _____

24. $y^2 - 4y = 21$ _____

25. $a(2a - 1) = 15$ _____

26. $x(x - 2) = 24$ _____

27. $\dfrac{x}{2} + 1 = \dfrac{4}{x}$ _____

28. $\dfrac{9}{t} = \dfrac{t}{4}$ _____

29. $x = \dfrac{30}{x - 1}$ _____

30. $\dfrac{y}{4} = \dfrac{5}{y - 8}$ _____

31. $\dfrac{5}{x} = \dfrac{x - 2}{2x - 7}$ _____

32. $\dfrac{3}{x} = \dfrac{12 - x}{x + 6}$ _____

33. $2x(x + 1) = 20 - x$ _____

34. $3x(x + 5) = 2(3 - x)$ _____

35. $4x + 5 = \dfrac{6}{x}$ _____

36. $5x + 17 = \dfrac{12}{x}$ _____

37. $2x - 1 = \dfrac{3 + 2x}{2x}$ _____

38. $3x + 7 = \dfrac{x - 6}{2x}$ _____

39. $(x - 2)^2 = 16$ _____

40. $(2x - 1)^2 = 25$ _____

Date _____

Name _____

Course/Section _____

Unit 6
Quadratic
Equations

Solving Quadratic
Equations

A. Solve by completing the square.

1. $x^2 + 4x = 0$ _____

2. $y^2 + 6y = 0$ _____

3. $p^2 + 8p = 9$ _____

4. $a^2 + 10a + 5 = 0$ _____

5. $x^2 = 2x + 1$ _____

6. $x^2 - 8x = 2$ _____

7. $2x^2 + 6x + 1 = 0$ _____

8. $3y^2 + 8y + 2 = 0$ _____

9. $x^2 = 1 + \dfrac{x}{2}$ _____

10. $1 + 7x = x^2$ _____

11. $10(x - 1) = x^2$ _____

12. $x^2 - 6 = 5x$ _____

B. Solve using the quadratic formula.

1. $x^2 - 3x - 10 = 0$ _____

2. $4x^2 + 11x - 3 = 0$ _____

3. $3x^2 - 17x = 6$ _____

4. $x^2 - x - 30 = 0$ _____

5. $2x^2 - 5x - 1 = 0$ _____

6. $a^2 + 3a + 1 = 0$ _____

7. $2x^2 + x - 1 = 0$ _____

8. $2x^2 + 3x = 0$ _____

9. $4x^2 - 3 = 0$ _____

10. $y^2 - 7y + 5 = 0$ _____

11. $p^2 + 8p + 9 = 0$ _____

12. $x^2 + 3x = 1$ _____

13. $x^2 = 4x - 1$ _____

14. $x^2 - x = 5$ _____

15. $(x - 4)^2 = 6$ _____

16. $(y + 2)^2 = 5$ _____

17. $(2x - 1)^2 = 3$ _____

18. $(3x + 2)^2 = 3$ _____

19. $3a^2 + 5a + 1 = 0$ _____

20. $2x^2 + 8x = 5$ _____

21. $2y^2 - 3y - 1 = 0$ _____

22. $a^2 - 2 = 5a$ _____

23. $2x^2 + 1 = 6x$ _____

24. $5x^2 - 2x - 1 = 0$ _____

25. $\dfrac{x^2}{2} + 3x - 1 = 0$ _____

26. $\dfrac{2x^2}{3} - 5x + 8 = 0$ _____

27. $2x = 4 + \dfrac{3}{x}$ _____

28. $3x - 1 = \dfrac{2}{x}$ _____

Date

29. $x^2 - \dfrac{11x}{5} + 1 = 0$ _____

30. $x^2 - \dfrac{2x}{3} = \dfrac{1}{2}$ _____

Name

Course/Section

31. $\dfrac{1}{x} + \dfrac{1}{x - 1} = 2$ _____

32. $\dfrac{2}{x} - \dfrac{1}{x - 2} = 3$ _____

33. $11x^2 - 7x - 9 = 0$ _____

34. $8x^2 + 9x - 10 = 0$ _____

35. $x^2 - \dfrac{2x}{3} - 2 = 0$ _____

36. $x^2 + \dfrac{3x}{5} = 3$ _____

37. $1 + \dfrac{7x}{2} = -x^2$ _____

38. $4 - \dfrac{8x}{3} = 2x^2$ _____

39. $(x + 1)(1 - x) = 3(x - 2)$ _____

40. $(x - 2)^2 = 2(x + 3)$ _____

Unit 6 Quadratic Equations

A. Graph each of the following equations and find its roots. If necessary, round the roots to the nearest tenth.

1. $y = \dfrac{x^2}{3}$ _____

2. $y = \dfrac{-x^2}{2}$ _____

3. $y = 1 - 2x^2$ _____

4. $x^2 - 3x = y$ _____

5. $y - 4x = x^2$ _____

6. $y = -x^2 - 7x - 10$ _____

7. $x^2 + 4 = y$ _____

8. $x^2 + 3x = y - 1$ _____

9. $y = 2x^2 + 3x + 3$ _____

10. $y = 2x^2 - x - 1$ _____

11. $y = 3x^2 + 5x - 1$ _____

12. $y = 5x - x^2 - 1$ _____

13. $x^2 + 2x - 5 = y$ _____

14. $x^2 - 3 - 2x = y$ _____

15. $y = \dfrac{x^2}{2} - x - 2$ _____

16. $y = 1 - \dfrac{x^2}{3} - \dfrac{x}{2}$ _____

B. Solve:

1. The product of two positive consecutive odd integers is 143. Find the integers.

   _____

2. The square of a number, decreased by 3, is equal to twice the number. Find the number.

   _____

3. The sum of the squares of two integers is 290. The sum of the same two integers is 24. Find the two integers.

   _____

4. The difference of two numbers is 6 and the sum of their squares is 90. Find the numbers.

   _____

5. The sum of a number and its reciprocal is $-\frac{17}{4}$. Find the number.

   _____

6. The length of a rectangular garden is twice its width. If the area of the garden is 882 square feet, find its dimensions.

   _____

Date _____

Name _____

Course/Section _____

7. The base of a triangle is four times its height. The area of the triangle is 98 square inches. Find its base and height. (*Hint:* Area of a triangle = $\frac{1}{2}$ · base · height)

_____

8. Find a in the triangle shown:

_____

9. Two motorists, traveling independently, each went a distance of 120 miles. One traveled at an average speed 15 mph faster than the other and took one hour less time to make the trip. What was the average speed of each car? (Round your answer to the nearest tenth mph.)

_____

10. Working together, Danielle and Trish can weed their vegetable garden in 4 hours. Working by herself, Trish can do the weeding in 6 hours less than Danielle. How many hours would each girl need to weed the garden if they worked alone?

_____

Every fraction is a division expressed in the form $\frac{a}{b}$, where a and b are real numbers—integers, decimals, or even other fractions—and b cannot equal zero.

$$a \div b = \frac{a}{b} \quad \underleftarrow{\text{Numerator}} \quad \underleftarrow{\text{Denominator}}$$

Some fractions: $\quad \dfrac{2}{3}, \dfrac{1}{10}, \dfrac{100}{1}, \dfrac{3.25}{5.18}, \dfrac{1.555\ldots}{2}, \dfrac{\frac{1}{2}}{\frac{1}{4}}$

Equivalent Fractions

Two fractions are said to be *equivalent* if they name the same number. For example,

$$\frac{1}{2} = \frac{2}{4} \quad \text{and} \quad \frac{1}{10} = \frac{10}{100}$$

or even $\quad \dfrac{2}{3} = \dfrac{4}{6} = \dfrac{8}{12} = \dfrac{10}{15} \quad$ and so on.

There are an infinite number of fractions equivalent to any given fraction.

To obtain a fraction equivalent to any given fraction, multiply the original numerator and denominator by the same non-zero number. For example,

$$\frac{1}{2} = \frac{1 \times 3}{2 \times 3} = \frac{3}{6} \quad \text{The fraction is multiplied by } \frac{3}{3} \text{ or 1.}$$

or $\quad \dfrac{2}{3} = \dfrac{2 \times 5}{3 \times 5} = \dfrac{10}{15} \quad$ The fraction is multiplied by $\frac{5}{5}$ or 1.

The value of the fraction has not changed, we have simply renamed it.

A. PRACTICE PROBLEMS

Find equivalent fractions as shown.

Example: $\quad \dfrac{4}{5} = \dfrac{?}{10} \qquad \dfrac{4 \times 2}{5 \times 2} = \dfrac{8}{10}$

1. $\dfrac{5}{6} = \dfrac{?}{42}$ 　　 2. $\dfrac{7}{16} = \dfrac{?}{48}$ 　　 3. $\dfrac{3}{7} = \dfrac{?}{56}$ 　　 4. $\dfrac{5}{3} = \dfrac{?}{12}$

5. $\dfrac{7}{8} = \dfrac{?}{16}$ 　　 6. $\dfrac{3}{4} = \dfrac{?}{12}$ 　　 7. $\dfrac{1}{2} = \dfrac{?}{78}$ 　　 8. $\dfrac{1}{9} = \dfrac{?}{63}$

9. $\dfrac{7}{9} = \dfrac{?}{45}$ 　　 10. $\dfrac{11}{12} = \dfrac{?}{72}$ 　　 11. $\dfrac{7}{10} = \dfrac{?}{50}$ 　　 12. $\dfrac{6}{7} = \dfrac{?}{35}$

The answers to all practice problems in the Fractions Review are on page 527.

It is often very helpful in both arithmetic and algebra to find the simplest equivalent form of a fraction. For example, the simplest fraction of all those equivalent to $\frac{10}{20}$ is $\frac{1}{2}$.

The numerator and denominator of a fraction are sometimes called the terms of the fraction and the process of finding the simplest equivalent fraction is known as *reducing the fraction to lowest terms*. To reduce a fraction to lowest terms, follow these three steps.

Step 1. Factor the numerator and denominator into their prime factors.

Example: $\frac{30}{42} = \frac{2 \times 3 \times 5}{2 \times 3 \times 7}$

Step 2. Find the factors common to both numerator and denominator and separate them out.

Example: $\frac{30}{42} = \boxed{\frac{2}{2}} \cdot \boxed{\frac{3}{3}} \cdot \frac{5}{7}$

Step 3. Replace all fractions of the form $\frac{a}{a}$ by 1.

Example: $\frac{30}{42} = 1 \cdot 1 \cdot \frac{5}{7}$

or $\frac{30}{42} = \frac{5}{7}$

In this example, the net result is that we have divided both numerator and denominator of the original fraction by 2×3 or 6, the factor common to both 30 and 42.

A fraction has been reduced to lowest terms when its numerator and denominator have no factors in common.

B. PRACTICE PROBLEMS

Reduce to lowest terms.

Example: $\frac{90}{105} = \frac{2 \times 3 \times 3 \times 5}{3 \times 5 \times 7} = \boxed{\frac{3}{3}} \cdot \boxed{\frac{5}{5}} \cdot \frac{2 \times 3}{7}$

$= 1 \cdot 1 \cdot \frac{2 \times 3}{7} = \frac{6}{7}$

Usually we abbreviate this process by writing

$\frac{90}{105} = \frac{2 \times \cancel{3} \times 3 \times \cancel{5}}{\cancel{3} \times \cancel{5} \times 7} = \frac{2 \times 3}{7} = \frac{6}{7}$

1. $\frac{15}{84}$ 2. $\frac{21}{35}$ 3. $\frac{4}{12}$ 4. $\frac{26}{30}$ 5. $\frac{12}{15}$ 6. $\frac{8}{10}$

7. $\frac{27}{54}$ 8. $\frac{18}{45}$ 9. $\frac{7}{42}$ 10. $\frac{42}{120}$ 11. $\frac{21}{56}$ 12. $\frac{54}{144}$

The answers to all practice problems in the Fractions Review are on page 527.

Fractions having the same denominator are called *like* fractions.

$\frac{3}{4}$, $\frac{1}{4}$, $\frac{7}{4}$, and $\frac{11}{4}$ are all like fractions. All have denominator 4.

It is easy to add or subtract like fractions: *first,* add or subtract the numerators to find the numerator of the sum or difference, then, *second,* use the denominator that the fractions have in common as the denominator of the sum or difference.

$$\frac{2}{9} + \frac{5}{9} = \frac{2+5}{9} = \frac{7}{9}$$

Add numerators

Same denominator

Always reduce to lowest terms.

C. PRACTICE PROBLEMS

Add or subtract and reduce your answer to lowest terms.

Example: $\dfrac{3}{5} + \dfrac{1}{5} = \dfrac{3+1}{5} = \dfrac{4}{5}$

1. $\dfrac{2}{8} + \dfrac{3}{8}$ 2. $\dfrac{7}{9} + \dfrac{5}{9}$ 3. $\dfrac{1}{2} + \dfrac{5}{2}$ 4. $\dfrac{1}{7} + \dfrac{4}{7} + \dfrac{5}{7}$

5. $\dfrac{5}{12} + \dfrac{7}{12}$ 6. $\dfrac{7}{16} + \dfrac{3}{16}$ 7. $\dfrac{3}{32} + \dfrac{11}{32}$ 8. $\dfrac{7}{9} - \dfrac{1}{9}$

9. $\dfrac{11}{12} - \dfrac{4}{12}$ 10. $\dfrac{19}{32} - \dfrac{7}{32}$ 11. $\dfrac{8}{10} - \dfrac{5}{10}$ 12. $\dfrac{12}{20} - \dfrac{9}{20}$

The answers are on page 527.

To add two fractions whose denominators are not the same, rewrite the fractions as equivalent fractions whose denominators are equal. For example, to add the fractions $\frac{2}{3}$ and $\frac{3}{4}$ rewrite them as

$$\frac{2}{3} = \frac{2 \times 4}{3 \times 4} = \frac{8}{12} \qquad \frac{3}{4} = \frac{3 \times 3}{4 \times 3} = \frac{9}{12}$$

Then

$$\frac{2}{3} + \frac{3}{4} = \frac{8}{12} + \frac{9}{12} = \frac{17}{12}$$

How do you know what number to use as the new denominator? In general, you cannot simply guess to find the best denominator, you need a method for finding it. Here is a good method to use.

Step 1. Write each denominator as a product of prime factors.

Example: Add $\dfrac{5}{60} + \dfrac{4}{18}$

$$60 = 2 \times 2 \times 3 \times 5$$
$$= 2^2 \times 3^1 \times 5^1$$
$$18 = 2 \times 3 \times 3 = 2^1 \times 3^2$$

Step 2. Write each prime that appears in either number.

$$2 \qquad 3 \qquad 5$$

Step 3. Attach to each prime the largest exponent that appears on it in either number.

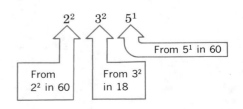

$$2^2 \qquad 3^2 \qquad 5^1$$

From 5^1 in 60

From 2^2 in 60

From 3^2 in 18

Step 4. Multiply to find the new denominator.

$$2^2 \times 3^2 \times 5^1$$
$$= 4 \times 9 \times 5 = 180$$

Now use this new denominator to form equivalent fractions.

$$\frac{5}{60} = \frac{?}{180} \qquad \frac{5 \times \boxed{3}}{60 \times \boxed{3}} = \frac{15}{180} \qquad \frac{4}{18} = \frac{?}{180} \qquad \frac{4 \times \boxed{10}}{18 \times \boxed{10}} = \frac{40}{180}$$

Add $\quad \dfrac{5}{60} + \dfrac{4}{18} = \dfrac{15}{180} + \dfrac{40}{180} = \dfrac{55}{180}$

Reduce to lowest terms $\quad \dfrac{55}{180} = \dfrac{\cancel{5} \times 11}{\cancel{5} \times 36} = \dfrac{11}{36}$

This new denominator is usually called the Least Common Denominator or LCD.

D. PRACTICE PROBLEMS

Find the LCD and add or subtract as shown.

Example: Add $\dfrac{5}{18} + \dfrac{5}{16}$

$$18 = 2 \times 3^2 \qquad 16 = 2^4 \qquad \text{LCD} = 2^4 \times 3^2 = 16 \times 9 = 144$$

$$\frac{5 \times \boxed{8}}{18 \times \boxed{8}} = \frac{40}{144} \qquad \frac{5 \times \boxed{9}}{16 \times \boxed{9}} = \frac{45}{144}$$

$$\frac{5}{18} + \frac{5}{16} = \frac{40}{144} + \frac{45}{144} = \frac{85}{144}$$

1. $\dfrac{2}{3} + \dfrac{1}{6}$ 2. $\dfrac{3}{7} + \dfrac{4}{5}$ 3. $\dfrac{3}{4} + \dfrac{7}{8}$ 4. $\dfrac{5}{10} + \dfrac{3}{8}$

5. $\dfrac{3}{4} + \dfrac{5}{6}$ 6. $\dfrac{7}{12} + \dfrac{3}{16}$ 7. $\dfrac{3}{4} - \dfrac{5}{8}$ 8. $\dfrac{5}{6} - \dfrac{1}{2}$

9. $\dfrac{7}{12} - \dfrac{1}{5}$ 10. $\dfrac{11}{18} - \dfrac{3}{12}$ 11. $\dfrac{7}{10} - \dfrac{1}{6}$ 12. $\dfrac{9}{20} - \dfrac{4}{15}$

The answers are on page 527.

Multiplying Fractions

Multiplication is the simplest arithmetic operation with fractions. The product of two fractions is a fraction whose numerator is the product of their numerators and whose denominator is the product of their denominators. For example,

$$\frac{5}{6} \times \frac{2}{3} = \frac{5 \times 2}{6 \times 3} = \frac{10}{18} \qquad \text{Multiply the numerators}$$
$$\text{Multiply the denominators}$$

Of course, we would reduce this answer to lowest terms.

$$\frac{10}{18} = \frac{\boxed{2} \times 5}{\boxed{2} \times 9} = \frac{5}{9}$$

We could also divide out the common factor 2 before multiplying:

$$\frac{5 \times \overset{1}{\cancel{2}}}{\underset{3}{\cancel{6}} \times 3} = \frac{5}{3 \times 3} = \frac{5}{9}$$

Multiplication of a fraction by an integer is easy if you recognize that for any integer a,

$a = \frac{a}{1}$. For example, $2 = \frac{2}{1}$, $4 = \frac{4}{1}$, and so on.

The product $3 \times \frac{2}{5}$ becomes $\frac{3}{1} \times \frac{2}{5} = \frac{3 \times 2}{1 \times 5} = \frac{6}{5}$.

E. PRACTICE PROBLEMS

Multiply and reduce to lowest terms.

Example: $\frac{3}{4} \times \frac{2}{5} = \frac{3 \times 2}{4 \times 5} = \frac{6}{20} = \frac{2 \times 3}{2 \times 10} = \frac{3}{10}$

1. $\frac{1}{2} \times \frac{1}{4}$ 2. $\frac{2}{3} \times \frac{5}{6}$ 3. $\frac{3}{8} \times \frac{1}{3}$ 4. $\frac{5}{16} \times \frac{8}{3}$

5. $\frac{3}{7} \times \frac{2}{6}$ 6. $\frac{8}{3} \times \frac{5}{12}$ 7. $3 \times \frac{2}{3}$ 8. $4 \times \frac{1}{5}$

9. $6 \times \frac{2}{3}$ 10. $\frac{5}{12} \times \frac{3}{4} \times \frac{8}{15}$ 11. $\left(\frac{1}{4}\right)^2$ 12. $\frac{16}{5} \times \frac{-15}{8}$

The answers are on page 527.

Dividing Fractions

The division $8 \div 4$ is read "8 divided by 4" and written $4\overline{)8}$ or $\frac{8}{4}$.

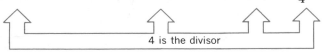

4 is the divisor

To divide by a fraction, invert the fraction and multiply by it. For example,

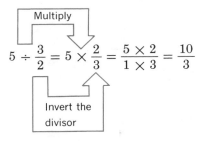

Multiply

$5 \div \frac{3}{2} = 5 \times \frac{2}{3} = \frac{5 \times 2}{1 \times 3} = \frac{10}{3}$

Invert the divisor

Another example: $\frac{3}{4} \div \frac{5}{6} = \frac{3}{4} \times \frac{6}{5} = \frac{3 \times 6}{4 \times 5} = \frac{18}{20}$ or $\frac{9}{10}$ reduced to lowest terms

To invert a fraction means to exchange numerator and denominator.
Inverting $\frac{3}{2}$ gives $\frac{2}{3}$. Inverting $\frac{1}{5}$ gives $\frac{5}{1}$ or 5. Inverting 7 gives $\frac{1}{7}$.
With fractions, every division problem is converted into a multiplication.

F. PRACTICE PROBLEMS

Divide and reduce to lowest terms.

Example: $\dfrac{2}{5} \div \dfrac{8}{3} = \dfrac{2}{5} \times \dfrac{3}{8} = \dfrac{2 \times 3}{5 \times 8} = \dfrac{6}{40} = \dfrac{3}{20}$ reduced to lowest terms

1. $\dfrac{7}{40} \div \dfrac{21}{25}$ 2. $3 \div \dfrac{2}{3}$ 3. $\dfrac{2}{3} \div 3$ 4. $4 \div \dfrac{1}{2}$

5. $\dfrac{1}{2} \div 4$ 6. $\dfrac{15}{4} \div \dfrac{5}{2}$ 7. $\dfrac{5}{6} \div \dfrac{1}{2}$ 8. $\dfrac{7}{20} \div \dfrac{4}{5}$

9. $\dfrac{1}{2} \div \dfrac{1}{3}$ 10. $\dfrac{3}{4} \div \dfrac{5}{16}$ 11. $\dfrac{2}{9} \div \dfrac{8}{3}$ 12. $\dfrac{3}{4} \div \dfrac{7}{8}$

The answers are on page 527.

Negative Fractions

Because the numerator and denominator of a fraction may be negative numbers, we often encounter fractions such as

$$-\dfrac{2}{3}, \ \dfrac{2}{-3}, \ \text{or} \ \dfrac{-2}{3}.$$

It is important to realize that these are all equivalent fractions.

$$-\dfrac{2}{3} = \dfrac{(-1)(2)}{3} = \dfrac{-2}{3}$$

$$\dfrac{2}{-3} = \dfrac{(-1)(2)}{(-1)(-3)} = \dfrac{-2}{3} \qquad \text{where we multiplied top and bottom of the fraction by } (-1).$$

In algebra, you will find it most helpful to write any negative fraction in the form $\dfrac{-a}{b}$.

Write $\quad -\dfrac{a}{b} = \dfrac{(-1)a}{b} \quad$ and $\quad \dfrac{a}{-b} = \dfrac{(-1)a}{b}$.

For example,

$$-\dfrac{-2}{5} = (-1)\dfrac{(-1)(2)}{5} = \dfrac{(-1)(-1)(2)}{5} = \dfrac{2}{5}$$

$$\dfrac{-3}{-4} = \dfrac{(-1)(3)}{(-1)(4)} = \dfrac{(-1)}{(-1)} \cdot \dfrac{3}{4} = \dfrac{3}{4}$$

$$-\dfrac{3}{-7} = (-1)\dfrac{(3)}{(-1)(7)} = \dfrac{(-1)}{(-1)} \cdot \dfrac{3}{7} = \dfrac{3}{7}$$

G. PRACTICE PROBLEMS

Rewrite each fraction in the form shown above.

Example: $\quad -\dfrac{3}{8} = \dfrac{-3}{8} \qquad -\dfrac{-4}{5} = \dfrac{4}{5}$

1. $-\dfrac{2}{5}$ 2. $\dfrac{-3}{-8}$ 3. $\dfrac{1}{-5}$ 4. $\dfrac{3}{-16}$

5. $-\dfrac{-5}{11}$ 6. $-\dfrac{4}{-7}$ 7. $-\dfrac{7}{16}$ 8. $\dfrac{-11}{-32}$

9. $-\dfrac{2}{-15}$ 10. $-\dfrac{(-3)}{(-11)}$ 11. $-\dfrac{4}{-9}$ 12. $\dfrac{3}{-8}$

The answers are on page 527.

Answers to practice problems in Fractions Review.

A. 1. $\dfrac{35}{42}$ 2. $\dfrac{21}{48}$ 3. $\dfrac{24}{56}$ 4. $\dfrac{20}{12}$ 5. $\dfrac{14}{16}$ 6. $\dfrac{9}{12}$

   7. $\dfrac{39}{78}$ 8. $\dfrac{7}{63}$ 9. $\dfrac{35}{45}$ 10. $\dfrac{66}{72}$ 11. $\dfrac{35}{50}$ 12. $\dfrac{30}{35}$

B. 1. $\dfrac{5}{28}$ 2. $\dfrac{3}{5}$ 3. $\dfrac{1}{3}$ 4. $\dfrac{13}{15}$ 5. $\dfrac{4}{5}$ 6. $\dfrac{4}{5}$

   7. $\dfrac{1}{2}$ 8. $\dfrac{2}{5}$ 9. $\dfrac{1}{6}$ 10. $\dfrac{7}{20}$ 11. $\dfrac{3}{8}$ 12. $\dfrac{3}{8}$

C. 1. $\dfrac{5}{8}$ 2. $\dfrac{4}{3}$ 3. 3 4. $\dfrac{10}{7}$ 5. 1 6. $\dfrac{5}{8}$

   7. $\dfrac{7}{16}$ 8. $\dfrac{2}{3}$ 9. $\dfrac{7}{12}$ 10. $\dfrac{3}{8}$ 11. $\dfrac{3}{10}$ 12. $\dfrac{3}{20}$

D. 1. $\dfrac{5}{6}$ 2. $\dfrac{43}{35}$ 3. $\dfrac{13}{8}$ 4. $\dfrac{7}{8}$ 5. $\dfrac{19}{12}$ 6. $\dfrac{37}{48}$

   7. $\dfrac{1}{8}$ 8. $\dfrac{1}{3}$ 9. $\dfrac{23}{60}$ 10. $\dfrac{13}{36}$ 11. $\dfrac{8}{15}$ 12. $\dfrac{11}{60}$

E. 1. $\dfrac{1}{8}$ 2. $\dfrac{5}{9}$ 3. $\dfrac{1}{8}$ 4. $\dfrac{5}{6}$ 5. $\dfrac{1}{7}$ 6. $\dfrac{10}{9}$

   7. 2 8. $\dfrac{4}{5}$ 9. 4 10. $\dfrac{1}{6}$ 11. $\dfrac{1}{16}$ 12. -6

F. 1. $\dfrac{5}{24}$ 2. $\dfrac{9}{2}$ 3. $\dfrac{2}{9}$ 4. 8 5. $\dfrac{1}{8}$ 6. $\dfrac{3}{2}$

   7. $\dfrac{5}{3}$ 8. $\dfrac{7}{16}$ 9. $\dfrac{3}{2}$ 10. $\dfrac{12}{5}$ 11. $\dfrac{1}{12}$ 12. $\dfrac{6}{7}$

G. 1. $\dfrac{-2}{5}$ 2. $\dfrac{3}{8}$ 3. $\dfrac{-1}{5}$ 4. $\dfrac{-3}{16}$ 5. $\dfrac{5}{11}$ 6. $\dfrac{4}{7}$

   7. $\dfrac{-7}{16}$ 8. $\dfrac{11}{32}$ 9. $\dfrac{2}{15}$ 10. $\dfrac{-3}{11}$ 11. $\dfrac{4}{9}$ 12. $\dfrac{-3}{8}$

ARITHMETIC REVIEW: SQUARE ROOTS

The square of any number is the product of that number with itself.

$2^2 = 2 \cdot 2 = 4$
$3^2 = 3 \cdot 3 = 9$
$4^2 = 4 \cdot 4 = 16$
and so on.

The reverse of this process is useful in algebra and we have devised special symbols for it.

$$\sqrt{4} = 2 \quad \text{since} \quad 2 \cdot 2 = 4$$
$$\sqrt{9} = 3 \quad \text{since} \quad 3 \cdot 3 = 9$$
$$\sqrt{16} = 4 \quad \text{since} \quad 4 \cdot 4 = 16 \quad \text{and so on.}$$

Radical Sign

The symbol $\sqrt{}$ is called a *radical sign* and is read "the square root of."

Notice that $2 \cdot 2 = 4$ and $(-2)(-2) = 4$. Every positive number has two square roots, one positive and one negative. When we speak of the square root of a number or use the symbol $\sqrt{}$ we are referring to the positive square root of that number.

$$\sqrt{16} = 4$$

and the negative square root of 16 is

$$-\sqrt{16} = -4$$

We can find the square root of any real number but, in general, the square root will not be a whole number, fraction, or an exact decimal. For example,

$$\sqrt{2} = 1.414213562\ldots$$
$$\sqrt{3} = 1.732050808\ldots$$
$$\sqrt{5} = 2.236067977\ldots$$

where the three dots indicate that the decimal number continues on forever with no pattern of digits. Any number of this sort that cannot be written as a finite decimal or expressed as the quotient of two integers in a fraction is called an *irrational* number.

Only positive numbers may appear under the radical sign—at least in basic algebra. The square root of a negative number is *not* a real number. For example, the quantity $\sqrt{-2}$ is not a real number because there is no real number x such that $x \cdot x = -2$. Numbers like $\sqrt{-2}$ are very useful in more advanced mathematics where they appear in the solution of certain equations. Such numbers, called *imaginary* numbers, will not be studied in this book.

A. PRACTICE PROBLEMS

Find these square roots using the table of square roots on page 533. Round each answer to three decimal places.

Example: $\sqrt{7} = 2.646$ rounded to three decimal places.

1. $\sqrt{4}$ 2. $\sqrt{9}$ 3. $\sqrt{64}$ 4. $\sqrt{16}$ 5. $\sqrt{49}$ 6. $\sqrt{36}$

7. $\sqrt{25}$ 8. $\sqrt{1}$ 9. $\sqrt{8}$ 10. $\sqrt{15}$ 11. $\sqrt{6}$ 12. $\sqrt{10}$

13. $\sqrt{50}$ 14. $\sqrt{65}$ 15. $\sqrt{185}$ 16. $\sqrt{125}$ 17. $\sqrt{12}$ 18. $\sqrt{110}$

The answers to all practice problems in this Arithmetic Review are on page 530.

Adding and Subtracting Radicals

We can perform all of the usual arithmetic operations with the irrational numbers resulting from the square root procedure. This ability to do arithmetic with radical numbers is necessary in algebra.

The expression $4 + \sqrt{2}$ represents the sum of the numbers 4 and $\sqrt{2}$ or $4 + 1.4142136\ldots$ or roughly 5.414, rounded to three decimal places.

$4 - \sqrt{2} \cong 4 - 1.4142136\ldots \cong 2.586$ rounded.

$3\sqrt{2}$ represents the product of the numbers 3 and $\sqrt{2}$ or $3(1.4142136\ldots)$ or approximately 4.243, rounded to three decimal places.

$-2\sqrt{5} = (-2)(\sqrt{5}) \cong (-2)(2.236068\ldots) \cong -4.472$ rounded.

B. PRACTICE PROBLEMS

Evaluate these radical expressions using the table of square roots and round to three decimal places.

Example: $3 + \sqrt{3} \cong 3 + 1.7321\ldots \cong 4.732$

1. $1 + \sqrt{3}$	2. $3 + \sqrt{5}$	3. $5 - \sqrt{10}$	4. $\sqrt{7} - 2$
5. $1 - \sqrt{5}$	6. $2 - \sqrt{8}$	7. $2\sqrt{3}$	8. $3\sqrt{3}$
9. $-\sqrt{5}$	10. $-4\sqrt{5}$	11. $\frac{1}{2}\sqrt{10}$	12. $\frac{2}{3}\sqrt{6}$
13. $3 + 2\sqrt{5}$	14. $-2 + 3\sqrt{2}$	15. $-1 - 3\sqrt{6}$	16. $5 - 2\sqrt{7}$
17. $\frac{1 + \sqrt{2}}{3}$	18. $\frac{1 + \sqrt{5}}{2}$	19. $\frac{3 - \sqrt{2}}{3}$	20. $\frac{1 - \sqrt{2}}{2}$
21. $\sqrt{3} - \sqrt{2}$	22. $\sqrt{3} - \sqrt{6}$	23. $2\sqrt{3} + 3\sqrt{2}$	24. $2\sqrt{5} - 3\sqrt{6}$

The answers to all practice problems in this Arithmetic Review are on page 530.

Factoring Radical
Numbers

Radical Expression

A number written in the form $\sqrt{6}$, $\sqrt{x}$, or $\sqrt{3ay^3}$ is called a *radical expression* and the number appearing under the radical sign is usually called the *radicand*. The radicand in $\sqrt{6}$ is 6; the radicand in $\sqrt{x}$ is x; the radicand in $\sqrt{3ay^3}$ is $3ay^3$.

If the radicand can be factored it may be possible to simplify the radical expression.

$$\begin{aligned} \sqrt{8} &= \sqrt{2 \cdot 2 \cdot 2} \\ &= \sqrt{2 \cdot 2} \cdot \sqrt{2} \\ &= 2\sqrt{2} \cong 2(1.4142) \cong 2.8284 \text{ rounded} \end{aligned}$$

or

$$\begin{aligned} \sqrt{18} &= \sqrt{2 \cdot 3 \cdot 3} \\ &= \sqrt{3 \cdot 3} \cdot \sqrt{2} \\ &= 3\sqrt{2} \cong 3(1.4142) \cong 4.2436 \text{ rounded} \end{aligned}$$

This simplification is possible because the square root of a product is equal to the product of the square roots of its factors.

$$\begin{aligned} \sqrt{252} &= \sqrt{4} \cdot \sqrt{9} \cdot \sqrt{7} \\ &= 2 \cdot 3 \cdot \sqrt{7} = 6\sqrt{7} \cong 6(2.6458) \cong 15.8748 \end{aligned}$$

This kind of factoring allows us to simplify many problems involving square roots and to find the value of some square roots beyond the range of the table.

Simplify by factoring.

Example: $\sqrt{40} = \sqrt{4} \cdot \sqrt{10} = 2\sqrt{10}$

1. $\sqrt{8}$	2. $\sqrt{12}$	3. $\sqrt{60}$	4. $\sqrt{80}$
5. $\sqrt{284}$	6. $\sqrt{549}$	7. $\sqrt{612}$	8. $\sqrt{175}$
9. $\sqrt{51300}$	10. $\sqrt{4949}$	11. $-\sqrt{704}$	12. $-\sqrt{1053}$

The answers to all practice problems in the Arithmetic Review are on page 530.

Answers to practice problems in Square Roots.

A.
1. 2	2. 3	3. 8	4. 4
5. 7	6. 6	7. 5	8. 1
9. 2.828	10. 3.873	11. 2.449	12. 3.162
13. 7.071	14. 8.062	15. 13.601	16. 11.180
17. 3.464	18. 10.488		

B.
1. 2.732	2. 5.236	3. 1.838	4. 0.646
5. −1.236	6. −0.828	7. 3.464	8. 5.196
9. −2.236	10. −8.944	11. 1.581	12. 1.633
13. 7.472	14. 2.243	15. −8.348	16. −0.292
17. 0.805	18. 1.618	19. 0.529	20. −0.207
21. 0.318	22. −0.717	23. 7.707	24. −2.876

C.
1. $2\sqrt{2}$	2. $2\sqrt{3}$	3. $2\sqrt{15}$	4. $4\sqrt{5}$
5. $2\sqrt{71}$	6. $3\sqrt{61}$	7. $6\sqrt{17}$	8. $5\sqrt{7}$
9. $30\sqrt{57}$	10. $7\sqrt{101}$	11. $-8\sqrt{11}$	12. $-9\sqrt{13}$

ARITHMETIC REVIEW:
FACTORING

It is very useful in arithmetic and in algebra to be able to write any whole number as a product of other numbers. For example, we can write

$6 = 2 \times 3$

Factors

The integers 2 and 3 are called the *factors* of 6.

We could also write

$6 = 1 \times 6$

The whole number factors of 6 are 1, 2, 3, and 6.

To *factor* a number means to write that number as a product of its positive integer factors. For example, the integer 12 may be written as

$12 = 1 \times 12$
$12 = 2 \times 6$

or

$12 = 3 \times 4$

so that the factors of 12 are 1, 2, 3, 4, 6, and 12.

Notice that zero is not a factor. There is no number x such that $0 \times x = 12$. The product of zero and any number is always zero. Zero is never written as a factor for any number.

Evenly Divisible

Any number is said to be *evenly divisible* by its factors, that is, every factor divides the number with zero remainder. For example,

18 is evenly divisible by the factor 3 since 3 divides 18 with zero remainder.

For some whole numbers the only factors are 1 and the number itself. For example, the only factors of 7 are 1 and 7.

$7 = 1 \times 7$

There are no other whole numbers that divide 7 with remainder zero. Numbers whose only factors are 1 and the number itself are known as *prime numbers*. Here is a list of the prime numbers less than 100:

2	3	5	7	11
13	17	19	23	29
31	37	41	43	47
53	59	61	67	71
73	79	83	89	97

Notice that 1 is not listed as a prime. All prime numbers have two distinct, unequal factors: 1 and the number itself. The number 1 has only one factor—itself. The number 1 is not a prime.

Prime Factors

The *prime factors* of any number are those factors that are prime numbers. The factors of 6 are 1, 2, 3, and 6. The prime factors of 6 are 2 and 3. The factors of 42 are 1, 2, 3, 6, 7, 14, 21, and 42. The prime factors of 42 are 2, 3, and 7. The other factors of 42 are not prime numbers.

It is particularly useful to be able to write any whole number as a product of its prime factors. For example, the prime factors of 30 are 2, 3, and 5 since

$30 = 2 \times 3 \times 5$.   30 is written here as a product of primes. Of course the order in which the prime factors are written is not important, we could write

$30 = 2 \times 5 \times 3$ or $30 = 3 \times 2 \times 5$ or any other arrangement. The important fact is that 2, 3, and 5 are the *only* primes whose product is 30.

Here is another example. The prime factors of 12 are 2 and 3; therefore 12 may be written as a product of primes like this:

$12 = 2 \times 2 \times 3$ or $12 = 2^2 \cdot 3$ using exponential notation.

Similarly,

$16 = 2 \times 2 \times 2 \times 2 = 2^4$ written as a product of primes
$9 = 3 \times 3 = 3^2$ written as a product of primes
$288 = 2 \times 2 \times 2 \times 2 \times 2 \times 3 \times 3 = 2^5 \cdot 3^2$ written as a product of primes

and

$7 = 7$ written as a product of primes

The ability to write any whole number as a product of primes is useful in arithmetic when working with fractions and it is very useful in algebra.

PRACTICE PROBLEMS

Write the following whole numbers as a product of primes.

Example: $28 = 2^2 \cdot 7$

1. 14	2. 20	3. 9	4. 170	5. 39
6. 32	7. 256	8. 81	9. 36	10. 56
11. 31	12. 252	13. 50	14. 136	15. 390
16. 24	17. 100	18. 40	19. 5	20. 288
21. 1369	22. 102	23. 96	24. 84	25. 1000

The answers are below.

Answers to practice problems in Factoring.

1. $2 \cdot 7$	2. $2^2 \cdot 5$	3. 3^2	4. $2 \cdot 5 \cdot 17$
5. $3 \cdot 13$	6. 2^5	7. 2^8	8. 3^4
9. $2^2 \cdot 3^2$	10. $2^3 \cdot 7$	11. 31	12. $2^2 \cdot 3^2 \cdot 7$
13. $2 \cdot 5^2$	14. $2^3 \cdot 17$	15. $2 \cdot 3 \cdot 5 \cdot 13$	16. $2^3 \cdot 3$
17. $2^2 \cdot 5^2$	18. $2^3 \cdot 5$	19. 5	20. $2^5 \cdot 3^2$
21. 37^2	22. $2 \cdot 3 \cdot 17$	23. $2^5 \cdot 3$	24. $2^2 \cdot 3 \cdot 7$
25. $2^3 \cdot 5^3$			

Table of Square Roots

Number	Square root	Number	Square root	Number	Square root	Number	Square root
1	1.0000	51	7.1414	101	10.0499	151	12.2882
2	1.4142	52	7.2111	102	10.0995	152	12.3288
3	1.7321	53	7.2801	103	10.1489	153	12.3693
4	2.0000	54	7.3485	104	10.1980	154	12.4097
5	2.2361	55	7.4162	105	10.2470	155	12.4499
6	2.4495	56	7.4833	106	10.2956	156	12.4900
7	2.6458	57	7.5498	107	10.3441	157	12.5300
8	2.8284	58	7.6158	108	10.3923	158	12.5698
9	3.0000	59	7.6811	109	10.4403	159	12.6095
10	3.1623	60	7.7460	110	10.4881	160	12.6491
11	3.3166	61	7.8102	111	10.5357	161	12.6886
12	3.4641	62	7.8740	112	10.5830	162	12.7279
13	3.6056	63	7.9373	113	10.6301	163	12.7671
14	3.7417	64	8.0000	114	10.6771	164	12.8062
15	3.8730	65	8.0623	115	10.7238	165	12.8452
16	4.0000	66	8.1240	116	10.7703	166	12.8841
17	4.1231	67	8.1854	117	10.8167	167	12.9228
18	4.2426	68	8.2462	118	10.8628	168	12.9615
19	4.3589	69	8.3066	119	10.9087	169	13.0000
20	4.4721	70	8.3666	120	10.9545	170	13.0384
21	4.5826	71	8.4261	121	11.0000	171	13.0767
22	4.6904	72	8.4853	122	11.0454	172	13.1149
23	4.7958	73	8.5440	123	11.0905	173	13.1529
24	4.8990	74	8.6023	124	11.1355	174	13.1909
25	5.0000	75	8.6603	125	11.1803	175	13.2288
26	5.0990	76	8.7178	126	11.2250	176	13.2665
27	5.1962	77	8.7750	127	11.2694	177	13.3041
28	5.2915	78	8.8318	128	11.3137	178	13.3417
29	5.3852	79	8.8882	129	11.3578	179	13.3791
30	5.4772	80	8.9443	130	11.4018	180	13.4164
31	5.5678	81	9.0000	131	11.4455	181	13.4536
32	5.6569	82	9.0554	132	11.4891	182	13.4907
33	5.7446	83	9.1104	133	11.5326	183	13.5277
34	5.8310	84	9.1652	134	11.5758	184	13.5647
35	5.9161	85	9.2195	135	11.6190	185	13.6015
36	6.0000	86	9.2736	136	11.6619	186	13.6382
37	6.0828	87	9.3274	137	11.7047	187	13.6748
38	6.1644	88	9.3808	138	11.7473	188	13.7113
39	6.2450	89	9.4340	139	11.7898	189	13.7477
40	6.3246	90	9.4868	140	11.8322	190	13.7840
41	6.4031	91	9.5394	141	11.8743	191	13.8203
42	6.4807	92	9.5917	142	11.9164	192	13.8564
43	6.5574	93	9.6437	143	11.9583	193	13.8924
44	6.6332	94	9.6954	144	12.0000	194	13.9284
45	6.7082	95	9.7468	145	12.0416	195	13.9642
46	6.7823	96	9.7980	146	12.0830	196	14.0000
47	6.8557	97	9.8489	147	12.1244	197	14.0357
48	6.9282	98	9.8995	148	12.1655	198	14.0712
49	7.0000	99	0.9499	149	12.2066	199	14.1067
50	7.0711	100	10.0000	150	12.2474	200	14.1421

Answers

UNIT 1 *Problem Set 1-1, Page 13*

A. 1. Expression, term, variable
    2. Expression, terms
    3. Factors
    4. Conditional equation, left member, expression
    5. Factor
    6. Inequality, variable
    7. Conditional equation, left member, right member, variable
    8. Expression, term
    9. Conditional equation, first term
    10. Right member, equation

B. 1. $\dfrac{x}{y}$ 2. $A - 3$ 3. $6xy$ 4. $a + b^2$

5. $3 + y$ 6. $x^2 y^2$ 7. $2(2a + b)$ 8. $\dfrac{a + 1}{b}$

9. $x - y$ 10. $3qst$ 11. $(a + b)(a - b)$ 12. $A + B + C$

13. $3a$ 14. $\dfrac{a}{2}$ 15. $1 - x$ 16. $t - q + 1$

17. $2 + y - t$ 18. $4(x - 2y)$ 19. $\dfrac{3p + t}{2x}$ 20. $\dfrac{x - 3}{2}$

21. $5A(A + 2)$ 22. $x + 2y - 3z$ 23. $(x - y + 1)(3x - 1)$
24. $(1 + 2a)(1 - 2b)$

C. 1. $<$ 2. $>$ 3. $<$ 4. $=$ 5. $<$
    6. $=$ 7. $<$ 8. $<$ 9. $>$ 10. $>$
    11. $=$ 12. $=$ 13. $=$ 14. $<$ 15. $>$
    16. $<$ 17. $=$ 18. $<$ 19. $>$ 20. $>$

UNIT 1 *Problem Set 1-2, Page 25*

A. 1. -3 2. -14 3. -19 4. 6
    5. -32 6. -6 7. 6 8. 40
    9. -14 10. 124 11. -81 12. -42
    13. -302 14. 0 15. -156 16. 148
    17. -2.4 18. -2.3 19. 1.1 20. -3.05

21. −13.92   22. 5.9   23. 29.96   24. −28.0
25. $-\frac{1}{2}$ 26. 2   27. $-6\frac{1}{4}$ 28. $-1\frac{2}{5}$
29. $-4\frac{1}{12}$ 30. $6\frac{11}{15}$ 31. $-13\frac{1}{2}$ 32. $-\frac{1}{12}$

B. 1. 12   2. −56   3. −45   4. 9
    5. −42   6. 81   7. 0   8. −1
    9. −28   10. −36   11. 42   12. −60
    13. −3   14. 7   15. −9   16. 4
    17. −8   18. −9   19. 9   20. −17
    21. 0.08   22. −0.72   23. −9   24. 4
    25. −4   26. 8   27. −0.4   28. −0.037
    29. $\frac{1}{3}$ 30. −6   31. −8   32. 1

C. 1. >   2. <   3. >   4. >   5. >   6. <   7. <
    8. >   9. <   10. <   11. >   12. >   13. <   14. <
    15. <   16. >   17. <   18. <   19. <   20. <   21. <

D. 1. −90   2. 1   3. $-\frac{1}{3}$ 4. 68
    5. 0   6. 5   7. −42   8. 132
    9. −5   10. 5.36   11. −1.37   12. −24.45
    13. 0.2702   14. 2   15. −8.9   16. −28.357
    17. 0.0001   18. −4012   19. $-7\frac{11}{12}$ 20. $\frac{1}{9}$
    21. $\frac{1}{120}$ 22. −1   23. −1   24. 0
    25. $\frac{1}{2}$ 26. −40

E. 1. 9th floor   2. −$10.94   3. A loss of 16 yards
    4. −2°F   5. 18,000 ft   6. $5\frac{3}{8}$

F. 1. (a) −1   (b) 5   (c) −5   (d) 1   (e) 6
    2. A decimal point
    3. (a)

    4. $4\frac{1}{4}$ in.

UNIT 1   *Problem Set 1-3, Page 39*

A. 1. 10^3 2. 5^5 3. 7^2
    4. c^7 5. $10a^3$ 6. $3^2x^2y^2$
    7. n^2p^3 8. e^2fg^3 9. $(x + y)^2$
    10. $(3 - b)^4$ 11. $6^2x^2(x - 2)^2$ 12. $7^3t^3(t - r)^2$
    13. $(d - c)^5$ 14. $(x + y)^2(x - y)^2$ 15. $10^4g^2(g - 2)^2$

B. 1. 32   2. 49   3. 1   4. 1,000,000
    5. 1   6. 243   7. 216   8. 8
    9. 10,000   10. 1024   11. 9   12. −1
    13. −125   14. 64   15. −1   16. 1
    17. $\frac{1}{27}$ 18. $\frac{1}{36}$ 19. $\frac{27}{64}$ 20. $\frac{16}{625}$
    21. $\frac{9}{4}$ 22. $\frac{25}{64}$ 23. $\frac{25}{4}$ 24. $\frac{27}{1000}$
    25. 0.81   26. 0.001   27. 11.56   28. 0.16
    29. −0.000027   30. −3.375   31. 0.0001   32. 4.2436

C. 1. $16x^4$ 2. $-125c^3$ 3. $49m^2n^2$ 4. $-27a^3$
    5. $64y^6$ 6. $\frac{d^2}{9}$ 7. $0.001y^3t^3$ 8. $100{,}000x^5y^5$

9. $-1.728r^3s^3$ 10. $\frac{9}{16}a^4c^4$ 11. $0.04n^4$ 12. $\frac{b^4}{81}$

13. y^7 14. t^{14} 15. e^{19} 16. $-x^5$

17. $2a^6$ 18. $\frac{c^{15}}{2}$ 19. $3n^6$ 20. $-2h^{10}$

21. b^6c 22. $3w^5x^4$ 23. y^{3+n} 24. e^2
25. t^7 26. m^8 27. k^2 28. $5y^4$
29. 3 30. $3c$ 31. $2a^2b$ 32. $-2s^2$

33. $81x^8y^8z^4$ 34. $\frac{p^4q^{12}t^8}{16}$ 35. $\frac{4a^6b^2c^{10}}{9}$ 36. $\frac{x^6}{y^9}$

37. $\frac{8}{x^3}$ 38. $\frac{9a^2b^4}{c^6}$ 39. x^{43} 40. a^{25}

41. p^{52} 42. x^{63} 43. x^{121} 44. a^{147}

45. $2^{10}x^{50}$ 46. 3^6a^{42} 47. $\frac{x^{176}}{y^{72}}$ 48. t^{23}

49. $3x^{34}$ 50. y^{38}

D. 1. $100,000,000$ 2. -27 3. $\frac{16}{81}$ 4. 0.125 5. 1.44
    6. -8 7. $\frac{1}{8}$ 8. $\frac{27}{64}$ 9. 15.625 10. $\frac{1}{4}$
   11. $\frac{1}{32}$ 12. $\frac{1000}{27}$ 13. 16 14. -16 15. 16

   16. $-\frac{1}{16}$ 17. $\frac{1}{16}$ 18. 1 19. $\frac{1}{x^4}$ 20. $\frac{1}{y^5}$

   21. $\frac{1}{a^8}$ 22. $\frac{1}{b^{18}}$ 23. $\frac{1}{x^{18}}$ 24. $\frac{1}{t^4}$ 25. $\frac{1}{x^6y^{10}}$

   26. $\frac{r^{12}}{t^6}$ 27. $\frac{1}{8x^3y^9}$ 28. $\frac{p^8}{q^6}$ 29. $\frac{y^{12}}{x^{20}}$ 30. $\frac{1}{9n^2p^8}$

   31. x^{-3} 32. a^4x^{-5} 33. $d^{-2}e^5$ 34. $2^{-2}x^2$ 35. $3^{-2}x^{-5}y^7$

   36. $5^{-1}x^{-1}y^3$ 37. $x^{-4}y^6$ 38. $2^{-1}x^{-8}y^3$ 39. $3^{-2}x^8y^{-5}$

E. 1. $0.0081y^4$ 2. $-\frac{b^5}{32}$ 3. $1,000,000c^6q^6$ 4. $\frac{p^2}{49}$

    5. f^{22} 6. $8y^5$ 7. x^4y^5 8. t^6m^7
    9. $2a^5$ 10. k^{15} 11. p^2 12. $4xy^2$
   13. $7b^3$ 14. $0.001728n^7$ 15. 6 16. $81q^2$
   17. $7x^2$ 18. $7.5h^4$ 19. $\frac{1}{2}a^{11}$ 20. $\frac{1}{2}$

   21. $a^{-4}b^{-6}$ or $\frac{1}{a^4b^6}$ 22. $\frac{1}{4x^4}$ 23. $\frac{1}{125x^3y^6}$ 24. $\frac{1}{64p^3q^9t^3}$

   25. x^3 26. 2 27. $\frac{3}{t^2}$ 28. $\frac{1}{y^2}$

   29. 6 30. $8p^2$

F. 1. c 2. $\frac{20}{3^8}$ 3. They are all true.
    4. The next line is $36^2 + 37^2 + 38^2 + 39^2 + 40^2 = 41^2 + 42^2 + 43^2 + 44^2$

UNIT 1   *Problem Set 1-4, Page 55*

A. 1. $8c$ 2. $17b$ 3. $18x$ 4. $-8r$
    5. $14n$ 6. $-9t$ 7. $-4h^2$ 8. $-12a^2b$
    9. $y - 5y^2$ 10. $6c^3d$ 11. $-4abc$ 12. $-15z$

13. $22ab$ 14. $-10cd$ 15. $28x^2$ 16. $-10vt$

17. $-11c^3de$ 18. $4j^2k + 7jk + 2jk^2$ 19. $9y$ 20. $4b^2c$

B. 1. $7b$ 2. $20a$ 3. $2p$ 4. $-27r$
5. 0 6. $25x$ 7. $-12a$ 8. $6d$
9. 0 10. 0 11. $-8x^2$ 12. $-0.3cd$
13. y^2z^2 14. $-6a^2b^2$ 15. $-0.6x^2y^2$ 16. $-4(a + b)$
17. $8d$ 18. $-11a^2$ 19. $-9rs^2$ 20. $-2x^2y$
21. $-14xy$

C. 1. $20a$ 2. $24c^2$ 3. m^5 4. $3y^3$ 5. $16rs$
6. $15z^4$ 7. $-a^2b^2c$ 8. $-18x^4y^8$ 9. $49d^2$ 10. $-7s^3t^3$
11. $-20d^4e$ 12. $-18z^6$ 13. $-8y^4$ 14. $-64b^3$ 15. $8r^5s$
16. $144c^6$ 17. $80n^2$ 18. $-8a^6$ 19. $-60x^2y^3z^4$ 20. $18b^3c^2$

D. 1. $8c$ 2. $-8d$ 3. $-5y$ 4. $-5a$ 5. $4p$

6. -7 7. $-8x^4$ 8. $-10t^2$ 9. $\dfrac{11}{m}$ 10. $-7r^3s$

11. $\dfrac{3b^2}{c^2}$ 12. $10d$ 13. $\dfrac{4p}{r^3}$ 14. $-a$ 15. $\dfrac{-5}{mn^2p^3}$

16. $\dfrac{-d^3z}{9}$ 17. 5 18. $-7x^5yz$ 19. $19a^4b^6$ 20. $125y$

21. $\dfrac{-12wx^5}{y^2}$

E. 1. $10c + 16d$ 2. $3x - 11y$ 3. $13ab - 6bc$
4. $7m - m^2$ 5. 8 6. $10a$

7. $27y$ 8. $\dfrac{2}{a^2c^2}$ 9. $-21d^2$

10. $3a^2y$ 11. $19 + 2b + 2b^2$ 12. 4
13. $3f - 2$ 14. $9c$ 15. $-2.7x^2y^3$
16. b^2z^6 17. $-6c^3d^4$ 18. $10 + 4b - 17ab$
19. $-4.22am$ 20. $-22y^2$ 21. $-8d^2$

22. $3q^4r^3s^2$ 23. $(9\frac{1}{3})x^2y^5$ or $\dfrac{28x^2y^5}{3}$ 24. $1.3b^2$

25. $2m^2 + n^2 - m^2n$ 26. $-7z^3$ 27. $18gh + 9g^2h + 12gh^2$
28. $2x^2$ 29. $7b^2$ 30. $12ef$

F. 1. $5a - 10b$ 2. $14x - 7z$ 3. $3y - x$
4. $t - 4a$ 5. $-2p - 8q$ 6. $-12x - 6y^2$
7. $16 - 4x$ 8. $5 - 5x^2$ 9. $2 - 4x$
10. $2 - 5p + 2q$ 11. $3 + 2m - m^2$ 12. $8 - 3x + 4y$
13. $2x + 4$ 14. $15m - 12$ 15. $q^2 - 3p - 2$
16. $6t - 1 + t^3$ 17. $x^2 + 2x$ 18. $a^2 - 2a$
19. $9y - 2y^2$ 20. $9x - 3x^2$ 21. $12x - 9y$
22. $12b - 8a$ 23. -4 24. $2 - 5b$
25. $1 - x$ 26. $4 - y$ 27. $10 - 8x$
28. $2 + 3p$ 29. $3 - a + a^2 - a^3$ 30. $x^3 - x^2 - 2x - 4$

G. 1. $20x + 8y$
2. There are at least three possible answers:
(a) 1 (The "I" of the rhyme.)
(b) 2 (The "I" of the rhyme and the man he met.)

3. $\dfrac{a^3c^6}{11b^4}$

UNIT 1   *Problem Set 1-5, Page 65*

A. 1. $+$ 2. $+$ 3. -4 4. x^3 5. $\times$ 6. $\times$
    7. $a-2$ 8. $=$ 9. $+$ 10. $+3$ 11. $=$ 12. $-$
    13. $\div$ 14. $\times$ 15. $\frac{2}{3}\times$ 16. $<$ 17. $\times 2$ 18. $=$
    19. $>$ 20. $+5$ 21. $+4$ 22. $b-a$ 23. a^2 24. -2
    25. $=$ 26. $=$ 27. $\div$ 28. $=$ 29. -5 30. $=$

B. 1. $a \div b$ or $\dfrac{a}{b}$ 2. $x+y$ 3. $p+4$

    4. $4t$ 5. $y \div 4$ or $\dfrac{y}{4}$ 6. $d-e$

    7. rm 8. $K+10$ 9. $z-5$

    10. $9B$ 11. $h-g$ or $g-h$ 12. $M-2$

    13. $u+5$ or $5+u$ 14. $x+\frac{1}{2}y$ or $x+\dfrac{y}{2}$ 15. $3a \div 5b$
                                                                           or $5b \div 3a$

    16. $2y \div 2$ or $\dfrac{2y}{2}$ 17. $3n$ 18. $8-g$

    19. n^3 20. W^2 21. $4n^2-7$

    22. a^2b^2 23. a^2+b^2 24. $\dfrac{qt}{rs}$

    25. $\dfrac{qt}{q+t}$ 26. M^2-PQ 27. $5AB$

    28. $4(D+E^2)$ 29. $\dfrac{2x^2t^2}{x+t}$ 30. $\dfrac{x+1}{(x-1)^2}$

    31. mv 32. $\frac{1}{2}gt^2$ 33. $4x-10$
    34. $L-6S$ 35. $2D+3$ 36. $G+W$
    37. $S-D+15$ 38. $\frac{1}{2}mv^2$ 39. $2(x-y)$
    40. xy^2

C. 1. $3x+4=6$ 2. $a+b=31$ 3. $s=c-5$

    4. $AB=10C$ 5. $5N-7=88$ 6. $S-J=2D$
                                                                 or $J-S=2D$

    7. $4x-3=2y$ 8. $J-4=2(N+3)$ 9. $6x+x^2=27$
    10. $A=\pi R^2$ 11. $V=\frac{1}{4}\pi HD^2$ 12. $B=3+P+N$
    13. $x+10=16$ 14. $P=2L+2W$ 15. $V=E^3$
    16. $x+y-4=2xy$ 17. $F=32+1.8C$ 18. $AB^2=6$
    19. $\frac{1}{2}(A+B)=7$ 20. $x^2+y^2=4+x+y$ 21. $F+10=3(D-6)$

    22. $160=A+B+C$ 23. $V=\dfrac{\pi D^3}{8}$ 24. $I=PRT$
         $C=2A$

    25. $IQ=\dfrac{100M}{C}$ 26. $D=Vt$ 27. $A=\frac{1}{4}\pi D^2$

    28. $L=6+2S$ 29. $L=20+2W$ 30. $N+9=4N$
    31. $A+B=\frac{1}{2}T$ 32. $G+S=\frac{1}{3}P$

D. 1. $\dfrac{12 + 144 + 20 + 3\sqrt{4}}{7} + 5 \cdot 11 = 9^2 + 0$

Problem Set 1-6, Page 75

A. 1. 6   2. 9   3. 12   4. 30   5. 2   6. 8
7. 7   8. 5   9. 11   10. 3   11. 46   12. 3
13. 6   14. 5   15. 4   16. 3   17. $13\frac{1}{2}$ 18. 12
19. 13   20. 16   21. 2   22. 24   23. 0   24. -33
25. 9   26. -3 27. 98   28. 20   29. 9   30. 2

B. 1. 22   2. 28   3. 46   4. 116   5. 96   6. 30
7. 4   8. -70 9. 405   10. $\frac{3}{7}$ 11. 550   12. 27
13. 100   14. 1188   15. 242   16. 26   17. 483   18. 15
19. 280   20. $\frac{14}{13}$ 21. 67   22. -144 23. 6   24. 108
25. 78   26. 108   27. -84 28. -45 29. 12   30. 4

C. 1. (a) 14   (b) 10   (c) 12   (d) 12   (e) 6   (f) 21
2. 339.12 ft^3 3. 16   4. 490 m   5. 60° C
6. 19 cm^2 7. 40   8. \$480   9. 904.32 in.^{3}

10. $0.8\,\dfrac{\text{kg}}{\text{m}^3}$ 11. 1, 3, 6, 10, 55   12. 2835.2 g

D. 1. (a) 153   (b) 38   (c) 183
2. (a) 88   (b) 14   (c) 76
3. 90,000,000,000 joules or 9×10^{10} joules

---

1. 6   2. 8.4   3. 4.2   4. -0.000008
5. $2x^3y^2z^2$ 6. x^7 7. $6xy + 2x^2$ 8. $6p^2q$
9. $-18x^6$ 10. $16ab^4$ 11. $-72c^5$ 12. $\dfrac{8n}{5m}$
13. $\dfrac{-5}{y}$ 14. $3b^2$ 15. $-3x^3$ 16. $-3st - 2st^2$
17. $2 - 2x$ 18. $13 - 6y$ 19. $5a - 1$ 20. $2x - 2$
21. $x^2 - pq$ 22. $5n + n^2 = 24$ 23. $6x - 5 = 2y$ 24. -88
25. 12   26. 28.26   27. 40   28. 24
29. 2000

Problem Set 2-1, Page 105

A. 1. Yes   2. No   3. No   4. Yes   5. No   6. No   7. No   8. Yes
9. Yes   10. No

B. 1. 5   2. 8   3. 3   4. 0   5. 43   6. 87   7. 24   8. 455
9. 58   10. 29   11. 27   12. -26 13. 52   14. 174   15. 0   16. 108
17. 4   18. 6   19. 12   20. 0   21. 7   22. 33   23. 11   24. 26

C. 1. -8 2. 35   3. 13   4. 37   5. 4   6. -5
7. 24   8. 0   9. 0   10. -10 11. -1 12. -5
13. 7   14. 7   15. $\dfrac{-3}{4}$ 16. 1.5   17. 5.6   18. 25
19. 12   20. -1.7 21. -24 22. 11   23. -16 24. 0
25. -3 26. -21 27. -2 28. 28   29. 8   30. -2.4
31. 11   32. -7 33. -13 34. 9   35. 0   36. 1
37. -2 38. 5   39. -10 40. 7   41. -5 42. 11
43. -5 44. -6 45. 3   46. 7   47. 6   48. -1
49. 4   50. 6

D. 1. $n - 37 = 17$, $n = 54$ 2. $24 + x = 42$, $x = 18$
     3. $w + 18 = 105$, $w = 87$ 4. $B - 13 = 6\frac{1}{2}$, $B = 19\frac{1}{2}$
     5. $t - 5.74 = 2.46$, $t = \$8.20$ 6. $s - 145 = 220$, $s = 365$ mph
     7. $974 = 116 + x$, $x = 858$ 8. $29 + b = 78$, $b = \$49$
     9. $t - 42 = 185$, $t = \$227$ 10. $24.90

E. 1. 23   2. $2
     3. $3x^2 + 3x = 3(x)(x + 1)$ This is always an even integer because x and $x + 1$ are consecutive integers and one of them must be even.

Mental Math Quiz, Page 110

1. 4   2. 7   3. 6   4. 5   5. 3   6. 6   7. −4
8. −7   9. 7   10. 7   11. 2   12. −3   13. 6   14. 5
15. 5   16. 12   17. 6   18. 12   19. 6   20. 5   21. −8
22. −4   23. −5   24. −10   25. −4   26. −3   27. −10   28. −3
29. 2   30. 2   31. 6   32. 2   33. 4   34. 8   35. 10
36. 13   37. 1   38. 5   39. −16   40. 10

UNIT 2   *Problem Set 2-2, Page 117*

A. 1. 6   2. 9   3. 9   4. 0   5. 39   6. −9   7. 8
    8. 8   9. $\frac{1}{2}$ 10. $3\frac{3}{4}$ 11. 0.4   12. 12   13. 56   14. $1\frac{1}{4}$
    15. 20   16. 1   17. −12   18. −6   19. 0   20. 3   21. −4
    22. $-\frac{3}{5}$ 23. 3   24. 20   25. $-\frac{4}{3}$ 26. 0   27. $2\frac{2}{3}$ 28. $7\frac{1}{2}$
    29. $\frac{1}{10}$ 30. $\frac{1}{12}$

B. 1. −0.2   2. 22   3. 1   4. 2   5. −10
    6. 23   7. $1\frac{3}{4}$ 8. 0   9. −0.25   10. −2
    11. 15   12. −81   13. 119   14. −13   15. 42
    16. 7   17. 3   18. −4   19. −7   20. −12
    21. −0.1875   22. $-2\frac{2}{5}$ 23. $-2\frac{1}{2}$ 24. 0.2   25. −0.04
    26. 0.4   27. 0.7   28. −6   29. 0   30. $6\frac{2}{3}$

C. 1. 0   2. 1.8   3. $\frac{-1}{27}$ 4. 7.8125   5. −0.2   6. 70
    7. 0.12   8. 4.5   9. 16   10. $\frac{1}{4}$ 11. 8   12. 6
    13. −9   14. −8.5   15. −11   16. 40   17. 16   18. −9
    19. 15   20. 4   21. −4   22. 2   23. −2   24. $-\frac{3}{8}$
    25. $\frac{2}{3}$ 26. $\frac{1}{5}$ 27. 3   28. $\frac{4}{3}$ 29. −1   30. 3

D. 1. 12, 48   2. 20, 15   3. 13, 39   4. $36, $24
    5. 4   6. 112, 70   7. 20   8. 83, 249
    9. Lot is worth $3600; cabin is worth $19,800.
    10. 2097

E. 1. $a = -1$, $b = 5$ 2. $\dfrac{2x + 6}{2} - x = 3$ 3. 84 years old

---

Mental Math Quiz, Page 126

1. 3   2. 3   3. 2   4. 2   5. 4   6. 4   7. 5   8. 2
9. 5   10. 6   11. 6   12. 2   13. 2   14. 4   15. 3   16. 7
17. 3   18. 3   19. 0   20. 16   21. 0   22. −2   23. −2   24. −3
25. 2   26. 1   27. −3   28. 12   29. 10   30. 8   31. 30   32. $-\frac{2}{3}$

UNIT 2   *Problem Set 2-3, Page 129*

A. 1. 4   2. 3   3. 5   4. -3 5. 6
    6. 15   7. -2 8. 9   9. 6   10. 5
    11. 11   12. 16   13. $-8\frac{1}{2}$ 14. -2 15. 2
    16. 6   17. $\frac{3}{7}$ 18. 1   19. 0   20. $\frac{1}{2}$

B. 1. 30   2. 16   3. $-42\frac{2}{3}$ 4. -7 5. 20
    6. 30   7. $-3\frac{1}{6}$ 8. -60 9. 1   10. -10
    11. 30   12. 1   13. $-\frac{1}{2}$ 14. $\frac{2}{5}$ 15. $\frac{1}{2}$
    16. -4 17. 0   18. $\frac{3}{4}$ 19. $\frac{5}{2}$ 20. $-2\frac{4}{5}$
    21. $-\frac{9}{2}$ 22. -0.8 23. 1   24. $\frac{2}{3}$ 25. 1
    26. $3\frac{1}{3}$ 27. 2   28. 25   29. $\frac{5}{2}$ 30. -2

C. 1. -9 2. $-\frac{1}{4}$ 3. $-\frac{4}{3}$ 4. $4\frac{1}{2}$ 5. $\frac{2}{3}$

    6. -1 7. -1 8. $\frac{1}{2}$ 9. $1\frac{1}{2}$ 10. 1

    11. 2   12. 5   13. $-\frac{1}{4}$ 14. 1   15. $-\frac{8}{5}$

    16. $-\frac{1}{4}$ 17. 1   18. $\frac{1}{2}$ 19. $\frac{1}{3}$ 20. $\frac{1}{3}$

    21. -5 22. $-\frac{1}{6}$ 23. $\frac{3}{4}$ 24. 1   25. 1

    26. 1   27. -1 28. 3   29. 2   30. $-\frac{1}{5}$

D. 1. 6, 12   2. 10, 3   3. $-4°F$ 4. 22   5. $-\dfrac{B + C}{2}$

    6. 33   7. 96   8. $8\frac{1}{2}$ in.   9. 7 ft by 35 ft
    10. Bill earned \$35.50; Pam earned \$16.00

E. 1. $1\frac{1}{9}$ 2. $\dfrac{2b - a}{2b^2 - a^2}$ 3. $a + 2b$

    4. 18   5. 10   6. $\frac{40}{11}$ hr or 3 hr 38 min

UNIT 2   *Problem Set 2-4, Page 139*

A. 1. $x = \dfrac{A}{3}$ 2. $x = B - 4$ 3. $x = \dfrac{Y}{2b}$

    4. $x = \dfrac{Rd}{a}$ 5. $x = 5 - Q$ 6. $x = \dfrac{C - 1}{2\pi}$

    7. $x = \dfrac{3c}{2a}$ 8. $x = \dfrac{b^2}{a}$ 9. $x = T - A - 3$

    10. $x = \dfrac{B - A}{c}$ 11. $x = \dfrac{2bK}{ay}$ 12. $x = -E - 3$

    13. $x = \dfrac{2A + 4}{3}$ 14. $x = \dfrac{B - A}{3}$ 15. $x = \dfrac{y}{2} - 1$

    16. $x = 4A - 2$ 17. $x = P - 2$ 18. $x = -3A$

    19. $x = \dfrac{2t + 4}{5}$ 20. $x = -3y$

B. 1. $a = px^2 - 2$ 2. $a = \dfrac{2x - 3}{2}$ 3. $a = -x$

4. $b = \dfrac{x - y}{t}$ 5. $a = \dfrac{c}{2d}$ 6. $b = \dfrac{d}{3mt^2}$

7. $a = \dfrac{cx + d - 1}{d}$ 8. $a = \dfrac{-p}{qc}$ 9. $b = \dfrac{t + p}{2m}$

10. $a = \dfrac{s + p}{t}$ 11. $b = \dfrac{3x - e}{3}$ 12. $b = \dfrac{3s + 2p}{p}$

13. $b = \dfrac{3(x + y)}{2}$ 14. $a = \dfrac{3x + 2}{4}$ 15. $a = 2c - 1$

16. $a = \dfrac{2cd}{3}$ 17. $a = \dfrac{8y + 5}{2}$ 18. $a = \dfrac{1}{t}$

19. $a = \dfrac{-x}{2}$ 20. $a = \dfrac{d^2}{2}$ 21. $a = 2y$

22. $a = -t - d$ 23. $a = \dfrac{5x + 6}{7}$ 24. $a = \dfrac{2p^2 + 3p}{5}$

25. $b = 2x$ 26. $a = \dfrac{3 + 3y}{8}$ 27. $a = \dfrac{xc}{d}$

28. $a = \dfrac{xd}{c}$ 29. $a = \dfrac{2t}{p}$ 30. $a = \dfrac{2x^2}{2y - 1}$

31. $a = \dfrac{y - x}{y}$ 32. $b = \dfrac{2 - t}{t}$ 33. $a = \dfrac{3p - 1}{p}$

34. $b = \dfrac{x - 2}{7}$ 35. $a = \dfrac{x}{2x - 1}$ 36. $a = \dfrac{x}{x^2 + 1}$

37. $a = \dfrac{2}{y + 1}$ 38. $b = \dfrac{2}{3 - t}$ 39. $a = \dfrac{6}{2x - 3}$

40. $b = \dfrac{6}{3y + 2}$

C. 1. $a = \dfrac{p - 2b}{2}$ 2. $m = \dfrac{E}{c^2}$ 3. $x = \dfrac{-c - By}{A}$

4. $v = \dfrac{-Ax - c}{B}$ 5. $L = \dfrac{100B}{C}$ 6. $F = \dfrac{ab}{a + b}$

7. $r = \dfrac{s - a}{s - t}$ 8. $t = \dfrac{a + sr - s}{r}$ 9. $R^2 = \dfrac{3V + \pi h^3}{3\pi h}$

10. $g = \dfrac{2S}{t^2}$ 11. $R = \dfrac{P}{i^2}$ 12. $h = \dfrac{A}{4\pi R^2}$

13. $m = \dfrac{2E}{v^2}$ 14. $h = \dfrac{P}{mg}$ 15. $R = \dfrac{v^2}{a}$

16. $P = \dfrac{nRT}{V}$ 17. $T = \dfrac{PV}{nR}$ 18. $R = \dfrac{Vr}{2KL}$

19. $a = \dfrac{bKQ}{bV + KQ}$ 20. $a = \dfrac{E - IR}{I}$

D. 1. $a = \dfrac{180L}{\pi R}$ 2. $a = \dfrac{360A}{\pi R^2}$ 3. (a) $V_1 = \dfrac{V_2 P_2}{P_1}$

3. (b) $V_2 = \dfrac{V_1 P_1}{P_2}$ 3. (c) $P_1 = \dfrac{P_2 V_2}{V_1}$ 3. (d) $P_2 = \dfrac{V_1 P_1}{V_2}$

4. $N \cong \dfrac{3.78P}{t^2 d}$ 5. $D = \dfrac{C(A + 12)}{A}$ 6. $R = \dfrac{rF(n - 1)}{r - F(n - 1)}$

7. (a) $L = \dfrac{6V}{\pi T^2}$ 8. $R_1 = \dfrac{V - R_2 i}{i}$

Answers

Copyright © 1982 by John Wiley & Sons, Inc.

A. 1. $x < 15$

2. $x > -4$

3. $a > -1$

4. $d < 2$

5. $y \leq 0$

6. $x \leq 10$

7. $x > 4$

8. $y > 7$

9. $a \leq -1$

10. $x \geq -3$

11. $Z < 7$

12. $Q < -1$

13. $x < -4$

14. $P < -3$

15. $t \leq -14$

16. $x \leq 2$

B. 1. $x < 5\frac{1}{2}$ 2. $a > 3\frac{2}{3}$ 3. $a < 2$ 4. $x > -\frac{2}{3}$

5. $x < -6$ 6. $t > 20$ 7. $y \geq 3\frac{1}{2}$ 8. $a \geq -2\frac{1}{3}$

9   $E < -\frac{3}{4}$ 10. $K > -1\frac{2}{3}$ 11. $G > 3$ 12. $m < -2$

13. $X \geq 3$ 14. $x \geq -1\frac{1}{2}$ 15. $x < -3\frac{1}{2}$ 16 $y \leq -2$

17. $m \leq \frac{1}{2}$ 18. $a > 1\frac{4}{5}$ 19. $x < -1$ 20. $Z < 3$

21. $X < 1\frac{5}{9}$ 22. $Y < 2$ 23. $X < -6$ 24. $Q > -1$

25. $a \leq -15$ 26. $m \geq -12$ 27. $x > 3$ 28. $a < 3$

29. $G > 1\frac{3}{4}$ 30. $p < 1\frac{3}{4}$

C. 1. 24   2. $W < 2\frac{1}{2}\,\text{ft}$ 3. $n_1 < 10\frac{1}{2},\ n_2 < 3\frac{1}{2}$

4. $S_1 = 4,\ S_2 < 4\frac{2}{3},\ S_3 < 11\frac{1}{3}$ 5. $S \geq 95$ 6. $x \geq -\frac{1}{3}$

7. Not possible   8. Eric's age < 12

---

1. 8 half dollars, 64 quarters, 76 dimes   2. 10 lb of protein powder
3. Approximately 111 minutes; Jan ran $14\frac{6}{7}\,\text{mi}$; Paul ran $11\frac{1}{7}\,\text{mi}$
4. 4.8 hr   5. 3 gallons   6. 15 gallons   7. 2.5 gallons   8. 1528
9. $6\frac{2}{3}$ hours   10. $4\frac{4}{9}$ hr   11. 30 hr
12. 1 pint wheatgerm oil, 5 pints carrot juice, 4 pints cider vinegar
13. 15 miles   14. 345 miles   15. $7\frac{1}{2}$ minutes

Self-Test 2
Unit 2, Page 179

1. 39   2. $2\frac{1}{2}$ 3. $1\frac{1}{2}$ 4. -3 5. -6

6. 3   7. 1   8. $-6\frac{2}{3}$ 9. $\dfrac{D}{3M^2}$ 10. $R = \dfrac{E}{I} - S$

11. $\dfrac{S}{2(A-1)}$ 12. $\dfrac{2a}{B}$ 13. $1\frac{1}{4}$ 14. $x > 2$

15. $a \geq \frac{3}{4}$ 16. $y < -3$

17. $R > \frac{1}{2}$ 18. 33, 34, and 35

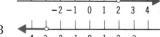

19. $\frac{4}{7}$ hr   20. 32, 34, and 36   21. $6\frac{2}{3}$ lb   22. 4 hr
23. 6 hr   24. 15 pints   25. $1\frac{1}{5}$ days

Mental Math Quiz,
Page 197

1. $2a^2$ 2. $6x^2$ 3. $6ap$ 4. $9x^3$ 5. $4x^2y$
6. $6p^2t$ 7. $5a^2b$ 8. $3t^3$ 9. $3a^2$ 10. $4xy^2$
11. $6t$ 12. $6y^3$ 13. a^4 14. x^5 15. x^4y^3
16. p^7 17. $10x^7$ 18. abx^5 19. a^3t^3 20. x^2y^4
21. $-3x$ 22. $-2x^2$ 23. $6y$ 24. $10x^2$ 25. $10ax$
26. $6ap^3$ 27. $12t^4$ 28. $-4k^4$ 29. b^3t^2 30. pt^7
31. $-6atx^2$ 32. $6x^3$ 33. $8y^2$ 34. $-10q^2$ 35. $-8p^3$
36. $3x^2$ 37. $-3m^3n$ 38. $-4s^2t^2$ 39. $-2xy^3$ 40. $-3at^2$

Mental Math Quiz,
Page 204

1. $2(x+2)$ 2. $3(y-3)$ 3. $5(p+3)$ 4. $4(2-q)$
5. $x(2+a)$ 6. $y(3-p)$ 7. $y(1-a)$ 8. $x(1-a)$
9. $a(1-a)$ 10. $k(k-2)$ 11. $p(p^2-3)$ 12. $t^2(t-1)$
13. $d(2d-3)$ 14. $a(2a-3)$ 15. $2t(t^3-2)$ 16. $x(1-y)$
17. $q(1+q)$ 18. $b(a+b)$ 19. $mn(m-1)$ 20. $ab(a-1)$
21. $3(a^3-5)$ 22. $-y(2+y^2)$ 23. $-4p(1+2p^2)$ 24. $-3b(a+3b)$
25. $3(x-y)$ 26. $b(2a-3)$ 27. $xy(5-y^4)$ 28. $a(bc+1)$
29. $xy(z-1)$ 30. $a(2xt-1)$ 31. $y(3x^2-1)$ 32. $ab(bx+y)$
33. $-(x+y)$ 34. $-(ab+t)$ 35. $-(2x+3)$ 36. $-(3t^2+2)$
37. $-(2+x^2)$ 38. $x(y-1)$ 39. $-x(2+3yz)$ 40. $3(t-2k)$

UNIT 3   *Problem Set 3-1, Page 205*

A.   1. $2t+2$ 2. $2x^2+3x$ 3. $15-5y$ 4. $-2p-6$
5. $-ax-3a$ 6. $-2b+b^2$ 7. $px-p^3$ 8. $4c-4c^2$
9. $6x^2-2x$ 10. $6n^2+6n$ 11. $-y+y^2$ 12. $-2t^2+2t^3$
13. $21pq+14p^2$ 14. $-3x^2+6x$ 15. $8a^2x-6ax^2$ 16. x^2-x^3
17. $2pt^2-2p^2t$ 18. $m^3n+m^2n^2$ 19. $3x^2-5x$ 20. $2y^3-12y^2$

B.   1. $3x+3y-3$ 2. $2x^2-4x+10$
3. $15a^2-10a+5$ 4. $8b^2-4b-12$
5. x^3-x^2-x 6. $x^3y+2xy-2y$
7. $2a^2b+2ab^2-2ab$ 8. $3p^3q-3pq^3-3pq^2$
9. $-4a^3x-4a^2x+4ax^2$ 10. $-2x^3-2ax^2+4x^2$
11. $-a^2b^2-2ab^3+b^4$ 12. $q^2x+q^3+q^2$
13. $x^3+x^4-2x^5$ 14. $6y^3-4y^4+2y^5$
15. $3x^3-4x^2y+2x^3y$ 16. $y-2xy+4y^2$
17. $5rt^2-5t^3+5t^4+5rt^3-15t^2$ 18. $a^2b+ab^2-a^2b^2+a^3b^2-a^2b^3$
19. $-a^2+a^3+a^4+a^5-a^7$ 20. $-x^4+x^3-x^2+x$

C.   1. $x(x-1)$ 2. $a(1-2a)$ 3. $3(2y-1)$
4. $4(2+t)$ 5. $2(a+2b)$ 6. $3b(2a^2c+3b^2)$
7. $4x(2y^2-3x^2)$ 8. $2b(a-2c)$ 9. $9tx(3t+2x)$
10. $7ab(3bc+2a)$ 11. $-3a(a^2+2)$ 12. $-2b(5b^2+c)$
13. $2x(2x+1)$ 14. $4a^2(2a+1)$ 15. $10a^2(3a^2-1)$
16. $12t^2(2t^3-3)$ 17. $x(x^2+x+2)$ 18. $m(2m^2-3m+5)$
19. $2(a^2-a+3)$ 20. $3(b^3-b+3)$ 21. $4x(-5x^3+x-3)$
22. $-2n(3n+2)$ 23. $-5p(4q+3p)$ 24. $-2y(6y^2+2y+3)$
25. $2a^2b^2c(3a+5b^3c)$ 26. $4xy(4y^2t^2-3x^3z^3)$
27. $7tr^2(3tr+4)$ 28. $3p^2q^2r(3p^3-4q^2r)$
29. $a(a^3-a^2+ac^2-c^3)$ 30. $x(x^6-4x^2-2x-3)$
31. $x(1-2t)$ 32. $x^2(a+b-c)$

Answers   Copyright © 1982 by John Wiley & Sons, Inc.

33. $4p(2x - 3)$ 34. $x(x - 3)(x + 3)$

35. $y(x - y)(y - 5)$ 36. $2a(a + 2)(2a - 7)$

D. 1. $x + (x + 1) + (x + 2) = 3x + 3 = 3(x + 1)$
 2. $2n + (2n + 2) + (2n + 4) = 6n + 6 = 6(n + 1)$
 3. The original quantity is 9.

UNIT 3   *Problem Set 3-2, Page 217*

A. 1. $a^2 + 5a + 6$ 2. $x^2 - 5x + 6$
 3. $t^2 + t - 12$ 4. $p^2 + p - 20$
 5. $x^2 - 5x + 4$ 6. $-y^2 - 3y + 10$
 7. $6R^2 - R - 1$ 8. $-15Q^2 + 7Q + 2$
 9. $2a^2 + 2a - 4$ 10. $3c^2 - 10c + 3$
 11. $4x^2 + 6ax - 4a^2$ 12. $5t^2 - 17pt + 6p^2$
 13. $-8x^2 - 4ax + 2a + 4x$ 14. $-6t^2 - t + 15$
 15. $ax + ay + bx + by$ 16. $ax + ay - bx - by$
 17. $ax - ay - bx + by$ 18. $ax - ay + bx - by$
 19. $2a^2x + 4ax^2 - ax - 2x^2$ 20. $-6p^2 + q^2 + pq$
 21. $x^3 - 2x^2 + x$ 22. $3a^3 + 8a^2 + 4a$
 23. $2t^4 + t^3 - 6t^2$ 24. $5m^2n^2 - 2m^4 + 3n^4$
 25. $2a^2x - 2ax - a + 1$ 26. $8R + 2R^2t - 3Rt - 12$
 27. $6x^2 - 7x + 2$ 28. $3G^2 - 10FG + 3F^2$
 29. $c^3 - c^2 - c + 1$ 30. $n^5 + n^3 - n^2 - 1$

B. 1. $x^2 - 2x + 1$ 2. $Q^2 + 4Q + 4$
 3. $T^2 - 6T + 9$ 4. $q^2 + 2q + 1$
 5. $p^2 + 2pq + q^2$ 6. $m^2 - 2mn + n^2$
 7. $4a^2 - 4ab + b^2$ 8. $9x^2 + 12xy + 4y^2$
 9. $x^2 + 4x + 4$ 10. $c^2 + 6c + 9$
 11. $d^2 - 8d + 16$ 12. $x^2 - 10x + 25$
 13. $4m^2 + 12mn + 9n^2$ 14. $m^4 + 2m^2n^2 + n^4$
 15. $9a^4 - 12a^3x + 4a^2 x^2$ 16. $4c^4 - 20bc^3 + 25b^2c^2$
 17. $2x^2 - 16x + 32$ 18. $3y^2 - 18y + 27$
 19. $a^3 + 2a^2 + a$ 20. $4p^3 + 12p^2 + 9p$
 21. $2xy + 6x - 4y - 12$ 22. $10a^2 - 15a + 5$
 23. $-2t^3 + 3t^2 - t$ 24. $6x^3 - ax^2 - a^2x$
 25. $a^2 - 1$ 26. $4 - x^2$
 27. $4t^2 - 9x^2$ 28. $4t^2 - 9x^2$
 29. $4p^2 - 4q^2$ 30. $a^2x^4 - y^4$
 31. $2p^2 - 2$ 32. $p^3 - 4p$

C. 1. $x^3 - x^2 - 11x - 10$ 2. $y^3 - 4y^2 + y + 6$
 3. $-a^3 + 5a^2 - 5a - 3$ 4. $4c^3 + 5c^2 - 4c + 4$
 5. $4x^2 + 4x - 2y - y^2$ 6. $x^3 + 2x^2 + 2x + 1$
 7. $x^3 + 1$ 8. $2n^3 + 3n^2 + n$
 9. $n^4 + 2n^3 + n^2$ 10. $a^3 - x^3$
 11. $a^3 + 3a^2x + 3ax^2 + x^3$ 12. $p^3 + 6p^2 + 12p + 8$
 13. $a^3 + a^2x + ax^2 + x^3$ 14. $a^3 + a^2x + ax^2 + bx^2 + abx + a^2b$
 15. $2x^3 - 5x^2y - 10xy^2 + 3y^3$ 16. $-a^3 - a^2b + 9ab^2 - 6b^3$
 17. $x^3 + 5x^2 + 8x + 4$ 18. $t^3 - 4t^2 + 5t - 2$
 19. $1 - x^4$ 20. $a^4 + 2a^2x^2 + x^4 - 2a^2x$
 21. $x^3 - 5x - 2$ 22. $2a^3 - 5a^2b + ab^2 + 2b^3$
 23. $y^3 + 3y^2 - 4y$ 24. $4x^3 - 16x^2 + 15x$
 25. $x^3 - 2x^2 - 5x + 6$ 26. $a^3 - a^2 - 14a + 24$
 27. $x^5 - x^4 - x^3 + x^2 - x + 1$ 28. $6t^4 - 3t^3 - 2t^2 + 5t - 2$
 29. $12a^3 - 20a^2 + 11a - 2$ 30. $20x^3 - 72x^2 + 81x - 27$

D. 1. $-a + ac + ad$ 3. (a) $\dfrac{2n^3 + 3n^2 + n}{6}$ 3. (b) 42,925

4. $a^2 + 2ab + b^2 + 2bc + c^2 + 2ac$, $a^2 + 2ab + b^2 - 2bc + c^2 - 2ac$,
    $a^2 - 2ab + b^2 + 2bc + c^2 - 2ac$, $x^4 + 4x^3 + 6x^2 + 4x + 1$

8. $x^2 - y^2 = (x + y)(x - y)$ But if $x - y = 1$, then $x^2 - y^2 = x + y$

9. $a = 5,\ b = -17$

UNIT 3   *Problem Set 3-3, Page 235*

A. 1. $(x + 2)(x + 8)$ 2. $(a + 1)(a + 1)$ 3. $(p + 2)(p + 4)$
    4. $(y - 2)(y - 6)$ 5. $(c - 2)(c - 3)$ 6. $(x - 3)(x + 2)$
    7. $(q + 4)(q + 5)$ 8. $(a + 7)(a + 8)$ 9. $(k + 8)(k - 5)$
   10. $(x + 2)(x - 11)$ 11. $(y + 3)(y + 12)$ 12. $(p + 1)(p + 6)$
   13. $(m + 5)(m - 10)$ 14. $(y - 4)(y + 15)$ 15. $(x - 2y)(x - 3y)$
   16. $(e - 5)(e - 7)$ 17. $(a - b)(a - b)$ 18. $(x + y)(x - 2y)$
   19. $(a - b)(a + b)$ 20. $(x - 2y)(x + 2y)$ 21. $(t + 3a)(t - 2a)$
   22. $(x + 4)(x + 4)$ 23. $(p - 2)(p - 10)$ 24. $(t + 8)(t + 9)$

B. 1. $(2a + 1)(a + 3)$ 2. $2(x - 1)(x + 4)$
    3. $2(t + 1)(t + 4)$ 4. $(3c + 1)(c + 5)$
    5. $(3x + 1)(x + 1)$ 6. $(2p + 1)(p + 1)$
    7. $(2q - 1)(q - 1)$ 8. $(3m + 1)(m + 2)$
    9. $(a - 1)(3a + 5)$ 10. $2(y + 3)(y - 1)$
   11. $(3t + 1)(2t - 1)$ 12. $(4q - 1)(q - 1)$
   13. $(4c + 1)(4c - 5)$ 14. $(2z - 1)(2z - 1)$
   15. $(2a - 5)(3a - 7)$ 16. $(x - 1)(3x + 4)$
   17. $(3x - y)(3x + 4y)$ 18. $(2x - y)(x + 3y)$
   19. $(2p + 5q)(2p + 3q)$ 20. $(5t - a)(3t + 4a)$
   21. $(2x - 3y)(2x + 3y)$ 22. $(4a - 1)(4a + 1)$
   23. $(3y - 2)(3y + 2)$ 24. $(5ta - 2b)(5ta + 2b)$

C. 1. $3(x + 4)(x + 5)$ 2. $2(a - 6)(a - 5)$
    3. $x(x^2 + 2x + 4)$ 4. $t(t - 6)(t - 7)$
    5. $y(y + 4)(y - 4)$ 6. $p(2p - 3)(2p + 3)$
    7. Cannot be factored   8. $2x(x - 1)(3x + 5)$
    9. $3q(2q - 5)(2q + 1)$ 10. $2b(a + 6)(a + 1)$
   11. $3a^2(t + 7)(t + 1)$ 12. Cannot be factored
   13. $xy(2x - 5)(x + 3)$ 14. $2y(3x - 8)(2x + 5)$
   15. $(2y + 3ab)(2y - 3ab)$ 16. $4q(2p^2 - q^2)$
   17. Cannot be factored   18. $ab(a + b)(a - b)$
   19. $5t(5p + 7)(2p - 3)$ 20. $-2(2x^2 + 3x - 1)$

D. 1. 3   2. $\frac{6}{5}$ 3. -4 4. 2   5. $\frac{1}{2}$ 6. -1 7. $\dfrac{3}{2}$
    8. 8   9. $\frac{2}{3}$ 10. $\frac{1}{2}$

E. 1. $x + \dfrac{5}{2}$ 2. $\dfrac{3x}{4} - 1$ 3. $ab - ab^2$

    4. $2x + y^2$ 5. $-1 + \dfrac{3x}{y}$ 6. $\dfrac{y}{x} + 2x$

    7. $\dfrac{b}{a} + 3b + 1$ 8. $xy^2 - \dfrac{1}{y^2}$ 9. $\dfrac{y}{2x} - \dfrac{2y^2}{3x}$

   10. $-\dfrac{y}{2x} + \dfrac{z}{4x} - \dfrac{3}{4z}$ 11. $x^2 - 4x - 6$

   12. $2x + 7$ 13. $2x^2 + x - 7 + \dfrac{20}{x + 5}$

   14. $3y^2 + 3y - 1 + \dfrac{4}{y - 1}$ 15. $x^3 - x + 1$

16. $x^2 - 2x + 3 - \dfrac{5}{x + 2}$ 17. $x^3 + x^2 + x + 1 - \dfrac{1}{x - 1}$

18. $x^4 - x^3 + x^2 - x + 1 + \dfrac{2}{x + 1}$

19. $x^3 - \dfrac{x^2}{2} - \dfrac{5x}{4} + \dfrac{5}{8} + \dfrac{3}{8(2x + 1)}$

20. $t^2 + 2t + 1 + \dfrac{2}{2t + 1}$ 21. $x^2 + ax + a^2$

22. $x^4 - ax^3 + a^2x^2 - a^3x + a^4 - \dfrac{2a^5}{x + a}$

23. $x + 2$ 24. $x - 3 + \dfrac{-9x + 23}{x^2 - x + 1}$

25. $x^2 - 2x + 1 - \dfrac{2}{x + 3}$ 26. $x^3 - x + 1 + \dfrac{8}{x - 5}$

27. $x^3 + x^2 + x + 1 + \dfrac{3}{x - 1}$ 28. $x^2 + x + 1 + \dfrac{2}{x - 1}$

F. 1. 10 quarters, 53 dimes, 212 nickels
    2. $(x + 2)(x + 4) - x(x + 6) = 8$ or $x^2 + 6x + 8 - x^2 - 6x = 8$
    3. Bill is 6 years old.   6. $x^4 + 4 = (x^2 + 2x + 2)(x^2 - 2x + 2)$

    8. $\dfrac{81a^6b^6c^{12}d^2}{16}$ 9. $2a^2(m^2 - 1) + 2b^2(n^2 - 1)$

1. $1 - x^2$ 2. $a^2 + 6a + 9$ 3. $2x^2 + 5xy - 3y^2$
4. $6q^3 + 9q^2 - 6q$ 5. $a^3 + ab^2 + a^2b + b^3$ 6. $4c^2 - 12ck + 9k^2$
7. $q^2 - 4$ 8. $x^3 - 6x^2 + 11x - 6$ 9. $3x(x - 2)$
10. $x(ax - a + x)$ 11. $3s(-p + 2 - ps)$ 12. $-x(2 + x)$
13. $(y + 2)(y + 5)$ 14. $(1 + x)(1 - x)$ 15. $(3x + 2)(2x - 1)$
16. $c(2a + b)(2a - b)$ 17. $(b + 2a)(2b - a)$ 18. $(1 - x)(a + b)$

19. $x = 1$ 20. $p = \frac{3}{4}$ 21. $t = -\frac{2}{3}$

22. $8, 10, 12$ 23. $7, 9, 11$ 24. 20 pennies
25. 50 cookies, 40 cupcakes, 30 doughnuts

26. $2x^2 - 3 + \dfrac{1}{2x}$ 27. $2x^2 + x - 2 + \dfrac{2}{x - 4}$

28. $x^4 + x^3 + x^2 + x + 1 + \dfrac{3}{x - 1}$

Answers to box on Page 257

1. $y = 1 - x$ 2. $y = \dfrac{x - 3}{2}$ 3. $y = \dfrac{2 + x}{2}$

4. $y = \dfrac{3}{2 + x}$ 5. $y = 2x - 6$

A. 1.

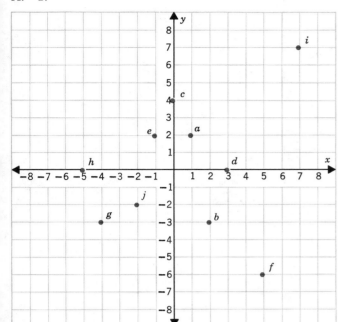

2.

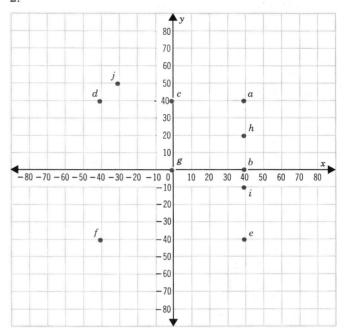

B. 1. (a) $(-2, 7)$ (b) $(0, 5)$ (c) $(8, 7)$ (d) $(-5, 4)$
        (e) $(1, 2)$ (f) $(8, 3)$ (g) $(-3, 0)$ (h) $(0, 0)$
        (i) $(7, 0)$ (j) $(-5, -3)$ (k) $(-3, -3)$ (l) $(3, -2)$
        (m) $(-4, -6)$ (n) $(1, -4)$ (o) $(4, -4)$ (p) $(0, -8)$

C. 1.

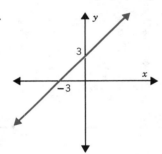

2.

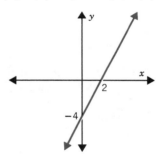

3.

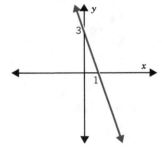

4.

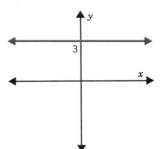

5.

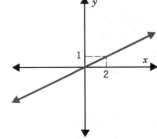

6.

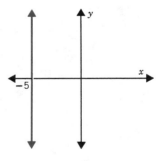

7.

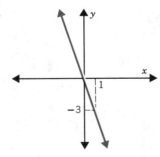

8.

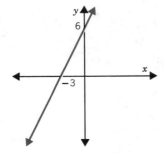

9.

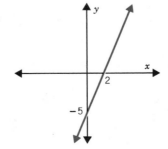

10.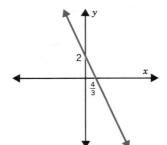

11.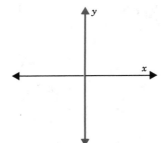

12.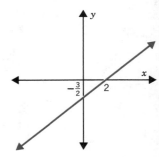

13.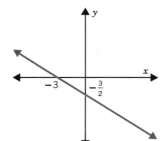

14.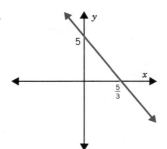

15.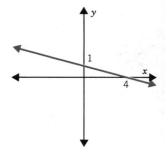

16.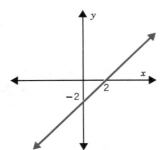

17.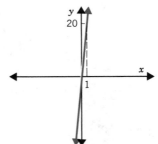

18.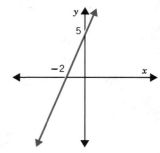

19.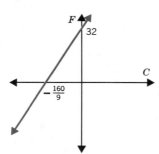

20.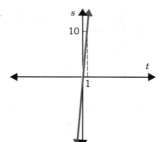

21.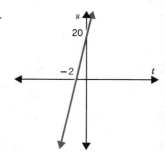

22.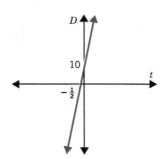

23.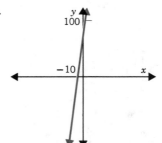

24.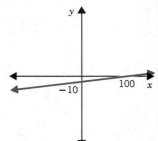

D. 1. $m = 1$, $b = 3$, $y = x + 3$ 2. $m = 2$, $b = -1$, $y = 2x - 1$
    3. $m = \frac{1}{2}$, $b = -5$, $y = \frac{1}{2}x - 5$ 4. $m = \frac{1}{2}$, $b = \frac{1}{2}$, $y = \frac{1}{2}x + \frac{1}{2}$
    5. $m = \dfrac{1}{3}$, $b = \dfrac{2}{3}$, $y = \dfrac{x}{3} + \dfrac{2}{3}$ 6. $m = 1$, $b = -4$, $y = x - 4$
    7. $m = -\frac{1}{2}$, $b = -2$, $y = -\frac{1}{2}x - 2$ 8. $m = 2$, $b = 0$, $y = 2x + 0$

9. $m = -5, b = 0, y = -5x + 0$
10. $m = 0, b = 6, y = 0 \cdot x + 6$
11. $m = -\frac{3}{2}, b = 30, y = \frac{-3x}{2} + 30$
12. $m = \frac{4}{5}, b = -4, y = \frac{4x}{5} - 4$
13. $m = \frac{9}{5}, b = 32, F = \frac{9}{5}C + 32$
14. $m = 10, b = 0, s = 10t + 0$
15. $m = 0, b = -4, y = 0 \cdot x - 4$
16. $m = \frac{2}{5}, b = -\frac{1}{5}, y = \frac{2}{5}x - \frac{1}{5}$
17. $m = \frac{1}{3}, b = -\frac{5}{3}, y = \frac{1}{3}x - \frac{5}{3}$
18. $m = -\frac{1}{2}, b = 2, y = -\frac{1}{2}x + 2$
19. $m = 1, b = -3, y = x - 3$
20. $m = 1, b = 0, y = x + 0$
21. $b = -\frac{1}{5}, m = \frac{2}{5}, y = \frac{2}{5}x - \frac{1}{5}$
22. $b = 3, m = -\frac{2}{3}, y = -\frac{2}{3}x + 3$
23. $b = -\frac{4}{3}, m = 2, y = 2x - \frac{4}{3}$
24. $b = -2, m = \frac{1}{2}, y = \frac{1}{2}x - 2$
25. $b = \frac{2}{5}, m = \frac{4}{5}, y = \frac{4}{5}x + \frac{2}{5}$
26. $b = \frac{1}{4}, m = -1, y = -x + \frac{1}{4}$
27. $b = \frac{1}{12}, m = -\frac{1}{6}, y = -\frac{1}{6}x + \frac{1}{12}$
28. $b = \frac{3}{2}, m = 6, y = 6x + \frac{3}{2}$
29. $b = \frac{3}{2}, m = 1, y = x + \frac{3}{2}$
30. $b = 3, m = -4, y = -4x + 3$

E.
1. $y = -2x + 9$
2. $y = -x + 3$
3. $5y = -9x + 7$
4. $7y = 6x - 11$
5. $3y = -x + 4$
6. $y = 2x$
7. $y = x - 2$
8. $2y = -x - 1$
9. $2y = -2x + 3$
10. $4y = -8x + 5$
11. $y = 3x - 3$
12. $y = -x$
13. $y = -x - 13$
14. $y = 7x + 23$
15. $y = -2x + 6$
16. $y = 3x - 9$
17. $y = 3x + 1$
18. $y = 2x - 10$
19. $y = 4x - 14$
20. $y = x - 1$
21. $y = -3x + 12$
22. $y = -4x + 19$
23. $y = -2x + 7$
24. $y = -x - 7$
25. $y = \frac{1}{2}x - 1$
    or $2y = x - 2$
26. $y = \frac{2}{5}x - 12$
    or $5y = 2x - 60$
27. $y = -\frac{3}{4}x - \frac{23}{2}$
    or $4y = -3x - 46$
28. $y = -\frac{2}{3}x - 11$
    or $3y = -2x - 33$
29. $y = -x + \frac{1}{4}$
30. $y = 2x + 2$

F. 1.

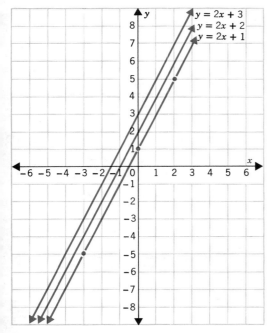

2.

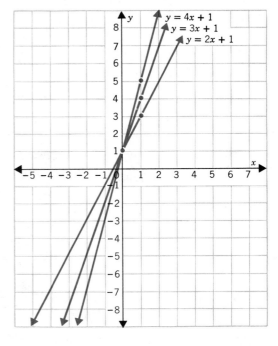

3.

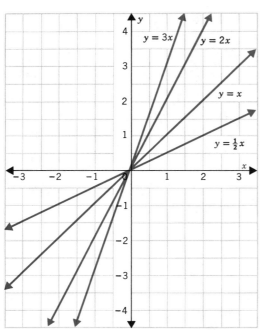

4.

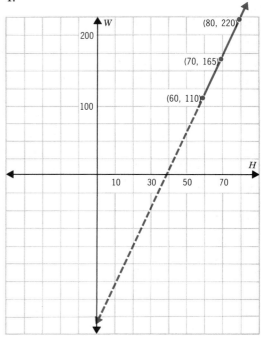

5.

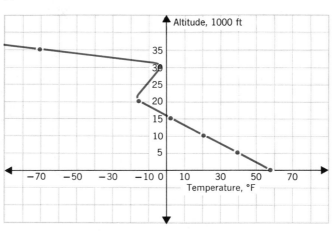

A.   **1.**

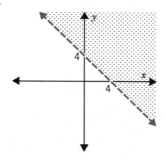

2.

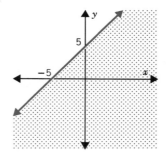

3.

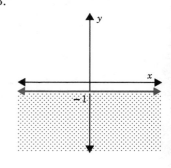

4.

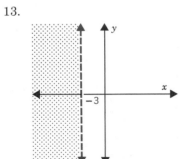

5.

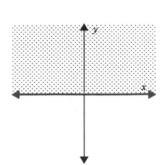

6.

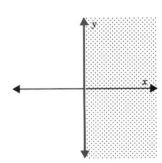

7.

8.

9.

10.

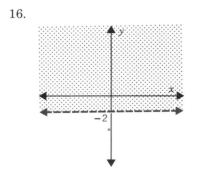

11.

12.

13.

14.

15.

16.

17.

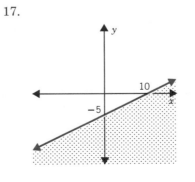

18.

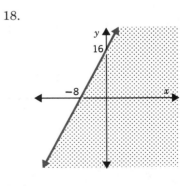

19.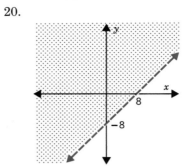

20.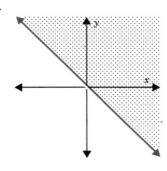

B. **1.**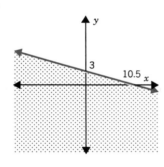

2.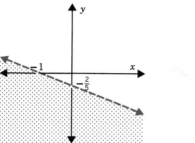

3.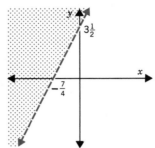

4.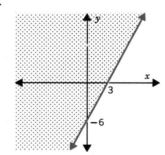

5.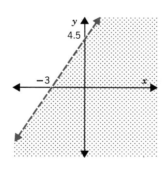

6.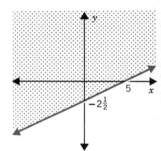

7.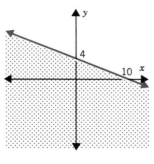

8.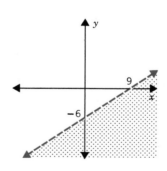

9.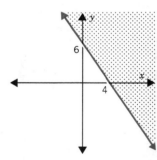

10.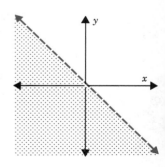

11.

12.

13.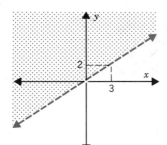

14.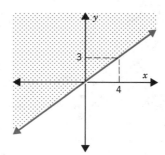

15.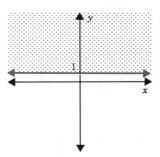

16.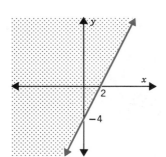

17.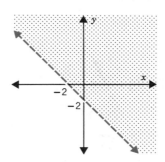

18.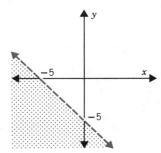

19.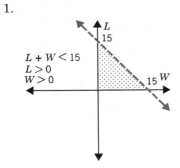

20.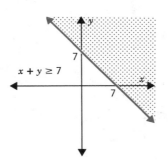

C. 1.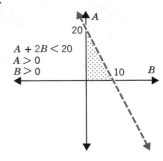

$L + W < 15$
$L > 0$
$W > 0$

2.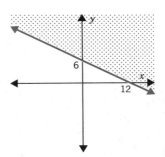

$x + y \geq 7$

3.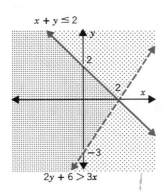

$A + 2B < 20$
$A > 0$
$B > 0$

4.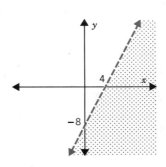

$x + y \leq 2$

$2y + 6 > 3x$

5.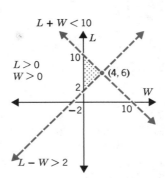

$L + W < 10$

$L > 0$
$W > 0$

$L - W > 2$

6.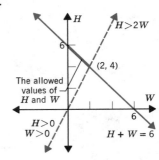

$H > 2W$

$(2, 4)$

The allowed values of H and W

$H > 0$
$W > 0$

$H + W = 6$

The height of the shortest possible frammis is 4 Mars bars.

A. 1.

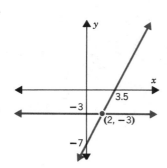

2.

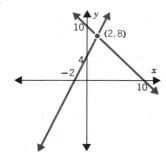

3.

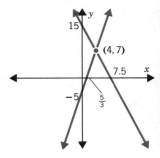

4.

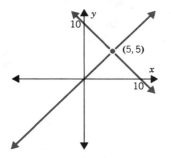

5.

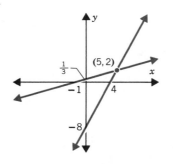

6.

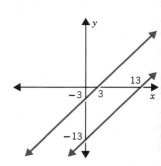

7.

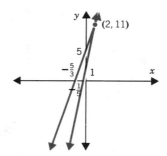

8.

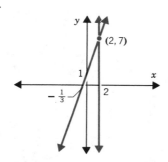

9.

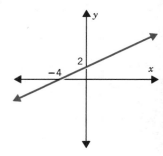

10.

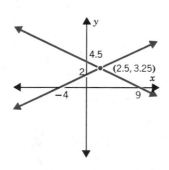

6. is inconsistent.
9. is dependent.
All others are consistent.

B. 1. $(2, 8)$ 2. $(2, -3)$ 3. $(4, 7)$ 4. $(2, 11)$
    5. $(7, 6)$ 6. $(4, 5)$ 7. No solution   8. $(2, 4)$

    9. $(13, -3)$ 10. $(-\frac{1}{9}, -\frac{1}{9})$ 11. $(7, -6)$ 12. $(-5, 2)$

    13. $(-1, 6)$ 14. $(3, -5)$ 15. $(-4, -2)$ 16. $(2, -2)$
    17. $(-1, 4)$ 18. $(-3, -7)$

C. 1. $(9, -4)$ 2. $(2, 5)$ 3. $(4, 1)$ 4. $(2, -1)$
    5. No solution   6. $(6, 2)$ 7. No solution   8. $(-1, 1)$
    9. $(-2, -3)$ 10. Dependent, no unique solution
    11. $(-3, 1)$ 12. $(4, -2)$

13. $(-1, -1.5)$ 14. $(-\frac{2}{3}, 1)$
15. $(4, 7)$ 16. $(-2, 3)$
17. Dependent, no unique solution   18. $(2, -2)$
19. $\left(\dfrac{a + b}{2}, \dfrac{a - b}{2}\right)$ 20. $(1, 1)$
21. $(\frac{1}{2}, \frac{3}{4})$ 22. $(\frac{2}{3}, -\frac{1}{3})$
23. $(-3, \frac{1}{2})$ 24. $(-\frac{3}{4}, 2)$

D. 1. $x = 3, y = 2, z = 8$ 2. $(10, 0)$

3. (a) $\left(\dfrac{a - b}{3}, \dfrac{2a + b}{3}\right)$ (b) $\left(\dfrac{de - bf}{ad - bc}, \dfrac{af - ce}{ad - bc}\right)$

4. $(3, 2)$ 5. 62¢ per dozen for eggs, 51¢ per loaf for bread
6. 212 lb, 106 lb, 53 lb

UNIT 4   *Problem Set 4-4, Page 321*

A. 1. 16 and 23   2. 3 and 11   3. $2\frac{3}{4}$ and $5\frac{3}{4}$
4. 18 nickels,   5. $-10\frac{1}{2}$ and $-31\frac{1}{2}$ 6. 8 and 12
    36 dimes
7. 21 and 29   8. \$2   9. 6 pennies,
                                                11 nickels
10. 112 thirteen cent   11. 3 ft by 4 ft   12. 9 dimes,
    stamps, 88 twenty   2 nickels
    cent stamps
13. 5 and 2   14. 22 ft by 28 ft   15. 21 and 36
16. 3 tens,   17. 3 lb gingersnaps,   18. 30¢ for peas,
    9 twenties   2 lb sugar cookies   35¢ for corn
19. 7 mph   20. plane speed 540 mph,   21. $53\frac{1}{3}$ mph
                        wind speed 60 mph
22. 1120 mi   23. Steve is 19,   24. $6\frac{2}{3}$ liter of 40%,
                        Tom is 7 $13\frac{1}{3}$ liter of 25%
25. 67   26. 12 mi   27. 342 adults,
                                                206 children
28. 8 min   29. 10 mi   30. 5 gal

Self-Test 4
Unit 4, Page 325   1. 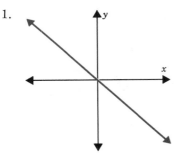   2. 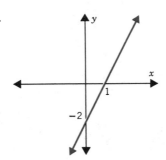   3.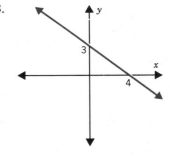

4. 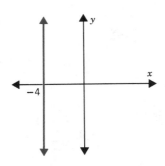   5. $m = -\frac{7}{5}, b = 7$ 6. $m = -\frac{15}{4}, b = -7\frac{1}{2}$

7.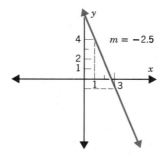
   $m = -2.5$

8.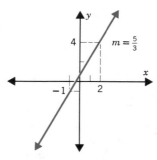
   $m = \frac{5}{3}$

9. $y = 5x - 4$

10. $3y = x + 6$

11. $y = -\frac{1}{2}x + 2$

12. $y = \frac{2}{3}x - 2$

13.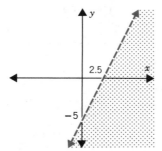

14.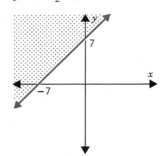

15.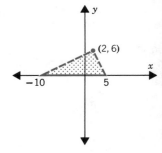

16. $(2, -1)$

17. $(4, 0)$

18. $(-4, 7)$

19. Dependent, no unique solution

20. $(4, 1)$

21. $(3\frac{1}{2}, 2\frac{1}{2})$

22. 4 and 17

23. Speed of boat = 11 mph, speed of current = 3 mph

24. 1.5 liter of 50% and 3.5 liter of 20%

25. 9 lb and 26 lb

UNIT 5   *Problem Set 5-1, Page 349*

A. 1. $\dfrac{2}{x}$ 2. a^3 3. $\dfrac{1}{3x}$ 4. $\dfrac{3}{2y^2}$

5. $\dfrac{1}{4x^4}$ 6. $\dfrac{-1}{3x^2}$ 7. $\dfrac{-4}{y^3}$ 8. $\dfrac{1}{ab^2}$

9. xy^2 10. ab^2 11. $-a$ 12. $\dfrac{2ab^2}{c}$

13. $\dfrac{xy}{7a}$ 14. $\dfrac{ac^2}{3}$ 15. $\dfrac{t}{q}$ 16. 2

17. 2   18. $\dfrac{3}{a + 3}$ 19. $\dfrac{2x - 3}{x^2}$ 20. $\dfrac{a + 2b}{3}$

21. 6   22. $\frac{1}{4}$ 23. $\dfrac{x - y}{x + y}$ 24. 1

25. $a + b$ 26. $\dfrac{x}{x + 2}$ 27. $\dfrac{-1}{2(a - 2)}$ 28. $\dfrac{x}{x - 1}$

29. -1 30. $\dfrac{x - y}{x}$ 31. $\dfrac{a + b}{2b - 1}$ 32. $\dfrac{p + q^2}{pq}$

33. $\dfrac{2x + 1}{1 - 2x}$ 34. $\dfrac{2}{x + 4}$ 35. $\dfrac{a + 4}{a + 1}$ 36. -1

37. $\dfrac{2(x + 2)(x + 2)}{(x - 2)(x + 3)}$ 38. 2   39. $\dfrac{x^2 + x + 2}{x^2 + x + 3}$ 40. $\dfrac{x - 2}{x + 3}$

B. 1. 20   2. abx^2 3. $2ax$ 4. $4x$

5. $4ab$ 6. $3x^2y$ 7. xy 8. $2y^2$

9. $-4x$ 10. $-12x^2$ 11. $-2a^2$ 12. $-8x^2$
13. $-a^3x$ 14. $-2x^2y^2$ 15. $-xy$ 16. $2x^4y$
17. $x + 2$ 18. $6(x + y)$ 19. $5x(x + 1)$ 20. $-6x(x + 1)$
21. $2x(x - y)$ 22. $3(x + 1)$ 23. $-(x + 2)$ 24. $-8a(2a + b)$
25. $x(x - 2)$ 26. $2(1 + y)$ 27. $-4(1 + p)$ 28. $-(x + 1)^2$
29. $x^2 - 1$ 30. $-(a + b)$ 31. $2y(y - 3)$ 32. $-x^2(x + 1)$
33. $2x(x - 1)$ 34. $(x - 1)(x + 1)$ 35. $3x^2(3x + 2)$ 36. $(x + 1)^4$
37. $x(x + 1)^2$ 38. $-a(a - 3)(a + 2)$ 39. $-(x + 1)$ 40. $y(1 - y)$

Mental Math Quiz, Page 354

1. $\dfrac{3}{x}$ 2. $\dfrac{5}{a}$ 3. $\dfrac{x + y}{4}$ 4. $\dfrac{3a}{2}$ 5. $\dfrac{7}{3a}$

6. x 7. $\dfrac{6}{7y}$ 8. $\dfrac{4p}{q}$ 9. $\dfrac{2}{x}$ 10. $\dfrac{3t}{4}$

11. $\dfrac{2}{xy}$ 12. $\dfrac{x - y}{a}$ 13. $\dfrac{b - 2}{x}$ 14. $\dfrac{2 - y}{x}$ 15. $\dfrac{7}{3z}$

16. $\dfrac{3}{x - 1}$ 17. $\dfrac{6k}{t}$ 18. $\dfrac{3}{x^2}$ 19. $\dfrac{3x}{ab^2}$ 20. $\dfrac{xy + x - y}{ab}$

UNIT 5 *Problem Set 5-2, Page 359*

A. 1. $\dfrac{4x}{7}$ 2. $\dfrac{10}{x}$ 3. $\dfrac{a}{x}$ 4. $\dfrac{2 + 3a}{5y}$

5. $\dfrac{2y}{x}$ 6. $\dfrac{y + 4}{x^2}$ 7. $\dfrac{4y}{y - 1}$ 8. $\dfrac{6c + 7}{5c}$

9. $\dfrac{7x - 4y}{xy}$ 10. $\dfrac{4x + 4y}{x - y}$ 11. $\dfrac{3 + 3t}{2t^2}$ 12. $\dfrac{15a - 5}{4 - y}$

13. $\dfrac{3t + 1}{p + 2}$ 14. $\dfrac{2a}{x - b}$ 15. $\dfrac{6a - b}{a^2 + a + 1}$ 16. $\dfrac{7y + 3x}{y - 2x}$

17. $\dfrac{10}{w - 2}$ 18. $\dfrac{4x + 4}{x^2 + 5}$ 19. 0 20. $\dfrac{3a + b + ab}{x}$

21. $\dfrac{3x + 2}{y}$ 22. 1 23. $\dfrac{4x + 4}{y - 1}$

B. 1. $\dfrac{-1}{3b}$ 2. $\dfrac{-4x}{7}$ 3. $\dfrac{6a}{x}$ 4. $\dfrac{-2b}{a}$

5. 1 6. $\dfrac{a - b}{x}$ 7. $\dfrac{2p + 3q}{pq}$ 8. $\dfrac{a - 9}{3(a - 2)}$

9. $\dfrac{a + 8c}{a - c}$ 10. $\dfrac{x + 7}{x + 2y}$ 11. $\dfrac{3 - 2x}{x}$ 12. $\dfrac{7 - 2w}{wx}$

13. $\dfrac{2x - 6}{x + 1}$ 14. $\dfrac{2a - 5}{a + b}$ 15. $\dfrac{3 - x}{x^2 + 3x + 2}$ 16. $\dfrac{x - 6y}{x + y}$

17. $\dfrac{4p - q}{p + q}$ 18. $\dfrac{-9}{3y - 4}$ 19. $\dfrac{4x - 8}{x + 2}$ 20. $\dfrac{2a - b}{a^2}$

21. $\dfrac{a + 2}{x}$ 22. $\dfrac{1 - 4x}{y}$ 23. -1 24. $\dfrac{2y + 2}{y + 4}$

C. 1. $\dfrac{2b + 3a}{ab}$ 2. $\dfrac{3x + 20}{4x}$ 3. $\dfrac{3p + 4}{p^2}$ 4. $\dfrac{3}{x}$

5. $\dfrac{a^2 + b^2}{ab}$ 6. $\dfrac{5x + 4y}{2xy}$ 7. $\dfrac{2x + 25}{5x}$ 8. $\dfrac{2a + 4b}{ab}$

9. $\dfrac{x^2 + 6}{2x^2}$ 10. $\dfrac{38}{5t}$ 11. $\dfrac{5c + d}{c^2d^2}$ 12. $\dfrac{x + 1}{6x^2}$

13. $\dfrac{11t^2 + 2}{st^2}$ 14. $\dfrac{25d}{18}$ 15. $\dfrac{2x}{13}$ 16. $\dfrac{8a + 35ac}{14bc}$

17. $\dfrac{3r + 4}{r^2}$ 18. $\dfrac{3a + 2}{abc}$ 19. $\dfrac{25x - 27}{24}$ 20. $\dfrac{8w + 4}{15w}$

21. $\dfrac{2x - y}{x(x - y)}$ 22. $\dfrac{20x + 4}{(x - 4)(3x + 2)}$ 23. $\dfrac{2a}{(a + b)(a - b)}$

24. $\dfrac{4x + 2}{x(x - 3)}$ 25. $\dfrac{2(x^2 + y^2)}{(x^2 - y^2)}$ 26. $\dfrac{2(x^2 + y^2)}{(x^2 - y^2)}$

27. $\dfrac{x + 1}{2(x^2 - 9)}$ 28. $\dfrac{8x + 1}{2x}$ 29. $\dfrac{x + 1}{x}$

30. $\dfrac{x^2 + 2x + 2}{x + 1}$ 31. $\dfrac{a^2 + 4}{a - 2}$ 32. $\dfrac{-x}{x - 3}$

33. $\dfrac{13}{2a + 6}$ 34. $\dfrac{a^2 + 17a + 4}{4a(a + 1)}$ 35. $\dfrac{5x^2 - 5x + 2}{(x - 3)(x - 2)(2x - 1)}$

36. $\dfrac{x^2 + x + 4}{2x(x + 1)(x - 2)}$ 37. $\dfrac{xy^2 + xy + x}{y^3}$ 38. $\dfrac{5x^2 - x + 4}{4x(x - 1)}$

39. $\dfrac{9a^2 - 19a}{(a - 1)(a + 2)(a - 3)}$ 40. $\dfrac{4x + 1}{x(x + 2)}$

41. $\dfrac{6x^2 + 17x + 6}{x(x + 2)(x + 3)}$ 42. $\dfrac{3y^3 - 2y^2 - 6y + 4}{y(y - 1)(y - 2)}$

43. $\dfrac{2x - 9}{x^2 - 5x}$ 44. $\dfrac{10 - x}{8x - 2x^2}$

45. $\dfrac{2y^2 - 2y - 5}{y(y + 1)(y - 5)}$ 46. $\dfrac{2a^2 - 1}{a^2(a - 1)}$

47. $\dfrac{x^2y + xy^2 - y^2 + x^2}{xy(x - y)}$ 48. $\dfrac{x^3 + 2a^2x - a^3}{ax(x - a)}$

49. $\dfrac{x^2 + x - 1}{x - 1}$ 50. $\dfrac{2x^2 + 9x + 10}{x + 3}$

D. 1. $\dfrac{3a}{8}$ 2. $\dfrac{x}{3}$ 3. $\dfrac{4xy}{45}$ 4. $\dfrac{5t - 4p}{20}$

5. $\dfrac{3x - 10y}{30}$ 6. $\dfrac{xy - a^2}{ay}$ 7. $\dfrac{y}{3}$ 8. $\dfrac{3 - 4x}{x^2}$

9. $\dfrac{7 - 4a^2}{a^3}$ 10. $\dfrac{-7x - 8}{12}$ 11. $\dfrac{4a + 45}{15a}$ 12. $\dfrac{-11n + 7}{20n}$

13. $\dfrac{53 - 3b}{8}$ 14. $\dfrac{-10c - 2}{5}$ 15. $\dfrac{17x + 2}{2x}$ 16. $\dfrac{23 - y}{15}$

17. $\dfrac{-8b^2 + 7b - 1}{b^2}$ 18. $\dfrac{41a + 14}{(a + 1)(5a + 2)}$ 19. $\dfrac{-16}{d^2 - 16}$

20. $\dfrac{-3x - 6}{x^2 - 16}$ 21. $\dfrac{3xy - 3y^2 - 5y}{x^2 - y^2}$ 22. $\dfrac{8}{a^2 - ab}$

23. $\dfrac{4b^2 + 7ab - 3a^2}{4b(a - b)}$ 24. $\dfrac{x^2 - 3x + 6}{3x^2}$ 25. $\dfrac{11x^2 + 30xy - 13y^2}{2(x^2 - y^2)}$

26. $\dfrac{-3t^3 + 12t - 23}{5t - 5}$ 27. $\dfrac{5q^2 + 7q + 2}{3q(q - 1)}$ 28. $\dfrac{-x^2 + 2xy + y^2}{x(x^2 - y^2)}$

29. $\dfrac{x^2 + x - 1}{x + 1}$ 30. $\dfrac{-y^2 + 7y + 3}{y^2 - 9}$ 31. $\dfrac{-2}{(2x + 1)(x + 1)(x - 1)}$

32. $\dfrac{x^2 - 3x - 2}{2(x + 5)(x - 1)}$ 33. $\dfrac{-x^2 + x - 3}{x^2(x - 3)}$ 34. $\dfrac{2}{(y - 1)(y - 3)}$

35. $\dfrac{x^2 - xy - x - y}{x(x + y)}$ 36. $\dfrac{xy - y^2 - x}{x(x - y)}$ 37. $\dfrac{x^2 - 3x}{x - 2}$

38. $\dfrac{-y^2}{y + 2}$ 39. $\dfrac{3}{(x - 2)(x + 1)(x - 6)}$

40. $\dfrac{y - 1}{(y - 3)(y + 4)(y - 5)}$

Answers to box on Page 364, Negative Fractions

1. $\dfrac{-2}{3}$ 2. $\dfrac{-2x}{3}$ 3. $\dfrac{-5x^2}{3y}$ 4. $\dfrac{-(x + 2)}{3x}$

5. $\dfrac{3y}{2a}$ 6. $\dfrac{4x^2}{5y}$ 7. $\dfrac{pq}{2a}$ 8. $\dfrac{-3x^2}{11y}$

9. $\dfrac{x + 1}{x + 2}$ 10. 1 11. 1 12. $\dfrac{2t}{t + 1}$

13. -1 14. -2 15. 1 16. -1

UNIT 5 *Problem Set 5-3, Page 373*

A. 1. $\dfrac{3x}{2y}$ 2. $\dfrac{2a^2}{3}$ 3. $\dfrac{5x}{3y}$ 4. $3b$ 5. $\dfrac{10}{3}$

6. $\dfrac{3y}{8x}$ 7. $\dfrac{3a^2}{10}$ 8. $\dfrac{7}{2p}$ 9. $\dfrac{3}{x}$ 10. $\dfrac{5x^2}{3}$

11. 4 12. $\dfrac{3b}{a}$ 13. $\dfrac{p}{q^3}$ 14. $\dfrac{8a^2c}{b^2}$ 15. $\dfrac{8xt}{3y}$

16. $4x^2$ 17. $\dfrac{a}{3c}$ 18. $\dfrac{2x^2}{3y}$ 19. $\dfrac{t^2}{2}$ 20. $\dfrac{13xz^3}{y^3}$

21. $\dfrac{-4}{x^2}$ 22. $-2xy$ 23. $\dfrac{-5a}{2b}$ 24. $\dfrac{-3x}{2y}$ 25. $\dfrac{-4xy^3}{z}$

26. $\dfrac{-2a^4}{3}$ 27. $\dfrac{-4x^4}{a^4}$ 28. $\dfrac{-5x^5}{4a^3}$ 29. $\dfrac{2x^3y^2}{a^4}$ 30. $\dfrac{4p^3q^3}{t^5}$

B. 1. $\dfrac{15}{4}$ 2. $\dfrac{4x}{3(x - 1)}$ 3. $\dfrac{5(x - 1)}{4(x + 1)}$ 4. $\dfrac{-x(x - 4)}{(x + 5)}$

5. $\dfrac{2a}{b}$ 6. $\dfrac{8(p - a)}{p(p + a)}$ 7. $\dfrac{2x^2}{x + 2}$ 8. $\dfrac{x + 1}{x}$

9. $\dfrac{x(x + 3)}{2(x + 2)}$ 10. 1 11. $\dfrac{2x(x + 2)}{3(x + 3)}$ 12. $\dfrac{b}{2}$

13. $\dfrac{a}{b(a + 2)}$ 14. $\dfrac{-2}{x}$ 15. $\dfrac{x^2y}{x + 1}$ 16. 3

17. $\dfrac{y^2 - y}{x + 2}$ 18. $\dfrac{x^2 - 1}{x^2 - 4}$ 19. $\dfrac{x^2 - 8x + 12}{x^2 + x - 2}$ 20. $\dfrac{3x^2 - 24x}{2xy + 8y}$

21. $\dfrac{2b - 4}{3ab + 6b}$

C.   1. 1   2. $\dfrac{x^2}{2}$ 3. $\dfrac{x^2}{y^2}$ 4. a 5. $\dfrac{1}{xy}$

     6. $2x$ 7. $\dfrac{a}{3b}$ 8. $2p$ 9. $\dfrac{9}{x^2}$ 10. $\dfrac{2}{3p}$

    11. a^3 12. $\dfrac{1}{a^3}$ 13. $\dfrac{y}{2}$ 14. $\dfrac{2x}{ab}$ 15. $\dfrac{ad}{bc}$

    16. x 17. $\dfrac{8}{3a^2b}$ 18. $\dfrac{2}{y^2}$ 19. $2xy$ 20. 36

    21. $\dfrac{-2x}{5}$ 22. $-b^2$ 23. $\dfrac{-x^3}{y}$ 24. $\dfrac{-1}{2q}$ 25. $\dfrac{-ad}{bc}$

    26. $\dfrac{ad}{bc}$ 27. $\dfrac{-ad}{bc}$ 28. $\dfrac{ad}{bc}$ 29. $\dfrac{4}{q}$ 30. $\dfrac{2}{a^2x}$

    31. $\dfrac{3}{y^3}$ 32. $\dfrac{3y}{5}$ 33. $\dfrac{4}{xy^2}$ 34. $\dfrac{x}{y}$ 35. $\dfrac{x^3}{y^3}$

    36. $2b^3$ 37. x^3y 38. $\dfrac{2}{x^2}$ 39. $\dfrac{y}{9x^2}$ 40. $16x^3y^3$

    41. $4a^3b^4c^2$ 42. $\dfrac{4}{3x}$

D.   1. $\dfrac{x+1}{x+3}$ 2. $\dfrac{3}{2a}$ 3. $\dfrac{2x(x+1)}{y(x-1)}$ 4. $\dfrac{x-4}{2ax}$ 5. $\dfrac{6}{x}$

    6. $\dfrac{ad}{bc}$ 7. $\dfrac{(x+1)(x-4)}{(x+2)(x-2)}$ 8. $\dfrac{1-a}{1-b}$ 9. 1 10. $\dfrac{2(x+4)}{3x}$

    11. $\dfrac{2x-1}{x-4}$ 12. $\dfrac{p+2}{p}$ 13. a^2-ab 14. $\dfrac{1}{3x+6}$ 15. $\dfrac{x-1}{x^2+1}$

    16. $\dfrac{x+2}{x^2+2}$ 17. $-\dfrac{x-2}{x-4}$ 18. $-\dfrac{x-5}{x-3}$ 19. $\dfrac{x+1}{x-1}$ 20. $\dfrac{-2}{a}$

    21. $\dfrac{x^2+1}{2}$ 22. $a+1$ 23. a 24. $\dfrac{-a}{b}$ 25. $\dfrac{y-x}{y+x}$

    26. $\dfrac{x+1}{x}$

E.   1. $\dfrac{1}{x-1}$ 2. $674{,}321{,}870$ The expression is $\dfrac{A}{(A+1)^2 - A(A+2)} = A$

                  where $A = 674{,}321{,}870$

    3. (a) $\dfrac{7}{5}$

       (c) $1 + \dfrac{1}{2} = \dfrac{3}{2} = 1.5$ (d) The approximations are
3, $\frac{22}{7} \cong 3.142857$, $\frac{333}{106} \cong 3.1415094$,
$\frac{355}{113} \cong 3.141593$, $\frac{103993}{33102} \cong 3.14159265$
The number is π.

$$1 + \cfrac{1}{2 + \dfrac{1}{2}} = \dfrac{7}{5} = 1.4$$

$$1 + \cfrac{1}{2 + \cfrac{1}{2 + \dfrac{1}{2}}} = \dfrac{17}{12} = 1.4166\ldots$$

      (e) 1, 2, $\frac{3}{2} = 1.5$, $\frac{5}{3} \cong 1.666$, $\frac{8}{5} = 1.600$,
$\frac{13}{8} = 1.6250$, $\frac{21}{13} \cong 1.61538\ldots$
Each numerator is the sum of the previous two numerators and each denominator is the sum of the previous two denominators. The numbers in the sequence approach closer and closer to $1.61803398\ldots$ or
$\dfrac{1+\sqrt{5}}{2}$.

4. $\left(1 + \dfrac{1}{n}\right)(n + 1) = (n + 2) + \dfrac{1}{n}$ and $\left(1 + \dfrac{1}{n}\right) + (n + 1) = (n + 2) + \dfrac{1}{n}$

5. $v = \dfrac{2c}{3}$

UNIT 5   *Problem Set 5-4, Page 399*

A. 1. 8   2. 2   3. $1\frac{1}{3}$ 4. $13\frac{1}{2}$ 5. $\frac{15}{16}$ 6. -1
    7. 2   8. 4   9. 2   10. 0   11. 5   12. -4
    13. 2   14. 9   15. -1 16. 3   17. -2 18. $-\frac{4}{3}$
    19. 40   20. -1 21. $\frac{1}{3}$ 22. 3   23. $-\frac{1}{2}$ 24. $-\frac{1}{3}$

B. 1. $\dfrac{ab}{a - b}$ 2. $\dfrac{at}{c}$ 3. $\dfrac{Qr + b}{a + Q}$ 4. $\dfrac{At}{n - Bt}$

   5. $\dfrac{2a}{a - 1}$ 6. $\dfrac{a(a - b)}{b}$ 7. $\dfrac{c}{a - b}$ 8. $\dfrac{A}{1 + tr}$

   9. $\dfrac{2S}{L + a}$ 10. $\dfrac{dn + a}{1 + d}$

C. 1. $(4, 1)$ 2. $(6, 10)$ 3. $(3, 5)$

   4. $(-4, -5)$ 5. $\left(\dfrac{12}{5}, \dfrac{16}{5}\right)$ 6. $(0, -1)$

D. 1. -6 2. 3   3. $\frac{3}{4}$ 4. 3 and 12   5. \$3060 and \$24,480
    6. 868 tickets   7. 2 and 4   8. \$4 and \$16
    9. 20 nickels, 60 dimes, 12 quarters   10. 6, 8, and 10
    11. 10 mph   12. $1\frac{5}{7}$ days   13. 40 walnuts, 60 raisins, 240 peanuts
    14. $13\frac{1}{3}$ 15. $93\frac{1}{3}$¢ or 93¢ rounded   16. (a) 182   (b) 63
    16. (c) 42   (d) 27   17. (a) 10   (b) $-\frac{1}{2}$ (c) 11   (d) -2
    18. 31 minutes   19. 240   20. $11\frac{1}{4}$ in.   21. 56
    22. 4   23. 17.5 lb

   24. (a) $S = kD$ (b) $D = kt^2$ (c) $C = \dfrac{k}{A}$ (d) $A = kH$

      (e) $v = \dfrac{k}{s}$ (f) $H = kT$

E. 1. (a) F decreases   (b) F is $\frac{1}{4}$ as large   (c) F is 4 times as large.
    2. (a) $\frac{133}{8}$ (b) $\frac{32}{3}$ (c) 12   (d) $17\frac{1}{2}$ (e) 9   (f) $\frac{1386}{109}$

   3. 10   4. $\dfrac{2x^2y^2 + 2}{xy}$

   5. (a) 1   (b) $x + y$ (c) $\dfrac{4pq}{p^2 - q^2}$ (d) y

   6. 2 mi + $\frac{2}{3}$ mi + $\frac{1}{3}$ mi or 3 mi

UNIT 5   *Problem Set 5-5, Page 427*

A. 1. $2\sqrt{7}$ 2. $3\sqrt{3}$ 3. $3\sqrt{5}$ 4. $4\sqrt{3}$ 5. $-3\sqrt{7}$
    6. $5\sqrt{5}$ 7. $10\sqrt{2}$ 8. $-7\sqrt{2}$ 9. $6\sqrt{3}$ 10. $8\sqrt{5}$
    11. $-2\sqrt{11}$ 12. $3\sqrt{13}$ 13. $-5\sqrt{2}$ 14. $10\sqrt{7}$ 15. $9\sqrt{7}$
    16. $p\sqrt{p}$ 17. $-x^2\sqrt{x}$ 18. n 19. $2x$ 20. $2\sqrt{y}$
    21. $-3\sqrt{ab}$ 22. $\dfrac{2\sqrt{x}}{a}$ 23. $2x\sqrt{5x}$ 24. $2y^2\sqrt{6y}$ 25. $-a\sqrt{2b}$
    26. $2p^3\sqrt{3}$ 27. $3a^2\sqrt{5b}$ 28. $xy\sqrt{xz}$ 29. $-bc\sqrt{ac}$ 30. $6pq\sqrt{q}$
    31. $-4\sqrt{x}$ 32. $9a^2\sqrt{a}$ 33. $-4a^2\sqrt{3a}$ 34. $3a\sqrt{a}$ 35. $4x^2\sqrt{y}$
    36. $3(x + y)\sqrt{2}$ 37. $2(a - b)\sqrt{2(a - b)}$ 38. $2x + 1$
    39. $x\sqrt{x + y}$ 40. $-x\sqrt{2 - x}$

B.
1. $5\sqrt{5}$ 2. $2\sqrt{3}$ 3. $-3\sqrt{6}$ 4. $2\sqrt{5}$ 5. $5\sqrt{6}$
6. $-\sqrt{3}$ 7. $\sqrt{x}$ 8. $-\sqrt{a}$ 9. $3\sqrt{pq}$ 10. $-2\sqrt{y}$
11. $-3\sqrt{x}$ 12. $3\sqrt{2a}$ 13. $(a-b)\sqrt{b}$ 14. $(k-1)\sqrt{x}$ 15. $3\sqrt{3t}$
16. $4\sqrt{2}$ 17. $4\sqrt{3}$ 18. $\sqrt{5}$ 19. $5\sqrt{x}$ 20. $\sqrt{a}$
21. $4\sqrt{2x}$ 22. $(a+1)\sqrt{ab}$ 23. $-4\sqrt{2x}$ 24. $(2x-a)\sqrt{x}$
25. $4\sqrt{3p}$ 26. $(x+y)\sqrt{3}$ 27. $-p\sqrt{2}$ 28. $2\sqrt{2}$
29. 0 30. $(1+2a+3b)\sqrt{ab}$ 31. $\dfrac{1-\sqrt{3}}{3}$ 32. $\dfrac{2+\sqrt{2}}{a}$
33. $\dfrac{\sqrt{3}-\sqrt{2}}{x}$ 34. $\dfrac{2-5\sqrt{2}}{10}$ 35. $\dfrac{3b-a\sqrt{5}}{ab}$ 36. $\dfrac{6+x\sqrt{x}}{3x}$
37. $\dfrac{3\sqrt{3}-5}{15}$ 38. $\dfrac{4\sqrt{3}-3\sqrt{2}}{6}$ 39. $\dfrac{2\sqrt{3}-6}{3}$ 40. $\dfrac{2\sqrt{5}+6}{3}$

C.
1. $\sqrt{15}$ 2. $2\sqrt{3}$ 3. $2\sqrt{6}$ 4. $\sqrt{14}$ 5. $3\sqrt{2}$
6. 4 7. $\sqrt{35}$ 8. $\sqrt{30}$ 9. $\sqrt{30}$ 10. $6\sqrt{5}$
11. $5\sqrt{6}$ 12. $4\sqrt{15}$ 13. $\sqrt{xy}$ 14. a 15. $b\sqrt{c}$
16. $x\sqrt{2y}$ 17. $a\sqrt{6}$ 18. $ab\sqrt{6}$ 19. $2pq\sqrt{3q}$ 20. $3ab\sqrt{2b}$
21. $4y^2\sqrt{2}$ 22. $8t\sqrt{6}$ 23. $5k\sqrt{2k}$ 24. $6a\sqrt{6}$ 25. $6x\sqrt{x}$
26. $a^2\sqrt{15}$ 27. x^2y^2 28. $x^5\sqrt{x}$ 29. $2+2\sqrt{6}$
30. $3\sqrt{2}+3\sqrt{3}$ 31. $5\sqrt{3}-5$ 32. $4\sqrt{2}+2\sqrt{6}$ 33. $2\sqrt{2}+2$
34. $\sqrt{6}+3$ 35. $\sqrt{6}-3\sqrt{2}$ 36. $2\sqrt{5}+5\sqrt{2}$ 37. $3-\sqrt{15}$
38. $2-2\sqrt{3}$ 39. $x-\sqrt{xy}$ 40. $ab\sqrt{ab}-a^2b$ 41. $3+2\sqrt{2}$
42. -1 43. $6+4\sqrt{2}$ 44. $5+2\sqrt{6}$ 45. $-\sqrt{3}$
46. $\sqrt{5}+\sqrt{10}-\sqrt{2}-2$ 47. a^2-b 48. $x^2+y+2x\sqrt{y}$
49. $4+9x+12\sqrt{x}$ 50. $9+\sqrt{15}$

D.
1. $\dfrac{\sqrt{6}}{6}$ 2. $\dfrac{\sqrt{5}}{5}$ 3. $\dfrac{4\sqrt{y}}{y}$ 4. $\dfrac{\sqrt{2}}{2x}$
5. $\dfrac{\sqrt{2y}}{2xy}$ 6. $\sqrt{2}$ 7. $\dfrac{\sqrt{14}}{4}$ 8. $\dfrac{2\sqrt{3}}{3}$
9. $\dfrac{\sqrt{15}}{3}$ 10. $\dfrac{\sqrt{x}}{x}$ 11. $\dfrac{2\sqrt{2q}}{pq}$ 12. $\dfrac{\sqrt{6}}{2}$
13. $\dfrac{\sqrt{10}}{2}$ 14. $\dfrac{\sqrt{30}}{5}$ 15. $\sqrt{14}$ 16. $2\sqrt{3}$
17. $2x\sqrt{3}$ 18. $\dfrac{\sqrt{6t}}{t}$ 19. $\sqrt{3a}$ 20. $\sqrt{2x}$
21. $\dfrac{2x\sqrt{y}}{y}$ 22. $\dfrac{2\sqrt{abc}}{ab}$ 23. $\dfrac{5\sqrt{3x}}{2x}$ 24. $\dfrac{3\sqrt{6pt}}{2p^2}$
25. $-2(1+\sqrt{2})$ 26. $-2(1-\sqrt{3})$ 27. $3(2-\sqrt{2})$
28. $2(\sqrt{6}+1)$ 29. $\dfrac{a(1-\sqrt{a})}{1-a}$ 30. $\dfrac{\sqrt{x}-x}{1-x}$
31. $\dfrac{1+x+2\sqrt{x}}{1-x}$ 32. $\dfrac{x+y+2\sqrt{xy}}{x-y}$

E.
1. $\sqrt{7}$ 2. $\sqrt[3]{6}$ 3. $2\sqrt{3}$ 4. $2\sqrt[3]{2}$
5. $4\sqrt{x}$ 6. $\sqrt[3]{17y^2}$ 7. $\dfrac{\sqrt{5}}{2}$ 8. $\dfrac{\sqrt[3]{17}}{2}$
9. $\dfrac{\sqrt{5}}{5}$ 10. $\dfrac{\sqrt[3]{36}}{6}$ 11. $\sqrt[3]{4}$ 12. $\sqrt[4]{27}$

13. $16\sqrt{2}$ 14. $3\sqrt[3]{3}$ 15. $\sqrt{xy}$ 16. $\sqrt[3]{ab^2}$

17. $x\sqrt[3]{xy}$ 18. $\sqrt[4]{p^3q}$ 19. $b\sqrt{a}$ 20. $y\sqrt[3]{x^2}$

21. $3^{1/2}$ 22. $11^{1/3}$ 23. $\dfrac{2^{1/2}}{3^{1/2}}$ 24. $\dfrac{7^{1/2}}{5^{1/2}}$

25. $x^{1/2}y^{1/2}$ 26. $a^{1/2}b^{3/2}$ 27. $\dfrac{p^{1/2}}{q^{1/2}}$ 28. $\dfrac{a^{1/2}}{b^{3/2}}$

29. $3^{1/4}t^{1/4}$ 30. $2^{1/3}x^{2/3}$ 31. $x^{4/3}$ 32. $y^{1/3}$

33. $\dfrac{1}{x^{1/3}}$ 34. $\dfrac{1}{t^{1/4}}$ 35. $\dfrac{3}{y^{1/2}}$ 36. $\dfrac{5}{x^{2/3}}$

37. $x^{2/3}$ 38. $2x^{3/4}$ 39. $3x^2y^{2/3}$ 40. $a^3b^{3/4}$

41. x^2 42. $y^{16/15}$ 43. a 44. $\dfrac{1}{x^{5/3}}$

45. $\dfrac{1}{x^{5/2}}$ 46. $\dfrac{1}{x^4}$ 47. $\dfrac{1}{x^{3/5}y^{6/5}}$ 48. $\dfrac{x^{3/4}}{y^{1/2}}$

49. $\dfrac{1}{x^{9/10}}$ 50. $y^{7/2}$ 51. $\dfrac{b^{4/3}}{a^8}$ 52. $\dfrac{x^{2/3}}{y^{5/3}}$

53. $\dfrac{1}{x^{5/12}}$ 54. $y^{23/12}$ 55. $a^{13/10}$ 56. $x^{19/12}$

57. $\dfrac{4y^{3/2}}{x^4z}$ 58. $\dfrac{b^9}{27a^{3/4}c^2}$ 59. $\dfrac{a^{3/2}c^2}{b^{3/4}}$ 60. $\dfrac{y^8}{xz^{4/3}}$

F. 1. $\sqrt{a + \dfrac{a}{a^2 - 1}} = \sqrt{\dfrac{a(a^2 - 1) + a}{a^2 - 1}}$

$$= \sqrt{\dfrac{a^3 - a + a}{a^2 - 1}}$$

$$= \sqrt{\dfrac{a^3}{a^2 - 1}}$$

$$= a\sqrt{\dfrac{a}{a^2 - 1}}$$

2. $\sqrt{a^2 + \dfrac{a^2}{a^2 - 1}} = \sqrt{\dfrac{a^2(a^2 - 1) + a^2}{a^2 - 1}}$

$$= \sqrt{\dfrac{a^4 - a^2 + a^2}{a^2 - 1}}$$

$$= \sqrt{\dfrac{a^4}{a^2 - 1}}$$

$$= a^2\sqrt{\dfrac{1}{a^2 - 1}}$$

3. (a) Square both sides: $19 + 6\sqrt{2} = (1 + 3\sqrt{2})^2$
   (b) Square both sides: $75 + 12\sqrt{21} = (3\sqrt{7} + 2\sqrt{3})^2 = 63 + 12\sqrt{21} + 12$

4. (a) $L = 10\sqrt{99} \cong 99.5$ ft   (b) $L = 60$ ft   (c) $L = 0$ ft

5. 228.2 miles

6. (a) No. The diagonal of the box is 18.97 cm.
   (b) Yes. The diagonal of the cube is 19.05 cm.

7. 36,660 ft/sec or about 25,000 mph

8. $(x - y) = (x^{1/2} + y^{1/2})(x^{1/2} - y^{1/2}) = (\sqrt{x} + \sqrt{y})(\sqrt{x} - \sqrt{y})$

Self-Test 5
Unit 5, Page 433

1. $\frac{2}{3}$ 2. $\dfrac{2 - 3x}{x - 1}$ 3. $-4a^3b$ 4. $\dfrac{x^2 + 4}{x + 1}$ 5. $\dfrac{x}{1 - y}$

6. $\dfrac{11a}{15}$ 7. $\dfrac{3x - y}{3x^2y}$ 8. $\dfrac{x}{x-1}$ 9. $\dfrac{2a^2}{5}$ 10. $\dfrac{x-1}{x+1}$

11. $\dfrac{xy}{3}$ 12. $2y\sqrt{5xy}$ 13. $3xy\sqrt{2x}$ 14. $\dfrac{a\sqrt{ab}}{b}$ 15. $2\sqrt{3x}$

16. $\dfrac{y - \sqrt{y}}{y-1}$ 17. $6 - x + \sqrt{x}$ 18. $3\sqrt{2x}$ 19. $x^2\sqrt{ab}$ 20. 36

21. $-4\frac{1}{4}$ 22. -12 23. 8 mph and 20 mph   24. 14 25. 21

26. (a) $x^{3/5}$ (b) $\dfrac{2y^{2/5}}{x^{4/3}z^3}$ (c) $\dfrac{x^{2/3}}{y^{8/9}}$ (d) $\dfrac{1}{x^{7/12}}$ (e) $\dfrac{x^9}{2^3y^{9/2}z^{3/4}}$

UNIT 6   *Problem Set 6-1, Page 461*

A. 1. $1, -1$ 2. $\sqrt{6}, -\sqrt{6}$ 3. $3\sqrt{2}, -3\sqrt{2}$
    4. $2\sqrt{3}, -2\sqrt{3}$ 5. $\sqrt{5}, -\sqrt{5}$ 6. $2\sqrt{2}, -2\sqrt{2}$
    7. $2, -2$ 8. $\sqrt{3}, -\sqrt{3}$ 9. No real roots
    10. $\dfrac{\sqrt{7}}{2}, \dfrac{-\sqrt{7}}{2}$ 11. $\dfrac{2\sqrt{3}}{3}, \dfrac{-2\sqrt{3}}{3}$ 12. No real roots
    13. $\pm\dfrac{\sqrt{6}}{3}$ 14. $\pm\dfrac{\sqrt{2}}{2}$ 15. $\pm\dfrac{\sqrt{6}}{6}$
    16. $\pm\sqrt{2}$ 17. ±2 18. $\pm\dfrac{\sqrt{6}}{2}$

B. 1. $c = 10$ 2. $c = 25$ 3. $b = 60$ 4. $a = 15$
    5. $c = \sqrt{2}$ 6. $c = 73$ 7. $a = 35$ 8. $c = 41$
    9. $a = \sqrt{105}$ 10. $b = 2\sqrt{11}$ 11. $c = \sqrt{89}$ 12. $a = 4\sqrt{15}$
    13. $a = 2\sqrt{30}$ 14. $x = \sqrt{29}$ 15. $a = 8\sqrt{3}$ 16. $c = \dfrac{\sqrt{13}}{6}$

C. 1. $3, -1$ 2. $8, -2$ 3. $4, 3$ 4. $2, \frac{5}{2}$
    5. $\frac{1}{3}, -\frac{1}{2}$ 6. $7, -3$ 7. $6, -3$ 8. $6, -1$
    9. $3, -\frac{1}{2}$ 10. $\frac{3}{2}, -1$ 11. $1, \frac{2}{3}$ 12. $\frac{1}{3}, -3$
    13. $-\frac{1}{2}, -\frac{1}{3}$ 14. $-\frac{2}{3}, -\frac{3}{2}$ 15. $\frac{2}{5}, -\frac{1}{5}$ 16. $\frac{5}{3}, -\frac{2}{3}$
    17. $\frac{1}{2}, -\frac{5}{2}$ 18. $\frac{4}{3}, -\frac{1}{2}$ 19. $-\frac{3}{2}$ 20. $5, -\frac{3}{2}$
    21. $1, -7$ 22. $4, -\frac{1}{2}$ 23. $1, 2$ 24. $\frac{1}{3}, \frac{2}{3}$
    25. $8, -3$ 26. $\frac{3}{2}, -1$ 27. $-2, -4$ 28. $4, 6$
    29. $3, 11$ 30. $4, -\frac{5}{2}$ 31. $\frac{1}{3}$ 32. $\frac{3}{4}, -1$
    33. $4, -1$ 34. $7, -2$ 35. $0, 1$ 36. 0
    37. $-1, 4$ 38. $3, -4$ 39. $0, -3$ 40. $-3, 5$

D. 1. $\dfrac{\sqrt{2}}{2}, \dfrac{-\sqrt{2}}{2}$ 2. $\dfrac{5}{2}, \dfrac{-5}{2}$ 3. $\dfrac{\sqrt{6}}{2}, \dfrac{-\sqrt{6}}{2}$ 4. $\frac{1}{2}, -\frac{1}{2}$
    5. $1, -3$ 6. $1, -4$ 7. 2 8. $2, 4$
    9. $\dfrac{3}{2}, \dfrac{-4}{3}$ 10. $4, \dfrac{-5}{2}$ 11. $1, 2$ 12. $3, 5$
    13. $5, -2$ 14. $8, -4$ 15. $5, -9$ 16. $-2, 3$
    17. $1, -2$ 18. 0 19. $2, -\frac{1}{2}$ 20. $5, 10$
    21. $1, -5$ 22. $-1, 7$ 23. $-1, \frac{2}{3}$ 24. $-2, \frac{4}{5}$
    25. $1, \frac{3}{2}$ 26. $-2, \frac{4}{3}$ 27. $\frac{3}{2}, -\frac{1}{5}$ 28. $2, -\frac{3}{7}$
    29. $4, \frac{1}{2}$ 30. $\frac{1}{4}, -\frac{2}{3}$

E. 1. $1, -1, -2$ 2. $0, a + b$ 3. $3, \dfrac{-1}{3}$ 4. $\pm\sqrt{a^2 - ab + b^2}$

5. (a) $\dfrac{\sqrt{5}}{5}$ (b) $\dfrac{2\sqrt{5}}{5}$ (c) $\dfrac{a\sqrt{5}}{5}$ (d) $3\sqrt{2}$ (e) $x = \sqrt{2}$

6. Originally there were 24 rows of 25 plants.
7. $c^2 = a^2 + b^2$ gives $b^2 = c^2 - a^2 = (c - a)(c + a)$

But if $c = a + 2$ then $c - a = 2$ and $b^2 = 2(c + a)$ or $b = \sqrt{2c + 2a}$

8. (a) In addition to the eight in the table, $(21, 20, 29)$, $(63, 16, 65)$, $(33, 56, 65)$, $(55, 48, 73)$, $(13, 84, 85)$, $(77, 36, 85)$, $(39, 80, 89)$, and $(65, 72, 97)$.

(b) $b + c = \dfrac{m^2 - n^2}{2} + \dfrac{m^2 + n^2}{2} = \dfrac{2m^2}{2} = m^2$

(c) $a + c = mn + \dfrac{m^2 + n^2}{2} = \dfrac{m^2 + 2mn + n^2}{2} = \dfrac{(m + n)^2}{2} = 2\left[\dfrac{m + n}{2}\right]^2$

Answers to box on Page 481, Equations Containing Radicals

(a) 7   (b) 2   (c) 5   (d) 12   (e) 5   (f) $\frac{1}{2}$ (g) 5, 3   (h) $\dfrac{-1 + \sqrt{13}}{2}$

UNIT 6   *Problem Set 6-2, Page 485*

A. 1. 2   2. $3 \pm 2\sqrt{7}$ 3. $\dfrac{-3 \pm \sqrt{29}}{2}$ 4. $1 \pm \sqrt{7}$

5. 1, 3   6. $3 \pm 3\sqrt{2}$ 7. $-1 \pm \sqrt{3}$ 8. $\dfrac{1 \pm \sqrt{33}}{2}$

9. $\dfrac{-5 \pm 3\sqrt{5}}{2}$ 10. $\frac{3}{2}, -2$ 11. $\frac{1}{3}, -2$ 12. $\dfrac{3 \pm \sqrt{41}}{4}$

13. $\frac{1}{2}, \frac{2}{3}$ 14. $\dfrac{2 \pm \sqrt{14}}{2}$ 15. $\dfrac{10 \pm \sqrt{110}}{2}$ 16. $\dfrac{2 \pm \sqrt{10}}{3}$

17. $1 \pm \sqrt{5}$ 18. $-2 \pm \sqrt{3}$ 19. $-1, -3$ 20. $\dfrac{3 \pm \sqrt{5}}{2}$

B. 1. 3, 4   2. $1, \frac{2}{5}$ 3. $\dfrac{-1 \pm \sqrt{33}}{4}$ 4. $\dfrac{1 \pm 3\sqrt{5}}{2}$

5. $\dfrac{-1 \pm \sqrt{41}}{4}$ 6. 0, 7   7. $\pm\sqrt{6}$ 8. $\dfrac{-3 \pm \sqrt{29}}{2}$

9. $\dfrac{1 \pm \sqrt{6}}{5}$ 10. $-1, 10$ 11. $-\frac{1}{2}, -3$ 12. 1, 3

13. $\dfrac{7 \pm \sqrt{269}}{10}$ 14. $-1 \pm \sqrt{3}$ 15. $\frac{1}{5}, -1$ 16. $\dfrac{3 \pm \sqrt{29}}{2}$

17. $4 \pm 2\sqrt{6}$ 18. $\dfrac{-1 \pm \sqrt{21}}{10}$ 19. $1, \frac{1}{4}$ 20. $\dfrac{2 \pm \sqrt{10}}{2}$

21. 1, 2   22. $\dfrac{3 \pm \sqrt{37}}{2}$ 23. $\dfrac{2 \pm \sqrt{2}}{2}$ 24. $\dfrac{5 \pm \sqrt{13}}{6}$

25. 0, -4 26. $\dfrac{1 \pm \sqrt{3}}{2}$ 27. $\frac{1}{2}, -2$ 28. $\dfrac{-5 \pm \sqrt{37}}{6}$

29. $-1 \pm \sqrt{3}$ 30. $\dfrac{-5 \pm \sqrt{89}}{8}$ 31. $1 \pm \sqrt{7}$ 32. $\dfrac{5 \pm \sqrt{65}}{4}$

33. $\dfrac{3 \pm \sqrt{5}}{2}$ 34. $\dfrac{-5 \pm \sqrt{21}}{2}$ 35. $3 \pm \sqrt{13}$ 36. $\dfrac{-3 \pm \sqrt{113}}{6}$

37. $\dfrac{1 \pm \sqrt{5}}{2}$ 38. $\dfrac{-1 \pm \sqrt{33}}{2}$ 39. $-1 \pm \sqrt{6}$ 40. $\dfrac{-1 \pm \sqrt{17}}{2}$

C. 1. (a) $\dfrac{6 \pm \sqrt{3}}{3}$ (b) $\frac{3}{4}, \frac{5}{8}$ 2. $-4, -7$

3. (a) $53, -49$ (b) $-53, -49$

4. (a) $1 + \sqrt{2} \pm \sqrt{3}$ (b) $-3\sqrt{2} \pm \sqrt{13}$

6. (a) $\dfrac{1 + \sqrt{5}}{2}$ (b) $\sqrt{2}$ 7. (a) 2 (b) 4 (c) $\dfrac{-1 + \sqrt{5}}{2}$

   (d) $\dfrac{1 + \sqrt{1 + 4n}}{2}$ (e) $\dfrac{-1 + \sqrt{1 + 4n}}{2}$

8. (a) $G = \dfrac{1 + \sqrt{5}}{2} \cong 1.62$

   (c) The pattern is $G^n = A_n + B_n \cdot G$, where A_n is equal to B_{n-1} from the previous equation, and $B_n = A_{n-1} + B_{n-1}$ from the previous equation.

   (d) The ratio of the areas is G.

9. $X = \dfrac{AY \pm A\sqrt{Y^2 - 4}}{2}$

UNIT 6   *Problem Set 6-3, Page 507*

A. 1. 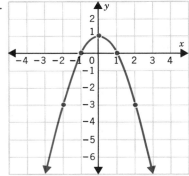 Roots $= \pm 1$

2. 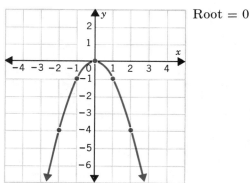 Root $= 0$

3. 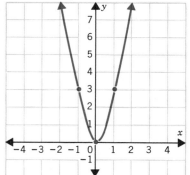 Root $= 0$

4. 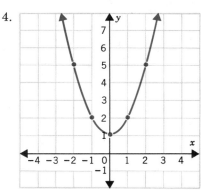 No real roots

5. 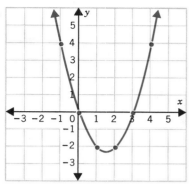 Roots = 0, 3

6. 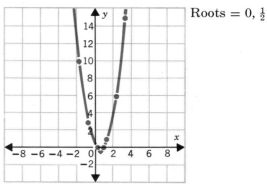 Roots = 0, $\frac{1}{2}$

7. 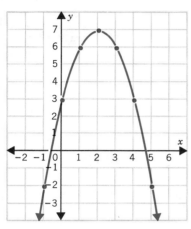 Roots $\cong$ 4.6, -0.6

Rounded to the nearest tenth.

8. 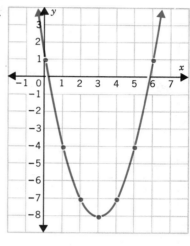 Roots $\cong$ 5.8, 0.2

Rounded to the nearest tenth.

9. 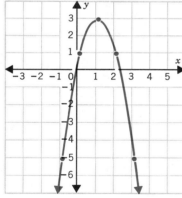 Roots $\cong$ 2.2, -0.2

Rounded to the nearest tenth

10. 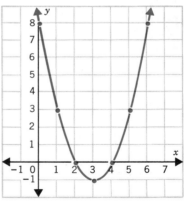 Roots = 2, 4

11. 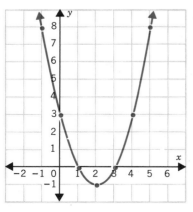 Roots = 1, 3

12. 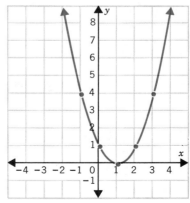 Root = 1

13. 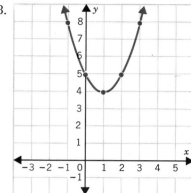 No real roots

14. 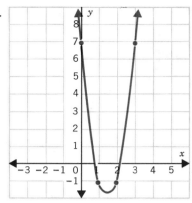 Roots $\cong$ 2.2, 0.8

Rounded to the nearest tenth.

15. 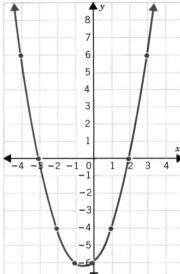 Roots = -3, 2

16. 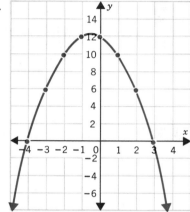 Roots = -4, 3

17. 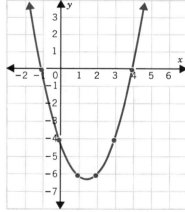 Roots = -1, 4

18. 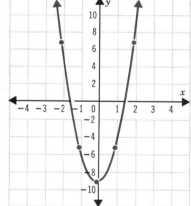 Roots = ± 1.5

19. 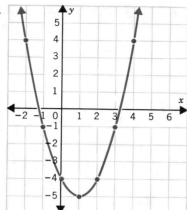 Roots ≅ 3.2, −1.2

Rounded to the nearest tenth.

20. 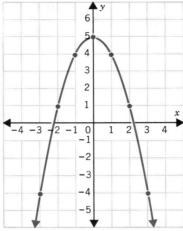 Roots ≅ 2.2, −2.2

Rounded to the nearest tenth.

B.
1. $6, -5$
2. $2 \pm 2\sqrt{6}$
3. 3
4. $4, 6$
5. $20, 21$
6. $3 \pm \sqrt{11}$
7. $\dfrac{7 \pm \sqrt{33}}{4}$
8. $7, 9, 11$ or $-3, -1, 1$
9. $8, 10$
10. $4, 6$
11. $\sqrt{10}$ by $2\sqrt{10}$ ft
12. 13 ft by 21 ft
13. $2\sqrt{3}$ ft and $4 + 2\sqrt{3}$ ft
14. 5 ft
15. 4 ft, 14 ft
16. 2.3 in.
17. 8
18. $\dfrac{6 + 2\sqrt{39}}{3}$
19. $\dfrac{-1 + \sqrt{31}}{3}$
20. $4 + 2\sqrt{6}$
21. 3.5 in., 7.0 in.
22. 6 m
23. $-15 + 5\sqrt{29}$
24. 7 in., 24 in.
25. $9, 40, 41$ in.
26. $20, 21, 29$ in.
27. Dan runs at 8.8 mph; George runs at 10.8 mph
28. Joe takes 6 hours; Alice takes 12 hours.
29. 2.9 hours   30. 2 mph   31. 50 mph   32. 6 bags per shekel

C.
1. The maximum is at $x = 50$ ft, $y = 2500$ sq ft
2. $n = 17$ 3. $4\frac{1}{2}$ 4. $5 \pm \sqrt{-15}$ This was the first time that complex numbers were used in the solution of a quadratic equation.
5. (a) $\dfrac{26 \pm 2\sqrt{7}}{9}$ (b) $\dfrac{6 \pm \sqrt{3}}{3}$

Self-Test 6
Unit 6, Page 513

1. $0, \frac{3}{2}$
2. $7, -2$
3. $11, -4$
4. -6
5. $\frac{1}{3}, -\frac{5}{2}$
6. $-\frac{3}{2}, -7$
7. $-2 \pm \sqrt{11}$
8. $-4 \pm 2\sqrt{3}$
9. $\dfrac{3 \pm \sqrt{13}}{2}$
10. $\dfrac{7 \pm \sqrt{57}}{4}$
11. $7, -9$
12. $0, 7$
13. $\dfrac{3 \pm \sqrt{7}}{2}$
14. $1 \pm \sqrt{11}$
15. $\dfrac{4 \pm \sqrt{21}}{5}$
16. $\dfrac{-3 \pm \sqrt{33}}{6}$

17. 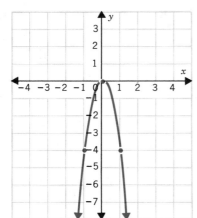 Root = 0

18. 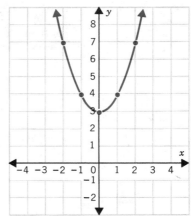 No real roots

19. 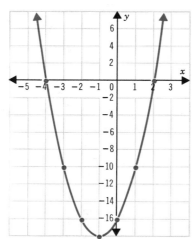 Roots = −4, 2

20. 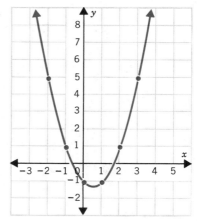 Roots = $\dfrac{1 \pm \sqrt{5}}{2}$

$\cong 1.6$ and -0.6

21. 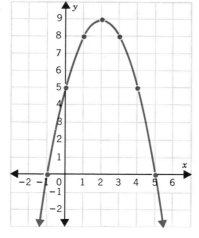 Roots = −1, 5

22. 8, 13   23. 12, 14

24. $\dfrac{6 \pm \sqrt{11}}{5}$

25. Width = $-1 + \sqrt{13} \cong 2.6$ ft
Length = $2 + 2\sqrt{13} \cong 9.2$ ft

Index